Commutative Algebra

Marco Fontana • Sophie Frisch • Sarah Glaz
Editors

Commutative Algebra

Recent Advances in Commutative Rings,
Integer-Valued Polynomials, and Polynomial
Functions

 Springer

Editors
Marco Fontana
Dipartimento di Matematica
Università degli Studi Roma Tre
Roma, Italy

Sophie Frisch
Mathematics Department
Graz University of Technology
Graz, Austria

Sarah Glaz
Department of Mathematics
University of Connecticut
Storrs, CT, USA

ISBN 978-1-4939-4741-6   ISBN 978-1-4939-0925-4 (eBook)
DOI 10.1007/978-1-4939-0925-4
Springer New York Heidelberg Dordrecht London

Mathematics Subject Classification (2010): 13-06, 13Axx, 13Bxx, 13Cxx, 13Dxx, 13Exx, 13Fxx, 13Gxx, 13Hxx, 13Jxx

Printed on acid-free paper

Springer is part of Springer Science+Business Media (www.springer.com)

Preface

This volume is mainly the outcome of a series of mini-courses and a conference on *Commutative rings, integer-valued polynomials and polynomial functions* at Technische Universität Graz, Austria, December 16–18 (mini-courses) and December 19–22 (conference), 2012. It also contains a small collection of invited articles by some of the leading experts in the area, carefully selected for the impact of their research on the major themes of the conference.

The aim of this meeting was to present recent progress in the area of commutative algebra, with primary emphasis on commutative ring theory and integer-valued polynomials along with connections to algebraic number theory, algebraic geometry and homological algebra. The wide range of topics is reflected in the table of contents of this volume. Some of the invited speakers who gave mini-courses have supplied surveys of the state of the art in newly emerging subfields.

At the conference, we had the good fortune to see that our field attracts excellent young mathematicians (who submitted good work, both individually and in collaboration with the old guard) and the not so good fortune to see that none of the young researchers have permanent jobs. May the first trend remain in full force and the second one be remedied in the near future!

Among the people and organizations who helped to make the conference and this volume of proceedings possible, our special thanks go to the departmental secretary Hermine Panzenböck and the doctoral student Roswitha Rissner, who, between the two of them, shared all the hard work of organizing, from designing the conference poster and implementing the website to applications for subsidies and the painstaking work of bookkeeping and balancing the accounts. Without their efforts, the conference would not have taken place and this volume would not have seen the light of day.

We thank the sponsors of the conference: the province of Styria, whose subsidy allowed us to sponsor the travel expenses of some graduate students and conference participants from low-income countries, and the faculty of mathematics and physics of Technische Universität Graz and the joint graduate school of natural sciences "NAWI Graz" of Technische Universität Graz and Karl-Franzens Universität Graz, who together paid the travel expenses of all the invited speakers. Last, but not least, we thank the editorial staff of Springer, in particular Elizabeth Loew, for their cooperation, hard work and assistance with the present volume.

Rome, Italy   Marco Fontana
Graz, Austria   Sophie Frisch
Storrs, Connecticut, USA   Sarah Glaz
December 2013

Contents

Weak Global Dimension of Prüfer-Like Rings

Khalid Adarbeh and Salah-Eddine Kabbaj

Abstract In 1969, Osofsky proved that a chained ring (i.e., local arithmetical ring) with zero divisors has infinite weak global dimension; that is, the weak global dimension of an arithmetical ring is 0, 1, or ∞. In 2007, Bazzoni and Glaz studied the homological aspects of Prüfer-like rings, with a focus on Gaussian rings. They proved that Osofsky's aforementioned result is valid in the context of coherent Gaussian rings (and, more generally, in coherent Prüfer rings). They closed their paper with a conjecture sustaining that "the weak global dimension of a Gaussian ring is 0, 1, or ∞." In 2010, the authors of Bakkari et al. (J. Pure Appl. Algebra 214:53–60, 2010) provided an example of a Gaussian ring which is neither arithmetical nor coherent and has an infinite weak global dimension. In 2011, the authors of Abuihlail et al. (J. Pure Appl. Algebra 215:2504–2511, 2011) introduced and investigated the new class of fqp-rings which stands strictly between the two classes of arithmetical rings and Gaussian rings. Then, they proved the Bazzoni-Glaz conjecture for fqp-rings. This paper surveys a few recent works in the literature on the weak global dimension of Prüfer-like rings making this topic accessible and appealing to a broad audience. As a prelude to this, the first section of this paper provides full details for Osofsky's proof of the existence of a module with infinite projective dimension on a chained ring. Numerous examples—arising as trivial ring extensions—are provided to illustrate the concepts and results involved in this paper.

Keywords Weak global dimension • Arithmetical ring • fqp-ring • Gaussian ring • Prüfer ring • Semihereditary ring • Quasi-projective module • Trivial extension

Mathematics Subject Classification 13F05, 13B05, 13C13, 16D40, 16B50

K. Adarbeh • S.-E. Kabbaj (✉)
Department of Mathematics and Statistics, KFUPM, Dhahran 31261, Saudi Arabia
e-mail: khalidwa@kfupm.edu.sa; kabbaj@kfupm.edu.sa

M. Fontana et al. (eds.), *Commutative Algebra: Recent Advances in Commutative Rings,*
Integer-Valued Polynomials, and Polynomial Functions, DOI 10.1007/978-1-4939-0925-4_1,
© Springer Science+Business Media New York 2014

1 Introduction

All rings considered in this paper are commutative with identity element and all modules are unital. Let R be a ring and M an R-module. The weak (or flat) dimension (resp., projective dimension) of M, denoted w. $\dim_R(M)$ (resp., p. $\dim_R(M)$)), measures how far M is from being a flat (resp., projective) module. It is defined as follows: Let n be an integer ≥ 0. We have w. $\dim_R(M) \leq n$ (resp., p. $\dim_R(M) \leq n$) if there is a flat (resp., projective) resolution

$$0 \to E_n \to E_{n-1} \to \ldots \to E_1 \to E_0 \to M \to 0.$$

If n is the least such integer, w. $\dim_R(M) = n$ (resp., p. $\dim_R(M) = n$). If no such resolution exists, w. $\dim_R(M) = \infty$ (resp., p. $\dim_R(M) = \infty$). The weak global dimension (resp., global dimension) of R, denoted by w. gl. $\dim(R)$ (resp., gl. $\dim(R)$), is the supremum of w. $\dim_R(M)$ (resp., p. $\dim_R(M)$), where M ranges over all (finitely generated) R-modules. For more details on all these notions, we refer the reader to [6, 13, 23].

A ring R is called coherent if every finitely generated ideal of R is finitely presented, equivalently, if $(0 : a)$ and $I \cap J$ are finitely generated for every $a \in R$ and any two finitely generated ideals I and J of R [13]. Examples of coherent rings are Noetherian rings, Boolean algebras, von Neumann regular rings, and semihereditary rings.

Gaussian rings belong to the class of Prüfer-like rings which has recently received much attention from commutative ring theorists. A ring R is called Gaussian if for every $f, g \in R[X]$, one has the content ideal equation $c(fg) = c(f)c(g)$ where $c(f)$, the content of f, is the ideal of R generated by the coefficients of f [25]. The ring R is said to be a chained ring (or valuation ring) if its lattice of ideals is totally ordered by inclusion; and R is called arithmetical if R_m is a chained ring for each maximal ideal m of R [11, 18]. Also R is called semihereditary if every finitely generated ideal of R is projective [8]; and R is Prüfer if every finitely generated regular ideal of R is projective [7, 16]. In the domain context, all these notions coincide with the concept of Prüfer domain. Glaz, in [14], constructs examples which show that all these notions are distinct in the context of arbitrary rings. More examples, in this regard, are provided via trivial ring extensions [1, 3].

The following diagram of implications puts the notion of Gaussian ring in perspective within the family of Prüfer-like rings [1, 4, 5]:

Semihereditary ring

$\Downarrow$

Ring with weak global dimension ≤ 1

$\Downarrow$

Arithmetical ring

$\Downarrow$

fqp-Ring

$\Downarrow$

Gaussian ring
⇓
Prüfer ring

In 1969, Osofsky proved that a local arithmetical ring (i.e., chained ring) with zero divisors has infinite weak global dimension [22]. In view of [13, Corollary 4.2.6], this result asserts that the weak global dimension of an arithmetical ring is 0, 1, or ∞.

In 2007, Bazzoni and Glaz proved that if R is a coherent Prüfer ring (and, a fortiori, a Gaussian ring), then w. gl. $\dim(R) = 0$, 1, or ∞ [5, Proposition 6.1]. And also they proved that if R is a Gaussian ring admitting a maximal ideal m such that the nilradical of the localization R_m is a nonzero nilpotent ideal, then w. gl. $\dim(R)$ = ∞ [5, Theorem 6.4]. At the end of the paper, they conjectured that "the weak global dimension of a Gaussian ring is 0, 1, or ∞" [5]. In two preprints [9, 10], Donadze and Thomas claim to prove this conjecture (see the end of Sect. 3).

In 2010, the authors of [3] proved that if $(A, \mathfrak{m})$ is a local ring, E is a nonzero $\frac{A}{\mathfrak{m}}$-vector space, and $R := A \ltimes E$ is the trivial extension of A by E, then:

- R is a total ring of quotients and hence a Prüfer ring.
- R is Gaussian if and only if A is Gaussian.
- R is arithmetical if and only if $A := K$ is a field and $\dim_K E = 1$.
- w. gl. $\dim(R) \geqq 1$. If, in addition, m admits a minimal generating set, then w. gl. $\dim(R) = \infty$.

As an application, they provided an example of a Gaussian ring which is neither arithmetical nor coherent and has an infinite weak global dimension [3, Example 2.7]; which widened the scope of validity of the above conjecture beyond the class of coherent Gaussian rings.

In 2011, the authors of [1] investigated the correlation of fqp-rings with well-known Prüfer conditions; namely, they proved that the class of fqp-rings stands between the two classes of arithmetical rings and Gaussian rings [1, Theorem 3.1]. They also examined the transfer of the fqp-property to trivial ring extensions in order to build original examples of fqp-rings. Also they generalized Osofsky's result (mentioned above) and extended Bazzoni-Glaz's result on coherent Gaussian rings by proving that the weak global dimension of an fqp-ring is equal to 0, 1, or ∞ [1, Theorem 3.11]; and then they provided an example of an fqp-ring that is neither arithmetical nor coherent [1, Example 3.9].

Recently, several papers have appeared in the literature investigating the weak global dimension of various settings subject to Prüfer conditions. This survey paper plans to track and study these works dealing with this topic from the very origin, that is, 1969 Osofsky's proof of the existence of a module with infinite projective dimension on a local arithmetical ring. Precisely, we will examine all main results published in [1, 3, 5, 15, 22].

Our goal is to make this topic accessible and appealing to a broad audience; including graduate students. For this purpose, we present complete proofs of all main results via ample details and simplified arguments along with exact references.

Further, numerous examples—arising as trivial ring extensions—are provided to illustrate the concepts and results involved in this paper. We assume familiarity with the basic tools used in the homological aspects of commutative ring theory, and any unreferenced material is standard as in [2, 6, 8, 13, 17, 19, 23, 27].

2 Weak Global Dimension of Arithmetical Rings

In this section, we provide a detailed proof for Osofsky's Theorem that the weak global dimension of an arithmetical ring with zero divisors is infinite. In fact, this result enables one to state that the weak global dimension of an arithmetical ring is 0, 1, or ∞. We start by recalling some basic definitions.

Definition 2.1. Let R be a ring and M an R-module. Then:

(1) The weak dimension of M, denoted by w. dim(M), measures how far M is from being flat. It is defined as follows: Let n be a positive integer. We have w. dim$(M) \leq n$ if there is a flat resolution

$$0 \to E_n \to E_{n-1} \to \cdots \to E_1 \to E_0 \to M \to 0.$$

If no such resolution exists, w. dim$(M) = \infty$; and if n is the least such integer, w. dim$(M) = n$.

(2) The weak global dimension of R, denoted by w. gl. dim(R), is the supremum of w. dim(M), where M ranges over all (finitely generated) R-modules.

Definition 2.2. Let R be a ring. Then:

(1) R is said to be a chained ring (or valuation ring) if its lattice of ideals is totally ordered by inclusion.

(2) R is called an arithmetical ring if R_m is a chained ring for each maximal ideal m of R.

Fields and $\mathbb{Z}_{(p)}$, where $\mathbb{Z}$ is the ring of integers and p is a prime number, are examples of chained rings. Also, $\mathbb{Z}/n^2\mathbb{Z}$ is an arithmetical ring for any positive integer n. For more examples, see [3]. For a ring R, let $Z(R)$ denote the set of all zero divisors of R.

Next we give the main theorem of this section.

Theorem 2.3. *Let R be an arithmetical ring. Then* w. gl. dim$(R) = 0, 1,$ *or* ∞.

To prove this theorem we make the following reductions:

(1) We may assume that R is a chained ring since w. gl. dim(R) is the supremum of w. gl. dim(R_m) for all maximal ideal m of R [13, Theorem 1.3.14 (1)].

(2) We may assume that R is a chained ring with zero divisors. Then we prove that w. gl. dim$(R) = \infty$ since if R is a valuation domain, then w. gl. dim$(R) \leq 1$ by [13, Corollary 4.2.6].

(3) Finally, we may assume that $(R, \mathfrak{m})$ is a chained ring with zero divisors such that $Z(R) = \mathfrak{m}$, since $Z(R)$ is a prime ideal, $Z(R_{Z(R)}) = Z(R)R_{Z(R)}$, and w. gl. $\dim(R_{Z(R)}) \le$ w. gl. $\dim(R)$.

So our task is reduced to prove the following theorem.

Theorem 2.4 ([22, Theorem]). *Let $(R, \mathfrak{m})$ be a chained ring with zero divisors such that $Z(R) = \mathfrak{m}$. Then* w. gl. $\dim(R) = \infty$.

To prove this theorem we first prove the following lemmas. Throughout, let $(R, \mathfrak{m})$ be a chained ring with $Z(R) = \mathfrak{m}$, M an R-module, $I = \{x \in R \mid x^2 = 0\}$, and for $x \in M$, $(0 : x) = \{y \in R \mid yx = 0\}$. One can easily check that I is a nonzero ideal since R is a chained ring with zero divisors.

Lemma 2.5 ([22, Lemma 1]). $I^2 = 0$, *and for all* $x \notin R$, $x \notin I \Rightarrow (0 : x) \subseteq I$.

Proof. To prove that $I^2 = 0$, it suffices to prove that $ab = 0$ for all $a, b \in I$. So let $a, b \in I$. Then either $a \in bR$ or $b \in aR$, so that $ab \in a^2R = 0$ or $ab \in b^2R = 0$.

Now let $x \in R \setminus I$ and $y \in (0 : x)$. Then either $x \in yR$ or $y \in xR$. But $x \in yR$ implies that $x^2 \in xyR = 0$, absurd. Therefore $y \in xR$, so that $y^2 \in xyR = 0$. Hence $y \in I$. □

Lemma 2.6 ([22, Lemma 2]). *Let* $0 \ne x \in Z(R)$ *such that* $(0 : x) = yR$. *Then* w. gl. $\dim(R) = \infty$.

Proof. We first prove that $(0 : y) = xR$. The inclusion $(0 : y) \supseteq xR$ is trivial since $xy = 0$. Now to prove the other inclusion let $z \in (0 : y)$. Then either $z = xr$ for some $r \in R$ and in this case we are done, or $x = zj$ for some $j \in R$. We may assume $j \in \mathfrak{m}$. Otherwise, j is a unit and then we return to the first case. Since $x \ne 0$, $j \notin (0 : z)$, so $jR \nsubseteq (0 : z)$ which implies $(0 : z) \subseteq jR$, and hence $y = jk$ for some $k \in \mathfrak{m}$. But then $0 = zy = zjk = xk$, so $k \in (0 : x) = yR$, and hence $k = yr$ for some $r \in R$. Hence $y = kj = yrj$, and as $j \in \mathfrak{m}$ we have the equality $y = y(1 - rj)(1 - rj)^{-1} = 0$, which contradicts the fact that x is a zero divisor. Hence $z \in xR$, and therefore $(0 : y) = xR$.

Now let m_x (resp., m_y) denote the multiplication by x (resp., y). Since $(0 : x) = yR$ and $(0 : y) = xR$ we have the following infinite flat resolution of xR with syzygies xR and yR:

$$\cdots \longrightarrow R \xrightarrow{m_y} R \xrightarrow{m_x} R \xrightarrow{m_y} \cdots \xrightarrow{m_y} R \xrightarrow{m_x} xR \longrightarrow 0$$

We claim that xR and yR are not flat. Indeed, recall that a projective module over a local ring is free [23]. So no projective module is annihilated by x or y. Since xR is annihilated by y and yR is annihilated by x, both xR and yR are not projective. Further, xR and yR are finitely presented in view of the exact sequence $0 \to yR \to R \to xR \to 0$. It follows that xR and yR are not flat (since a finitely presented flat module is projective [23, Theorem 3.61]). □

Corollary 2.7 ([22, Corollary]). *If $I = \mathfrak{m}$, then I is cyclic and R has infinite weak global dimension.*

Proof. Assume that $I = \mathfrak{m}$. Then $\mathfrak{m}^2 = 0$. Now let $0 \neq a \in \mathfrak{m}$. We claim that $\mathfrak{m} = aR$. Indeed, let $b \in \mathfrak{m}$. Since R is a chained ring, either $b = ra$ for some $r \in R$ and in this case we are done, or $a = rb$ for some $r \in R$. In the later case, either r is a unit and then $b = r^{-1}a \in aR$, or $r \in \mathfrak{m}$ which implies $a = rb = 0$, which contradicts the assumption $a \neq 0$. Thus $\mathfrak{m} = aR$, as claimed. Moreover, we have $(0 : a) = aR$. Indeed, $(0 : a) \supseteq aR$ since $a \in I$; if $x \in (0 : a)$, then $x \in Z(R) = \mathfrak{m} = aR$. Hence $(0 : a) = aR$. It follows that R satisfies the conditions of Lemma 2.6 and hence the weak global dimension of R is ∞. □

Throughout, an element x of an R-module M is said to be regular if $(0 : x) = 0$.

Lemma 2.8 ([22, Lemma 3]). *Let F be a free module and $x \in F$. Then x is contained in zR for some regular element z of F.*

Proof. Let $\{y_\alpha\}$ be a basis for F and let $x := \sum_{i=1}^{n} y_i r_i \in F$, where $r_i \in R$. Since R is a chained ring, there is $j \in \{1, 2, \ldots, n\}$ such that $\sum_{i=1}^{n} r_i R \subseteq r_j R$. So that for each $i \in \{1, 2, \ldots, n\}$, $r_i = r_j s_i$ for some $s_i \in R$ with $s_j = 1$. Hence $x = r_j(\sum_{i=1}^{n}(y_i s_i))$. We claim that $z := \sum_{i=1}^{n} y_i s_i$ is regular. Suppose not and let $t \in R$ such that $t(\sum_{i=1}^{n} y_i s_i) = 0$. Then $t s_i = 0$ for all $i \in \{1, 2, \ldots, n\}$. In particular $t = t s_j = 0$, absurd. Therefore z is regular and $x = r_j z$, as desired. □

Note, for convenience, that in the proof of Theorem 2.4 (below), we will prove the existence of a module M satisfying the conditions (1) and (2) of the next lemma, which will allow us to construct—via iteration—an infinite flat resolution of M.

Lemma 2.9 ([22, Lemma 4]). *Assume that $(0 : r)$ is infinitely generated for all $0 \neq r \in \mathfrak{m}$. Let M be an R-submodule of a free module N such that:*

(1) $M = M_1 \bigcup M_2 \bigcup M_3$, where $M_1 = \bigcup_{\substack{x \in M \\ x \text{ regular}}} xR$, $M_2 = \bigcup_{i=0}^{\infty} yu_i R$, with y regular in N, $u_i R \subsetneqq u_{i+1} R$, and yu_i is not in M_1, and $M_3 = \sum v_j R$.
(2) $yu_0 R \cap xR$ is infinitely generated for some regular $x \in M$.

Let F be a free R-module with basis $\{y_x \mid x \text{ regular} \in M\} \cup \{z_i \mid i \in \omega\} \cup \{w_j\}$, and let $v : F \longrightarrow N$ be the map defined by: $v(y_x) = x$, $v(z_i) = yu_i$, and $v(w_j) = v_j$. Then $K = \text{Ker}(v)$ has properties (1), (2), and M is not flat.

Proof. First the map v exists by [19, Theorem 4.1]. (1) by (2), there exist $r, s \in R$ such that $yu_0 r = xs \neq 0$. Here $r \in \mathfrak{m}$; otherwise, $yu_0 = xsr^{-1} \in M_1$, contradiction. Since $Z(R) = \mathfrak{m}$, the expression for any regular element in terms of a basis for N has one coefficient a unit. Indeed, let $(n_\alpha)_{\alpha \in \Delta}$ be a basis for N and z a regular element in N with $z = \sum_{i=0}^{i=k} c_i n_i$ where $c_i \in R$. As R is a chained ring, there exists $j \in \{0, \ldots, k\}$ such that for all $i \in \{0, \ldots, k\}$, there exists $d_i \in R$ with $c_i = c_j d_i$ and $d_j = 1$. We claim that c_j is a unit. Suppose not. Then $c_j \in Z(R)$. So there is a nonzero $d \in R$ with $dc_j = 0$, and hence $dz = dc_j \sum_{i=0}^{i=k} d_i n_i = 0$. This is absurd since z is regular.

Now, let $x = \sum_{\substack{i \in I \\ I \text{ finite}}} a_i n_i$ and $y = \sum_{\substack{i \in I \\ I \text{ finite}}} b_i n_i$. Then $b_i u_0 r = a_i s$ for all $i \in I$. Let $i_0 \in I$ such that a_{i_0} is a unit. So $s = u_0 r t$, where $t = b_{i_0} a_{i_0}^{-1} \in R$. Note that $b_{i_0} \neq 0$ since $xs \neq 0$. Clearly, $z_0 - y_x u_0 t$ is regular in F (since z_0, y_x are part of the basis of F) and is not in K [otherwise, $v(z_0 - y_x u_0 t) = 0$ yields $yu_0 = xu_0 t$, which contradicts (1)] and $(z_0 - y_x u_0 t)r \in K$. We claim that $(z_0 - y_x u_0 t)r$ is not in $K_1 := \bigcup_{\substack{x' \in K \\ x' \text{ regular}}} x'R$. Suppose not and assume that $r(z_0 - u_0 t y_x) = r'x'$ with $r' \in R$ and x' regular in K. Then $r' \neq 0$ since $r \neq 0$ and as $x' \in K \subseteq F$, there are $a, b, a_i \in R$ such that $x' = az_0 - by_x + x''$, where $x'' = \sum_{\substack{y_x \neq f_i \\ z_0 \neq f_i}} a_i f_i$. Thus $r = r'a$, $ru_0 t = r'b$, and $r'x'' = 0$. Since x' is regular in F and $r'x'' = 0$, a or b is unit. We claim that a is always a unit. Indeed, if b is a unit, then $r(1 - ab^{-1}u_0 t) = 0$, so if $a \in \mathfrak{m}$, then $(1 - ab^{-1}u_0 t)$ is a unit which implies $r = 0$, absurd. So $a^{-1}x' = z_0 - a^{-1}by_x + a^{-1}x''$, $r' = a^{-1}r$, and $ru_0 t = ra^{-1}b$ which implies $z_0 - u_0 t y_x + (u_0 t - a^{-1}b)y_x + a^{-1}x'' = a^{-1}x' \in K$. By Lemma 2.8 $(u_0 t - a^{-1}b)y_x + a^{-1}x'' = pq$, fore some q regular in F and $p \in R$. But clearly since $r = r'a$, $ru_0 t = r'b$, and $r'x'' = 0$, then $rpq = 0$. Hence $rp = 0$. It follows that $(z_0 - y_x u_0 t + qp) \in K$, where q is regular in F and $p \in (0 : r)$. Thus by applying v we obtain $yu_0 - xu_0 t + pv(q) = 0$. But R is a chained ring, so p and $u_0 t$ are comparable and since $u_0 t r \neq 0$, $p = u_0 t h$ for some $h \in R$. Hence $yu_0 = (x - hv(q))u_0 t$; we show that $(x - hv(q))$ is regular in M which contradicts property (1). First clearly $(x - hv(q)) \in M$ since $x, v(q) \in M$. Now suppose that $a(x - hv(q)) = 0$ for some $a \in \mathfrak{m}$. Either $u_0 t = a'a$ for some $a' \in R$, this yields $yu_0 = (x - hv(q))aa' = 0$ also impossible, or $a = u_0 tm$ for some $m \in R$, and this yields $mu_0 y = (x - hv(q))a = 0$, so $mu_0 = 0$ as y is regular, and hence $a = mu_0 t = 0$. We conclude that $(x - hv(q))$ is regular in M and hence $yu_0 \in M_1$, the desired contradiction.

Last, let $yu_0 R \cap xR = \langle x_0, x_1, \ldots, x_n, \ldots \rangle$, where

$$\langle x_0, x_1, \ldots, x_i \rangle \subsetneqq \langle x_0, x_1, \ldots, x_i, x_{i+1} \rangle.$$

For any integer $i \geq 0$, let $x_i = yu_0r_i$ for some $r_i \in R$. It is clear that $r_0R \subsetneqq r_1R \subsetneqq \ldots \subsetneqq r_iR \subsetneqq r_{i+1}R \subsetneqq \ldots$. Now, let $y' := z_0 - y_xu_0t$, $u'_i := r_i$ for each $i \in \mathbb{N}$. Then

$$K = K_1 \bigcup K_2 \bigcup K_3, \text{ where } K_1 := \bigcup_{\substack{x' \in K \\ x' \text{ regular}}} x'R, \ K_2 := \bigcup_{i=0}^{\infty} y'u'_iR \text{ with } y' \text{ regular}$$

in F and $u'_iR \subsetneqq u'_{i+1}R$, and $K_3 := K \setminus (K_1 \bigcup K_2)$. Thus K satisfy Property (1).

(2) Since $u_0R \subsetneqq u_1R$, $u_0 = u_1m'$ for some $m' \in \mathfrak{m}$. Hence $x' := z_0 - z_1m'$ is regular in K since $v(x') = v(z_0 - z_1m') = yu_0 - yu_1m' = 0$ and z_0, z_1 are basis elements. We claim that $(z_0 - z_1m')R \cap (z_0 - y_xu_0t)r_0R = z_0(0 : m')$. Indeed, since z_0, z_1, y_x are basis elements, then $(z_0 - z_1m')R \cap (z_0 - y_xu_0t)r_0 \subseteq z_0R$. Also $(z_0 - z_1m')R \cap z_0R = z_0(0 : m')$. For, let $l \in (z_0 - z_1m')R \cap z_0R$. Then $l = (z_0 - z_1m')a = z_0a'$ for some $a, a' \in R$. Hence $a = a'$ and $am' = 0$, whence $l = az_0$ with $am' = 0$. So $l \in z_0(0 : m')$. The reverse inclusion is straightforward. Consequently, $(z_0 - z_1m')R \cap (z_0 - y_xu_0t)r_0R \subseteq z_0(0 : m')$. To prove the reverse inclusion, let $k \in (0 : m')$. Then either $k = r_0k'$ or $r_0 = kk'$, for some $k' \in R$. The second case is impossible since $r_0u_0 \neq 0$. Hence $z_0k = (z_0 - y_xu_0t)r_0k' \in (z_0 - y_xu_0t)r_0R$. Further, $z_0k \in (z_0 - z_1m')R$. Therefore our claim is true. But z_0 is regular, so $z_0(0 : m') \cong (0 : m')$ which is infinitely generated by hypothesis. Therefore $y'u'_0R \cap x'R$ is infinitely generated, as desired.

Finally, M is not flat. Suppose not, then by [23, Theorem 3.57], there is an R-map $\theta : F \longrightarrow K$ such that $\theta((z_0 - y_xu_0t)r_0) = (z_0 - y_xu_0t)r_0$. Assume that $\theta(z_0) = az_0 + by_x + Z_1$ for some $a, b \in R$ and $\theta(y_x) = a'z_0 + b'y_x + Z_2$ for some $a', b' \in R$. Then $r_0a - r_0u_0ta' = r_0$, $r_0b - r_0u_0tb' = -r_0u_0t$, and $r_0Z_1 - r_0u_0tZ_2 = 0$. Hence $r_0(1 - a + u_0ta') = 0$ and since $r_0 \neq 0$, a or a' is a unit. Suppose that a is a unit and without loss of generality we can assume that $a = 1$. Thus we have the equation $z_0 - u_0ty_x - u_0ta'z_0 + (u_0t - u_0tb' + b)y_x + Z_1 - u_0tZ_2 = \theta(z_0) - u_0t\theta(Z_2) \in K$. By Lemma 2.8, $-u_0ta'z_0 + (u_0t - u_0tb' + b)y_x + Z_1 - u_0tZ_2 = pq$, where q is regular in F and, clearly, $r_0p = 0$ since $r_0u_0ta' = 0$. Thus $z_0 - u_0ty_x + pq \in K$, which is absurd (as seen before in the second paragraph of the proof of Lemma 2.9).

Now we are able to prove Theorem 2.4.

Proof of Theorem 2.4. If $(0 : r)$ is cyclic for some $r \in \mathfrak{m}$, then R has infinite weak global dimension by Lemma 2.6. Next suppose that $(0 : r)$ is not cyclic, for all $0 \neq r \in \mathfrak{m}$, which is equivalent to assume that $(0 : r)$ is infinitely generated for all $0 \neq r \in \mathfrak{m}$, since R is a chained ring.

Let $0 \neq a \in I$ and $b \in \mathfrak{m} \setminus I$. Note that b exists since $I \neq \mathfrak{m}$ by the proof of Corollary 2.7. Let N be a free R-module on two generators y, y' and let $M := (y - y'b)R + y(0 : a)$. Then:

(A) $M_1 := \bigcup_{\substack{x \in M \\ x \text{ regular}}} xR = \{(yt - y'b)r \mid 1 - t \in (0 : a), r \in R\}$. To show this equality, let c be a regular element in M. Then $c = (r_1 + r_2)y - r_1by'$ for some $r_1 \in R, r_2 \in (0 : a)$. We claim that r_1 is a unit. Suppose not. So either $r_1 \in (r_2)$ hence $ac = 0$, or $r_2 = nr_1$ for some $n \in R$ and since $r_1 \in \mathfrak{m} = Z(R)$, there is $r'_1 \neq 0$ such that $r_1 r'_1 = 0$, so $r'_1 c = 0$. In both cases there is a contradiction with the fact that c is regular. Thus, r_1 is a unit. It follows that $c = (1 + r_1^{-1} r_2) y r_1 - b y' r_1 \in \{(yt - y'b) r \mid 1 - t \in (0 : a), r \in R\}$. Now let $c = yt - y'b$, where $(1 - t) \in (0 : a)$. Then c is regular. Indeed, if $rc = 0$ for some $r \in R$, then $rt = 0$. Moreover, either $r = na$ for some $n \in R$, and in this case $r(1 - t) = na(1 - t) = 0$, so $r = rt = 0$ as desired, or $a = nr$ for some $n \in R$, so $a = at = nrt = 0$, absurd.

(B) There exists a countable chain of ideals $u_0 R \subsetneq u_1 R \subsetneq \dots$ where $u_i \in (0 : a) \setminus (0 : b)$. Since $0 \neq a \in I$ and $b \in \mathfrak{m} \setminus I$, $(a) \subseteq (b)$. Thus $(0 : b) \subseteq (0 : a)$. Moreover $(0 : b) \subsetneq (0 : a)$; otherwise, $a \in (0 : a) = (0 : b)$, and hence $ab = 0$. Hence $b \in (0 : a) = (0 : b) \subseteq I$ by Lemma 2.5, absurd. Now let $u_0 \in (0 : a) \setminus (0 : b)$. Since $(0 : a)$ is infinitely generated, there are $u_1, u_2, \dots$ such that $(u_0) \subsetneq (u_0, u_1) \subsetneq \dots \subseteq (0 : a)$. So $u_0 R \subsetneq u_1 R \subsetneq \dots$ and necessarily $u_i \notin (0 : b)$ for all $i \geq 1$ since $u_0 \notin (0 : b)$.

Note that $y u_i \in M$ [since $u_i \in (0 : a)$]. Also $y u_i \notin M_1$; otherwise, if $y u_i = y t r - y' b r$ with $1 - t \in (0 : a)$ and $r \in R$, then $u_i = t r$ and $b r = 0$. Hence $b u_i = b t r = 0$ and thus $u_i \in (0 : b)$, contradiction. Also note that y is regular in N (part of the basis) and $y \notin M$; if $y = (y - y'b) r_1 + r_2 y$ with $r_1 \in R$ and $r_2 \in (0 : a)$, then $r_1 b = 0$ and $r_1 + r_2 = 1$. So $r_1 \in \mathfrak{m}$, $a r_1 = a$, and hence $a = 0$, absurd.

(A) and (B) imply that (1) of Lemma 2.9 holds.

Let us show that $y u_0 R \cap (y - y'b) R = y(0 : b)$. Indeed, if $c = y u_0 r = (y - y'b) r'$ where $r, r' \in R$, then $u_0 r = r'$ and $r' b = 0$. Hence $c \in y(0 : b)$. If $c = ry$ where $rb = 0$, then $r = u_0 t$ for some $t \in R$ as $u_0 \in (0 : a) \setminus (0 : b)$. Thus $c = r(y - y'b)$. Now $y(0 : b) \cong (0 : b)$ is infinitely generated. Therefore (2) of Lemma 2.9 holds.

Since K satisfies the properties of M we can consider it as a new module M, and then there is a free module F_1 and a map $v_1 : F_1 \longrightarrow F$ such that $K_1 = \text{Ker}(v_1)$ satisfies the same conditions of K and K_1 is not flat. We can repeat this iteration above to get the infinite flat resolution of M:

$$\dots \to F_n \to F_{n-1} \to \dots \to F_1 \to F_0 \to M \to 0.$$

with none of the syzygies $K, K_1, K_2, \dots$ is flat. Therefore R has an infinite weak global dimension. □

3 Weak Global Dimension of Gaussian Rings

In 2005, Glaz proved that if R is a Gaussian coherent ring, then w. gl. $\dim(R) = 0, 1,$ or ∞ [15]. In this section, we will see that the same conclusion holds for the larger class of Prüfer coherent rings and fore some contexts of Gaussian rings. We start by recalling the definitions of Gaussian, Prüfer, and coherent rings.

Definition 3.1. Let R be a ring. Then:

(1) R is called a Gaussian ring if for every $f, g \in R[X]$, one has the content ideal equation $c(fg) = c(f)c(g)$, where $c(f)$, the content of f, is the ideal of R generated by the coefficients of f.
(2) R is called a Prüfer ring if every nonzero finitely generated regular ideal is invertible (or, equivalently, projective)
(3) R is called a coherent ring if every finitely generated ideal of R is finitely presented, equivalently, if $(0 : a)$ and $I \cap J$ are finitely generated for every $a \in R$ and any two finitely generated ideals I and J of R.

Recall that Arithmetical ring $\Rightarrow$ Gaussian ring $\Rightarrow$ Prüfer ring. To see the proofs of the above implications and that they cannot be reversed, in general, we refer the reader to [5, 14, 15] and Sect. 5 of this paper.

Noetherian rings, valuation domains, and $K[x_1, x_2, \ldots]$ where K is a field are examples of coherent rings. For more examples, see [13].

Let $Q(R)$ denote the total ring of fractions of R and $\text{Nil}(R)$ its nilradical. The following proposition is the first main result of this section.

Proposition 3.2 ([5, Proposition 6.1]). *Let R be a coherent Prüfer ring. Then the weak global dimension of R is equal to 0, 1, or ∞.*

The proof of this proposition relies on the following lemmas. Recall that a ring R is called regular if every finitely generated ideal of R has a finite projective dimension and von Neumann regular if every R-module is flat.

Lemma 3.3 ([13, Corollary 6.2.4]). *Let R be a coherent regular ring. Then $Q(R)$ is a von Neumann regular ring.*

Lemma 3.4 ([15, Lemma 2.1]). *Let R be a local Gaussian ring and $I = (a_1, \ldots, a_n)$ be a finitely generated ideal of R. Then $I^2 = (a_i^2)$, for some $i \in \{1, 2, \ldots, n\}$.*

Proof. We first assume that $I = (a, b)$. Let $f(x) := ax + b$, $g(x) := ax - b$, and $h(x) := bx + a$. Since R is Gaussian, $c(fg) = c(f)c(g)$, so that $(a, b)^2 = (a^2, b^2)$, also $c(fh) = c(f)c(h)$ which implies that $(a, b)^2 = (ab, a^2 + b^2)$. Hence $(a^2, b^2) = (ab, a^2 + b^2)$, whence $a^2 = rab + s(a^2 + b^2)$, for some r and s in R. That is, $(1 - s)a^2 + rab + sb^2 = 0$. Since R is a local ring, either s or $1 - s$ is a unit in R. If s is a unit in R, then $b^2 + rs^{-1}ab + (s^{-1} - 1)a^2 = 0$. Next we show that $ab \in (a^2)$. Let $k(x) := (b + \alpha a)x - a$, where $\alpha := rs^{-1}$. Then $c(hk) = c(h)c(k)$ implies that $(b(b + \alpha a), \alpha a^2, -a^2) = (a, b)((b + \alpha a), a)$. But clearly $(b(b + \alpha a), \alpha a^2, -a^2) = ((s^{-1} - 1)a^2, \alpha a^2, -a^2) = (a^2)$. Thus $(a^2) = (a, b)((b + \alpha a), a)$. In particular, $ab \in (a^2)$ and so does b^2. If $1 - s$ is unit, similar arguments imply that ab, and hence $a^2 \in (b^2)$. Thus for any two elements a and b, $ab \in (b^2)$ or (a^2). It follows that $I^2 = (a_1, \ldots, a_n)^2 = (a_1^2, \ldots, a_n^2)$. An induction on n leads to the conclusion. $\square$

Recall that a ring R is called reduced if it has no nonzero nilpotent elements.

Lemma 3.5 ([15, Theorem 2.2]). *Let R be a ring. Then* w. gl. $\dim(R) \leq 1$ *if and only if R is a Gaussian reduced ring.*

Proof. Assume that w. gl. $\dim(R) \leq 1$. By [13, Corollary 4.2.6], R_p is a valuation domain for every prime ideal p of R. As valuation domains are Gaussian, R is locally Gaussian, and therefore Gaussian. Further, R is reduced. For, let $x \in R$ such that x is nilpotent. We claim that $x = 0$. Suppose not and let $n \geq 2$ be an integer such that $x^n = 0$. Then there exists a prime ideal q in R such that $x \neq 0$ in R_q [2, Proposition 3.8]. It follows that $x^n = 0$ in R_q, a contradiction since R_q is a domain.

Conversely, since R is Gaussian reduced, R_p is a local, reduced, Gaussian ring for any prime ideal p of R. We claim that R_p is a domain. Indeed, let a and b in R_p such that $ab = 0$. By Lemma 3.4, $(a, b)^2 = (b)^2$ or (a^2). Say $(a, b)^2 = (b^2)$. Then $a^2 = tb^2$ for some $t \in R_p$. Thus $a^3 = tb(ab) = 0$. Since R_p is reduced, $a = 0$, and R_p is a domain. Therefore R_p is a valuation domain for all prime ideals p of R. So w. gl. $\dim(R) \leq 1$ by [13, Corollary 4.2.6]. $\qquad \square$

Lemma 3.6 ([5, Theorem 3.3]). *Let R be a Prüfer ring. Then R is Gaussian if and only if $Q(R)$ is Gaussian.*

Lemma 3.7 ([5, Theorem 3.12(ii)]). *Let R be a ring. Then* w. gl. $\dim(R) \leq 1$ *if and only if R is a Prüfer ring and* w. gl. $\dim(Q(R)) \leq 1$.

Proof. If w. gl. $\dim(R) \leq 1$, R is Prüfer and, by localization, w. gl. $\dim(Q(R)) \leq 1$. Conversely, assume that R is a Prüfer ring such that w. gl. $\dim(Q(R)) \leq 1$. By Lemma 3.5, $Q(R)$ is a Gaussian reduced ring. So R is reduced and, by Lemma 3.6, R is Gaussian. By Lemma 3.5, w. gl. $\dim(R) \leq 1$. $\qquad \square$

Proof of Proposition 3.2. Assume that w. gl. $\dim(R) = n < \infty$ and let I be any finitely generated ideal of R. Then I has a finite weak dimension. Since R is a coherent ring, I is finitely presented. Hence the weak dimension of I equals its projective dimension by [13, Corollary 2.5.5]. Whence, as I is an arbitrary finitely generated ideal of R, R is a regular ring. So, by [13, Corollary 6.2.4], $Q(R)$ is von Neumann regular. By Lemma 3.7, w. gl. $\dim(R) \leq 1$. $\qquad \square$

The following is an example of a coherent Prüfer ring with infinite weak global dimension.

Example 3.8. Let $R = \mathbb{R} \ltimes \mathbb{C}$. Then R is coherent by [20, Theorem 2.6], Prüfer by Theorem 4.2, and w. gl. $\dim(R) = \infty$ by Lemma 4.1.

In order to study the weak global dimension of an arbitrary Gaussian ring, we make the following reductions:

(1) We may assume that R is a local Gaussian ring since w. gl. $\dim(R)$ is the supremum of w. gl. $\dim(R_m)$ for all maximal ideal m of R [13, Theorem 1.3.14 (1)].
(2) We may assume that R is a non-reduced local Gaussian ring since every reduced Gaussian ring has weak global dimension at most 1 by Lemma 3.5.

(3) Finally, we may assume that $(R, \mathfrak{m})$ is a local Gaussian ring with the maximal ideal $\mathfrak{m}$ such that $\mathfrak{m} = \mathrm{Nil}(R)$, since the prime ideals of a local Gaussian ring R are linearly ordered, so that $\mathrm{Nil}(R)$ is a prime ideal and w. gl. dim$(R) \geq$ w. gl. dim$(R_{\mathrm{Nil}(R)})$.

Next we announce the second main result of this section.

Theorem 3.9 ([5, Theorem 6.4]). *Let R be a Gaussian ring with a maximal ideal $\mathfrak{m}$ such that $\mathrm{Nil}(R_{\mathfrak{m}})$ is a nonzero nilpotent ideal. Then* w. gl. dim$(R) = \infty$.

The proof of this theorem involves the following results:

Lemma 3.10. *Consider the following exact sequence of R-modules*

$$0 \longrightarrow M' \longrightarrow M \longrightarrow M'' \longrightarrow 0$$

where M is flat. Then either the three modules are flat or w. dim$(M'') =$ w. dim$(M') + 1$.

Proof. This is a classic result. We offer here a proof for the sake of completeness. Suppose that M'' is flat. Then by the long exact sequence theorem [23, Theorem 8.3] we get the exact sequence

$$0 = \mathrm{Tor}_2(M'', N) \longrightarrow \mathrm{Tor}_1(M', N) \longrightarrow \mathrm{Tor}_1(M, N) = 0$$

for any R-module N. Hence $\mathrm{Tor}_1(M', N) = 0$ which implies that M' is flat.

Next, assume that M'' is not flat. In this case, we claim that

$$\text{w. dim}(M'') = \text{w. dim}(M') + 1.$$

Indeed, let w. dim$(M') = n$. Then we have the exact sequence

$$0 = \mathrm{Tor}_{n+2}(M, N) \longrightarrow \mathrm{Tor}_{n+2}(M'', N) \longrightarrow \mathrm{Tor}_{n+1}(M', N) = 0$$

for any R-module N. Hence $\mathrm{Tor}_{n+2}(M'', N) = 0$ for any R-module N which implies

$$\text{w. dim}(M'') \leq n + 1 = \text{w. dim}(M') + 1.$$

Now let w. dim$(M'') = m$. Then we have the exact sequence

$$0 = \mathrm{Tor}_{m+1}(M'', N) \longrightarrow \mathrm{Tor}_m(M', N) \longrightarrow \mathrm{Tor}_m(M, N) = 0$$

for any R-module N. Hence $\mathrm{Tor}_m(M', N) = 0$ for any R-module N which implies that

$$\text{w. dim}(M'') = m \geq \text{w. dim}(M') + 1.$$

Consequently, $\mathrm{w. \dim}(M'') = \mathrm{w. \dim}(M') + 1$. ☐

Recall that an exact sequence of R-modules

$$0 \longrightarrow M' \longrightarrow M \longrightarrow M'' \longrightarrow 0$$

is pure if it remains exact when tensoring it with any R-module. In this case, we say that M' is a pure submodule of M [23].

Lemma 3.11 ([5, Lemma 6.2]). *Let $(R, \mathfrak{m})$ be a local ring which is not a field. Then $\mathrm{w. \dim}(R/\mathfrak{m}) = \mathrm{w. \dim}(\mathfrak{m}) + 1$.*

Proof. Consider the short exact sequence

$$0 \to \mathfrak{m} \to R \to R/\mathfrak{m} \to 0.$$

Assume that $R/\mathfrak{m}$ is flat. By [13, Theorem 1.2.15 (1,2,3)], $\mathfrak{m}$ is pure and $(aR)\,\mathfrak{m} = aR \cap \mathfrak{m} = aR$ for all $a \in \mathfrak{m}$. Hence $a\,\mathfrak{m} = aR$, for all $a \in \mathfrak{m}$, and so by Nakayama's Lemma, $a = 0$, absurd. By Lemma 3.10, $\mathrm{w. \dim}(R/\mathfrak{m}) = \mathrm{w. \dim}_R(\mathfrak{m}) + 1$. ☐

Proposition 3.12 ([5, Proposition 6.3]). *Let $(R, \mathfrak{m})$ be a local ring with nonzero nilpotent maximal ideal. Then $\mathrm{w. \dim}(\mathfrak{m}) = \infty$.*

Proof. Let n be the minimum integer such that $\mathfrak{m}^n = 0$. We claim that for all $1 \le k < n$, $\mathrm{w. \dim}(\mathfrak{m}^{n-k}) = \mathrm{w. \dim}(\mathfrak{m}) + 1$. Indeed, let $k = 1$. Then $\mathfrak{m}^{n-1}\,\mathfrak{m} = 0$, so $\mathfrak{m}^{n-1}$ is an $(R/\mathfrak{m})$-vector space; hence $0 \neq \mathfrak{m}^{n-1} \cong \bigoplus R/\mathfrak{m}$ implies that $\mathrm{w. \dim}_R(\mathfrak{m}^{n-1}) = \mathrm{w. \dim}(R/\mathfrak{m}) = \mathrm{w. \dim}(\mathfrak{m}) + 1$ by Lemma 3.11. Now let h be the maximum integer in $\{1, \ldots, n-1\}$ such that $\mathrm{w. \dim}(\mathfrak{m}^{n-k}) = \mathrm{w. \dim}(\mathfrak{m}) + 1$ for all $k \le h$. Assume by way of contradiction that $h < n - 1$. Then we have the exact sequence:

$$0 \to \mathfrak{m}^{n-h} \to \mathfrak{m}^{n-(h+1)} \to \mathfrak{m}^{n-(h+1)}/\mathfrak{m}^{n-h} \to 0, \qquad (*)$$

where $\mathfrak{m}^{n-(h+1)}/\mathfrak{m}^{n-h}$ is a nonzero $(R/\mathfrak{m})$-vector space. So by Lemma 3.11, we have $\mathrm{w. \dim}(\mathfrak{m}^{n-(h+1)}/\mathfrak{m}^{n-h}) = \mathrm{w. \dim}(\mathfrak{m}) + 1$. By hypothesis, $\mathrm{w. \dim}(\mathfrak{m}^{n-h}) = \mathrm{w. \dim}(\mathfrak{m}) + 1$. Let us show that $\mathrm{w. \dim}(\mathfrak{m}^{n-(h+1)}) = \mathrm{w. \dim}(\mathfrak{m}) + 1$. Indeed, if $l := \mathrm{w. \dim}(\mathfrak{m}) + 1$, then by applying the long exact sequence theorem to $(*)$, we get

$$0 = \mathrm{Tor}_{l+1}(\mathfrak{m}^{n-h}, N) \longrightarrow \mathrm{Tor}_{l+1}(\mathfrak{m}^{n-(h+1)}, N) \longrightarrow \mathrm{Tor}_{l+1}\left(\frac{\mathfrak{m}^{n-(h+1)}}{\mathfrak{m}^{n-h}}, N\right) = 0$$

for any R-module N. Hence $\mathrm{Tor}_{l+1}(\mathfrak{m}^{n-(h+1)}, N) = 0$ for any R-module N which implies

$$\mathrm{w. \dim}(\mathfrak{m}^{n-(h+1)}) \le l = \mathrm{w. \dim}(\mathfrak{m}) + 1.$$

Further, if $\text{w. dim}(\mathfrak{m}^{n-(h+1)}) \lneqq l$, then we have

$$0 = \text{Tor}_{l+1}\left(\frac{\mathfrak{m}^{n-(h+1)}}{\mathfrak{m}^{n-h}}, N\right) \longrightarrow \text{Tor}_l(\mathfrak{m}^{n-h}, N) \longrightarrow \text{Tor}_l(\mathfrak{m}^{n-(h+1)}, N) = 0$$

for any R-module N. Hence $\text{Tor}_l(\mathfrak{m}^{n-h}, N) = 0$ for any R-module N which implies that $\text{w. dim}(\mathfrak{m}^{n-h)}) \leq l - 1$, absurd. Hence $\text{w. dim}(\mathfrak{m}^{n-(h+1)}) = \text{w. dim}(\mathfrak{m}) + 1$, the desired contradiction. Therefore the claim is true and, in particular, for $k = n - 1$, we have $\text{w. dim}(\mathfrak{m}) = \text{w. dim}(\mathfrak{m}) + 1$, which yields $\text{w. dim}(\mathfrak{m}) = \infty$.   □

Proof of Theorem 3.9. Suppose that R is Gaussian and $\mathfrak{m}$ is a maximal ideal in R such that $\text{Nil}(R_\mathfrak{m})$ is a nonzero nilpotent ideal. Then $R_\mathfrak{m}$ is also Gaussian and $\text{Nil}(R_\mathfrak{m})$ is a prime ideal in R. Moreover $\text{Nil}(R_\mathfrak{m}) = pR_\mathfrak{m} \neq 0$ for some prime ideal p in R. Now, the maximal ideal pR_p of R_p is nonzero since $0 \neq pR_\mathfrak{m} \subseteq pR_p$. Also by assumption, there is a positive integer n such that $(pR_\mathfrak{m})^n = 0$, whence $p^n = 0$. So $(pR_p)^n = 0$ and hence pR_p is nilpotent. Therefore R_p is a local ring with nonzero nilpotent maximal ideal. By Proposition 3.12, $\text{w. gl. dim}(R_p) = \infty$. Since $\text{w. gl. dim}(R) \geq \text{w. gl. dim}(R_S)$ for any localization R_S of R, we get $\text{w. gl. dim}(R) = \infty$.   □

In the previous section, we saw that the weak global dimension of an arithmetical ring is 0, 1, or ∞. In this section, we saw that the same result holds if R is Prüfer coherent or R is a Gaussian ring with a maximal ideal $\mathfrak{m}$ such that $\text{Nil}(R_\mathfrak{m})$ is a nonzero nilpotent ideal.

The question of whether this result is true for an arbitrary Gaussian ring was the object of Bazzoni-Glaz conjecture which sustained that the weak global dimension of a Gaussian ring is 0, 1, or ∞. In a first preprint [9], Donadze and Thomas claim to prove this conjecture in all cases except when the ring R is a non-reduced local Gaussian ring with nilradical N satisfying $N^2 = 0$. Then in a second preprint [10], they claim to prove the conjecture for all cases.

4   Gaussian Rings via Trivial Ring Extensions

In this section, we will use trivial ring extensions to construct new examples of non-arithmetical Gaussian rings , non-Gaussian Prüfer rings, and illustrative examples for Theorems 2.4 and 3.9. Let A be a ring and M an R-module. The trivial ring extension of A by M (also called the idealization of M over A) is the ring $R := A \ltimes M$ whose underlying group is $A \times M$ with multiplication given by

$$(a, x)(a', x') = (aa', ax' + a'x).$$

Recall that if I is an ideal of A and M' is a submodule of M such that $IM \subseteq M'$, then $J := I \ltimes M'$ is an ideal of R; ideals of R need not be of this form [20, Example 2.5]. However, the form of the prime (resp., maximal) ideals of R is $p \ltimes M$, where p is a prime (resp., maximal) ideal of A [17, Theorem 25.1(3)]. Suitable background on trivial extensions is [13, 17, 20].

The following lemma is useful for the construction of rings with infinite weak global dimension.

Lemma 4.1 ([3, Lemma 2.3]). *Let K be a field, E a nonzero K-vector space, and $R := K \ltimes E$. Then* w. gl. $\dim(R) = \infty$.

Proof. First note that $R^{(I)} \cong A^{(I)} \ltimes E^{(I)}$. So let us identify $R^{(I)}$ with $A^{(I)} \ltimes E^{(I)}$ as R-modules. Now let $\{f_i\}_{i \in I}$ be a basis of E and $J := 0 \ltimes E$. Consider the R-map $u : R^{(I)} \longrightarrow J$ defined by $u((a_i, e_i)_{i \in I}) = (0, \sum_{i \in I} a_i f_i)$. Then we have the following short exact sequence of R-modules

$$0 \longrightarrow \mathrm{Ker}(u) \longrightarrow R^{(I)} \overset{u}{\longrightarrow} J \longrightarrow 0$$

But $\mathrm{Ker}(u) = 0 \ltimes E^{(I)}$. Indeed, clearly $0 \ltimes E^{(I)} \subseteq \mathrm{Ker}(u)$. Now suppose $u((a_i, e_i)) = (0, 0)$. Then $\sum_{i \in I} a_i f_i = 0$; hence $a_i = 0$ for each i as $\{f_i\}_{i \in I}$ is a basis for E and we have the equality. Therefore the above exact sequence becomes

$$0 \longrightarrow 0 \ltimes E^{(I)} \longrightarrow R^{(I)} \overset{u}{\longrightarrow} J \longrightarrow 0 \qquad (*)$$

We claim that J is not flat. Suppose not. Then $0 \ltimes E^{(I)} \cap JR^{(I)} = (0 \ltimes E^{(I)})J$ by [23, Theorem 3.55]. But $(0 \ltimes E^{(I)})J = 0$. We use the above identification to obtain $0 = 0 \ltimes E^{(I)} \cap JR^{(I)} = (J)^{(I)} \cap J^{(I)} = J^{(I)} = 0 \ltimes E^{(I)}$, absurd (since $E \neq 0$).

Now, by Lemma 3.10, w. $\dim(J) = $ w. $\dim(J^{(I)}) + 1 = $ w. $\dim(J) + 1$. It follows that w. gl. $\dim(R) = $ w. $\dim(J) = \infty$. $\qquad \square$

Next, we announce the main result of this section.

Theorem 4.2 ([3, Theorem 3.1]). *Let $(A, \mathfrak{m})$ be a local ring, E a nonzero $\frac{A}{\mathfrak{m}}$-vector space, and $R := A \ltimes E$ the trivial ring extension of A by E. Then:*

(1) R is a total ring of quotients and hence a Prüfer ring.
(2) R is Gaussian if and only if A is Gaussian.
(3) R is arithmetical if and only if $A := K$ is a field and $\dim_K(E) = 1$.
(4) w. gl. $\dim(R) \gneqq 1$. If $\mathfrak{m}$ admits a minimal generating set, then w. gl. $\dim(R)$ is infinite.

Proof.

(1) Let $(a, e) \in R$. Then either $a \in \mathfrak{m}$ in which case we get $(a, e)(0, e) = (0, ae) = (0, 0)$ or $a \notin \mathfrak{m}$ which implies a is a unit and hence $(a, e)(a^{-1}, -a^{-2}e) = (1, 0)$, the unity of R. Therefore R is a total ring of quotients and hence a Prüfer ring.

(2) Suppose that R is Gaussian. Then, since $A \cong \frac{R}{0 \ltimes E}$ and the Gaussian property is stable under factor rings, A is Gaussian.

Conversely, assume that A is Gaussian and let $F := \sum(a_i, e_i)X^i$ be a polynomial in $R[X]$. If $a_i \notin \mathfrak{m}$ for some i, then (a_i, e_i) is invertible since we have $(a_i, e_i)(a_i^{-1}, -a^{-2}e_i) = (1, 0)$. We claim that F is Gaussian. Indeed, for any $G \in R[X]$, we have $c(F)c(G) = Rc(G) = c(G) \subseteq c(FG)$. The reverse inclusion always holds. If $a_i \in \mathfrak{m}$ for each i, let $G := \sum(a_j', e_j')X^j \in R[X]$. We may assume, without loss of generality, that $a_j' \in \mathfrak{m}$ for each j (otherwise, we return to the first case) and let $f := \sum a_i X^i$ and $g := \sum a_j' X^j$ in $A[X]$. Then $c(FG) = c(fg) \ltimes c(fg)E$. But since E is an $\frac{A}{\mathfrak{m}}$-vector space, $\mathfrak{m} E = 0$ yields $c(FG) = c(fg) \ltimes 0 = c(f)c(g) \ltimes 0 = c(F)c(G)$, since A is Gaussian. Therefore R is Gaussian, as desired.

(3) Suppose that R is arithmetical. First we claim that A is a field. On the contrary, assume that A is not a field. Then $\mathfrak{m} \neq 0$, so there is $a \neq 0 \in \mathfrak{m}$. Let $e \neq 0 \in E$. Since R is a local arithmetical ring (i.e., chained ring), either $(a, 0) = (a', e')(0, e) = (0, a'e)$ for some $(a', e') \in R$ which contradicts $a \neq 0$ or $(0, e) = (a'', e'')(a, 0) = (a'a, 0)$ for some $(a'', e'') \in R$ which contradicts $e \neq 0$. Hence A is a field. Next, we show that $\dim_K(E) = 1$. Let e, e' be two nonzero vectors in E. We claim that they are linearly dependent. Indeed, since R is a local arithmetical ring, either $(0, e) = (a, e'')(0, e') = (0, ae')$ for some $(a, e'') \in R$, hence $e = ae'$ or similarly if $(0, e') \in (0, e)R$. Consequently, $\dim_K(E) = 1$.

Conversely, let J be a nonzero ideal in $K \ltimes K$ and let (a, b) be a nonzero element of J. So $(0, a^{-1})(a, b) = (0, 1) \in J$. Hence $0 \ltimes K \subseteq J$. But $0 \ltimes K$ is maximal since 0 is the maximal ideal in K. So the ideals of $K \ltimes K$ are $(0, 0)K \ltimes K$, $0 \ltimes K = R(0, 1)$, and $K \ltimes K$. Therefore $K \ltimes K$ is a principal ring and hence arithmetical.

(4) First w. gl. $\dim(R) \gneqq 1$. Let $J := 0 \ltimes E$ and $\{f_i\}_{i \in I}$ be a basis of the $\frac{A}{\mathfrak{m}}$-vector space E. Consider the map $u : R^{(I)} \longrightarrow J$ defined by $u((a_i, e_i)_{i \in I}) = (0, \sum_{i \in I} a_i f_i)$. Here we are using the same identification that has been used in Lemma 4.1. Then clearly $\mathrm{Ker}(u) = (\mathfrak{m} \ltimes E)^{(I)}$. Hence we have the short exact sequence of R-modules

$$0 \longrightarrow (\mathfrak{m} \ltimes E)^{(I)} \longrightarrow R^{(I)} \xrightarrow{u} J \longrightarrow 0. \tag{1}$$

We claim that J is not flat. Otherwise, by [23, Theorem 3.55], we have

$$J^{(I)} = (\mathfrak{m} \ltimes E)^{(I)} \cap JR^{(I)} = J(\mathfrak{m} \ltimes E^{(I)}) = 0.$$

Hence, by [23, Theorem 2.44], w. gl. $\dim(R) \gneqq 1$.

Next, assume that $\mathfrak{m}$ admits a minimal generating set. Then $\mathfrak{m} \ltimes E$ admits a minimal generating set (since E is a vector space). Now let $(b_i, g_i)_{i \in L}$ be a minimal generating set of $\mathfrak{m} \ltimes E$. Consider the R-map $v : R^{(L)} \longrightarrow \mathfrak{m} \ltimes E$ defined by $v((a_i, e_i)_{i \in L}) = \sum_{i \in L}(a_i, e_i)(b_i, g_i)$. Then we have the exact sequence

$$0 \longrightarrow \mathrm{Ker}(v) \longrightarrow R^{(L)} \overset{v}{\longrightarrow} \mathfrak{m} \ltimes E \longrightarrow 0 \qquad (2)$$

We claim that $\mathrm{Ker}(v) \subseteq (\mathfrak{m} \ltimes E)^{(L)}$. On the contrary, suppose that there is $x = ((a_i, e_i)_{i \in L}) \in \mathrm{Ker}(v)$ and $x \notin (\mathfrak{m} \ltimes E)^{(L)}$. Then $\sum_{i \in L} (a_i, e_i)(b_i, g_i) = 0$ and as $x \notin (\mathfrak{m} \ltimes E)^{(L)}$, there is (a_j, e_j) with $a_j \notin \mathfrak{m}$. So that (a_j, e_j) is a unit, which contradicts the minimality of $(b_i, g_i)_{i \in L}$. It follows that

$$\mathrm{Ker}(v) = V \ltimes E^{(L)} = (V \ltimes 0) \bigoplus (0 \ltimes E^{(L)}) = (V \ltimes 0) \bigoplus J^{(L)}$$

where $V := \{(a_i)_{i \in L} \in \mathfrak{m}^i \mid \sum_{i \in L} a_i b_i = 0\}$. Indeed, if $x \in \mathrm{Ker}(v)$, then $x = (a_i, b_i)_{i \in L}$ where $a_i \in \mathfrak{m}$, $b_i \in E$, with $\sum_{i \in L} a_i b_i = 0$, hence $\mathrm{Ker}(v) \subseteq V \ltimes E^{(L)}$. The other inclusion is trivial. Now, by Lemma 3.10 applied to (1), we get

$$\mathrm{w.\,dim}(J) = \mathrm{w.\,dim}((\mathfrak{m} \ltimes E)^I) + 1 = \mathrm{w.\,dim}(\mathfrak{m} \ltimes E) + 1.$$

On the other hand, from (2) we obtain

$$\mathrm{w.\,dim}(J) \leq \mathrm{w.\,dim}(V \ltimes 0 \oplus J^L) = \mathrm{w.\,dim}(\mathrm{Ker}(v)) \leq \mathrm{w.\,dim}(\mathfrak{m} \ltimes E).$$

It follows that

$$\mathrm{w.\,dim}(J) \leq \mathrm{w.\,dim}(J) - 1.$$

Consequently, $\mathrm{w.\,gl.\,dim}(R) = \mathrm{w.\,dim}(J) = \infty$. $\qquad \square$

Next, we give examples of non-arithmetical Gaussian rings.

Example 4.3.

(1) Let p be a prime number. Then $(\mathbb{Z}_{(p)}, p\mathbb{Z}_{(p)})$ is a non-trivial valuation domain. Hence $\mathbb{Z}_{(p)} \ltimes \frac{\mathbb{Z}}{p\mathbb{Z}}$ is a non-arithmetical Gaussian total ring of quotients by Theorem 4.2.
(2) Since $\dim_{\mathbb{R}}(\mathbb{C}) = 2 \gneq 1$, $\mathbb{R} \ltimes \mathbb{C}$ is a non-arithmetical Gaussian total ring of quotient. In general, if K is a field and E is a K-vector space with $\dim_K(E) \gneq 1$, then $R := K \ltimes E$ is a non-arithmetical Gaussian total ring of quotients by Theorem 4.2.

Next, we provide examples of non-Gaussian total rings of quotients and hence non-Gaussian Prüfer rings.

Example 4.4. Let $(A, \mathfrak{m})$ be a non-valuation local domain. By Theorem 4.2, $R := A \ltimes \frac{A}{\mathfrak{m}}$ is a non-Gaussian total ring of quotients, hence a non-Gaussian Prüfer ring.

The following is an illustrative example for Theorem 2.4.

Example 4.5. Let $R := \mathbb{R} \ltimes \mathbb{R}$. Then R is a local ring with maximal ideal $0 \ltimes \mathbb{R}$ and $Z(R) = 0 \ltimes \mathbb{R}$. Further, R is arithmetical by Theorem 4.2. By Osofsky's Theorem (Theorem 2.4) or by Lemma 4.1, w. gl. $\dim(R) = \infty$.

Now we give an example of a non-coherent local Gaussian ring with nilpotent maximal ideal and infinite weak global dimension (i.e., an illustrative example for Theorem 3.9).

Example 4.6. Let K be a field and X an indeterminate over K and let $R := K \ltimes K[X]$. Then:

(1) R is a non-arithmetical Gaussian ring since K is Gaussian and $\dim_K(K[X]) = \infty$ by Theorem 4.2.
(2) R is not a coherent ring since $\dim_K(K[X]) = \infty$ by [20, Theorem 2.6].
(3) R is local with maximal ideal $\mathfrak{m} = 0 \ltimes K[X]$ by [17, Theorem 25.1(3)]. Also $\mathfrak{m}$ is nilpotent since $\mathfrak{m}^2 = 0$. Therefore, by Theorem 3.9, w. gl. $\dim(R) = \infty$.

5 Weak Global Dimension of fqp-Rings

Recently, Abuhlail, Jarrar, and Kabbaj studied commutative rings in which every finitely generated ideal is quasi-projective (fqp-rings). They investigated the correlation of fqp-rings with well-known Prüfer conditions; namely, they proved that fqp-rings stand strictly between the two classes of arithmetical rings and Gaussian rings [1, Theorem 3.2]. Also they generalized Osofsky's Theorem on the weak global dimension of arithmetical rings (and partially resolved Bazzoni-Glaz's related conjecture on Gaussian rings) by proving that the weak global dimension of an fqp-ring is 0, 1, or ∞ [1, Theorem 3.11]. In this section, we will give the proofs of the above mentioned results. Here too, the needed examples in this section will be constructed by using trivial ring extensions. We start by recalling some definitions.

Definition 5.1.

(1) Let M be an R-module. An R-module M' is M-projective if the map $\psi : \mathrm{Hom}_R(M', M) \longrightarrow \mathrm{Hom}_R(M', \frac{M}{N})$ is surjective for every submodule N of M.
(2) M' is quasi-projective if it is M'-projective.

Definition 5.2. A commutative ring R is said to be an fqp-ring if every finitely generated ideal of R is quasi-projective.

The following theorem establishes the relation between the class of fqp-rings and the two classes of arithmetical and Gaussian rings.

Theorem 5.3 ([1, Theorem 3.2]). *For a ring R, we have*

$$R \text{ arithmetical } \Rightarrow R \text{ fqp-ring } \Rightarrow R \text{ Gaussian}$$

where the implications are irreversible in general.

The proof of this theorem needs the following results.

Lemma 5.4 ([1, Lemma 2.2]). *Let R be a ring and let M be a finitely generated R-module. Then M is quasi-projective if and only if M is projective over $\frac{R}{\text{Ann}(M)}$.*

Lemma 5.5 ([12, Corollary 1.2]). *Let $M_{i1 \leq i \leq n}$ be a family of R-modules. Then $\bigoplus_{i=1}^{n} M_i$ is quasi-projective if and only if M_i is M_j-projective $\forall \; i, j \in \{1, 2, \ldots, \}$.*

Lemma 5.6 ([1, Lemma 3.6]). *Let R be an fqp-ring. Then $S^{-1}R$ is an fqp-ring, for any multiplicative closed subsets of R.*

Proof. Let J be a finitely generated ideal of $S^{-1}R$. Then $J = S^{-1}I$ for some finitely generated ideal I of R. Since R is an fqp-ring, I is quasi-projective and hence, by Lemma 5.4, I is projective over $\frac{R}{\text{Ann}(I)}$. By [23, Theorem 3.76], $J :=$ $S^{-1}I$ is projective over $\frac{S^{-1}R}{S^{-1}\text{Ann}(I)}$. But $S^{-1}\text{Ann}(I) = \text{Ann}(S^{-1}I) = \text{Ann}(J)$ by [2, Proposition 3.14]. Therefore $J := S^{-1}I$ is projective over $\frac{S^{-1}R}{\text{Ann}(S^{-1}I)}$. Again by Lemma 5.4, J is quasi-projective. It follows that $S^{-1}R$ is an fqp-ring. $\square$

Lemma 5.7 ([1, Lemma 3.8]). *Let R be a local ring and a, b two nonzero elements of R such that (a) and (b) are incomparable. If (a, b) is quasi-projective, then $(a) \cap (b) = 0$, $a^2 = b^2 = ab = 0$, and $\text{Ann}(a) = \text{Ann}(b)$.*

Proof. Let $I := (a, b)$ be quasi-projective. Then by [26, Lemma 2], there exist f_1, $f_2 \in \text{End}_R(I)$ such that $f_1(I) \subseteq (a)$, $f_2(I) \subseteq (b)$, and $f_1 + f_2 = 1_I$. Now let $x \in (a) \cap (b)$. Then $x = r_1 a = r_2 b$ for some r_1, $r_2 \in R$. But $x = f_1(x) + f_2(x) = f_1(r_1 a) + f_2(r_2 b) = r_1 f_1(a) + r_2 f_2(b) = r_1 a' a + r_2 b' b = a' x + b' x$ where a', $b' \in R$. We claim that a' is a unit. Suppose not. Since R is local, $1 - a'$ is a unit. But $a = f_1(a) + f_2(a) = a' a + f_2(a)$. Hence $(1 - a')a = f_2(a) \subseteq (b)$ which implies that $a \in (b)$. This is absurd since (a) and (b) are incomparable. Similarly, b' is a unit. It follows that $(a' - (1 - b'))$ is a unit. But $x = a' x + b' x$ yields $(a' - (1 - b'))x = 0$. Therefore $x = 0$ and $(a) \cap (b) = 0$.

Next, we prove that $a^2 = b^2 = ab = 0$. Obviously, $(a) \cap (b) = 0$ implies that $ab = 0$. So it remains to prove that $a^2 = b^2 = 0$. Since $(a) \cap (b) = 0$, $I = (a) \oplus (b)$. By Lemma 5.5, (b) is (a)-projective. Let $\varphi : (a) \longrightarrow \frac{(a)}{a\,\text{Ann}(b)}$ be the canonical map and $g : (b) \longrightarrow \frac{(a)}{a\,\text{Ann}(b)}$ be defined by $g(rb) = r\bar{a}$. If $r_1 b = r_2 b$, then $(r_1 - r_2)b = 0$. Hence $r_1 - r_2 \in \text{Ann}(b)$ which implies that $(r_1 - r_2)\bar{a} = 0$. So $g(r_1 b) = g(r_2 b)$. Consequently, g is well defined. Clearly g is an R-map. Now, since (b) is (a)-projective, there exists an R-map $f : (b) \longrightarrow (a)$ with $\varphi \circ f = g$. For b, we have $f(b) \in (a)$; hence $f(b) = ra$ for some $r \in R$. Also $(\varphi \circ f)(b) = g(b)$. Hence $f(b) - a \in a\,\text{Ann}(b)$. Whence $ra - a = at$ for some $t \in \text{Ann}(b)$ which implies that $(t + 1)a = ra$. By multiplying the last equality by a we obtain, $(t + 1)a^2 = ra^2$. But $ab = 0$ implies $0 = f(ab) = af(b) = ra^2$. Hence $(t + 1)a^2 = 0$. Since $t \in \text{Ann}(b)$ and R is local, $(t + 1)$ is a unit. It follows that $a^2 = 0$. Likewise $b^2 = 0$.

Last, let $x \in \mathrm{Ann}(b)$. Then $f(xb) = xra = 0$. The above equality $(t+1)a = ra$ implies $(t + 1 - r)a = 0$. But $t + 1$ is a unit and R is local. So r is a unit $(b \neq 0)$. Hence $xa = 0$. Whence $x \in \mathrm{Ann}(a)$ and $\mathrm{Ann}(b) \subseteq \mathrm{Ann}(a)$. Similarly we can show that $\mathrm{Ann}(a) \subseteq Ann(b)$. Therefore $\mathrm{Ann}(a) = \mathrm{Ann}(b)$.   □

Proof of Theorem 5.3. R arithmetical $\Rightarrow R$ fqp-ring.

Let R be an arithmetical ring, I a nonzero finitely generated ideal of R, and p a prime ideal of R. Then $I_p := I R_p$ is finitely generated. But R is arithmetical; hence, R_p is a chained ring and I_p is a principal ideal of R_p. By [21], I_p is quasi-projective. By [29, 19.2] and [28], it suffices to prove that $(\mathrm{Hom}_R(I, \ I))_p \cong \mathrm{Hom}_{R_p}(I_p, \ I_p)$. But $\mathrm{Hom}_{R_p}(I_p, \ I_p) \cong \mathrm{Hom}_R(I, \ I_p)$ by the adjoint isomorphisms theorem [23, Theorem 2.11] (since $\mathrm{Hom}_{S^{-1}R}(S^{-1}N, S^{-1}M) \cong \mathrm{Hom}(N, S^{-1}M)$ where $S^{-1}N \cong N \otimes_R S^{-1}R$ and $S^{-1}M \cong \mathrm{Hom}_{S^{-1}R}(S^{-1}R, S^{-1}M))$. So let us prove that

$$(\mathrm{Hom}_R(I, \ I))_p \cong \mathrm{Hom}_R(I, \ I_p).$$

Let

$$\phi : (\mathrm{Hom}_R(I, \ I))_p \longrightarrow \mathrm{Hom}_R(I, \ I_p)$$

be the function defined by $\frac{f}{s} \in (\mathrm{Hom}_R(I, \ I))_p$, $\phi(\frac{f}{s}) : I \longrightarrow I_p$ with $\phi(\frac{f}{s})(x) = \frac{f(x)}{s}$, for each $x \in I$. Clearly ϕ is a well-defined R-map. Now suppose that $\phi(\frac{f}{s}) = 0$. I is finitely generated, so let $I = (x_1, x_2, \ldots, x_n)$, where n is an integer. Then for every $i \in \{1, 2, \ldots, n\}$, $\phi(\frac{f}{s})(x_i) = \frac{f(x_i)}{s} = 0$, whence there exists $t_i \in R \setminus p$ such that $t_i f(x_i) = 0$. Let $t := t_1 t_2 \ldots t_n$. Clearly, $t \in R \setminus p$ and $t f(x) = 0$, for all $x \in I$. Hence $\frac{f}{s} = 0$. Consequently, ϕ is injective. Next, let $g \in \mathrm{Hom}_R(I, \ I_p)$. Since I_p is principal in R_p, $I_p = aR_p$ for some $a \in I$. But $g(a) \in I_p$. Hence $g(a) = \frac{ca}{s}$ for some $c \in R$ and $s \in R \setminus p$. Let $x \in I$. Then $\frac{x}{1} \in I_p = aR_p$. Hence $\frac{x}{1} = \frac{ra}{u}$ for some $r \in R$ and $u \in R \setminus p$. So there exists $t \in R \setminus p$ such that $tux = tra$. Now, let $f : I \longrightarrow I$ be the multiplication by c. (i.e., for $x \in I$, $f(x) = cx$). Then $f \in \mathrm{Hom}_R(I, \ I)$ and we have

$$\phi\left(\frac{f}{s}\right)(x) = \frac{f(x)}{s} = \frac{cx}{s} = \frac{c}{s}\frac{x}{1} = \frac{cra}{su} = \frac{r}{u}g(a) = \frac{1}{tu}g(tra) = \frac{1}{tu}g(tux) = g(x).$$

Therefore ϕ is surjective and hence an isomorphism, as desired.

R fqp-ring $\Rightarrow R$ Gaussian

Recall that, if $(R, \mathfrak{m})$ is a local ring with maximal ideal $\mathfrak{m}$, then R is a Gaussian ring if and only if for any two elements a, b in R, $(a, b)^2 = (a^2)$ or (b^2) and if $(a, b)^2 = (a^2)$ and $ab = 0$, then $b^2 = 0$ [5, Theorem 2.2 (d)].

Let R be an fqp-ring and let P be any prime ideal of R. Then by Lemma 5.6 R_p is a local fqp-ring. Let $a, b \in R_P$. We investigate two cases. The first case is $(a, b) = (a)$ or (b), say (b). So $(a, b)^2 = (b^2)$. Now assume that $ab = 0$.

Since $a \in (b)$, $a = cb$ for some $c \in R$. Therefore $a^2 = cab = 0$. The second case is $I := (a, b)$ with $I \neq (a)$ and $I \neq (b)$. Necessarily, $a \neq 0$ and $b \neq 0$. By Lemma 5.7, $a^2 = b^2 = ab = 0$. Both cases satisfy the conditions that were mentioned at the beginning of this proof (The conditions of [5, Theorem 2.2 (d)]). Hence R_p is Gaussian. But p being an arbitrary prime ideal of R and the Gaussian notion being a local property, then R is Gaussian.

To prove that the implications are irreversible in general, we will use the following theorem to build examples for this purpose. □

Theorem 5.8 ([1, Theorem 4.4]). *Let $(A, \mathfrak{m})$ be a local ring and E a nonzero $\frac{A}{\mathfrak{m}}$-vector space. Let $R := A \ltimes E$ be the trivial ring extension of A by E. Then R is an fqp-ring if and only if $\mathfrak{m}^2 = 0$.*

The proof of this theorem depends on the following lemmas.

Lemma 5.9 ([24, Theorem 2]). *Let R be a local fqp-ring which is not a chained ring. Then $(\mathrm{Nil}(R))^2 = 0$.*

Lemma 5.10 ([1, Lemma 4.5]). *Let R be a local fqp-ring which is not a chained ring. Then $Z(R) = \mathrm{Nil}(R)$.*

Proof. We always have $\mathrm{Nil}(R) \subseteq Z(R)$. Now, let $s \in Z(R)$. Then there exists $t \neq 0 \in R$ such that $st = 0$. Since R is not chained, there exist nonzero elements $x, y \in R$ such that (x) and (y) are incomparable. By Lemma 5.7, $x^2 = xy = y^2 = 0$. Either (x) and (s) are incomparable and hence, by Lemma 5.7, $s^2 = 0$, whence $s \in \mathrm{Nil}(R)$, or (x) and (s) are comparable. In this case, either $s = rx$ for some $r \in R$ which implies that $s^2 = r^2x^2 = 0$ and hence $s \in \mathrm{Nil}(R)$. Or $x = sx'$ for some $x' \in R$. Same arguments applied to (s) and (y) yield either $s \in \mathrm{Nil}(R)$ or $y = sy'$ for some $y' \in R$. Since (x) and (y) are incomparable, (x') and (y') are incomparable. Hence, by Lemma 5.7, $(x') \cap (y') = 0$. If (x') and (t) are incomparable, then by Lemma 5.7 $\mathrm{Ann}(x') = \mathrm{Ann}(t)$, so that $s \in \mathrm{Ann}(x')$ which implies that $x = sx' = 0$, absurd. If $(t) \subseteq (x')$, then $(t) \cap (y') \subseteq (x') \cap (y') = 0$. So (t) and (y') are incomparable, whence similar arguments as above yield $y = 0$, absurd. Last, if $(x') \subseteq (t)$, then $x' = r't$ for some $r' \in R$. Hence $x = sx' = str' = 0$, absurd. Therefore all the possible cases lead to $s \in \mathrm{Nil}(R)$. Consequently, $Z(R) = \mathrm{Nil}(R)$. □

Lemma 5.11 ([1, Lemma 4.6]). *Let $(R, \mathfrak{m})$ be a local ring such that $\mathfrak{m}^2 = 0$. Then R is an fqp-ring.*

Proof. Let I be a nonzero proper finitely generated ideal of R. Then $I \subseteq \mathfrak{m}$ and $\mathfrak{m}I = 0$. Hence $\mathfrak{m} \subseteq \mathrm{Ann}(I)$, whence $\mathfrak{m} = \mathrm{Ann}(I)$ $(I \neq 0)$. So that $\frac{R}{\mathrm{Ann}(I)} \cong \frac{A}{\mathfrak{m}}$ which implies that I is a free $\frac{R}{\mathrm{Ann}(I)}$-module, hence projective over $\frac{R}{\mathrm{Ann}(I)}$. By Lemma 5.4, I is quasi-projective. Consequently, R is an fqp-ring. □

Proof of Theorem 5.8. Assume that R is an fqp-ring. We may suppose that A is not a field. Then R is not a chained ring since $((a, 0)$ and $((0, e))$ are incomparable where $a \neq 0 \in \mathfrak{m}$ and $e = (1, 0, 0, \ldots) \in E$. Also R is local with maximal $\mathfrak{m} \ltimes E$.

By Lemma 5.10, $Z(R) = \mathrm{Nil}(R)$. But $\mathfrak{m} \ltimes E = Z(R)$. Let $(a, e) \in \mathfrak{m} \ltimes E$. Since E is an $\frac{A}{\mathfrak{m}}$-vector space, $(a, e)(0, e) = (0, ae) = (0, 0)$. Hence $\mathfrak{m} \ltimes E \subseteq Z(R)$. The other inclusion holds since $Z(R)$ is an ideal. Hence $\mathfrak{m} \ltimes E = \mathrm{Nil}(R)$. By Lemma 5.9, $(\mathrm{Nil}(R))^2 = 0 = (\mathfrak{m} \ltimes E)^2$. Consequently, $\mathfrak{m}^2 = 0$.

Conversely, $\mathfrak{m}^2 = 0$ implies $(\mathfrak{m} \ltimes E)^2 = 0$ and hence by Lemma 5.11, R is an fqp-ring.   □

Now we can use Theorem 5.8 to construct examples which prove that the implications in Theorem 5.3 cannot be reversed in general. The following is an example of an fqp-ring which is not an arithmetical ring

Example 5.12. $R := \frac{\mathbb{R}[X]}{(X^2)} \ltimes \mathbb{R}$ is an fqp-ring by Theorem 5.8, since R is local with a nilpotent maximal ideal $\frac{(X)}{(X^2)} \ltimes \mathbb{R}$. Also, since $\frac{\mathbb{R}[X]}{(X^2)}$ is not a field, R is not arithmetical by Theorem 4.2.

The following is an example of a Gaussian ring which is not an fqp-ring.

Example 5.13. $R := \mathbb{R}[X]_{(X)} \ltimes \mathbb{R}$ is Gaussian by Theorem 4.2. Also, by Theorem 5.8, R is not an fqp-ring.

Now the natural question is what are the values of the weak global dimension of an arbitrary fqp-ring? The answer is given by the following theorem.

Theorem 5.14 ([1, Theorem 3.11]). *Let R be an fqp-ring. Then* $\mathrm{w. gl. dim}(R) = 0, 1,$ *or* ∞.

Proof. Since $\mathrm{w. gl. dim}(R) = \sup\{\mathrm{w. gl. dim}(R_p) \mid p \text{ prime ideal of } R\}$, one can assume that R is a local fqp-ring. If R is reduced, then $\mathrm{w. gl. dim}(R) \leq 1$ by Lemma 3.5. If R is not reduced, then $\mathrm{Nil}(R) \neq 0$. By Lemma 5.9, either $(\mathrm{Nil}(R))^2 = 0$, in this case $\mathrm{w. gl. dim}(R) = \infty$ by Theorem 3.9 (since an fqp-ring is Gaussian) or R is a chained ring with zero divisors ($\mathrm{Nil}(R) \neq 0$), in this case $\mathrm{w. gl. dim}(R) = \infty$ by Theorem 2.3. Consequently, $\mathrm{w. gl. dim}(R) = 0, 1,$ or ∞.   □

It is clear that Theorem 5.14 generalizes Osofsky's Theorem on the weak global dimension of arithmetical rings (Theorem 2.3) and partially resolves Bazzoni-Glaz conjecture on Gaussian rings.

References

1. J. Abuihlail, M. Jarrar, S. Kabbaj, Commutative rings in which every finitely generated ideal is quasi-projective. J. Pure Appl. Algebra **215**, 2504–2511 (2011)
2. M.F. Atiyah, I.G. Macdonald, *Introduction to Commutative Algebra* (Westview Press, New York, 1969)
3. C. Bakkari, S. Kabbaj, N. Mahdou, Trivial extensions defined by Prüfer conditions. J. Pure Appl. Algebra **214**, 53–60 (2010)

4. S. Bazzoni, S. Glaz, *Prüfer Rings, Multiplicative Ideal Theory in Commutative Algebra* (Springer, New York, 2006), pp. 263–277
5. S. Bazzoni, S. Glaz, Gaussian properties of total rings of quotients. J. Algebra **310**, 180–193 (2007)
6. N. Bourbaki, *Commutative Algebra*, Chapters 1–7. (Springer, Berlin, 1998)
7. H.S. Butts, W. Smith, Prüfer rings. Math. Z. **95**, 196–211 (1967)
8. H. Cartan, S. Eilenberg, *Homological Algebra* (Princeton University Press, Princeton, 1956)
9. G. Donadze, V.Z. Thomas, On a conjecture on the weak global dimension of Gaussian rings. arXiv:1107.0440v1 (2011)
10. G. Donadze, V.Z. Thomas, Bazzoni-Glaz conjecture. arXiv:1203.4072v1 (2012)
11. L. Fuchs, Über die Ideale Arithmetischer Ringe. Comment. Math. Helv. **23**, 334–341 (1949)
12. K.R. Fuller, D.A. Hill, On quasi-projective modules via relative projectivity. Arch. Math. (Basel) **21**, 369–373 (1970)
13. S. Glaz, *Commutative Coherent Rings*. Lecture Notes in Mathematics, vol. 1371 (Springer, Berlin, 1989)
14. S. Glaz, *Prüfer Conditions in Rings with Zero-Divisors*. Series of Lectures in Pure and Applied Mathematics, vol. 241 (CRC Press, Boca Raton, 2005), pp. 272–282
15. S. Glaz, The weak dimension of Gaussian rings. Proc. Am. Math. Soc. **133**(9), 2507–2513 (2005)
16. M. Griffin, Prüfer rings with zero-divisors. J. Reine Angew. Math. **239/240**, 55–67 (1969)
17. J.A. Huckaba, *Commutative Rings with Zero-Divisors* (Dekker, New York, 1988)
18. C.U. Jensen, Arithmetical rings. Acta Math. Hungar. **17**, 115–123 (1966)
19. S. Lang, *Algebra, Graduate Texts in Mathematics* (Springer, New York, 2002)
20. S. Kabbaj, N. Mahdou, Trivial extensions defined by coherent-like conditions. Comm. Algebra **32**(10), 3937–3953 (2004)
21. A. Koehler, Rings for which every cyclic module is quasi-projective. Math. Ann. **189**, 311–316 (1970)
22. B. Osofsky, Global dimension of commutative rings with linearly ordered ideals. J. London Math. Soc. **44**, 183–185 (1969)
23. J.J. Rotman, *An Introduction to Homological Algebra* (Academic, New York, 1979)
24. S. Singh, A. Mohammad, Rings in which every finitely generated left ideal is quasi-projective. J. Indian Math. Soc. **40**(1–4), 195–205 (1976)
25. H. Tsang, Gauss's lemma, Ph.D. thesis, University of Chicago, Chicago, 1965
26. A. Tuganbaev, Quasi-projective modules with the finite exchange property. Communications of the Moscow Mathematical Society. Russian Math. Surveys **54**(2), 459–460 (1999)
27. W.V. Vasconcelos, *The Rings of Dimension Two*. Lecture Notes in Pure and Applied Mathematics, vol. 22 (Dekker, New York, 1976)
28. R. Wisbauer, Local-global results for modules over algebras and Azumaya rings. J. Algebra **135**, 440–455 (1990)
29. R. Wisbauer, *Modules and Algebras: Bimodule Structure and Group Actions on Algebras* (Longman, Harlow, 1996)

Quasi-complete Semilocal Rings and Modules

Daniel D. Anderson

Abstract Let R be a (commutative Noetherian) semilocal ring with Jacobon radical J. Chevalley has shown that if R is complete, then R satisfies the following condition: given any descending chain of ideals $\{A_n\}_{n=1}^{\infty}$ with $\bigcap_{n=1}^{\infty} A_n = 0$, for each positive integer k there exists an s_k with $A_{s_k} \subseteq J^k$. A finitely generated R-module M is said to be *(weakly) quasi-complete* if for any descending chain $\{A_n\}_{n=1}^{\infty}$ of R-submodules of M (with $\bigcap_{n=1}^{\infty} A_n = 0$) and $k \geq 1$, there exists an s_k with $A_{s_k} \subseteq (\bigcap_{n=1}^{\infty} A_n) + J^k M$. An easy modification of Chevalley's proof shows that a finitely generated R-module over a complete semilocal ring is quasi-complete. However, the converse is false as any DVR is quasi-complete. In this paper we survey known results about (weakly) quasi-complete rings and modules and prove some new results.

Keywords Quasi-complete rings • Quasi-complete modules • Noether lattices

Subject Classifications: 13E05, 13H10, 13A15, 06F10.

1 Introduction

Throughout this paper all rings are commutative with identity and all modules are unitary. Local rings and semilocal rings carry the Noetherian hypothesis.

The following result is proved by Chevalley [6, Lemma 7].

D.D. Anderson (✉)
Department of Mathematics, University of Iowa, Iowa City, IA 52242, USA
e-mail: dan-anderson@uiowa.edu

M. Fontana et al. (eds.), *Commutative Algebra: Recent Advances in Commutative Rings,*
Integer-Valued Polynomials, and Polynomial Functions, DOI 10.1007/978-1-4939-0925-4_2,
© Springer Science+Business Media New York 2014

Theorem 1. *Let R be a complete semilocal ring, and let (b_n) be a sequence of ideals in R such that $b_{n+1} \subset b_n$ $(1 \le n < \infty)$ and $\bigcap_{n=1}^{\infty} b_n = \{0\}$. If $p_1, \ldots, p_k$ are the maximal prime ideals of R, we have $b_n \subset (p_1 \cdots p_k)^{m(n)}$, where $m(n)$ is an exponent which increases indefinitely with n.*

An equivalent formation may be found in Nagata [17, Theorem 30.1].

Theorem 2. *Let R be a semilocal ring with Jacobson radical J. If a_n $(n = 1, 2, 3, \ldots)$ are ideals of R such that $a_{n+1} \subseteq a_n$ for any n and such that $\bigcap_{n=1}^{\infty} a_n = 0$, then given any natural number n, there exists a natural number $m(n)$ such that $a_{m(n)} \subseteq J^n$.*

A proof of the theorem in the local case may also be found in Northcott [18, Theorem 1, page 86]. All three authors use the theorem to show that if R is a semilocal ring with Jacobson radical J and R' is a semilocal extension ring of R with Jacobson radical J', then R is a subspace of R' if and only if $J = J' \cap R$ [6, Proposition 4], [17, Corollary 30.2], [18, Theorem 2, page 88]. By this we mean that the J-adic topology on R is the subspace topology induced by the J'-adic topology on R'. We make the following fundamental definition.

Definition 1. Let R be a semilocal ring with Jacobson radical J and let M be a finitely generated R-module. Then M is *(weakly) J-quasi-complete* if for any descending chain $\{A_n\}_{n=1}^{\infty}$ of submodules of M (with $\bigcap_{n=1}^{\infty} A_n = 0$) and $k \ge 1$, there exists an s_k with $A_{s_k} \subseteq (\bigcap_{n=1}^{\infty} A_n) + J^k M$.

When the context is clear we will often just say (weakly) quasi-complete. Now an easy modification of any of the three previously mentioned proofs shows that a finitely generated module over a complete semilocal ring is weakly quasi-complete and hence is actually quasi-complete as seen by passing to the R-module $M/\bigcap_{n=1}^{\infty} A_n$. We state this as a theorem and for completeness give its proof.

Theorem 3. *Let M be a finitely generated module over a complete semilocal ring. Then M is quasi-complete (and hence is weakly quasi-complete).*

Proof. Let R be a complete semilocal ring with Jacobson radical J. Let $\{A_n\}_{n=1}^{\infty}$ be a descending chain of submodules of M and put $A = \bigcap_{n=1}^{\infty} A_n$. Then $\bar{M} = M/A$ is again a finitely generated module over the complete semilocal ring R. Put $\bar{A}_n = A_n/A$; so $\{\bar{A}_n\}_{n=1}^{\infty}$ is a descending chain of submodules of $\bar{M}$ with $\bigcap_{n=1}^{\infty} \bar{A}_n = 0$. If we show that $\bar{M}$ is weakly quasi-complete, then for each $k \ge 1$, there is an s_k with $\bar{A}_{s_k} \subseteq J^k \bar{M}$ and hence $A_{s_k} \subseteq A + J^k M$. Thus it suffices to show that each finitely generated R-module is weakly quasi-complete. So let $\{A_n\}_{n=1}^{\infty}$ be a descending chain of submodules of M with $\bigcap_{n=1}^{\infty} A_n = 0$. Let $k \ge 1$. Put $A'_n = A_n + J^k M$. So $\{A'_n\}_{n=1}^{\infty}$ is a descending chain of submodules of M which stabilizes since $M/J^k M$ is Artinian; say $A'_{s_k} = A'_{s_k+1} = \cdots$. We may assume that the sequence $\{s_k\}_{k=1}^{\infty}$ is strictly increasing. Put $C_k = A_{s_k}$, so $\{C_k\}_{k=1}^{\infty}$ is a descending chain of R-submodules with $\bigcap_{k=1}^{\infty} C_k = 0$. Also, $C_k + J^k M = C_{k+1} + J^k M$. We show that $C_k \subseteq J^k M$ for each $k \ge 1$. Let $x \in C_k$. Now $x \in C_k \subseteq C_{k+1} + J^k M$, so $x = y_{k+1} + a_k$ where $y_{k+1} \in C_{k+1}$ and $a_k \in J^k M$. Now $y_{k+1} \in C_{k+1} \subseteq C_{k+2} + J^{k+1}M$. Continuing we get sequences $y_{k+1}, y_{k+2}, \ldots$ and $a_k, a_{k+1}, \ldots$ where $y_{k+j} \in C_{k+j}$ and $a_{k+j} \in J^{k+j}M$ with $x = y_{k+n} + a_k + a_{k+1} + \cdots + a_{k+n-1}$ for all n. Since $\lim_{n \to \infty} a_{k+n} = 0$, the sequence $t_n = a_k + a_{k+1} + \cdots + a_{k+n-1}$ converges, say to a. Hence $\lim_{n \to \infty} y_{k+n} = x - a$. Since $t_n \in J^k M$, $a \in J^k M$. Since $y_{k+p}, y_{k+p+1}, \ldots$ are all in C_{k+p}, we have $x - a \in C_{k+p}$ for every $p \geq 0$. Hence $x - a = 0$, so $x = a \in J^k M$. □

Now it turns out that the notion of completeness cannot be given solely in ideal-theoretic terms. For example, $\mathbb{Z}_{(p)}$ (p a prime) and its completion the p-adic integers $\hat{\mathbb{Z}}_{(p)}$ have isomorphic lattices of ideals, yet $\hat{\mathbb{Z}}_{(p)}$ is complete, while $\mathbb{Z}_{(p)}$ is not. (Throughout the paper we will use $\hat{M}$ to denote the completion of M.) In fact, $\mathbb{Z}_{(p)}$ is quasi-complete but not complete. Actually, any DVR, or more generally any one-dimensional analytically irreducible local domain is quasi-complete (Corollary 2), but need not be complete.

As noted in the previous paragraph the notion of completeness cannot be given in lattice-theoretic terms. Dilworth [7] introduced the Noether lattice as the abstraction of the lattice of ideals of a Noetherian ring. E.W. Johnson and J.A. Johnson have developed a theory of completions for lattice modules over semilocal Noether lattices (see the references). Basically, a lattice module is defined to be complete if it satisfies Definition 1 stated in lattice-theoretic terms. This is a reasonable definition as the map $L_R(M) \to L_{\hat{R}}(\hat{M})$ given by $N \to \hat{R} \otimes N = \hat{R}N$ from the lattice of R-submodule of M (R is a semilocal ring and M a finitely generated R-module) to the lattice of $\hat{R}$-submodules of $\hat{M}$ is a multiplicative lattice module isomorphism if and only if M is quasi-complete. This is explained in greater detail later in the paper (see especially the two paragraphs after the proof of Corollary 2).

In Sect. 2 we consider the case of quasi-complete local rings. Theorem 5 gives 15 additional characterizations of quasi-complete local rings. Many of these are extended to quasi-complete modules over semilocal rings in Sect. 3. Theorem 6 states that a quasi-complete semilocal ring is a finite direct product of quasi-complete local rings and thus effectively reduces the semilocal case to the local case.

2 Quasi-complete Local Rings

In this section we consider the local case of quasi-completeness. As we shall see in the next section, the general case readily reduces to local case (Theorem 6). For the reader's convenience we repeat the definition of (weak) quasi-completeness in the case of a local ring.

Let (R, M) be a local ring with maximal ideal M (here local includes the Noetherian hypothesis). We say R is (weakly) quasi-complete [8] if for each descending sequence $\{A_n\}_{n=1}^{\infty}$ of ideals of R (with $\bigcap_{n=1}^{\infty} A_n = 0$) and each $k \geq 1$, there exists an $s_k \geq 1$ with $A_{s_k} \subseteq (\bigcap_{n=1}^{\infty} A_n) + M^k$ ($A_{s_k} \subseteq M^k$). It is easily seen that R is quasi-complete if and only if R/A is weakly quasi-complete for each proper ideal A of R. Thus a homomorphic image of a quasi-complete local ring is quasi-complete. It is well known that a complete local ring is weakly quasi-complete (see, for instance [17, Theorem 30.1]) and hence quasi-complete since a homomorphic image of a complete local ring is again complete. A module-theoretic generalization was given in Theorem 3. We next give alternative characterizations of (weak) quasi-completeness. As usual, $\hat{R}$ denotes the M-adic completion of R.

Theorem 4. *For a local ring (R, M), the following conditions are equivalent.*

1. R *is (weakly) quasi-complete.*
2. *For each descending sequence $\{Q_n\}_{n=1}^{\infty}$ of M-primary ideals of R (with $\bigcap_{n=1}^{\infty} Q_n = 0$) and each $k \geq 1$, there exists an $s_k \geq 1$ with $Q_{s_k} \subseteq (\bigcap_{n=1}^{\infty} Q_n) + M^k$ ($Q_{s_k} \subseteq M^k$).*
3. *For ideals $A \subsetneq B$ of $\hat{R}$ (with $A = 0$), $R \cap A \subsetneq R \cap B$.*
4. *For any decreasing sequence $\{A_n\}_{n=1}^{\infty}$ of ideals of R (with $\bigcap_{n=1}^{\infty} A_n = 0$) and any finitely generated R-module N, $\bigcap_{n=1}^{\infty} A_n N = (\bigcap_{n=1}^{\infty} A_n) N$.*
5. *Condition (4) with N cyclic.*
6. *For any decreasing sequence $\{A_n\}_{n=1}^{\infty}$ of ideals of R (with $\bigcap_{n=1}^{\infty} A_n = 0$), we have $\bigcap_{n=1}^{\infty} \hat{R} A_n = \hat{R}(\bigcap_{n=1}^{\infty} A_n)$.*

Proof. (1)$\Rightarrow$(2) Clear. (2)$\Rightarrow$(3) Suppose that $A \subseteq B$ are ideals of $\hat{R}$ with $R \cap A = R \cap B$. Put $Q'_n = B + \hat{M}^n$; so Q'_n is $\hat{M}$-primary and $\bigcap_{n=1}^{\infty} Q'_n = B$. Let $Q_n = R \cap Q'_n$; so $\{Q_n\}_{n=1}^{\infty}$ is a descending sequence of M-primary ideals of R with $\bigcap_{n=1}^{\infty} Q_n = \bigcap_{n=1}^{\infty} (R \cap Q'_n) = R \cap (\bigcap_{n=1}^{\infty} Q'_n) = R \cap B = R \cap A$. Hence for each $k \geq 1$, there exists an $s_k \geq 1$ with $Q_{s_k} \subseteq (R \cap A) + M^k$. We can assume that $s_k \geq k$. Thus $Q'_{s_k} = \hat{R} Q_{s_k} \subseteq \hat{R}(R \cap A + M^k) = \hat{R}(R \cap A) + \hat{M}^k \subseteq A + \hat{M}^k$. So $B = \bigcap_{k=1}^{\infty} Q'_{s_k} \subseteq \bigcap_{k=1}^{\infty} (A + \hat{M}^k) = A$; hence $A = B$. (3)$\Rightarrow$(1) Let $\{A_n\}_{n=1}^{\infty}$ be a decreasing sequence of ideals of R. Then $\{\hat{R} A_n\}_{n=1}^{\infty}$ is a decreasing sequence of ideals of $\hat{R}$. Note that $R \cap (\bigcap_{n=1}^{\infty} \hat{R} A_n) = \bigcap_{n=1}^{\infty} (R \cap \hat{R} A_n) = \bigcap_{n=1}^{\infty} A_n = R \cap (\hat{R} \bigcap_{n=1}^{\infty} A_n)$; so $\bigcap_{n=1}^{\infty} \hat{R} A_n = \hat{R}(\bigcap_{n=1}^{\infty} A_n)$. Now for $k \geq 1$, there exists an s_k with $\hat{R} A_{s_k} \subseteq (\bigcap_{n=1}^{\infty} \hat{R} A_n) + \hat{M}^k = \hat{R}(\bigcap_{n=1}^{\infty} A_n) + \hat{R} M^k = \hat{R}(\bigcap_{n=1}^{\infty} A_n + M^k)$. Thus $A_{s_k} = R \cap \hat{R} A_{s_k} \subseteq R \cap (\hat{R}(\bigcap_{n=1}^{\infty} A_n + M^k)) = (\bigcap_{n=1}^{\infty} A_n) + M^k$. (1)$\Rightarrow$(4) Let $A = \bigcap_{n=1}^{\infty} A_n$. First suppose that R is weakly quasi-complete and $A = 0$. For $k \geq 1$, there is an $s_k \geq 1$ with $A_{s_k} \subseteq M^k$. Hence $A_{s_k} N \subseteq M^k N$. So $\bigcap_{n=1}^{\infty} A_n N \subseteq \bigcap_{k=1}^{\infty} A_{s_k} N \subseteq \bigcap_{k=1}^{\infty} M^k N = 0$. Hence $\bigcap_{n=1}^{\infty} A_n N = (\bigcap_{n=1}^{\infty} A_n) N$. Next suppose that R is quasi-complete and A is not necessarily 0. Now R/A is weakly quasi-complete and N/AN is a finitely generated R/A-module. So by the case $A = 0$, $\bigcap_{n=1}^{\infty} ((A_n/A)(N/AN)) = (\bigcap_{n=1}^{\infty} A_n/A)(N/AN)$. But this translates to $\bigcap_{n=1}^{\infty} \left(\frac{A_n N + AN}{AN} \right) = {}^{AN}/_{AN}$ or $\bigcap_{n=1}^{\infty} A_n N = AN$, which is what we needed to prove. (4)$\Rightarrow$(5) Clear. (5)$\Rightarrow$(1) Let $N = R/J$ where J is an ideal of R. Let $\{A_n\}_{n=1}^{\infty}$ be a decreasing sequence of ideals of R with $A = \bigcap_{n=1}^{\infty} A_n$. So $\bigcap_{n=1}^{\infty} A_n (R/J) = (\bigcap_{n=1}^{\infty} A_n)(R/J)$ or $\bigcap_{n=1}^{\infty} (A_n + J) = (\bigcap_{n=1}^{\infty} A_n) + J$. Suppose that $J = M^k$ where $k \geq 1$. Since R/M^k is Artinian, there exists an

$s_k \geq 1$ with $A_{s_k} + M^k = A_\ell + M^k$ for $\ell \geq s_k$. So $A_{s_k} \subseteq A_{s_k} + M^k = \bigcap_{n=1}^{\infty} (A_n + M^k) = (\bigcap_{n=1}^{\infty} A_n) + M^k$. So R is quasi-complete (weakly quasi-complete if we only assume that $A = 0$). (3)$\Rightarrow$(6) This is given in the first three sentences of the proof of (3)$\Rightarrow$(1). (6)$\Rightarrow$(1) This follows from the last two sentences of the proof of (3)$\Rightarrow$(1). □

Corollary 1. *Let (R, M) be a local ring and $\varphi: L(R) \rightarrow L(\hat{R})$ be given by $\varphi(A) = \hat{R}A$ where $L(R)$ [resp., $L(\hat{R})$] is the lattice of ideals of R (resp., $\hat{R}$). Then the following conditions are equivalent:*

1. *R is quasi-complete.*
2. *φ is surjective.*
3. *φ is a (multiplicative) lattice isomorphism.*
4. *Every (principal) ideal B of $\hat{R}$ has the form $B = \hat{R}A$ for some (necessarily principal) ideal A of R.*
5. *For each $x \in \hat{R}$, $x = ur$ for some $u \in U(\hat{R})$, the group of units of $\hat{R}$, and $r \in R$.*
6. *$L(R)$ and $L(\hat{R})$ are isomorphic as multiplicative lattices.*

Proof. (1)$\Rightarrow$(2) Let B be an ideal of $\hat{R}$. Then $\hat{R}(R \cap B) \subseteq B$ and $R \cap (\hat{R}(R \cap B)) = R \cap B$. So by Theorem 4, $\hat{R}(R \cap B) = B$. So φ is surjective. (2)$\Rightarrow$(3) As $\hat{R}A \cap R = A$ for each ideal A of R, φ is always injective. Thus by (2) φ is a bijection. Clearly φ and φ^{-1} are order preserving; so φ is a lattice isomorphism, even a multiplicative lattice isomorphism since $\varphi(AB) = \hat{R}AB = \hat{R}A\hat{R}B = \varphi(A)\varphi(B)$. (3)$\Rightarrow$(1) Suppose that $A \subsetneq B$ for ideals A and B of $\hat{R}$. Then $R \cap A = \varphi^{-1}(A) \subsetneq \varphi^{-1}(B) = R \cap B$. By Theorem 4, R is quasi-complete. (3)$\Rightarrow$(4) Clear. Note that if $\hat{R}x = \hat{R}B$ for some ideal B of R, then writing $B = \Sigma Rb_\alpha$ gives $\hat{R}x = \Sigma \hat{R}b_\alpha$. Since $\hat{R}x$ is principal and hence completely join-irreducible, $\hat{R}x = \hat{R}b_{\alpha_0}$ for some $b_{\alpha_0} \in R$. (4)$\Rightarrow$(2) This is clear since if each principal ideal of $\hat{R}$ is an extension of an ideal of R, then so is every ideal of $\hat{R}$. (5)$\Rightarrow$(4) Let $x \in \hat{R}$, so $x = ur$ where $u \in U(\hat{R})$ and $r \in R$. Then $\hat{R}x = \hat{R}ur = \hat{R}r = \hat{R}(Rr)$. (4)$\Rightarrow$(5) Let $x \in \hat{R}$, so $\hat{R}x = \hat{R}r$ for some $r \in R$. Hence $x = ur$ for some $u \in U(\hat{R})$. (3)$\Rightarrow$(6) Clear. (6)$\Rightarrow$(1) Let $\psi: L(R) \rightarrow L(\hat{R})$ be a multiplicative lattice isomorphism. First note that $\psi(M) = \hat{M}$ and so $\psi(M^n) = \hat{M}^n$. Let $\{A_n\}_{n=1}^{\infty}$ be a decreasing sequence of ideals of R. Then $\{\psi(A_n)\}_{n=1}^{\infty}$ is a decreasing sequence of ideals of $\hat{R}$. So for $k \geq 1$, there exists an s_k with $\psi(A_{s_k}) \subseteq \bigcap_{n=1}^{\infty} \psi(A_n) + \hat{M}^k = \psi(\bigcap_{n=1}^{\infty} A_n + M^k)$. Hence $A_{s_k} \subseteq \bigcap_{n=1}^{\infty} A_n + M^k$. So R is quasi-complete. □

Corollary 2. *Let R be a local integral domain.*

1. *Then R is weakly quasi-complete if and only if for each nonzero prime ideal P of $\hat{R}$, $P \cap R \neq 0$.*
2. *A weakly quasi-complete local domain is analytically irreducible.*
3. *Suppose further that dim $R = 1$. Then the following are equivalent.*

   a. *R is quasi-complete.*
   b. *R is weakly quasi-complete.*
   c. *R is analytically irreducible.*

Proof. (1) ($\Rightarrow$) Suppose that R is weakly quasi-complete. Then for any nonzero
ideal A of $\hat{R}$, prime or not, $A \cap R \neq 0$ by Theorem 4. ($\Leftarrow$) Suppose the R is
not weakly quasi-complete. So by Theorem 4 there is a nonzero ideal A of $\hat{R}$
with $A \cap R = 0$. Since $\hat{R}$ is Noetherian, we can suppose that A is maximal
with respect to this property. We claim that A is prime. Suppose that $xy \in A$,
but $x \notin A$ and $y \notin A$. Then $A \subsetneq (A, x)$ and $A \subsetneq (A, y)$; so $(A, x) \cap R \neq 0$
and $(A, y) \cap R \neq 0$. Now $((A, x) \cap R)((A, y) \cap R) \subseteq ((A, x)(A, y)) \cap R \subseteq$
$A \cap R = 0$. But R is an integral domain, so $((A, x) \cap R)((A, y) \cap R) \neq 0$.

(2) Suppose that R is a weakly quasi-complete integral domain. Suppose that $\hat{R}$ is
not a domain. Let P be a minimal prime of $\hat{R}$, so $P \subseteq Z(\hat{R})$. Let $0 \neq r \in$
$P \cap R$. Now $r \notin Z(R)$, and hence $r \notin Z(\hat{R})$ since $\hat{R}$ is a flat R-module. But
$r \in P \subseteq Z(\hat{R})$, a contradiction.

(3) We always have (a)$\Rightarrow$(b) and (b)$\Rightarrow$(c) follows from (2). (c)$\Rightarrow$(a) Now suppose
$\dim R = 1$. For a proper nonzero ideal A of R, $\dim R/A = 0$ and hence R/A
is complete. Thus R/A is (weakly) quasi-complete. Next, suppose that $A = 0$.
Now R is a one-dimensional analytically irreducible integral domain. So $\hat{M}$ is
the only nonzero prime ideal of $\hat{R}$. Certainly $\hat{M} \cap R = M \neq 0$. So by (1),
R is weakly quasi-complete. So every homomorphic image of R is (weakly)
quasi-complete and hence R is quasi-complete. $\qquad\square$

Now (1)$\Rightarrow$(4) and (1)$\Rightarrow$(5) of Corollary 1 have been obtained by E.W. Johnson
[8] from a different perspective. We outline his approach.

Let (R, M) be a local ring. For ideals A and B of R set $S(A, B) =$
$\sup \{i \,|\, A + M^i = B + M^i\}$; so $S(A, B) \geq 0$ and $S(A, B) = \infty$ if and only if
$A = B$. Let $d(A, B) = 1/2^{S(A,B)}$; so d is a metric on $L(R)$. He proved that $L(R)$
is complete if and only if R is quasi-complete and that the d-completion of $L(R)$
is isometric to $L(\hat{R})$. Later, J.A. Johnson [15] gave the equivalence of (1)–(6) of
Corollary 1. Also see [11,12] for the theory of completions of multiplicative lattices
and lattice modules. Note that for a decreasing sequence $\{A_n\}_{n=1}^{\infty}$ of ideals of R
with $A = \bigcap_{n=1}^{\infty} A_n$, $A = \lim_{n \to \infty} A_n$ if and only if for each $k \geq 1$, there exists an s_k
with $d(A_\ell, A) \leq 1/2^k$ for each $\ell \geq s_k$, i.e., $A_\ell + M^k = A + M^k$, or equivalently,
$A_\ell \subseteq A + M^k$. So R is (weakly) quasi-complete if and only if for each decreasing
sequence $\{A_n\}_{n=1}^{\infty}$ of ideals of R with $A = \bigcap_{n=1}^{\infty} A_n$ (with $A = 0$), $\lim_{n \to \infty} A_n = A$
[13].

We next list the various characterizations of quasi-complete local rings already
given and add a few more.

Theorem 5. *For a local ring (R, M) the following conditions are equivalent:*

1. *R is quasi-complete.*
2. *For each decreasing sequence $\{Q_n\}_{n=1}^{\infty}$ of M-primary ideals and each $k \geq 1$,
   there exits an $s_k \geq 1$ with $Q_{s_k} \subseteq (\bigcap_{n=1}^{\infty} Q_n) + M^k$.*
3. *Every homomorphic image of R is weakly quasi-complete.*

4. *Every homomorphic image* (S, N) *of* R *satisfies the following condition: if* $\{Q_n\}_{n=1}^{\infty}$ *is a decreasing sequence of* N-*primary ideals of* S *with* $\bigcap_{n=1}^{\infty} Q_n = 0$ *and* $k \geq 1$, *there exists an* $s_k \geq 1$ *so that* $Q_{s_k} \subseteq N^k$.

5. *For ideals* $A \subsetneqq B$ *of* $\hat{R}$, $R \cap A \subsetneqq R \cap B$.

6. *The map* $\varphi: L(R) \to L(\hat{R})$ *given by* $\varphi(A) = \hat{R}A$ *is surjective, i.e., each ideal* B *of* $\hat{R}$ *has the form* $B = \hat{R}A$ *for some ideal* A *of* R.

7. *The map* $\varphi: L(R) \to L(\hat{R})$ *is a multiplicative lattice isomorphism.*

8. $L(R)$ *and* $L(\hat{R})$ *are isomorphic as multiplicative lattices.*

9. $L(R)$ *is complete in the* d-*metric.*

10. *For each* $x \in \hat{R}$, *there exist* $u \in U(\hat{R})$ *and* $r \in R$ *with* $x = ur$.

11. *Given a decreasing sequence* $\{A_n\}_{n=1}^{\infty}$ *of ideals of* R *with* $\bigcap_{n=1}^{\infty} A_n = A$, $\lim_{n \to \infty} A_n = A$.

12. R *satisfies* $AB5^*$ (*the dual of* $AB5$), *i.e., for any ideal* B *of* R *and downward directed family* $\{B_\alpha\}$ *of ideals, we have* $B + (\bigcap B_\alpha) = \bigcap_\alpha (B + B_\alpha)$.

13. R *has a dual* R-*module* A, *i.e., there is an order-reversing lattice isomorphism* $\psi: L(R) \to L(A)$ *satisfying* $\psi(JN) = \psi(N):J$ *for all ideals* J *and submodules* N *of* R.

14. $E(R/M)$, *the injective envelope of* R/M, *is an* R-*module dual of* R.

15. *For any decreasing sequence* $\{A_n\}_{n=1}^{\infty}$ *of ideals of* R *and any finitely generated* (*or just cyclic*) R-*module* N, $\bigcap_{n=1}^{\infty} A_n N = (\bigcap_{n=1}^{\infty} A_n) N$.

16. *For any decreasing sequence* $\{A_n\}_{n=1}^{\infty}$ *of ideals of* R, *we have* $\bigcap_{n=1}^{\infty} \hat{R} A_n = \hat{R}(\bigcap_{n=1}^{\infty} A_n)$.

Proof. The equivalence of (1)–(11), (15), and (16) has already been given, while the equivalence of (1) and (12)–(14) may be found in [2]. □

We remark that in (12), the condition that $\{B_a\}$ is downward directed can be replaced by $\{B_a\}$ is a chain or even by $\{B_n\}_{n=1}^{\infty}$ is a countable descending chain, see [9, Lemma 3] or the proof of (5)$\Rightarrow$(1) of Theorem 4. Also, a "weak" version of any of these with $\bigcap B_a = 0$ characterizes weakly quasi-completeness.

Now for R a local ring we have R complete $\Rightarrow$ R is quasi-complete $\Rightarrow$ R is weakly quasi-complete. Now $Z_{(p)}$, $k[X]_{(X)}$ (k a field) or more generally any non-complete DVR (i.e., a one-dimensional regular local ring) is quasi-complete, but not complete. We do not know of an example of a weakly quasi-complete local ring that is not quasi-complete. We end this section with two examples. The first example shows that $k[X_1, \ldots, X_n]_{(X_1, \ldots, X_n)}$ need not be quasi-complete for $n \geq 2$. Thus a regular local ring need not be quasi-complete. The second example gives an example of a two-dimensional regular local ring that is quasi-complete but not complete.

Example 1. Let k be a countable field. Then $R_n = k[X_1, \ldots, X_n]_{(X_1, \ldots, X_n)}$ is weakly quasi-complete if and only if $n = 1$. Now $R_1 = k[X_1]_{(X_1)}$ is a DVR and hence is even quasi-complete. Suppose $n \geq 2$. By Corollary 2 it suffices to show there is a nonzero prime ideal P of $\hat{R}_n = k[[X_1, \ldots, X_n]]$ with $P \cap R_n = 0$, or equivalently, $P \cap k[X_1, \ldots, X_n] = 0$. It even suffices to show there is a nonzero prime ideal P of $\hat{R}_2$ with $P \cap k[X_1, X_2] = 0$. Suppose not. Now R_2 has countably infinite many height-one prime ideals while $\hat{R}_2 = k[[X_1, X_2]]$ has uncountably many height-one prime ideals. Note that if Q is a height-one prime ideal of $\hat{R}_2$, then $Q \cap k[X_1, X_2]$ is a height-one prime ideal of $k[X_1, X_2]$ (since we are assuming that $Q \cap k[X_1, X_2] \neq 0$). So there is some (necessarily principal) height-one prime ideal (p) of $k[X_1, X_2]$ having an uncountable set of height-one prime ideals $\{P_\alpha\}$ of $\hat{R}_2$ with $P_\alpha \cap k[X_1, X_2] = (p)$. So there are infinitely many prime ideals of $\hat{R}_2$ minimal over $\hat{R}_n\, p$, a contradiction.

Conjecture 1. For any field k, $k[X_1, \ldots, X_n]_{(X_1, \ldots, X_n)}$ is not weakly quasi-complete for $n \geq 2$.

Example 2. Suppose that $(D, (\pi))$ is a complete DVR. Then $D[X]_{(\pi, X)}$ is a two-dimensional regular local ring that is quasi-complete, but not complete. Now $\overline{D[X]_{(\pi, X)}} = D[[X]]$, so $D[X]_{(\pi, X)}$ is not complete. We show that $D[X]_{(\pi, X)}$ is quasi-complete using (1)$\Leftrightarrow$ (5) of Corollary 1. Let $0 \neq f \in D[[X]]$. So $f = \lambda \pi^n f'$ where λ is a unit of D, $n \geq 0$, and $f' \in D[[X]]$ has some coefficient a unit. Now by the Preparation Theorem (for example, see [5, Chap. VII, Sect. 3.8, Proposition 6, page 510]) $f' = \lambda' f''$ where λ' is a unit of $D[[X]]$ and $f'' \in D[X]$. Then $f = (\lambda \lambda')(\pi^n f'')$ where $\lambda \lambda'$ is a unit of $D[[X]]$ and $\pi^n f'' \in D[X]_{(\pi, X)}$.

Conjecture 2. For a DVR $(D, (\pi))$, $D[X]_{(\pi, X)}$ is quasi-complete if and only if D is complete.

3   The General Case

The notion of quasi-completeness for a local ring can be generalized to finitely generated modules over semilocal rings. (See [16] for a further generalization.) Let R be a semilocal ring (this includes the Noetherian hypothesis) with maximal ideals $M_1, \ldots, M_n$ and Jacobson radical $J = M_1 \cap \cdots \cap M_n = M_1 \cdots M_n$. Let M be a finitely generated R-module. Recall that M is said to be (*weakly*) J-*quasi-complete* if for any descending chain $\{A_n\}_{n=1}^\infty$ of R-submodules of M (with $\bigcap_{n=1}^\infty A_n = 0$) and $k \geq 1$, there exists an s_k with $A_{s_k} \subseteq (\bigcap_{n=1}^\infty A_n) + J^k M$. We next show that R is J-quasi-complete if and only if $R = R_{M_1} \times \cdots \times R_{M_n}$ and each R_{M_i} is quasi-complete.

Theorem 6. *Let R be a semilocal ring with maximal ideals $M_1, \ldots, M_n$ and let $J = M_1 \cap \cdots \cap M_n$. Then R is J-quasi-complete if and only if $R = R_{M_1} \times \cdots \times R_{M_n}$ and each R_{M_i} is $M_{i_{M_i}}$-quasi-complete.*

Proof. ($\Leftarrow$) With a change of notation $R = R_1 \times \cdots \times R_n$ where (R_i, M_i) is a quasi-complete local ring. Let $\{A_i\}_{i=1}^\infty$ be a descending sequence of ideals of R. So $A_i = A_{i1} \times \cdots \times A_{in}$ and $\bigcap_{i=1}^\infty A_i = (\bigcap_{i=1}^\infty A_{i1}) \times \cdots \times (\bigcap_{i=1}^\infty A_{in})$. Now each (R_j, M_j) is quasi-complete, so for $k \geq 1$, there exists a $k_j \geq 1$ so that

$A_{kjj} \subseteq (\bigcap_{i=1}^{\infty} A_{ij}) + M_j^k$. Let $k' = \max\{k_1, \ldots, k_n\}$, so $A_{k'} \subseteq ((\bigcap_{i=1}^{\infty} A_{i1}) + M_1^k) \times \cdots \times ((\bigcap_{i=1}^{\infty} A_{in}) + M_n^k) = (\bigcap_{i=1}^{\infty} A_i) + M_1^k \times \cdots \times M_n^k = (\bigcap_{i=1}^{\infty} A_i) + J^k$. So R is J-quasi-complete. ($\Rightarrow$) Let $Q_i = \bigcap_{k=1}^{\infty} M_i^k$. So $Q_1 \cap \cdots \cap Q_n \subseteq M_1^k \cap \cdots \cap M_n^k = J^k$ for each $k \geq 1$; and hence $Q_1 \cap \cdots \cap Q_n = 0$. Since R is J-quasi-complete, for $k = 1$ and $M_i \supseteq M_i^2 \supseteq \cdots$, there exists $k_i \geq 1$ with $M_i^{k_i} \subseteq Q_i + J$. So for $i \neq j$, $R = M_i^{k_i} + M_j^{k_j} \subseteq Q_i + Q_j + J$. Hence $Q_i + Q_j + J = R$ and therefore $Q_i + Q_j = R$. So $R \approx R/Q_1 \times \cdots \times R/Q_n$ by the Chinese Remainder Theorem. Moreover, R/Q_i is local with unique maximal ideal M_i/Q_i. For if $Q_i \subseteq M_j$ for $j \neq i$, $R = Q_i + Q_j \subseteq M_j$, a contradiction. So with a change of notation with $R = R_1 \times \cdots \times R_n$ where (R_i, M_i) is local and R is J-quasi-complete, it suffices to show that R_i is quasi-complete. Let $A_1 \supseteq A_2 \supseteq \cdots$ be a descending sequence of ideals of R_i and put $\bar{A}_i = R_1 \times \cdots \times R_{i-1} \times A_i \times R_{i+1} \times \cdots \times R_n$. So for $k \geq 1$, there exists n_k with $\bar{A}_{n_k} \subseteq (\bigcap_{i=1}^{\infty} \bar{A}_i) + J^k$ and hence as in the proof of ($\Leftarrow$), $A_{n_k} \subseteq (\bigcap_{i=1}^{\infty} A_i) + M^k$. So R_i is quasi-complete. $\square$

The proof of Theorem 6 can easily be modified to obtain the first part of the following result. The second part of Theorem 7 is well known.

Theorem 7. *Let $R_1, \ldots, R_n$ be semilocal rings. Then $R_1 \times \cdots \times R_n$ is (weakly) quasi-complete if and only if each R_i is (weakly) quasi-complete. Moreover, $R_1 \times \cdots \times R_n$ is complete if and only if each R_i is complete.*

Theorem 7 can be used to give examples of quasi-complete semilocal rings that are not complete.

Example 3. Let k be a field and $n \geq 1$. Then $R_n = k[[X_1, \ldots, X_n]] \times k[X]_{(X)}$ is an n-dimensional regular semilocal ring that is quasi-complete but not complete.

Theorem 6 essentially reduces the study of J-quasi-complete rings to the local case. Note that if $(R, M_1, \ldots, M_n)$ is J-quasi-complete, then $L(R) = L(R_{M_1}) \times \cdots \times L(R_{M_n}) = L(\widehat{R_{M_1}}) \times \cdots \times L(\widehat{R_{M_n}}) = L(\hat{R})$. We leave it to the reader to extend Theorem 5 to the semilocal case. Here "Q is M-primary" is replaced by "R/Q is Artinian." In fact, several of these characterizations already appear in this form in the literature: $(1) \Leftrightarrow (13) \Leftrightarrow (14)$ [2] and $(1) \Leftrightarrow (6) \Leftrightarrow (7) \Leftrightarrow (8) \Leftrightarrow (9) \Leftrightarrow (11)$ [12]. What is not entirely obvious in the semilocal case is that if $\hat{R}b = \hat{R}I$ for some ideal I of R, then we can take I to be principal and then $b = ua$ for some $u \in U(\hat{R})$ and $a \in R$. We next prove this for a finitely generated module over a semilocal ring.

Proposition 1. *Let R be a semilocal ring and M a finitely generated R-module. Let $m \in \hat{M}$. Suppose that $\hat{R}m = \hat{R}N$ for some submodule N of M. Then N is cyclic and hence $\hat{R}m = \hat{R}n$ for some $n \in N$. For any $n \in N$ with $\hat{R}m = \hat{R}n$, we have $m = un$ for some $u \in U(\hat{R})$. Hence if M is J-quasi-complete, for each $\hat{m} \in \hat{M}$, there exists $u \in U(\hat{R})$ and $m \in M$ with $\hat{m} = um$.*

Proof. It suffices to show that N is cyclic. For if R is any semiquasilocal ring and M any R-module, with $m_1, m_2 \in M$, $Rm_1 = Rm_2$ implies $m_2 = um_1$ for some $u \in U(R)$ [3, Corollary 13]. To show that N is cyclic, it suffices to show that

N is a multiplication module (i.e., for each submodule $K \subseteq N$, $K = (K{:}N)N$) since a multiplication module over a semiquasilocal ring is cyclic [4]. So suppose that K is a submodule of N. Now $\hat{R}N = \hat{R}m$ gives that $\hat{R}N$ is cyclic and hence a multiplication module over $\hat{R}$. So $\hat{K} \subseteq \hat{R}N$ gives $\hat{K} = (\hat{K}{:}\hat{R}N)\hat{R}N$, so $\hat{K} = (\hat{K}{:}\hat{R}N)\hat{R}N = \widetilde{(K{:}N)}\hat{R}N = \hat{R}(K{:}N)N$. But then $K = \hat{K} \cap M = \hat{R}((K{:}N)N) \cap M = (K{:}N)N$. $\qquad\square$

We next give several characterizations of J-quasi-complete modules.

Theorem 8. *For a finitely generated module M over a semilocal ring R, the following conditions are equivalent:*

1. *M is J-quasi-complete.*
2. *For each decreasing sequence $\{N_n\}_{n=1}^{\infty}$ of R-submodules of M with each M/N_n Artinian and for each $k \geq 1$, there exists an $s_k \geq 1$ with $N_{s_k} \subseteq (\bigcap_{n=1}^{\infty} N_n) + J^k M$.*
3. *Let $N_1 \subsetneq N_2$ be $\hat{R}$-submodules of $\hat{M}$. Then $N_1 \cap M \subsetneq N_2 \cap M$.*
4. *For each $\hat{R}$ submodule N of $\hat{M}$, $N = \hat{R}N'$ for some R-submodule N' of M.*
5. *The map $\varphi{:}L_R(M) \to L_{\hat{R}}(\hat{M})$ given by $\varphi(N) = \hat{R}N$ is a lattice module isomorphism, i.e., φ is a lattice isomorphism and $\varphi(IN) = I\varphi(N)$.*
6. *For $\hat{m} \in \hat{M}$, there exist $u \in U(\hat{R})$ and $m \in M$ with $\hat{m} = um$.*

Proof. The equivalence of (1)–(3) follows from the proof of Theorem 4, mutatis mutandis. Clearly (6)$\Rightarrow$(4) and (4)$\Rightarrow$(6) follows from Proposition 1. Also, clearly (5)$\Rightarrow$(4) and (4)$\Rightarrow$(5) since φ is always injective. So (4)–(6) are equivalent. Clearly (5)$\Rightarrow$(3) and (3)$\Rightarrow$(4) follows as in (1)$\Rightarrow$(2) of Corollary 2. $\qquad\square$

We can abstract condition (6) of Theorem 8. Let $f{:}R \to S$ be a ring homomorphism, M an R-module and $\tilde{M} = S \otimes_R M$. We say that $\tilde{M}$ is *S-unit M-generated* if for $\tilde{m} \in \tilde{M}$, there exists $u \in U(S)$ and $m \in M$ with $\tilde{m} = um$. So for R semilocal and M a finitely generated R-module with $S = \hat{R}$, $\tilde{M} = \hat{M}$ is $\hat{R}$-unit M-generated if and only if M is quasi-complete. Two cases where each module $\tilde{M}$ is S-unit M-generated are (1) $S = R_N$, N a multiplicatively closed set ($\tilde{M} = R_N \otimes_R M = M_N$, $\tilde{m} = m/n = (1/n)m$, $m \in M$, $n \in N$) and $S = R/I$ with f the natural map ($\tilde{M} = R/I \otimes_R M = M/IM$, $\tilde{m} = m + IM = (1 + I)m$). However, for the ring extension $R \to R[X]$ and nonzero R-module M, $\tilde{M} = R[X] \otimes_R M = M[X]$ is never $R[X]$-unit M-generated.

There is also a ring abstraction. Let $f{:}A \to B$ be a ring homomorphism. We call f a *U-homomorphism* (or a *U-extension* if f is the inclusion map) if for each $b \in B$, there exists an $a \in A$ and $u \in U(B)$ with $b = f(a)u$, or equivalently, for each $b \in B$, there exists a $u \in U(B)$ with $ub \in f(A)$. Examples include (a) $R \subseteq \hat{R}$ where R is quasi-complete, (b) the natural map $R \to R_S$ where S is a multiplicatively closed subset of R, and (c) any surjection. Consider the following conditions on a ring extension $A \subseteq B$: (1) $A \subseteq B$ is a U-extension, (2) for each principal ideal Bb of B, $Bb = BI$ for some principal ideal I of A, (3) for each principal ideal Bb of B, $Bb = BI$ for some ideal I of A, i.e., the map $\varphi{:}L(A) \to L(B)$ given by $\varphi(I) = BI$ is a surjection, and (4) the map $\varphi{:}L(A) \to L(B)$

given by $\varphi(I) = BI$ is a bijection and hence a multiplicative lattice isomorphism. Clearly (1)$\Rightarrow$(2)$\Rightarrow$(3) and (4)$\Rightarrow$(3). Note that taking $A = \mathbb{Z}$ and $B = \mathbb{Z}_{(2)}$ shows that (1)$\not\Rightarrow$(4) and hence (2)$\not\Rightarrow$(4) and (3)$\not\Rightarrow$(4). Also (4)$\not\Rightarrow$(2) and hence (3)$\not\Rightarrow$(2). Let D be Dedekind domain that is not a PID. Then the map $\varphi:L(D) \to L(D(X))$ given by $\varphi(I) = D(X)I$ is a lattice isomorphism [1, Theorem 8]. Suppose that $I = (a,b)$ is a nonprincipal ideal of D and let $f = a + bX$. So $D(X)I = D(X)f$, but $D(X)f \neq D(X)c$ for any $c \in D$. Now (2)$\Rightarrow$(1) is true if B is a strongly associate ring [3]. For if $Bb = Ba$ for $a \in A$, then $b = ua$ for some $u \in U(B)$.

Over a complete semilocal ring R all finitely generated R-modules are complete. We next show that a finitely generated module over a quasi-complete local ring (even a DVR) need not be quasi-complete. Also, given a finitely generated module M over a semilocal ring R and a submodule N of M, M is complete if and only if N and M/N are complete. We show that only the ($\Rightarrow$) implication carries over for quasi-complete modules.

Proposition 2. *Let R be a semilocal ring, M a finitely generated R-module and N a submodule of M.*

1. *If M is quasi-complete, then N and M/N are also quasi-complete.*
2. *$R \oplus M$ is quasi-complete if and only if R is quasi-complete and M is complete (as an R-module). Hence $R \oplus R$ is quasi-complete if and only if R is complete. Thus the converse of (1) is false.*
3. *If N is a complete R-module and M/N is quasi-complete, then M is quasi-complete.*

Proof. (1) Let $\hat{n} \in \hat{N}$; so M quasi-complete gives $\hat{n} = um$ for some unit $u \in \hat{R}$ and $m \in M$. But then $u^{-1}\hat{n} = m \in M \cap \hat{N} = N$. So N is quasi-complete. Next let $x \in \widehat{M/N} = \hat{M}/\hat{N}$; so $x = \hat{m} + \hat{N}$ where $\hat{m} \in \hat{M}$. So $\hat{m} = um$ where $u \in \hat{R}$ is a unit and $m \in M$. So $x = um + \hat{N} = u(m + \hat{N}) = u(m + N)$ where $m + N \in M/N$. So M/N is also quasi-complete.

(2) ($\Leftarrow$) This follows from (3). ($\Rightarrow$) Since $R \oplus M$ is quasi-complete, its homomorphic image R is quasi-complete. Let $\hat{m} \in \hat{M}$. Then for $(1, \hat{m}) \in \hat{R} \oplus \hat{M} = \widehat{R \oplus M}$, $(1, \hat{m}) = u(r, m)$ where $u \in U(\hat{R})$, $r \in R$, and $m \in M$. Now $1 = ur$ implies $r \in U(\hat{R}) \cap R = U(R)$. So $u = r^{-1} \in U(R) \subseteq R$. Thus $\hat{m} = um \in M$. So $\hat{M} = M$.

(3) Let $\hat{m} \in \hat{M}$. So $\hat{m} + \hat{N} = u(m + N)$ where $u \in U(\hat{R})$ and $m \in M$ since M/N is quasi-complete and $\widehat{M/N} = \hat{M}/\hat{N}$. So $\hat{m} - um \in \hat{N}$. Hence $n := u^{-1}(\hat{m} - um) \in \hat{N} = N$. So $\hat{m} = um + un = u(m + n)$. Thus M is quasi-complete. $\square$

Example 4. Let D be a semilocal PID and M a finitely generated D-module. So $M = D^n \oplus T$ where $n \geq 0$ and T is torsion. (1) Suppose that D has exactly one maximal ideal, i.e., D is a DVR. If D is complete, then M is complete and hence quasi-complete. Suppose that D is not complete. Now T is complete, so

$M = D^n \oplus T$ is quasi-complete for $n = 0, 1$ by Proposition 2 (since D is quasi-complete). If $n \geq 2$, then $D^n \oplus T = D \oplus (D^{n-1} \oplus T)$ is not quasi-complete since $D^{n-1} \oplus T$ is not complete (Proposition 2). (2) Suppose that D has more than one maximal ideal. So D is not quasi-complete by Theorem 8. But M quasi-complete and $n \geq 1$ gives that D is quasi-complete. Thus $n = 0$, i.e., $M = T$ is torsion. Hence M is complete and thus quasi-complete.

The next theorem allows us to construct quasi-complete local rings having many nilpotent elements.

Theorem 9. *Let $(R, \mathcal{M})$ be a one-dimensional analytically irreducible local domain and let M be a finitely generated R-module. Then the idealization $R (+) M$ is quasi-complete if and only if M is quasi-complete.*

Proof. ($\Rightarrow$) Let $\hat{m} \in \hat{M}$, so $(0, \hat{m}) \in \hat{R}(+)\hat{M} = \widehat{R(+)M}$ and hence $(0, \hat{m}) = (\hat{u}, a)(r, m)$ where $\hat{u} \in U(\hat{R})$, $a \in \hat{M}$, $r \in R$, and $m \in M$. Now $0 = \hat{u}r \Rightarrow r = 0$ so $\hat{m} = \hat{u}m$. Hence M is quasi-complete. ($\Leftarrow$) Let $(\hat{r}, \hat{m}) \in \hat{R}(+)\hat{M} = \widehat{R(+)M}$. Case $\hat{r} = 0$. Now M quasi-complete $\Rightarrow \hat{m} = um$ for some $u \in U(\hat{R})$ and $m \in M$. So $(0, \hat{m}) = (u, 0)(0, m)$. Case $\hat{r} \neq 0$. First suppose that $\hat{r}$ is a unit. Then $(\hat{r}, \hat{m}) \in U(\widehat{R(+)M})$ and $(\hat{r}, \hat{m}) = (\hat{r}, \hat{m})(1, 0)$. So suppose $\hat{r}$ is not a unit. Now $\hat{r} = \hat{u}r$ where $\hat{u} \in U(\hat{R})$ and $r \in R$ where necessarily $r \in \mathcal{M} - \{0\}$. Choose n with $\mathcal{M}^n \subseteq Rr$. Choose $b \in M$ with $\hat{u}^{-1}\hat{m} - b \in \mathcal{M}^n \hat{M} \subseteq r\hat{M}$, say $\hat{u}^{-1}\hat{m} - b = r\hat{a}$. So $\hat{m} = r\hat{u}\hat{a} + \hat{u}b$. Then $(\hat{r}, \hat{m}) = (\hat{u}, \hat{u}a)(r, b)$. $\square$

Note that the implication ($\Rightarrow$) of Theorem 9 does not use the hypothesis that R is a one-dimensional analytically irreducible local domain. However, the implication ($\Leftarrow$) uses the fact that R is quasi-complete and a one-dimensional domain ($\mathcal{M}^n \subseteq Rr$) and so R is analytically irreducible by Corollary 2. We end with the following example.

Example 5. Let R be a one-dimensional analytically irreducible local domain. Then $R[X]/(X^2) \approx R(+)R$ is quasi-complete. However, $R[X, Y]/(X, Y)^2 \approx R(+)(R \oplus R)$ is quasi-complete if and only if R is complete in which case $R[X, Y]/(X, Y)^2$ is actually complete since $R \oplus R$ is quasi-complete if and only if R is complete (Proposition 2).

References

1. D.D. Anderson, Multiplication ideals, multiplication rings, and the ring $R(X)$. Canad. J. Math. **27**, 760–768 (1976)
2. D.D. Anderson, The existence of dual modules. Proc. Am. Math. Soc. **55**, 258–260 (1976)
3. D.D. Anderson, M. Axtell, S.J. Forman, J. Stickles, When are associates unit multiples? Rocky Mount. J. Math. **34**, 811–823 (2004)
4. A. Barnard, Multiplication modules. J. Algebra **71**, 174–178 (1981)
5. N. Bourbaki, *Commutative Algebra* (Addison-Wesley Publishing Company, Reading, 1972)
6. C. Chevalley, On the theory of local rings. Ann. Math. **44**, 690–708 (1943)

7. R.P. Dilworth, Abstract commutative ideal theory. Pacific J. Math. **12**, 481–498 (1962)
8. E.W. Johnson, A note on quasi-complete local rings. Coll. Math. **21**, 197–198 (1970)
9. E.W. Johnson, Modules: duals and principally generated fake duals. Algebra Universalis **24**, 111–119 (1987)
10. E.W. Johnson, J.A. Johnson, The Hausdorff completion of the space of closed subsets of a module. Canad. Math. Bull. **38**, 325–329 (1995)
11. J.A. Johnson, a-adic completions of Noetherian lattice modules. Fund. Math. **66**, 347–373 (1970)
12. J.A. Johnson, Semi-local lattices. Fund. Math. **90**, 11–15 (1975)
13. J.A. Johnson, Quasi-complete ideal lattices. Coll. Math. **33**, 59–62 (1975)
14. J.A. Johnson, Completeness in semilocal ideal lattices. Czechoslovak Math. J. **27**, 378–387 (1977)
15. J.A. Johnson, Quasi-completeness in local rings. Math. Japon. **22**, 183–184 (1977)
16. C.-P. Lu, Quasi-complete modules. Indiana Univ. Math. J. **29**, 277–286 (1980)
17. M. Nagata, *Local Rings, Interscience Tract in Pure and Applied Mathematics*, vol. 13 (Interscience, New York, 1962)
18. D.G. Northcott, *Ideal Theory* (Cambridge University Press, Cambridge, 1953)

On the Total Graph of a Ring and Its Related Graphs: A Survey

Ayman Badawi

Abstract Let R be a (commutative) ring with nonzero identity and $Z(R)$ be the set of all zero divisors of R. The *total graph* of R is the simple undirected graph $T(\Gamma(R))$ with vertices all elements of R, and two distinct vertices x and y are adjacent if and only if $x + y \in Z(R)$. This type of graphs has been studied by many authors. In this paper, we state many of the main results on the total graph of a ring and its related graphs.

Keywords Total graph · Zero divisors · Diameter · Girth · Connected graph Genus · Generalized total graph · Dominating set · Clique · Chromatic number

MSC(2010) classification: 13A15, 13B99, 05C99.

1 Introduction

Over the past several years, there has been considerable attention in the literature to associating graphs with commutative rings (and other algebraic structures) and studying the interplay between ring-theoretic and graph-theoretic properties; see the recent survey articles [13, 32]. For example, as in [10], the *zero-divisor graph* of R is the (simple) graph $\Gamma(R)$ with vertices $Z(R) \setminus \{0\}$, and distinct vertices x and y are adjacent if and only if $xy = 0$; see the articles [6,11–12, 15–17, 19, 36]. The total graph (as in [7]) has been investigated in [2–5, 25, 32, 33, 35, 37]; and several

A. Badawi (✉)
Department of Mathematics and Statistics, American University of Sharjah, P.O. Box 26666
Sharjah, United Arab Emirates
e-mail: abadawi@aus.edu

M. Fontana et al. (eds.), *Commutative Algebra: Recent Advances in Commutative Rings, Integer-Valued Polynomials, and Polynomial Functions*, DOI 10.1007/978-1-4939-0925-4_3, © Springer Science+Business Media New York 2014

variants of the total graph have been studied in [1, 8, 9, 14, 16, 18, 21–24, 26, 27, 31]. The goal of this survey article is to enclose many of the main results on the total graph of a commutative ring and its related graphs.

Let G be a (simple) graph. We say that G is *connected* if there is a path between any two distinct vertices of G. At the other extreme, we say that G is *totally disconnected* if no two vertices of G are adjacent. For vertices x and y of G, we define $d(x, y)$ to be the length of a shortest path from x to y ($d(x, x) = 0$ and $d(x, y) = \infty$ if there is no such path). The *diameter* of G is diam$(G) = \sup\{d(x, y) \mid x$ and y are vertices of $G\}$. The *girth* of G, denoted by gr(G), is the length of a shortest cycle in G (gr$(G) = \infty$ if G contains no cycles). The eccentricity of a vertex x in G is the distance between x and the vertex which is at the greatest distance from x, $e(x) = \max\{d(x, y) \mid y$ is a vertex in $G\}$. The radius of the graph G, $r(G)$, is defined by $r(G) = \min\{e(x) \mid x$ is a vertex in $G\}$, and the center of the graph is the set of all of its vertices whose eccentricity is minimal, i.e., it is equal to the radius. So, the radius of the graph is equal to the smallest eccentricity and diameter to the largest eccentricity of a vertex in this graph. It is well known that for connected graphs of diameter d and radius r, one has $r \leq d \leq 2r$. Recall that a *clique* in a graph is a set of pairwise adjacent vertices. The *clique number* of a graph G, denoted by $\omega(G)$, is the order of a largest clique in G. Also, $\chi(G)$ denotes the chromatic number of G and is the minimum number of colors which is needed for a proper coloring of G, i.e., a coloring of the vertices of G such that adjacent vertices have distinct colors. We denote the complete graph on n vertices by K^n and the complete bipartite graph on m and n vertices by $K^{m,n}$ (we allow m and n to be infinite cardinals). We will sometimes call a $K^{1,n}$ a *star graph*. We say that two (induced) subgraphs G_1 and G_2 of G are *disjoint* if G_1 and G_2 have no common vertices and no vertex of G_1 (resp., G_2) is adjacent (in G) to any vertex not in G_1 (resp., G_2). By abuse of notation, we will sometimes write $G_1 \subseteq G_2$ when G_1 is a subgraph of G_2. A general reference for graph theory is [20].

Throughout this paper, all rings R are with $1 \neq 0$. Let R be a commutative ring with nonzero identity. Then $Z(R)$ denotes its set of zero divisors, Nil(R) denotes its ideal of nilpotent elements, Reg(R) denotes its set of nonzero divisors (i.e., Reg$(R) = R \setminus Z(R)$), and $U(R)$ denotes its group of units. For $A \subseteq R$, let $A^* = A \setminus \{0\}$. We say that R is *reduced* if Nil$(R) = \{0\}$, and dim(R) will always mean Krull dimension. As usual, $\mathbb{Z}$, $\mathbb{Q}$, $\mathbb{Z}_n$, and $\mathbb{F}_q$ will denote the integers, rational numbers, integers modulo n, and the finite field with q elements, respectively. General references for ring theory are [29, 30].

2 The Total Graph of a Ring

In [7], Anderson and I defined the *total graph* of R to be the (undirected) graph $T(\Gamma(R))$ with all elements of R as vertices, and two distinct vertices x and y are adjacent if and only if $x + y \in Z(R)$. Let Reg$(T((\Gamma(R)))$ be the (induced) subgraph of $T(\Gamma(R))$ with vertices Reg(R).

Theorem 2.1 ([7, Theorem 2.2]). *Let R be a commutative ring such that $Z(R)$ is an ideal of R, and let $|Z(R)| = \alpha$ and $|R/Z(R)| = \beta$.*

1. If $2 \in Z(R)$, then $\mathrm{Reg}(T(\Gamma(R))$ is the union of $\beta - 1$ disjoint $K^{\alpha'}s$.
2. If $2 \notin Z(R)$, then $\mathrm{Reg}(T(\Gamma(R))$ is the union of $(\beta - 1)/2$ disjoint $K^{\alpha,\alpha'}s$.

Theorem 2.2 ([7, Theorem 2.4]). *Let R be a commutative ring such that $Z(R)$ is an ideal of R. Then*

1. $\mathrm{Reg}(T(\Gamma(R))$ is complete if and only if either $R/Z(R) \cong \mathbb{Z}_2$ or $R \cong \mathbb{Z}_3$.
2. $\mathrm{Reg}(T(\Gamma(R))$ is connected if and only if either $R/Z(R) \cong \mathbb{Z}_2$ or $R/Z(R) \cong \mathbb{Z}_3$.
3. $\mathrm{Reg}(T(\Gamma(R))$ is totally disconnected if and only if R is an integral domain with $\mathrm{char}(R) = 2$.

Theorem 2.3 ([7, Theorem 2.9]). *Let R be a commutative ring such that $Z(R)$ is an ideal of R. Then the following statements are equivalent:*

1. $\mathrm{Reg}(T(\Gamma(R))$ is connected.
2. Either $x + y \in Z(R)$ or $x - y \in Z(R)$ for all $x, y \in \mathrm{Reg}(R)$.
3. Either $x + y \in Z(R)$ or $x + 2y \in Z(R)$ for all $x, y \in \mathrm{Reg}(R)$. In particular, either $2x \in Z(R)$ or $3x \in Z(R)$ (but not both) for all $x \in \mathrm{Reg}(R)$.
4. Either $R/Z(R) \cong \mathbb{Z}_2$ or $R/Z(R) \cong \mathbb{Z}_3$.

Theorem 2.4 ([7, Theorems 3.3, 3.4]). *Let R be a commutative ring such that $Z(R)$ is not an ideal of R. Then $T(\Gamma(R))$ is connected if and only if $1 = z_1 + \cdots + z_n$ for some $z_1, \ldots, z_n \in Z(R)$. Furthermore, suppose that $T(\Gamma(R))$ is connected and let n be the least integer $1 = z_1 + \cdots + z_n$ for some $z_1, \ldots, z_n \in Z(R)$. Then $\mathrm{diam}(T(\Gamma(R))) = n$. In particular, if R is a finite commutative ring and $Z(R)$ is not an ideal of R, then $\mathrm{diam}(T(\Gamma(R))) = 2$.*

In the following example, for each integer $n \geq 2$, we construct a commutative ring R_n such that $Z(R_n)$ is not an ideal of R_n and $T(\Gamma(R_n))$ is connected with $\mathrm{diam}(T(\Gamma(R))) = n$.

Example 2.5. Let $n \geq 2$ be an integer, $D = Z[X_1, X_2, \ldots, X_{n-1}]$, K be the quotient field of D, $P_0 = (X_1 + X_2 + \mathrm{A} \cdots + X_{n-1})$, $P_i = (X_i)$ for each integer i with $1 \leq i \leq n - 2$, and $P_{n-1} = (X_{n-1} + 1)$. Then $P_0, P_1, \ldots, P_{n-1}$ are distinct prime ideals of D. Let $F = P_0 \cup P_1 \mathrm{A} \cup \cdots \cup P_{n-1}$; then $S = D \ F$ is a multiplicative subset of D. Set $R_n = D(+)(K/D_S)$. Then $Z(R_n) = F(+)(K/D_S))$. Since $(1, 0) = (-X_1 - X_2 - \cdots - X_{n-1}, 0) + (X_1, 0) + (X_2, 0) + (X_3, 0) + \mathrm{A} \cdots + (X_{n-1} + 1, 0)$ is the sum of n zero divisors of R_n, by construction we conclude that n is the least integer $m \geq 2$ such that 1 is the sum of m zero divisors of R_n. Hence $T(\Gamma(R_n))$ is connected with $\mathrm{diam}(T(\Gamma(R_n))) = n$ by Theorems 2.4 above.

Theorem 2.6 ([7, Theorem 3.1]). *If $\mathrm{Reg}(\Gamma(R))$ is connected, then $T(\Gamma(R))$ is connected.*

The converse of Theorem 2.6 is not true. We have the following example.

Example 2.7. Let $R = \mathbb{Q}[X](+)(\mathbb{Q}(X)/\mathbb{Q}[X])$. Then one can easily show that $Z(R) = (\mathbb{Q}[X]\,\mathbb{Q}^*)(+)(\mathbb{Q}(X)/\mathbb{Q}[X])$ is not an ideal of R and $\mathrm{Reg}(R) = U(R) = \mathbb{Q}^*(+)(\mathbb{Q}(X)/\mathbb{Q}[X])$. Thus $T(\Gamma(R))$ is connected with $\mathrm{diam}(T(\Gamma(R))) = 2$ (by Theorems 2.4) since $(1,0) = (X,0)(+)(X+1,0)$ with $(X,0),(X+1,0) \in Z(R)$. However, $\mathrm{Reg}(\Gamma(R))$ is not connected since there is no path from $(1,0)$ to $(2,0)$ in $\mathrm{Reg}(\Gamma(R))$.

Theorem 2.8. *1. [7, Corollary 3.5] If $T(\Gamma(R))$ is connected, then diam $(T(\Gamma(R)) = d(0,1)$.*

2. [7, Corollary 3.5] If $T(\Gamma(R))$ is connected and $\mathrm{diam}(T(\Gamma(R)) = n$, then $\mathrm{diam}(\mathrm{Reg}(\Gamma(R))) \geq n - 2$.

3. [4, Corollary 1] If R is a commutative Noetherian ring and $T(\Gamma(R))$ is connected with diameter n, then $n - 2 \leq \mathrm{diam}(\mathrm{Reg}(\Gamma(R))) \leq n$.

Theorem 2.9 ([8, Theorem 4.4]). *Let R be a commutative ring.*

(1) If R is either an integral domain or isomorphic to $\mathbb{Z}_4$ or $\mathbb{Z}_2[X]/(X^2)$, then $\mathrm{gr}(T(\Gamma(R))) = \infty$.

(2) If R is isomorphic to $\mathbb{Z}_2 \times \mathbb{Z}_2$, then $\mathrm{gr}(T(\Gamma(R))) = 4$.

(3) Otherwise, $\mathrm{gr}(T(\Gamma(R))) = 3$.

Theorem 2.10 ([35, Theorem 2.1]). *Let R be a finite commutative ring with 1 such that $Z(R)$ is not an ideal of R. Then $r(T(\Gamma(R))) = 2$.*

Theorem 2.11 ([35, Theorem 2.2]). *Let R be a commutative ring with 1 such that $Z(R)$ is not an ideal of R, and let n be the smallest integer such that $1 = z_1 + \cdots + z_n$, for some $z_1, \ldots, z_n \in \mathrm{A}Z(R)$. Then $r(T(\Gamma(R))) = n$.*

Theorem 2.12 ([35, Theorem 3.2]). *Let R be a ring such that $Z(R)$ is not an ideal of R. Then $T(\Gamma(R[x]))$ is connected if and only if $T(\Gamma(R))$ is connected. Furthermore if $\mathrm{diam}(T(\Gamma(R))) = n$, then $\mathrm{diam}(T(\Gamma(R[x]))) = r(T(\Gamma(R[x]))) = n$.*

Theorem 2.13 ([35, Theorem 3.4]). *Let R be a reduced ring such that $Z(R)$ is not an ideal of R. Then $T(\Gamma(R[[x]]))$ is connected if and only if $T(\Gamma(R))$ is connected. Furthermore if $\mathrm{diam}(T(\Gamma(R))) = n$, then $\mathrm{diam}(T(\Gamma(R[[x]]))) = r(T(\Gamma(R[[x]]))) = n$.*

Let G be a simple undirected graph. Recall that a *Hamiltonian path* of G is a path in G that visits each vertex of G exactly once. A *Hamilton cycle (circuit)* of G is a Hamilton path that is a cycle. A graph G is called a *Hamilton graph* if it has a Hamilton cycle.

Theorem 2.14 ([4, Theorem 3]). *Let R be a finite commutative ring such that $Z(R)$ is not an ideal. Then the following statements hold:*

1. $T(\Gamma(R))$ is a Hamiltonian graph.

2. $\mathrm{Reg}(\Gamma(R))$ is a Hamiltonian graph if and only if R is isomorphic to none of the rings: $\mathbb{Z}_2^{n+1}, \mathbb{Z}_2^n \times \mathbb{Z}_3, \mathbb{Z}_2^n \times \mathbb{Z}_4, \mathbb{Z}_2^n \times \mathbb{Z}_2[X]/(X^2)$, where n is a natural number.

Theorem 2.15 ([25, Theorem 5.2]). *If R is a commutative ring and diam $(T(\Gamma(R))) = 2$, then $T(\Gamma(R))$ is Hamilton graph.*

Theorem 2.16 ([25, Corollary 5.3]). *If R is an Artinian ring, then $T(\Gamma(R))$ is Hamilton graph.*

Recall that a simple undirected graph is called a *planar graph* if it can be drawn on the plane in such way that no edges cross each other. Recall that a commutative ring R is called a *local (quasilocal) ring* if it has exactly one maximal ideal.

Theorem 2.17 ([33, Theorem 1.5]). *Let R be a finite commutative ring such that $T(\Gamma(R))$ is planar. Then the following statements hold:*

1. *If R is a local ring, then R is a field or R is isomorphic to one of the following rings:*
   $\mathbb{Z}_4, \mathbb{Z}_2[X]/(X^2), \mathbb{Z}_2[X]/(X^3), \mathbb{Z}_2[X,Y]/(X,Y)^2, \mathbb{Z}_4[X]/(2X, X^2)$
   $\mathbb{Z}_4[X]/(2X, X^2 - 2), \mathcal{Z}_8, \mathbb{F}_4[X](X^2), \mathbb{Z}_4[X]/(X^2 + X + 1)$, *where $\mathbb{F}_4$ is a field with exactly four elements.*
2. *If R is not a local ring, then R isomorphic to either $\mathbb{Z}_2 \times \mathbb{Z}_2$ or $\mathbb{Z}_6$.*

A simple undirected nonplanar graph G is called *toroidal* if the vertices of G can be placed on a torus such that no edges cross. The

Theorem 2.18 ([33, Theorem 1.6]). *Let R be a finite commutative ring such that $T(\Gamma(R))$ is toroidal. Then the following statements hold:*

1. *If R is a local ring, then R is isomorphic to either $\mathbb{Z}_9$ or $\mathbb{Z}_3/(x^2)$.*
2. *If R is not a local ring, then R is isomorphic to one of the following rings: $\mathbb{Z}_2 \times \mathbb{F}_4, \mathbb{Z}_3 \times \mathbb{Z}_3, \mathbb{Z}_2 \times \mathbb{Z}_4, \mathbb{Z}_2 \times \mathbb{Z}_2[X]/(X^2), \mathbb{Z}_2 \times \mathbb{Z}_2 \times \mathbb{Z}_2$, where $\mathbb{F}_4$ is a field with exactly four elements.*

Let S_k denote the sphere with k handles, where k is a nonnegative integer, that is, k is an oriented surface with k handles. The genus of a graph G, denoted $G(G)$, is the minimal integer n such that the graph can be embedded in S_n. Intuitively, G is embedded in a surface if it can be drawn in the surface so that its edges intersect only at their common vertices. Note that a graph G is a planar iff $g(G) = 0$ and G is toroidal iff $g(G) = 1$. Note that if x is a real number, then $\lceil x \rceil$ is the least integer that is greater than or equal to x.

Theorem 2.19 ([24, Theorem 3.2]). *Let R be a finite commutative ring with identity, I be an ideal contained in $Z(R)$, $|I| = n$ and $|R/I| = m$. Then the following statements are true:*

1. *If $2 \in I$, then $g(T(\Gamma(R))) \geq m \lceil \frac{(n-3)(n-4)}{12} \rceil$.*
2. *If $2 \notin I$, then $g(T(\Gamma(R))) \geq \lceil \frac{(n-3)(n-4)}{12} \rceil + (\frac{m-1}{2}) \lceil \frac{(n-2)^2}{4} \rceil$.*

Theorem 2.20 ([24, Corollary 3.4]). *Let R be a finite commutative ring with identity such that $Z(R)$ is an ideal of R, $|Z(R)| = n$ and $|R/Z(R)| = m$. Then the following statements hold:*

1. If $2 \in Z(R)$, then $g(T(\Gamma(R))) = m\lceil \frac{(n-3)(n-4)}{12} \rceil$.
2. If $2 \notin I$, then $g(T(\Gamma(R))) = \lceil \frac{(n-3)(n-4)}{12} \rceil + (\frac{m-1}{2})\lceil \frac{(n-2)^2}{4} \rceil$.

Theorem 2.21 ([24, Theorem 4.3]). *Let R be a finite commutative ring. Then $g(T(\Gamma(R))) = 2$ if and only if R is isomorphic to either $\mathbb{Z}_{10}$ or $\mathbb{Z}_3 \times \mathbb{F}_4$, where $\mathbb{F}_4$ is a field with four elements.*

Let v be a vertex of a simple undirected graph G. Then the degree of v is denoted by $\deg(v)$. We say $\deg(v) = k$ if there are exactly k (distinct) vertices in G where each vertex is connected to v by an edge. Let G be a simple undirected graph. We say that G is *Eulerian* if it is connected and its vertex degrees are all even.

Theorem 2.22. *1. [37, Theorem 3.3] Let R be a finite commutative ring. Then $T(\Gamma(R))$ is Eulerian if and only if R is isomorphic to a direct sum of two or more finite fields of even orders, i.e., $R \cong \bigoplus_{i=1}^{k} \mathbb{F}_{2^{t_i}}$ for some $k \geq 2$.*
2. [25, Lemma 5.1] Suppose that $Z(R)$ is not an ideal of R. Then $T(\Gamma(R))$ is Eulerian if and only if $2 \in Z(R)$ and $|Z(R)|$ is an odd integer.

Let G be a simple undirected graph with V as its set of vertices. A subset S of V is called a *dominating set* of G if for every $a \in V \setminus S$, there is a $b \in S$ such that $a - b$ is an edge of the graph G. The domination number $\gamma(G)$ is the minimum size of a dominating set of G.

Theorem 2.23 ([37, Theorem 4.1]). *Let R be a finite commutative ring and $n = \min\{|R/M| \mid M$ is a maximal ideal of $R\}$. Then $\gamma(T(\Gamma(R))) = n$, except when R is a (finite) field of an odd order, where $\gamma(T(\Gamma(R))) = \frac{n-1}{2} + 1$.*

Let $H = \{d \mid d$ is a dominating set of $T(\Gamma(R))\}$. The *intersection graph of dominating sets* denoted by $IT(R)$ is a simple undirected graph with vertex set H and two distinct vertices a and b in H are adjacent if an only if $a \cap b = \emptyset$ (see [26, 27]).

Theorem 2.24 ([26, Theorem 3.1]). *Let R be a commutative Artinian ring with $|R| \geq 4$ and let I be an annihilator ideal of R such that $|R/I|$ is finite. Then*

1. *$IT(R)$ is connected and $\operatorname{diam}(IT(R)) \leq 2$.*
2. *$\operatorname{gr}(IT(R))) \in \{3, 4\}$. In particular, $\operatorname{gr}(IT(R)) = 4$ if and only if either $R \cong \mathbb{Z}_4$ or $R \cong \mathbb{Z}_2[X]/(X^2)$.*

Theorem 2.25 ([26, Theorem 3.2]). *Let R be a commutative Artinian ring with $|R| \geq 4$ and let I be an annihilator ideal of R such that $|R/I|$ is finite. Then*

1. *$IT(R)$ is a regular graph (i.e., all vertices in $IT(R)$ have the same degree).*
2. *$IT(R)$ is a complete graph if and only if R is an integral domain.*
3. *$IT(R)$ is a bipartite graph if and only if either $R \cong \mathbb{Z}_4$ or $R \cong \mathbb{Z}_2[X]/(X^2)$.*
4. *$IT(R)$ is a cycle if and only if either $R \cong \mathbb{Z}_4$ or $R \cong \mathbb{Z}_2[X]/(X^2)$.*

Theorem 2.26 ([26, Theorem 5.4]). *Let R be a finite commutative ring. Then*

1. *$IT(R)$ is planar if and only if R is isomorphic to either $\mathbb{Z}_3$ or $\mathbb{Z}_4$ or $\mathbb{Z}_5$ or $\mathbb{Z}_2[x]/(X^2)$ or $\mathbb{Z}_2 \times \mathbb{Z}_2$ or $\mathbb{F}_{2^n}$ (a field with 2^n elements) for some positive integer $n \geq 1$.*

2. $IT(R)$) is toroidal if and only if $R \cong \mathbb{Z}_6$.

3. $g(IT(R)) = 2$ if and only if $R \cong \mathbb{Z}_7$.

Theorem 2.27 ([26, Theorem 5.5]). *If R is a finite commutative ring, then $g(IT(R)) \leq g(T(\Gamma(R)))$.*

Theorem 2.28 ([27, Theorem 2.1]). *Let R be a commutative Artinian ring with $|R| \geq 4$ and assume that I is the unique annihilator ideal of R such that $|R/I|$ is minimum. Then $IT(R)$ is Eulerian if and only if R is not a field.*

Theorem 2.29 ([27, Theorem 2.2]). *Let R be a commutative Artinian ring with $|R| \geq 4$ and assume that I is an annihilator ideal of R such that $|R/I|$ is minimum. Then $IT(R)$ is a Hamilton graph.*

We recall that a graph G with number of vertices equals $m \geq 3$ is called *pancyclic* if G contains cycles of all lengths from 3 to m. Also G is called *vertex-pancyclic* if each vertex v of G belongs to every cycle of length l for $3 \leq l \leq m$.

Theorem 2.30. *Let R be a commutative Artinian ring with $|R| \geq 4$ and assume that I is an annihilator ideal of R such that $|R/I|$ is minimum. Then*

1. *[27, Theorem 2.3] $IT(R)$ is pancyclic if and only if either $R \cong \mathbb{Z}_4$ or $R \cong \mathbb{Z}_2[X]/(X^2)$.*
2. *[27, Corollary 2.1] $IT(R)$ is vertex-pancyclic if and only if neither $R \cong \mathbb{Z}_4$ nor $R \cong \mathbb{Z}_2[X]/(X^2)$ (i.e., $IT(R)$ is not pancyclic).*

We recall that a *perfect graph* is a graph in which the chromatic number of every induced subgraph equals the size of the largest clique of that subgraph.

Theorem 2.31. *Let R be a finite commutative ring. Then:*

1. *[27, Theorem 4.1] $\chi(IT(R) = \omega(IT(R))$.*
2. *[27, Theorem 4.2] $IT(R)$ is perfect if and only if either R is an integral domain or R has a unique annihilator ideal I with $|R/I| = 2$ or $R \cong \mathbb{Z}_2 \times \mathbb{Z}_2$.*

Let $CT(\Gamma(R))$ denotes the complement of the total graph of a commutative ring R, i.e., $CT(\Gamma(R))$ is a simple undirected graph with R as its vertex set, and two distinct vertices x, y in $CT(\Gamma(R))$ are adjacent if $x + y \in \text{Reg}(R)$.

Recall that a *path graph* is a particularly simple example of a tree, namely a tree with two or more vertices that is not branched at all, that is, contains only vertices of degree 2 and 1. In particular, it has two terminal vertices (vertices that have degree 1), while all others (if any) have degree 2.

Theorem 2.32 ([25, Theorem 2.16]). *Let R be a commutative ring. Then the following statements are true:*

1. *$CT(\Gamma(R))$ is a path if and only if $R \cong \mathbb{Z}_2$.*
2. *$CT(\Gamma(R))$ is complete if and only if R is an integral domain and $\text{char}(R) = 2$.*
3. *$CT(\Gamma(R)))$ is a star if and only if either $R \cong \mathbb{Z}_2$ or $R\mathbb{Z}_3$.*

4. $CT(\Gamma(R))$ is a cycle if and only if either $R \cong \mathbb{Z}_4$ or $R \cong \mathbb{Z}_2[X]/(X^2)$ or $R \cong \mathbb{Z}_6$.

5. $CT(\Gamma(R))$ is a complete bipartite graph if and only if either R is a local ring [with maximal ideal $Z(R)$] such that $R/Z(R) \cong \mathbb{Z}_2$ or $R \cong \mathbb{Z}_3$.

Theorem 2.33. *Let R be a finite commutative ring. Then*

1. *[25, Corollary 4.5]* $\mathrm{gr}(CT(\Gamma(R))) = 3, 4, 6, \infty$.
2. *[25, Lemma 5.1]* Suppose $Z(R)$ is not an ideal of R. Then $CT(\Gamma(R))$ is Eulerian if and only if $2 \in Z(R)$ and $|\mathrm{Reg}(R)|$ is an even integer.

Let $C(R)$ represent a simple undirected graph with vertex set R and for distinct $x, y \in R$, the vertices x and y are adjacent if and only if $x - y \in Z(R)$. It is natural for one to ask when is $T(\Gamma(R))$ isomorphic to $C(R)$? We have the following result.

Theorem 2.34 ([37, Theorem 5.2]). *Let R be a finite commutative ring. Then the two graphs $T(\Gamma(R))$ and $C(R)$ are isomorphic if and only if at least one of the following conditions is true:*

1. $R \cong R_1 \oplus \cdots \oplus R_k, k \geq 1$, and each R_i is a local ring of an even order.
2. $R \cong R_1 \oplus \cdots \oplus R_k, k \geq 2$, and each R_i is a local ring such that $\min\{|R_i/M_i|$ where M_i is the maximal ideal of $R_i\} = 2$.

Let R be a noncommutative ring. Then one can define $T(\Gamma(R))$ and $\mathrm{Reg}(\Gamma(R))$ in the same way as for the commutative case. Let R be a ring. Then $M_n(R), GL_n(R)$, and $T_n(R)$ denote the set of $n \times n$ matrices over R, the set of $n \times n$ invertible matrices over R, and the set of $n \times n$ upper triangular matrices over R, respectively.

Theorem 2.35 ([35, Theorem 3.7]). *Let R be a commutative ring. The total graph $T(\Gamma(M_n(R)))$ is connected and $\mathrm{diam}(T(\Gamma(M_n(R)))) = 2$.*

Theorem 2.36 ([3, Theorem 1]). *Let F be a field with $\mathrm{char}(F) \neq 2$ and n be a positive integer. Then $\omega(\mathrm{Reg}(\Gamma(M_n(F)))) < \infty$, and moreover $\omega(\mathrm{Reg}(\Gamma(M_n(F)))) \leq \sum_{k=0}^{n} \frac{(n!)^2}{k![(n-k)!]^2}$.*

Theorem 2.37 ([3, Theorem 2]). *For every field F with $\mathrm{char}(F) \neq 2$, $\omega(\mathrm{Reg}(\Gamma(M_2(F)))) = 5$.*

Theorem 2.38 ([3, Theorem 3]). *For every division ring $D, \mathrm{char}(D) \neq 2$, $\mathrm{diag}(\pm 1, \ldots, \pm 1\} \ldots, \pm 1)$ (the set of all diagonal matrices with diagonal entries in the set $\{-1, 1\}$ forms a maximal clique for $\mathrm{Reg}(\Gamma(M_n(D)))$).*

Theorem 2.39 ([5, Theorem 1]). *If F is a field, $\mathrm{char}(F) \neq 2$ and n is a positive integer, then $\chi(\mathrm{Reg}(\Gamma(T_n(F)))) = \omega(\mathrm{Reg}(\Gamma(T_n(F)))) = 2^n$.*

Theorem 2.40 ([2, Theorem 1, Theorem 3]). *Let R be a ring (not necessarily commutative). Then $\mathrm{gr}(\mathrm{Reg}(\Gamma(R))), \mathrm{gr}(T(\Gamma(R))) \in \{3, 4, \infty\}$.*

Recall that a *tree* is an undirected graph in which any two vertices are connected by exactly one simple path. In other words, any connected graph without simple cycles is a tree. A *forest* is a disjoint union of trees.

Theorem 2.41 ([2, Theorem 2]). *Let R be a left Artinian ring and $\text{Reg}(\Gamma(R))$ be a tree. Then R is isomorphic to one of the following rings: $\mathbb{Z}_3, \mathbb{Z}_4, \mathbb{Z}_2[X]/(X^2), \mathbb{Z}_2^r, \mathbb{Z}_3 \times \mathbb{Z}_2^r, \mathbb{Z}_4 \times \mathbb{Z}_2^r, \mathbb{Z}_2[X]/(X^2) \times \mathbb{Z}_2^r, T_2(\mathbb{Z}_2), T_2(\mathbb{Z}_2) \times \mathbb{Z}_2^r$, where $T_2(\mathbb{Z}_2)$ denotes the ring of 2×2 upper triangular matrices over $\mathbb{Z}_2$ and r is a natural number.*

Theorem 2.42 ([2, Theorem 5]). *Let R be a finite ring (not necessarily commutative). Then $\text{Reg}(\Gamma(R))$ is regular (i.e., all vertices have the same degree).*

Theorem 2.43. *Let R be ring (not necessarily commutative). Then*

1. *[2, Theorem 7] If R is a left Artinian ring and $\text{Reg}(\Gamma(R))$ contains a vertex adjacent to all other vertices, then $\text{Reg}(\Gamma(R))$ is complete.*
2. *[2, Theorem 8] If $2 \notin Z(R)$ and $\text{Reg}(\Gamma(R))$ is a complete graph, then $J(R) = 0$ (where $J(R)$ is the Jacobson radical of R).*
3. *[2, Theorem 9] If R is a left Artinian ring and $2 \notin Z(R)$, then $\text{Reg}(\Gamma(R))$ is a complete graph, if and only if $R \cong \mathbb{Z}_3^r$, for some natural number r.*
4. *[2, Corollary 4] If R is a reduced left Noetherian ring and $2 \notin Z(R)$ such that $\text{Reg}(\Gamma(R))$ is a complete graph, then $R \cong \mathbb{Z}_3^r$, for some natural number r.*

3 The Total Graph of a Commutative Ring Without the Zero Element

In this section, we consider the (induced) subgraph $T_0(\Gamma(R))$ of $T(\Gamma(R))$ obtained by deleting 0 as a vertex. Specifically, $T_0(\Gamma(R))$ has vertices $R^* = R \setminus \{0\}$, and two distinct vertices x and y are adjacent if and only if $x + y \in Z(R)$.

Let $\text{d}_T(x, y)$ (resp., $\text{d}_{T_0}(x, y)$) denote the distance from x to y in $T(\Gamma(R))$ (resp., $T_0(\Gamma(R))$).

Theorem 3.1 ([8, Theorem 4.3]). *Let R be a commutative ring. Then $\text{diam}(T_0(\Gamma(R))) = \text{diam}(T(\Gamma(R)))$.*

Theorem 3.2 ([8, Theorem 4.5]). *Let R be a commutative ring.*

(1) If R is either an integral domain or isomorphic to $\mathbb{Z}_4$, $\mathbb{Z}_2[X]/(X^2)$, or $\mathbb{Z}_2 \times \mathbb{Z}_2$, then $\text{gr}(T_0(\Gamma(R))) = \infty$.
(2) If R is isomorphic to $\mathbb{Z}_9$ or $\mathbb{Z}_3[X]/(X^2)$, then $\text{gr}(T_0(\Gamma(R))) = 4$.
(3) Otherwise, $\text{gr}(T_0(\Gamma(R))) = 3$.

Let $x, y \in R^*$ be distinct. We say that $x - a_1 - \cdots - a_n - y$ is a *zero-divisor path* from x to y if $a_1, \ldots, a_n \in Z(R)^*$ and $a_i + a_{i+1} \in Z(R)$ for every $0 \le i \le n$ (let $x = a_0$ and $y = a_{n+1}$). We define $\text{d}_Z(x, y)$ to be the length of a shortest zero-divisor path from x to y ($\text{d}_Z(x, x) = 0$ and $\text{d}_Z(x, y) = \infty$ if there is no such path) and $\operatorname{diam}_Z(R) = \sup\{d_Z(x,y) \mid x,y \in R^*\}$. In particular, if $x,y \in R^*$ are distinct and $x + y \in Z(R)$, then $x - y$ is a zero-divisor path from x to y with $d(x,y) = 1$.

Let $\operatorname{Min}(R)$ denote the set of all minimal prime ideals of a commutative ring R. Recall that $U(R)$ denotes the set of all units of a commutative ring R.

Theorem 3.3 ([8, Theorem 5.1]). *Let R be a commutative ring that is not an integral domain. Then there is a zero-divisor path from x to y for every $x,y \in R^*$ if and only if one of the following two statements holds.*

(1) R is reduced, $|\operatorname{Min}(R)| \geq 3$, and $R = (z_1, z_2)$ for some $z_1, z_2 \in Z(R)^*$.
(2) R is not reduced and $R = (z_1, z_2)$ for some $z_1, z_2 \in Z(R)^*$.

Moreover, if there is a zero-divisor path from x to y for every $x,y \in R^*$, then $\operatorname{diam}_Z(R) \in \{2,3\}$ and R is not quasilocal.

Theorem 3.4 ([8, Theorem 5.2]). *Let R be a commutative ring. Then $\operatorname{diam}_Z(R) \in \{0,1,2,3,\infty\}$.*

Theorem 3.5 ([8, Theorem 5.3]). *Let $R = R_1 \times R_2$ for commutative local (quasilocal) rings R_1, R_2 with maximal ideals M_1, M_2, respectively, and $\operatorname{Nil}(R_2) \neq \{0\}$. If there are $a_1 \in U(R_1)$ and $a_2 \in U(R_2)$ such that $(2a_1, 2a_2) \in U(R)$ and $(a_1, a_2) + (2a_1, 2a_2) \notin Z(R)$, then $\operatorname{diam}_Z(R) = 3$.*

Let $x,y \in R^*$ be distinct. We say that $x - a_1 - \cdots - a_n - y$ is a *regular path* from x to y if $a_1, \ldots, a_n \in \operatorname{Reg}(R)$ and $a_i + a_{i+1} \in Z(R)$ for every $0 \leq i \leq n$ (let $x = a_0$ and $y = a_{n+1}$). We define $d_{\operatorname{reg}}(x,y)$ to be the length of a shortest regular path from x to y ($d_{\operatorname{reg}}(x,x) = 0$ and $d_{\operatorname{reg}}(x,y) = \infty$ if there is no such path), and $\operatorname{diam}_{\operatorname{reg}}(R) = \sup\{d_{\operatorname{reg}}(x,y) \mid x,y \in R^*\}$. In particular, if $x,y \in R^*$ are distinct and $x + y \in Z(R)$, then $x - y$ is a regular path from x to y with $d_{\operatorname{reg}}(x,y) = 1$. Note that $\operatorname{diam}_{\operatorname{reg}}(\mathbb{Z}_2) = 0$, $\operatorname{diam}_{\operatorname{reg}}(\mathbb{Z}_3) = 1$, and $\operatorname{diam}_{\operatorname{reg}}(R) = \infty$ for any other integral domain R. We also have $\max\{\operatorname{diam}(T(\Gamma(R))), \operatorname{diam}(\operatorname{Reg}(\Gamma(R)))\} \leq \operatorname{diam}_{\operatorname{reg}}(R)$.

Theorem 3.6 ([8, Theorem 5.6]). *Let R be a commutative ring with diam $(T_0(\Gamma((R))) = n < \infty$.*

(1) Let $u \in U(R)$, $s \in R^*$, and P be a shortest path from s to u of length $n-1$ in $T_0(\Gamma(R))$. Then P is a regular path from s to u.
(2) Let $u \in U(R)$, $s \in R^*$, and $P : s - a_1 - \cdots - a_n = u$ be a shortest path from s to u of length n in $T_0(\Gamma(R))$. Then either P is a regular path from s to u, or $a_1 \in Z(R)^*$ and $a_1 - \cdots - a_n = u$ is a regular path of length $n-1 = d_{T_0}(a_1,u)$.

Theorem 3.7 ([8, Theorem 5.7]). *Let R be a commutative ring.*

(1) If $s \in \operatorname{Reg}(R)$ and $w \in \operatorname{Nil}(R)^*$, then there is no regular path from s to w. In particular, if there is a regular path from x to y for every $x,y \in R^*$, then R is reduced.
(2) If R is reduced and quasilocal, then there is no regular path from any unit to any nonzero nonunit in R.

In particular, if there is a regular path from x to y for every $x,y \in R^*$, then R is reduced and not quasilocal.

Recall from [28] that a commutative ring R is a *p.p. ring* if every principal ideal of R is projective. For example, a commutative von Neumann regular ring is a p.p. ring, and $\mathbb{Z} \times \mathbb{Z}$ is a p.p. ring that is not von Neumann regular. It was shown in [34, Proposition 15] that a commutative ring R is a p.p. ring if and only if every element of R is the product of an idempotent element and a regular element of R (thus a commutative p.p. ring that is not an integral domain has nontrivial idempotents).

Theorem 3.8 ([8, Theorem 5.9, Corollary 5.10]). *Let R be a commutative p.p. ring that is not an integral domain. Then there is a regular path from x to y for every $x, y \in R^*$. Moreover,* $\mathrm{diam}_{\mathrm{reg}}(R) = 2$. *In particular, if R be a commutative von Neumann regular ring that is not a field, then there is a regular path from x to y for every $x, y \in R^*$ and* $\mathrm{diam}_{\mathrm{reg}}(R) = 2$.

Theorem 3.9 ([8, Theorem 5. 14]). *Let R be a commutative ring that is not an integral domain. Then there is a regular path from x to y for every $x, y \in R^*$ if and only if R is reduced,* $\mathrm{Reg}(\Gamma(R))$ *is connected, and for each $a \in Z(R)^*$ there is a $b \in Z(R)^*$ such that $d_z(a, b) > 1$ (it is possible that $d_z(a, b) = \infty$).*

Theorem 3.10 ([8, Corollary 5.15]). *Let R be a reduced commutative ring such that* $|\mathrm{Min}(R)| = 2$. *Then there is a regular path from x to y for every $x, y \in R^*$ if and only if* $\mathrm{Reg}(\Gamma(R))$ *is connected.*

4   Generalized Total Graph

A subset H of R becomes a *multiplicative-prime* subset of R if the following two conditions hold: (i) $ab \in H$ for every $a \in H$ and $b \in R$, and (ii) if $ab \in H$ for $a, b \in R$, then either $a \in H$ or $b \in H$. For example, H is multiplicative-prime subset of R if H is a prime ideal of R, H is a union of prime ideals of R, $H = Z(R)$, or $H = R \setminus U(R)$. In fact, it is easily seen that H is a multiplicative-prime subset of R if and only if $R \setminus H$ is a saturated multiplicatively closed subset of R. Thus H is a multiplicative-prime subset of R if and only if H is a union of prime ideals of R [30, Theorem 2]. Note that if H is a multiplicative-prime subset of R, then $\mathrm{Nil}(R) \subseteq H \subseteq R \setminus U(R)$; and if H is also an ideal of R, then H is necessarily a prime ideal of R. In particular, if $R = Z(R) \cup U(R)$ (e.g., R is finite), then $\mathrm{Nil}(R) \subseteq H \subseteq Z(R)$.

Let H be a multiplicative-prime subset of a commutative ring R. the *generalized total graph* of R, denoted by $GT_H(R)$, as the (simple) graph with all elements of R as vertices, and for distinct $x, y \in R$, the vertices x and y are adjacent if and only if $x + y \in H$. For $A \subseteq R$, let $GT_H(A)$ be the induced subgraph of $GT_H(R)$ with all elements of A as the vertices. For example, $GT_H(R \setminus H)$ is the induced subgraph of $GT_H(R)$ with vertices $R \setminus H$. When $H = Z(R)$, we have that $GT_H(R)$ is the so-called total graph of R as introduced in [7] and denoted there by $T(\Gamma(R))$. As to be expected, $GT_H(R)$ and $T(\Gamma(R))$ share many properties. However, the concept of generalized total graph, unlike the earlier concept of total graph, allows us to study graphs of integral domains.

Theorem 4.1 ([9, Theorem 4.1]). *Let H be a prime ideal of a commutative ring R, and let $|H| = \alpha$ and $|R/H| = \beta$.*

1. *If $2 \in H$, then $GT_H(R \setminus H)$ is the union of $\beta - 1$ disjoint K^{α}'s.*
2. *If $2 \notin H$, then $GT_H(R \setminus H)$ is the union of $(\beta - 1)/2$ disjoint $K^{\alpha,\alpha}$'s.*

Theorem 4.2 ([9, Theorem 4.2]). *Let H be a prime ideal of a commutative ring R.*

1. *$GT_H(R \setminus H)$ is complete if and only if either $R/H \cong \mathbb{Z}_2$ or $R \cong \mathbb{Z}_3$.*
2. *$GT_H(R \setminus H)$ is connected if and only if either $R/H \cong \mathbb{Z}_2$ or $R/H \cong \mathbb{Z}_3$.*
3. *$GT_H(R \setminus H)$ (and hence $GT_H(H)$ and $GT_H(R)$) is totally disconnected if and only if $H = \{0\}$ (thus R is an integral domain) and $\mathrm{char}(R) = 2$.*

The next theorem gives a more explicit description of the diameter and girth of $GT_H(R \setminus H)$ when H is a prime ideal of R.

Theorem 4.3 ([9, Theorem 4.4]). *Let H be a prime ideal of a commutative ring R.*

1.   a. *$\mathrm{diam}(GT_H(R \setminus H)) = 0$ if and only if $R \cong \mathbb{Z}_2$.*
   b. *$\mathrm{diam}(GT_H(R \setminus H)) = 1$ if and only if either $R/H \cong \mathbb{Z}_2$ and $R \not\cong \mathbb{Z}_2$ (i.e., $R/H \cong \mathbb{Z}_2$ and $|H| \geq 2$), or $R \cong \mathbb{Z}_3$.*
   c. *$\mathrm{diam}(GT_H(R \setminus H)) = 2$ if and only if $R/H \cong \mathbb{Z}_3$ and $R \not\cong \mathbb{Z}_3$ (i.e., $R/H \cong \mathbb{Z}_3$ and $|H| \geq 2$).*
   d. *Otherwise, $\mathrm{diam}(GT_H(R \setminus H)) = \infty$.*
2.   a. *$\mathrm{gr}(GT_H(R \setminus H)) = 3$ if and only if $2 \in H$ and $|H| \geq 3$.*
   b. *$\mathrm{gr}(GT_H(R \setminus H)) = 4$ if and only if $2 \notin H$ and $|H| \geq 2$.*
   c. *Otherwise, $\mathrm{gr}(GT_H(R \setminus H)) = \infty$.*
3.   a. *$\mathrm{gr}(GT_H(R)) = 3$ if and only if $|H| \geq 3$.*
   b. *$\mathrm{gr}(GT_H(R)) = 4$ if and only if $2 \notin H$ and $|H| = 2$.*
   c. *Otherwise, $\mathrm{gr}(GT_H(R)) = \infty$.*

The following examples illustrate the previous theorem.

Example 4.4 ([9, Example 4.5]). (a) Let $R = \mathbb{Z}$ and H be a prime ideal of R. Then $GT_H(R \setminus H)$ is complete if and only if $H = 2\mathbb{Z}$, and $GT_H(R \setminus H)$ is connected if and only if either $H = 2\mathbb{Z}$ or $H = 3\mathbb{Z}$. Moreover, $\mathrm{diam}(GT_H(R \setminus H)) = 1$ if and only if $H = 2\mathbb{Z}$, and $\mathrm{diam}(GT_H(R \setminus H)) = 2$ if and only if $H = 3\mathbb{Z}$. Let $p \geq 5$ be a prime integer and $H = p\mathbb{Z}$. Then $GT_H(R \setminus H)$ is the union of $(p-1)/2$ disjoint $K^{\omega,\omega}$'s; so $\mathrm{diam}(GT_H(R \setminus H)) = \infty$. Finally, $\mathrm{diam}(GT_H(R \setminus H)) = \infty$ when $H = \{0\}$.

Also, $\mathrm{gr}(GT_H(R \setminus H)) = \infty$ if $H = \{0\}$, $\mathrm{gr}(GT_H(R \setminus H)) = 3$ if $H = 2\mathbb{Z}$, and $\mathrm{gr}(GT_H(R \setminus H)) = 4$ otherwise. Moreover, $\mathrm{gr}(GT_{\{0\}}(R)) = \infty$ and $\mathrm{gr}(GT_H(R)) = 3$ for any nonzero prime ideal H of R.

(b) Let $R = \mathbb{Z}_{pm} \times R_1 \times \cdots \times R_n$, where $m \geq 2$ is an integer, p is a positive prime integer, and $R_1, \ldots, R_n$ are commutative rings. Then $H = p\mathbb{Z}_{pm} \times R_1 \times \cdots \times R_n$

is a prime ideal of R. The graph $GT_H(R \setminus H)$ is complete if and only if $p = 2$, and $GT_H(R \setminus H)$ is connected if and only if $p = 2$ or $p = 3$. Moreover, $\text{diam}(GT_H(R \setminus H)) = 1$ if and only if $p = 2$, and $\text{diam}(GT_H(R \setminus H)) = 2$ if and only if $p = 3$. Assume that $p \geq 5$. Then $GT_H(R \setminus H)$ is the union of $(p - 1)/2$ disjoint $K^{\alpha,\alpha}$'s, where $\alpha = m|R_1| \cdots |R_n|$; so $\text{diam}(GT_H(R \setminus H)) = \infty$.

Also, $\text{gr}(GT_H(R \setminus H)) = 3$ if $p = 2$ and $\text{gr}(GT_H(R \setminus H)) = 4$ otherwise. Moreover, $\text{gr}(GT_H(R)) = 3$ for any prime p.

Theorem 4.5 ([9, Theorem 4.7]). *Let H be a prime ideal of a commutative ring R. Then the following statements are equivalent.*

1. *$GT_H(R \setminus H)$ is connected.*
2. *Either $x + y \in H$ or $x - y \in H$ for every $x, y \in R \setminus H$.*
3. *Either $x + y \in H$ or $x + 2y \in H$ for every $x, y \in R \setminus H$. In particular, either $2x \in H$ or $3x \in H$ (but not both) for every $x \in R \setminus H$.*
4. *Either $R/H \cong \mathbb{Z}_2$ or $R/H \cong \mathbb{Z}_3$.*

Theorem 4.6 ([9, Theorem 5.1(3)]). *Let R be a commutative ring and H a multiplicative-prime subset of R that is not an ideal of R. If $GT_H(R \setminus H)$ is connected, then $GT_H(R)$ is connected.*

Theorem 4.7 ([9, Theorem 5.2, Theorem 5.3]). *Let R be a commutative ring and H a multiplicative-prime subset of R that is not an ideal of R. Then $GT_H(R)$ is connected if and only if $1 = z_1 + \cdots + z_n$, for some $z_1, \ldots, z_n \in H$. In particular, if H is not an ideal of R and either $\dim(R) = 0$ (e.g., R is finite) or R is an integral domain with $\text{diam}(R) = 1$, then $GT_H(R)$ is connected. Furthermore, suppose that $G_H(R)$ is connected. Let $n \geq 2$ be the least integer such that $1 = z_1 + \cdots + z_n$ for some $z_1, \ldots, z_n \in H$. Then $\text{diam}(GT_H(R)) = n$. In particular, if H is not an ideal of R and either $\dim(R) = 0$ (e.g., R is finite) or R is an integral domain with $\dim(R) = 1$, then $\text{diam}(GT_H(R)) = 2$.*

Theorem 4.8 ([9, Corollary 5.5]). *Let R be a commutative ring and H a multiplicative-prime subset of R that is not an ideal of R such that $GT_H(R)$ is connected.*

1. *$\text{diam}(GT_H(R)) = d(0, 1)$.*
2. *If $\text{diam}(GT_H(R)) = n$, then $\text{diam}(GT_H(R \setminus H)) \geq n - 2$.*

Theorem 4.9 ([9, Theorem 5.15]). *Let R be a commutative ring and H a multiplicative-prime subset of R that is not an ideal of R.*

1. *Either $\text{gr}(GT_H(H)) = 3$ or $\text{gr}(GT_H(H)) = \infty$. Moreover, if $\text{gr}(GT_H(H)) = \infty$, then $R \cong \mathbb{Z}_2 \times \mathbb{Z}_2$ and $H = Z(R)$; so $GT_H(H)$ is a $K^{1,2}$ star graph with center 0.*
2. *$\text{gr}(GT_H(R)) = 3$ if and only if $\text{gr}(GT_H(H)) = 3$.*
3. *$\text{gr}(GT_H(R)) = 4$ if and only if $\text{gr}(GT_H(H)) = \infty$ (if and only if $R \cong \mathbb{Z}_2 \times \mathbb{Z}_2$).*
4. *If $\text{char}(R) = 2$, then $\text{gr}(GT_H(R \setminus H)) = 3$ or ∞. In particular, $\text{gr}(GT_H(R \setminus H)) = 3$ if $\text{char}(R) = 2$ and $GT_H(R \setminus H)$ contains a cycle.*

5. $\mathrm{gr}(GT_H(R \setminus H)) = 3, 4,$ *or* ∞. *In particular,* $\mathrm{gr}(GT_H(R \setminus H)) \leq 4$ *if* $GT_H(R \setminus H)$ *contains a cycle.*

Let R be a commutative ring. Recall that a subset S of R is called a *multiplicatively closed subset* of R if S is closed under multiplication. A multiplicatively closed subset S of R is called *saturated* if $xy \in S$ implies that $x \in S$ and $y \in S$.

Let S be multiplicatively closed subset of a commutative ring R. The graph $\Gamma_S(R)$ is a simple undirected graph with all elements of R as vertices, and two distinct vertices x and y of R are adjacent if and only if $x + y \in S$.

Theorem 4.10 ([18, Corollary 1.6]). *Suppose that S is an ideal of R with $|S| = n$ and $|R/S| = m$.*

1. *If* $2 \in S$, *then* $\Gamma_S(R)$ *is the union of m disjoint K^n's.*
2. *If* $2x \notin S$ *for each* $x \in R$, *then* $\Gamma_S(R)$ *is the union of K^n with $(m-1)/2$ disjoint $K^{n,n}$'s.*

Theorem 4.11 ([18, Proposition 2.1]). *The graph $\Gamma_S(R)$ is complete if and only if $S = R$ or* ($\mathrm{char} R = 2$ *and* $S = R \setminus \{0\}$).

Theorem 4.12 ([18, Proposition 2.1]). *Let S be a saturated multiplicatively closed subset of R with $R \setminus S = \cup_{i=1}^n P_i$ such that $|R/P_i| = 2$ for some i. Then $\Gamma_S(R)$ is a bipartite graph. Furthermore, $\Gamma_S(R)$ is a complete bipartite graph if and only if $n = 1$.*

Theorem 4.13 ([18, Theorem 2.15]). *Let R be finite commutative ring and S be a saturated multiplicatively closed subset of R. Then $\mathrm{gr}(\Gamma_S(R)) \in \{3, 4, 6, \mathsf{A}\infty\}$.*

The following is an example of saturated multiplicatively closed sets, to show that each of the numbers $3, 4, 6,$ and ∞ given in the previous theorem can appear as the girth of some graphs.

Example 4.14 ([18, Example 2.16]). Let $R = \mathbb{Z}_6$. Then $\mathrm{gr}(\Gamma_{Z(R)}(R)) = 3$, $\mathrm{gr}(\Gamma_{U(R)}(R)) = 6$, and $\mathrm{gr}(\Gamma_S(R)) = 4$, where $S = \{1, 3, 5\}$. For the saturated multiplicatively closed subset $S = \{-1, 1\}$ of $\mathbb{Z}$, we have $\mathrm{gr}(\Gamma_S(R)) = \infty$.

Theorem 4.15 ([18, Theorem 2.17]). *Let R be finite and S be a saturated multiplicatively closed subset of R. Then $\mathrm{gr}(\Gamma_S(R)) = \mathsf{A}\infty$ if and only if one of the following statements holds:*

1. $R = \mathbb{Z}_3$.
2. $R = \mathbb{Z}_2 \times \cdots \times \mathbb{Z}_2$ *and* $|S| = 1$.

Theorem 4.16 ([18, Theorem 2.23]). *Let R be a finite commutative ring. For a saturated multiplicatively closed subset S of R, we have $\mathrm{diam}(\Gamma_S(R)) \in \{1, 2, 3, \mathsf{A}\infty\}$.*

References

1. A. Abbasi, S. Habib, The total graph of a commutative ring with respect to proper ideals. J. Korean Math. Soc. **49**, 85–98 (2012)
2. S. Akbari, F. Heydari, The regular graph of a non-commutative ring. Bull. Austral. Math. Soc. **89**, 132–140 (2013)
3. S. Akbari, M. Jamaali, S.A. Seyed Fakhari, The clique numbers of regular graphs of matrix algebras are finite. Linear Algebra Appl. **43**, 1715–1718 (2009)
4. S. Akbari, D. Kiani, F. Mohammadi, S. Moradi, The total graph and regular graph of a commutative ring. J. Pure Appl. Algebra **213**, 2224–2228 (2009)
5. S. Akbari, M. Aryapoor, M. Jamaali, Chromatic number and clique number of subgraphs of regular graph of matrix algebras. Linear Algebra Appl. **436**, 2419–2424 (2012)
6. D.F. Anderson, A. Badawi, On the zero-divisor graph of a ring. Comm. Algebra **36**, 3073–3092 (2008)
7. D.F. Anderson, A. Badawi, The total graph of a commutative ring. J. Algebra **320**, 2706–2719 (2008)
8. D.F. Anderson, A. Badawi, The total graph of a commutative ring without the zero element. J. Algebra Appl. **11**, (18 pages) (2012). doi:10.1142/S0219498812500740
9. D.F. Anderson, A. Badawi, The generalized total graph of a commutative ring. J. Algebra Appl. **12**, (18 pages) (2013). doi:10.1142/S021949881250212X
10. D.F. Anderson, P.S. Livingston, The zero-divisor graph of a commutative ring. J. Algebra **217**, 434–447 (1999)
11. D.F. Anderson, S.B. Mulay, On the diameter and girth of a zero-divisor graph. J. Pure Appl. Algebra **210**, 543–550 (2007)
12. D.D. Anderson, M. Winders, Idealization of a module. J. Comm. Algebra, **1** 3–56 (2009)
13. D.F. Anderson, M. Axtell, J. Stickles, Zero-divisor graphs in commutative rings, in *Commutative Algebra Noetherian and Non-Noetherian Perspectives*, ed. by M. Fontana, S.E. Kabbaj, B. Olberding, I. Swanson (Springer, New York, 2010), pp. 23–45
14. D.F. Anderson, J. Fasteen, J.D. LaGrange, The subgroup graph of a group. Arab. J. Math. **1**, 17–27 (2012)
15. N. Ashra, H.R. Maimani, M.R. Pournaki, S. Yassemi, Unit graphs associated with rings. Comm. Algebra **38**, 2851–2871 (2010)
16. S.E. Atani, S. Habibi, The total torsion element graph of a module over a commutative ring. An. Stiint. Univ.'Ovidius Constanta Ser. Mat. **19**, 23–34 (2011)
17. M. Axtel, J. Stickles: Zero-divisor graphs of idealizations. J. Pure Appl. Algebra **204**, 23–43 (2006)
18. Z. Barati, K. Khashyarmanesh, F. Mohammadi, K. Nafar, On the associated graphs to a commutative ring. J. Algebra Appl. **11**, (17 pages) (2012). doi:10.1142/S021949881105610
19. I. Beck, Coloring of commutative rings. J. Algebra **116**, 208–226 (1988)
20. B. Bollaboás, *Graph Theory, An Introductory Course* (Springer, New York, 1979)
21. T. Chelvam, T. Asir, Domination in total graph on $\mathbb{Z}_n$. Discrete Math. Algorithms Appl. **3**, 413–421 (2011)
22. T. Chelvam, T. Asir, Domination in the total graph of a commutative ring. J. Combin. Math. Combin. Comput. (to appear)
23. T. Chelvam, T. Asir, Intersection graph of gamma sets in the total graph. Discuss. Math. Graph Theory **32**, 339–354 (2012)
24. T. Chelvam, T. Asir, On the genus of the total graph of a commutative ring. Comm. Algebra **41**, 142–153 (2013)
25. T. Chelvam, T. Asir, On the total graph and its complement of a commutative ring. Comm. Algebra **41**, 3820–3835 (2013). doi:10.1080/00927872.2012.678956
26. T. Chelvam, T. Asir, The intersection graph of gamma sets in the total graph I. J. Algebra Appl. **12**, (18 pages), (2013). doi:10.1142/S0219498812501988

27. T. Chelvam, T. Asir, The intersection graph of gamma sets in the total graph II. J. Algebra Appl. **12**, (14 pages), (2013). doi:10.1142/S021949881250199X

28. S. Endo, Note on p.p. rings. Nagoya Math. J. **17**, 167–170 (1960)

29. J.A. Huckaba, *Commutative Rings with Zero Divisors* (Dekker. New York/Basel, 1988)

30. I. Kaplansky, *Commutative Rings* (University of Chicago Press, Chicago, 1974)

31. K. Khashyarmanesh, M.R. Khorsandi, A generalization of the unit and unitary Cayley graphs of a commutative ring. Acta Math. Hungar. **137**, 242–253 (2012)

32. H.R. Maimani, M.R. Pouranki, A. Tehranian, S. Yassemi, Graphs attached to rings revisited. Arab. J. Sci. Eng. **36**, 997–1011 (2011)

33. H.R. Maimani, C. Wickham, S. Yassemi, Rings whose total graphs have genus at most one. Rocky Mountain J. Math. **42**, 1551–1560 (2012)

34. W.W. McGovern, Clean semiprime f-rings with bounded inversion. Comm. Algebra **31**, 3295–3304 (2003)

35. Z. Pucanović, Z. Petrović, On the radius and the relation between the total graph of a commutative ring and its extensions. Publ. Inst. Math. (Beograd)(N.S.) **89**, 1–9 (2011)

36. P.K. Sharma, S.M. Bhatwadekar, A note on graphical representations of rings. J. Algebra **176**, 124–127 (1995)

37. M.H. Shekarriz, M.H. Shiradareh Haghighi, H. Sharif, On the total graph of a finite commutative ring. Comm. Algebra **40**, 2798–2807 (2012)

Prime Ideals in Polynomial and Power Series Rings over Noetherian Domains

Ela Celikbas, Christina Eubanks-Turner, and Sylvia Wiegand

Abstract In this article we survey recent results concerning the set of prime ideals in two-dimensional Noetherian integral domains of polynomials and power series. We include a new result that is related to current work of the authors [Celikbas et al., *Prime Ideals in Quotients of Mixed Polynomial-Power Series Rings*; see http://www. math.unl.edu/$\sim$swiegand1 (preprint)]: Theorem 5.4 gives a general description of the prime spectra of the rings $R[[x, y]]/P$, $R[x][[y]]/Q$ and $R[[y]][x]/Q'$, where x and y are indeterminates over a one-dimensional Noetherian integral domain R and P, Q, and Q' are height-one prime ideals of $R[[x, y]]$, $R[x][[y]]$, and $R[[y]][x]$, respectively. We also include in this survey recent results of Eubanks-Turner, Luckas, and Saydam describing prime spectra of simple birational extensions $R[x][f(x)/g(x)]$ of $R[x]$, where $f(x)$ and $g(x)$ are power series in $R[[x]]$ such that $f(x) \neq 0$ and is a prime ideal of $R[x][[y]]$—this is a special case of Theorem 5.4. We give some examples of prime spectra of homomorphic images of mixed power series rings when the coefficient ring R is the ring of integers $\mathbb{Z}$ or a Henselian domain.

Keywords Commutative ring • Noetherian ring • Integral domain • Polynomial ring • Power series ring • Prime ideals • Prime spectrum

1991 *Mathematics Subject Classification.* Primary 13B35, 13J10, 13A15

E. Celikbas
Department of Mathematics, University of Missouri, Columbia, MO 65211, USA
e-mail: celikbase@missouri.edu

C. Eubanks-Turner
Department of Mathematics, Loyola Marymount University, Los Angeles, CA 90045, USA
e-mail: ceturner@lmu.edu

S. Wiegand (✉)
Department of Mathematics, University of Nebraska–Lincoln, Lincoln, NE 68588, USA
e-mail: swiegand1@math.unl.edu

M. Fontana et al. (eds.), *Commutative Algebra: Recent Advances in Commutative Rings,*   55
Integer-Valued Polynomials, and Polynomial Functions, DOI 10.1007/978-1-4939-0925-4_4,
© Springer Science+Business Media New York 2014

1 Introduction

Prime ideals play a fundamental role in commutative ring theory, especially in the
theory of ideals and modules. By the primary decomposition theorem, every nonzero
ideal of a Noetherian ring has a unique set of associated prime ideals. Often if
a property can be demonstrated for prime ideals, then it holds for all ideals, for
example, finite generation, by Cohen's theorem [15, Theorem 3.4]. Murthy uses
prime ideals to show that a regular local ring is a UFD in [18]. For a Noetherian ring
R, the Grothendieck group of all finitely generated R-modules is generated by the
modules of the form R/P, where P is a prime ideal of R (see [2]). The Wiegands
demonstrate many connections between the set of prime ideals of a ring R and the
set of indecomposable R-modules in [28].

For R a commutative ring, we denote by Spec(R) the *prime spectrum* of R, that
is, the set of prime ideals of R, considered as a partially ordered set, or *poset*, under
inclusion. In 1950, Irving Kaplansky asked:

Question 1.1. Which partially ordered sets occur as Spec(R), for some Noetherian
ring R?

This problem remains open, although there have been many and varied results
related to Question 1.1:

(1) Hochster's characterization of the prime spectrum of a commutative ring as a
    topological space [9],
(2) Lewis' result that every finite poset is the prime spectrum of a commutative ring
    [12],
(3) Some properties of prime spectra of Noetherian rings [16, 27, 29],
(4) Examples of Noetherian rings such as those of Nagata, McAdam, and Heitmann
    that do not have other properties that might be expected of Noetherian rings
    [8, 17, 19, 21], and
(5) Characterizations of prime spectra of other specific classes of Noetherian rings
    or of particular Noetherian rings (see, for example, [6, 13, 24, 26]).

Many of these results are discussed in more detail in [29], along with other results.

In this article we focus on results over the past decade concerning prime spectra
for two-dimensional Noetherian integral domains of polynomials and power series.
We include background information related to this focus. In particular, our results
are related to S. Wiegand's theorem from the 1980s, Theorem 2.3, proved using
techniques developed by Heitmann and others; see [25]. Theorem 2.3 shows that
any finite amount of "misbehavior" is possible for prime ideals of a Noetherian
ring. Other results such as McAdam's Theorem 2.4 suggest that the converse is also
true: Perhaps, in some sense, the amount of such misbehavior is finite. Our current
and recent investigations of prime spectra show that certain finite subsets of these
spectra determine the partially ordered sets that are prime spectra for our rings; see
Theorem 5.4 and Definition 5.5.

The *characterization* of the prime spectrum of a particular ring requires (1) a list of axioms that describe the prime spectrum as a poset, and (2) a proof that any two posets satisfying these axioms are order-isomorphic. In order to characterize prime spectra for a class of Noetherian rings, the axioms of (1) may contain "genetic codes" that allow for some variety in the spectra of rings of the class; they depend upon cardinalities associated to the ring. In this case we require (2′). Each "genetic code" should determine a unique partially ordered set up to order-isomorphism and (3) examples to show that every poset fitting the axioms can be realized as a prime spectrum for some ring in the class.

For the remainder of this article, let x and y be indeterminates over a one-dimensional Noetherian domain R. Wiegand characterizes $\mathrm{Spec}(\mathbb{Z}[y])$, where $\mathbb{Z}$ is the ring of integers, in [26]; see Theorem 2.9. If R is a countable, semilocal one-dimensional Noetherian domain, Wiegand and Heinzer characterize $\mathrm{Spec}(R[y])$ in [6]; this characterization of course depends upon the number of maximal ideals of R. Shah and Wiegand extend this result to $\mathrm{Spec}(R[y])$, for R a semilocal one-dimensional Noetherian domain of any cardinality in [23,29]—this characterization depends upon the number of maximal ideals of R, the cardinality of R, and the cardinality of $R/\mathbf{m}$ for each maximal ideal $\mathbf{m}$ of R; see Theorem 2.13.

Several recent articles describe prime spectra for power series rings. In [7], Heinzer, Rotthaus, and Wiegand describe $\mathrm{Spec}(R[\![x]\!])$; see Theorem 2.15. In [5], Eubanks-Turner, Luckas, and Saydam describe prime spectra of simple birational extensions of $R[\![x]\!]$, that is, $\mathrm{Spec}(R[\![x]\!][g/f])$, where $g, f \in R[\![x]\!]$, $f \neq 0$, and either g, f is an $R[\![x]\!]$-sequence or $(g, f) = R[\![x]\!]$; see Theorem 6.1. In current work, the present authors describe prime spectra of rings of the form $R[\![x]\!][y]/Q$ or $R[y][\![x]\!]/Q$, where Q is a height-one prime ideal of the appropriate ring and $x \notin Q$; see Sect. 5 and [4].

In Sect. 2 we give notation and background results on prime spectra, and we mention some related items, such as the intriguing Conjecture 2.12 of Roger Wiegand. We give some general properties of mixed power series in Sect. 3. In Sect. 4 we characterize $\mathrm{Spec}(R[\![x]\!][\![y]\!]/Q)$, where R is a one-dimensional Noetherian domain and Q is a height-one prime ideal of $R[\![x, y]\!]$; see Theorem 4.1. In Sect. 5 we give new results related to the characterization of $\mathrm{Spec}(R[\![x]\!][y]/Q)$ and $\mathrm{Spec}(R[y][\![x]\!]/Q)$ from [4]; see Theorems 5.2 and 5.4. In Sect. 6 we give results from [5] concerning prime spectra of simple birational extensions of $R[\![x]\!]$; this yields a characterization in the case R is a countable Dedekind domain. In Sect. 7, we show two prime spectra examples of dimension two: $\mathrm{Spec}(\mathbb{Z}[y][\![x]\!]/Q)$, where Q is a specified prime ideal of $\mathbb{Z}[y][\![x]\!]$, and $\mathrm{Spec}(R[\![x]\!][y]/Q)$, where Q is a specified prime ideal of $R[\![x]\!][y]$ and R is a Henselian domain.

All rings are commutative with identity throughout the paper. Let $\mathbb{N}$ denote the natural numbers, let $\mathbb{Z}$ denote the integers, and let $\mathbb{R}$ denote the real numbers. Set $\mathbb{N}_0 := \mathbb{N} \cup \{0\}$ and $\aleph_0 := |\mathbb{N}|$.

2   Background

In this section, we give background information related to our focus on recent work concerning prime spectra of two-dimensional Noetherian integral domains of polynomials and power series. We refer the reader to [28, 29] for more general information concerning prime spectra in Noetherian rings.

We first introduce some notation.

Notation 2.1. Let U be a partially ordered set, sometimes abbreviated *poset*; let S be a subset of U and let $u, v \in U$. We define

$$u^{\uparrow(U)} = u^{\uparrow} := \{w \in U \mid u < w\}, \quad u^{\downarrow} := \{w \in U \mid w < u\}, \quad \mathrm{L}_e(S) := \{u \in U \mid u^{\uparrow} = S\};$$

$$\mathrm{max}(S) := \{\text{maximal elements of } S\}, \quad \text{and} \quad \mathrm{min}(S) := \{\text{minimal elements of } S\}.$$

For $u \in U$, the *height* of u, $\mathrm{ht}(u)$, is the length $t \in \mathbb{N}_0$ of a maximal length chain in U of form

$$u_0 < u_1 < u_2 < \cdots < u_t = u.$$

Set $\mathcal{H}_i(U) := \{u \in U \mid \mathrm{ht}(u) = i\}$, for each $i \in \mathbb{N}_0$. The *dimension* of U, $\dim(U)$, is the maximum of the heights of all elements of U.

We say v *covers* u and write $u \ll v$ if $u < v$ and there are no elements of U strictly between u and v. The *minimal upper bound set* of u and v, if $u \not\le v$ and $v \not\le u$, is the set $\mathrm{mub}(u, v) := \mathrm{min}(u^{\uparrow} \cap v^{\uparrow})$ and their *maximal lower bound set* is $\mathrm{Mlb}(u, v) := \mathrm{max}(u^{\downarrow} \cap v^{\downarrow})$.

Let R be a commutative ring. We use notation similar to that for the partially ordered set $U = \mathrm{Spec}(R)$. For example, if $P \in \mathrm{Spec}(R)$, $P^{\uparrow} = \{Q \in \mathrm{Spec}(R) \mid P \subsetneq Q\}$; $\mathrm{min}(R)$ is the set of minimal prime ideals of R; $\mathrm{max}(R)$ is the set of maximal ideals of R; and $\dim(R)$ is the supremum of the heights that occur for maximal ideals of R. We also use $V(S) := V_R(S) := \{\mathbf{q} \in \mathrm{Spec}(R) \mid S \subseteq \mathbf{q}\}$, for a subset S of R; for $a \in R$, put $V_R(a) := V_R(\{a\})$. For each $i \in \mathbb{N}_0$, we set $\mathcal{H}_i(R) := \{\mathbf{q} \in \mathrm{Spec}(R) \mid \mathrm{ht}(\mathbf{q}) = i\}$.

In Remarks 2.2 we establish that the rings we study are well behaved.

Remarks 2.2. (1) If a ring A is Cohen–Macaulay, $n, m \in \mathbb{N}_0$, and x_i and y_j are indeterminates over A, for $1 \le i \le n$, $1 \le j \le m$, then the *mixed polynomial-power series rings*, $A[\{x_i\}_{i=1}^n][[\{y_j\}_{j=1}^m]]$ and $A[[\{y_j\}_{j=1}^m][\{x_i\}_{i=1}^n]]$, are Cohen–Macaulay; see [15, Theorem 17.7]. Thus they are *catenary*: If $P \subseteq Q$ in $\mathrm{Spec}(R)$, then any two maximal chains of prime ideals from P to Q have the same length [15, Theorem 17.9].

(2) If R is a Noetherian integral domain of dimension one, then R is Cohen–Macaulay; see [15, Exercise 17.1, p. 139]. Thus every mixed polynomial-power series ring over a one-dimensional Noetherian domain R that involves a finite number of variables is catenary by item (1).

Theorem 2.3 was inspired by many examples of Noetherian prime spectra with finite amounts of "misbehavior" that were produced by Nagata, McAdam, Heitmann, and others; they show, for example, that Noetherian rings can be noncatenary and that there exist Noetherian rings containing two height-two prime ideals whose intersection contains no height-one prime ideal; see [8, 17, 20]. The proof of Theorem 2.3 uses techniques of these and other researchers. The statement of Theorem 2.3 summarizes the situation: All sorts of finite noncatenary or prescribed intersecting behavior is possible in the prime spectrum of some Noetherian rings. This idea is related to the later sections of this article where we show that the prime spectra that occur for our rings have a similar finite amount of "prescribed" discrepancy within a general form of the spectra; see Sects. 5 and 6. The difference here is that our rings are catenary by Remark 2.2.

Theorem 2.3 ([25, Theorem 1]). *Let F be an arbitrary finite poset. There exist a Noetherian ring A and a saturated order-embedding $\varphi : F \to \mathrm{Spec}(A)$ such that φ preserves minimal upper bound sets and maximal lower bound sets. In detail, for $u, v \in F$, we have*

(i) $u < v$ if and only if $\varphi(u) < \varphi(v)$;
(ii) v covers u if and only if $\varphi(v)$ covers $\varphi(u)$;
(iii) $\varphi(\mathrm{mub}_F(u,v)) = \mathrm{mub}(\varphi(u), \varphi(v))$; and
(iv) $\varphi(\mathrm{Mlb}_F(u,v)) = \mathrm{Mlb}(\varphi(u), \varphi(v))$.

A related theorem of Steve McAdam, Theorem 2.4, guarantees that noncatenary misbehavior cannot be too widespread in the prime spectrum of a Noetherian ring:

Theorem 2.4 ([16]). *Let P be a prime ideal of height n in a Noetherian ring. Then all but finitely many covers of P have height $n + 1$.*

Perhaps one might conjecture from Theorem 2.4 that in general prime spectra of Noetherian rings behave well, like the spectra of excellent rings, if a finite "bad" subset is removed.[1]

Corollary 2.5, which follows from Theorem 2.3, relates to our focus for this article because it describes exactly the countable posets that arise as prime spectra of two-dimensional semilocal Noetherian domains.

Corollary 2.5 ([25, Theorem 2]). *Let U be a countable poset of dimension two. Assume that U has a unique minimal element and $\max(U)$ is finite. Then $U \cong \mathrm{Spec}(R)$ for some countable Noetherian domain R if and only if $L_e(u)$ is infinite for each element u with $\mathrm{ht}(u) = 2$.*

Lemma 2.6 is useful for counting prime ideals in our rings.

---

[1]For the definition of "excellent ring" see [15, p. 260]. Basically "excellence" means the ring is catenary and has other nice properties that polynomial rings over a field possess.

Lemma 2.6 ([29, Lemma 4.2] and [5, Lemma 3.6, Remarks 3.7]). *Let T be a Noetherian domain, let y be an indeterminate, and let I be a proper ideal of T. Let $\beta = |T|$ and $\rho = |T/I|$. Then:*

(1) $|(T/I)[y]| = \rho \cdot \aleph_0 \le \beta \cdot \aleph_0 = |T[y]|$.
(2) $|T[y]| = \beta^{\aleph_0} = \rho^{\aleph_0}$.
(3) If $\beta \le \aleph_0$, then $|(T/I)[y]| = \aleph_0 = |T[y]|$.
(4) If $\beta = \aleph_0$ and $\max(T)$ is infinite, then $\beta = |\max(T)| = |T/I| \cdot \aleph_0$.
(5) If k is a field, y, y' are indeterminates, and c is an irreducible element of $k[y]$, then

$$|(k[y]/ck[y])[y']| = |k| \cdot \aleph_0 = |\max(k[y])|.$$

2.1 Prime Ideals in Polynomial Rings

This subsection includes basic facts, previous results, and technical lemmas concerning $\mathrm{Spec}(A[y])$, where A is a Noetherian domain and y is an indeterminate over A.

In Remarks 2.7, we give some basic facts.

Remarks 2.7. Let A be a Noetherian domain of dimension d and let y be an indeterminate over A.

(1) If P is a prime ideal of $A[y]$, then $\mathrm{ht}(P \cap A) \le \mathrm{ht}(P) \le \mathrm{ht}(P \cap A) + 1$; see [15, Theorem 15.1].
(2) If M is a prime ideal of $A[y]$ of height $d + 1$, then M is a maximal ideal of $A[y]$, the prime ideal $\mathbf{m} = M \cap A$ is a maximal ideal of A of height d, and $M = (\mathbf{m}, h(y))A[y]$, where $\overline{h(y)}$ is irreducible in $\overline{A[y]} = A[y]/(\mathbf{m}[y]) \cong (A/\mathbf{m})[y]$. This follows from item (1) and [10, Theorem 28, p. 17].
(3) If I is a nonzero ideal of $A[y]$ such that $I \cap A = (0)$, then $I = h(y)K[y] \cap A[y]$, where K is the field of fractions of A and $h(y) \in A[y]$ with $\deg(h(y)) \ge 1$. This follows since $K[y] = (A \setminus \{0\})^{-1}A[y]$ is a principal ideal domain (PID). If P is a prime ideal of $A[y]$ such that $P \cap A = (0)$, then $\mathrm{ht}(P) = 1$. The set of prime ideals P of $A[y]$ such that $P \cap A = (0)$ is in one-to-one correspondence with the set of height-one prime ideals of $K[y]$, via $P \mapsto PK[y] \mapsto PK[y] \cap A[y]$.

The proof of Lemma 2.8 is straightforward and follows from material in [10] on G-domains; see [4]. A *G-domain* is an integral domain A such that $A[y]$ contains a maximal ideal that intersects A in (0).

Lemma 2.8 ([4, 10]). *Let A be a Noetherian domain. If Q is a maximal ideal of $A[y]$ of height one, then*

(1) $Q \cap A = (0)$;
(2) $\dim(A) \le 1$ and $|\max(A)| < \infty$; say $\max(A) = \{\mathbf{m}_1, \ldots, \mathbf{m}_t\}$; and
(3) Q contains an element of form $h(y) = yg(y) + 1$, where $0 \ne g(y) \in (\cap_{i=1}^{t}\mathbf{m}_i)[y]$.

Moreover, if A is one-dimensional and semilocal with maximal ideals $\mathbf{m}_1, \ldots, \mathbf{m}_t$, *and Q is a prime ideal of A[y] that is minimal over an element of form* $h(y) = yg(y) + 1$, *where* $g(y) \in (\cap_{i=1}^{t}\mathbf{m}_i)[y]$, *then Q is a height-one maximal ideal of A[y].*

Theorem 2.9, due to Roger Wiegand, characterizes Spec($\mathbb{Z}[y]$), the spectrum of the ring of polynomials in the variable y over the integers $\mathbb{Z}$. The most important distinguishing feature of Spec($\mathbb{Z}[y]$) is Axiom RW.

Theorem 2.9 ([26, Theorem 2]). *Let* $U = \text{Spec}(\mathbb{Z}[y])$, *the partially ordered set of prime ideals of the ring of polynomials in one variable over the integers. Then U is characterized by the following axioms:*

(P1) U is countable and has a unique minimal element.
(P2) U has dimension two.
(P3) For each element u of height-one, $u^{\uparrow}$ is infinite.
(P4) For each pair u, v of distinct elements of height-one, $u^{\uparrow} \cap v^{\uparrow}$ is finite.
(RW) Every pair (S, T) of finite subsets S and T of U such that $\emptyset \neq S \subseteq \mathcal{H}_1(U)$
       and $T \subseteq \mathcal{H}_2(U)$ has a "radical element" in U. A "radical element" for such
       a pair (S, T) is a height-one element $w \in U$ such that $s^{\uparrow} \cap w^{\uparrow} \subseteq T \subseteq w^{\uparrow}$,
       for every $s \in S$.

A partially ordered set U satisfies the axioms of Theorem 2.9 if and only if U is order-isomorphic to Spec($\mathbb{Z}[y]$).

Theorem 2.9 leads to Question 2.10:

Question 2.10. For which two-dimensional Noetherian domains A is Spec(A) $\cong$ Spec($\mathbb{Z}[y]$)?

Remarks 2.11. (1) The following is known about rings that fit Question 2.10:

(a) Let k be a field and let z be another indeterminate. Then Spec($k[z, y]$) is order-isomorphic to Spec($\mathbb{Z}[y]$) $\iff$ k is an algebraic extension of a finite field. The ($\Leftarrow$) direction is due to Wiegand in [26, Theorem 2]; for the ($\Rightarrow$) direction, see [29].

(b) Let D be an order in an algebraic number field; that is, D is the ring of algebraic integers in a field K that is a finite extension of the rational numbers. Roger Wiegand shows Spec($D[y]$) is order-isomorphic to Spec($\mathbb{Z}[y]$) in [26, Theorem 1].

(c) In their 1998 article Li and Wiegand prove that if $B := \mathbb{Z}[y][\frac{g_1}{f}, \ldots, \frac{g_m}{f}]$, where f is nonzero and $f, g_1, \ldots, g_m \in \mathbb{Z}[y]$, then Spec($B$) is order-isomorphic to Spec($\mathbb{Z}[y]$); see [14].

(d) Saydam and Wiegand extend the result of Li and Wiegand in 2001 to show, for D an order in an algebraic number field and for B a finitely generated extension of $D[y]$ contained in the field of fractions of $D[y]$, that Spec(B) $\cong$ Spec($\mathbb{Z}[y]$) in [22].

(2) The prime spectrum of $R[y]$ is not known in general, for y an indeterminate over a one-dimensional Noetherian domain R with infinitely many maximal ideals. In fact $\mathrm{Spec}(R[y])$ is barely known beyond the examples of item (1) above and the rings of Theorem 2.13 below; see [29].

(3) The prime spectrum of $\mathbb{Q}[z, y]$, where $\mathbb{Q}$ is the field of rational numbers, is unknown, but Wiegand shows that it is *not* order-isomorphic to $\mathrm{Spec}(\mathbb{Z}[y])$ in [26]; see also [29, Remark 2.11.3].

In relation to Question 2.10, Wiegand's 1986 conjecture is still open:

Conjecture 2.12 (Wiegand [26]). For every two-dimensional Noetherian integral domain D that is finitely generated as a $\mathbb{Z}$-algebra, $\mathrm{Spec}(D) \cong \mathrm{Spec}(\mathbb{Z}[y])$.

The next theorem, Theorem 2.13, was first proved by Heinzer and Wiegand in case R is countable. Later Shah, Wiegand, and Wiegand proved it for cardinalities; see also [11]. By Theorem 2.13, the prime spectrum of a polynomial ring over a semilocal one-dimensional Noetherian domain is dependent upon whether or not the coefficient ring is Henselian.[2] For example, complete local rings, such as power series rings over a field, are Henselian.

Theorem 2.13 ([6, Theorem 2.7], [23, Theorem 2.4], and [29, Theorem 3.1]). *Let R be a semilocal one-dimensional Noetherian domain, let $\mathbf{m}_1, \ldots, \mathbf{m}_n$ be the maximal ideals of R where $n \in \mathbb{N}$, let y be an indeterminate, let $\beta = |R[y]|$, and let $\gamma_i = |(R/\mathbf{m}_i)[y]|$, for each i with $1 \leq i \leq n$. Then there exist exactly two possibilities for $U = \mathrm{Spec}(R[y])$ up to cardinality, depending upon whether or not R is Henselian and, if R is not Henselian, depending upon the number n of maximal ideals of R.*

- *In case R is **not Henselian**, U satisfies these axioms:*

(I_β) $|U| = \beta$ and U has a unique minimal element $u_0 = (0)$.

(II_β) $|\mathcal{H}_1(U) \cap \max(U)| = \beta$.

(III_γ) $\dim(U) = 2$.

(IV_n) There exist exactly n height-one elements $u_1, \ldots, u_n \in U$ such that $u_i^\uparrow$ is infinite. Also:

    (i) $u_1^\uparrow \cup \cdots \cup u_n^\uparrow = \mathcal{H}_2(U)$.

    (ii) $u_i^\uparrow \cap u_j^\uparrow = \emptyset$ if $i \neq j$.

    (iii) $|u_i^\uparrow| = \gamma_i$, for $1 \leq i \leq n$.

(V_n) If $v \in U$, v is not maximal, $\mathrm{ht}(v) = 1$ and $v \notin \{u_1, \ldots, u_n\}$, then $v^\uparrow$ is finite.

(VI_β) For every nonempty finite subset T of $\mathcal{H}_2(U)$, we have $|L_e(T)| = \beta$.

---

[2]Essentially a "Henselian" ring is one that satisfies Hensel's Lemma; see the definition in [20].

• If R is **Henselian**, then $n = 1$ and U satisfies Axioms $I_\beta, II_\beta, III_\gamma, IV_1$ and the
adjusted axioms V_1^h and VI_β^h below:

(V_1^h) If $v \in U$, v is not maximal, $\mathrm{ht}(v) = 1$ and $v \neq u_1$, then $|v^\uparrow| = 1$.
(VI_β^h) For every nonempty finite subset T of $\mathcal{H}_2(U)$, we have $|L_e(T)| = \beta \iff$
$\qquad |T| = 1$, $\quad$ and $\quad L_e(T) = \emptyset \iff |T| > 1$.

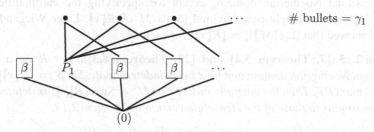

Diagram 2.13.h: Spec($R[y]$), R Henselian

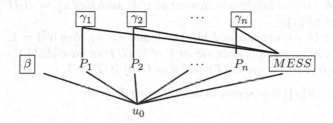

Diagram 2.13.nh: Spec($R[y]$), R non-Henselian

These diagrams show Spec($R[y]$) for the two cases of the theorem, where P_i is
the prime ideal of $R[y]$ corresponding to u_i, for each i with $1 \leq i \leq n$, and each
block $\boxed{\beta}$ represents β primes in that position.
The relations satisfied by the MESS box in Diagram 2.13 are too complicated to
show. They are described in Axiom VI_β.

2.2 Prime Ideals in Power Series Rings

In this subsection we describe prime ideals in power series rings over a Noetherian
domain. In the remainder of the paper we use the following straightforward remarks,
particularly Remark 2.14(1).

Remarks 2.14. Let x be an indeterminate over a Noetherian domain A. Then

(1) Every maximal ideal of $A[\![x]\!]$ has the form $(\mathbf{m}, x)A[\![x]\!]$, where $\mathbf{m}$ is a maximal
ideal of A; see [20, Theorem 15.1] (Nagata). Thus x is in every maximal ideal
of $A[\![x]\!]$.

(2) If $\mathbf{p}$ is a prime ideal of A, then $\mathbf{p}A[\![x]\!] \in \mathrm{Spec}(A[\![x]\!])$ and $\mathrm{ht}(\mathbf{p}A[\![x]\!]) = \mathrm{ht}(\mathbf{p})$; see [3, Theorem 4] or [1, Theorem 4].

(3) Thus every maximal ideal of $A[\![x]\!]$ of maximal possible height in a Noetherian catenary domain has the form $(\mathbf{m}, x)A[\![x]\!]$, where $\mathbf{m}$ is a maximal ideal of A with $\mathrm{ht}(\mathbf{m}) = \dim(A)$.

Heinzer, Rotthaus, and Wiegand almost characterized $\mathrm{Spec}(R[x])$ for R a one-dimensional Noetherian domain, except for specifying the cardinalities of the $L_e(\{M\})$ sets of height-two maximal ideals M of $R[x]$. Later Wiegand and Wiegand showed that $|L_e(\{M\})| = |R[x]|$ for each M.

Theorem 2.15 ([7, Theorem 3.4] and [29, Theorem 4.3]). *Let R be a one-dimensional Noetherian domain and let x be an indeterminate. Set $\beta := |R[x]|$ and set $\alpha := |\max(R)|$. Then the partially ordered set $U := \mathrm{Spec}(R[x])$ is determined by axioms similar to those of the Henselian version of Theorem 2.13:*

(I_β) $|U| = \beta$ *and U has a unique minimal element $u_0 = (0)$.*

(II_0) $\mathcal{H}_1(U) \cap \max(U) = \emptyset$.

(III_α) $\dim(U) = 2$, $|\mathcal{H}_2(U)| = \alpha$.

(IV_1) *There exists a height-one element $u_1 \in U$ such that $u_1^{\uparrow} = H_2(U)$, namely, $u_1 = xR[x]$.*

(V_1^h) *If $v \in U$, v is not maximal, $\mathrm{ht}(v) = 1$, and $v \neq u_1$, then $|v^{\uparrow}| = 1$.[3]*

(VI_β^h) *For every nonempty finite subset T of $\mathcal{H}_2(U)$, we have $|L_e(T)| = \beta$ if and only if $|T| = 1$, and $L_e(T) = \emptyset$ if and only if $|T| > 1$.*

Thus $\mathrm{Spec}(R[x])$ is as shown in the following diagram:

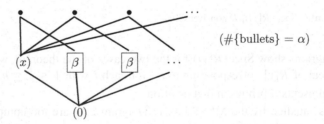

$(\#\{\text{bullets}\} = \alpha)$

Diagram 2.15.0: $\mathrm{Spec}(R[x])$

In Diagram 2.15.0, the cardinality of the set of bullets equals the cardinality of $\max(R)$ since the set of height-two maximal ideals of $R[x]$ is in one-to-one correspondence with the set of maximal ideals of the coefficient ring R by Remark 2.14(1). The boxed β beneath each maximal ideal of $R[x]$ means that there are exactly β prime ideals in that position (beneath that maximal ideal and

---

[3] Since axiom II_0 holds, this axiom could be stated here without saying "v is not maximal."

no other). For each value of α and β, any two posets described by Diagram 2.15.0 are order-isomorphic.

Remark 2.16. From Diagrams 2.15.0 and 2.13.h, we see that $\text{Spec}(\mathbb{Q}[y][[x]]) \not\cong \text{Spec}(\mathbb{Q}([x][[y]]))$, where $\mathbb{Q}$ is the field of rational numbers. Moreover the difference between the prime spectrum of a power series ring over a one-dimensional Noetherian domain, such as $\mathbb{Q}[y][[x]]$, and that of a polynomial ring over a Henselian ring, such as $\mathbb{Q}[[x]][y]$, is that the partially ordered set described in the Henselian case of Theorem 2.13 has β height-one maximal elements, whereas the other partially ordered set has no height-one maximal elements.

In our characterizations of prime spectra, we identify those prime ideals that are an intersection of maximal ideals, such as the prime ideal (x) in Diagram 2.15.0 and the prime ideal P_1 in Diagram 2.13.h. These are called *j-prime ideals.*

Definitions 2.17. (1) Let A be a commutative ring.

- A *j-prime (ideal)* of A is a prime ideal of A that is an intersection of maximal ideals of A.
- The *j-spectrum* of A is $j\text{-Spec}(A) := \{j\text{-primes} \in \text{Spec}(A)\}$.

(2) For U a partially ordered set, we say that $u \in U$ is a *j-element* if u is a maximal element of U or if $\min(u^\uparrow)$ is infinite. Then $j\text{-set}(U) := \{j\text{-elements of } U\}$.

Thus, if A is a two-dimensional integral domain, $\{j\text{-elements of } U = \text{Spec}(A)\} = \{j\text{-prime ideals of } A\}$.

Examples 2.18. We show $j\text{-Spec}(\mathbb{Q}[y][[x]])$ and $j\text{-Spec}(\mathbb{Q}[x][[y]])$, respectively, in Diagram 2.18.0; they are parts of Diagrams 2.15.0 and 2.13.h.

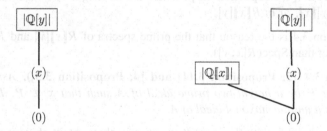

Diagram 2.18.0: $j\text{-Spec}(\mathbb{Q}[y][[x]])$ and $j\text{-Spec}(\mathbb{Q}[x][[y]])$

3 Properties of Mixed Polynomial-Power Series Rings

In this section we give some properties of prime spectra of three-dimensional Noetherian mixed polynomial-power series rings. We use the following setting:

Setting 3.1. Let x and y be indeterminates over a one-dimensional Noetherian domain R. Let A be either $R[y][[x]]$, $R[[x]][y]$, or $R[[x, y]]$. Let $A_1 = R[y]$ if

$A = R[y][\![x]\!]$ or $R[\![x]\!][y]$, and let $A_1 = R[y]$ if $A = R[\![x, y]\!]$. Then $A/xA \cong A_1$ and, depending on which A_1 we have, for every $\mathbf{m} \in \max(R)$.

$$A/(\mathbf{m}, x)A \cong A_1/\mathbf{m}A_1 \cong (R/\mathbf{m})[y] \quad \text{or} \quad A/(\mathbf{m}, x)A \cong (R/\mathbf{m})[\![y]\!].$$

Proposition 3.2 gives a description of the maximal ideals of A having maximal height, that is, height three.

Proposition 3.2 ([4]). *Assume Setting 3.1 and let $\mathcal{M}$ be a height-three maximal ideal of A. Then*

(1) $\mathcal{M} = (\mathbf{m}, x, h(y))A$, *for some* $\mathbf{m} \in \max(R)$ *and some* $h(y) \in A_1$ *with* $\overline{h(y)}$ *irreducible in* $\overline{A_1} = A_1/(\mathbf{m}A_1) \cong (R/\mathbf{m})[y]$ *or* $\overline{h(y)}$ *irreducible in* $\overline{A_1} \cong (R/\mathbf{m})[\![y]\!]$.

(2) *Conversely, the ideals* $(\mathbf{m}, x, h(y))A$ *are maximal and have height three, for every* $\mathbf{m} \in \max(R)$ *and for every* $h(y) \in A_1$ *such that* $\overline{h(y)}$ *is irreducible in* $\overline{A_1}$.

(3) *If* $A = R[\![x, y]\!]$, *then every maximal ideal of* $R[\![x, y]\!]$ *has height 3; there are* $|\max(R)|$ *maximal ideals in* $R[\![x, y]\!]$; *and* $\max(R[\![x, y]\!]) = \{(\mathbf{m}, x, y)R[\![x, y]\!]$, *where* $\mathbf{m} \in \max(R)\}$.

(4) *For* $A = R[\![x]\!][y]$ *or* $A = R[y][\![x]\!]$, *there are* $|(R/\mathbf{m})| \cdot \aleph_0$ *height-three maximal ideals that contain* $\mathbf{m}$, *for each fixed* $\mathbf{m} \in \max(R)$.

Proof. Item (3) follows from Remark 2.14(1). For the remaining items, see [4, Proposition 4.2].   □

Proposition 3.3 is also straightforward to prove using Remark 2.14(1) and Lemma 2.8; see [4].

Proposition 3.3 ([4, Propostion 4.3]). *There are no height-one maximal ideals in* $R[\![x, y]\!]$, $R[y][\![x]\!]$, *or in* $R[\![x]\!][y]$.

Proposition 3.4 is the reason that the prime spectra of $R[\![x]\!][y]$ and $R[y][\![x]\!]$ is much simpler than $\operatorname{Spec}(R[\![x, y]\!])$.

Proposition 3.4 ([5, Proposition 3.11] and [4, Proposition 3.3]). *Assume Setting 3.1. Let P be a height-two prime ideal of A such that $x \notin P$. Then P is contained in a unique maximal ideal of A.*

In Proposition 3.5, with Setting 3.1, we observe that certain obvious conditions on a height-one prime ideal Q of A are equivalent to saying that Q is not contained in any height-three maximal ideal of A. For $A = R[\![x, y]\!]$, these conditions never occur; see Proposition 3.2(3) or Theorem 4.1.

Proposition 3.5 ([4, Proposition 3.8]). *Assume Setting 3.1, so that R is a one-dimensional Noetherian domain and A is $R[\![x]\!][y]$, $R[y][\![x]\!]$, or $R[\![x, y]\!]$. Let Q be a height-one prime ideal of A. Then statements 1–4 are equivalent:*

(1) *Every prime ideal of A containing $(Q, x)A$ is a maximal ideal.*

(2) *For every $\mathbf{m} \in \max(R)$, every prime ideal of A containing $(Q, \mathbf{m})A$ is maximal.*

(3) Q is contained in no height-three maximal ideal of A.
(4) $\dim(A/Q) = 1$.

Moreover,

- *If $(Q, x)A = A$, then item (1) holds.*
- *If $\mathbf{m} \in \max(R)$ and $(Q, \mathbf{m})A = A$, then every prime ideal containing $(Q, \mathbf{m})A$ is maximal.*
- *Thus either of the conditions*
  (i) $(Q, x)A = A$ or
  (ii) $(Q, \mathbf{m})A = A$, for every $\mathbf{m} \in \max(R)$,

  implies (4) $\dim(A/Q) = 1$.

Proposition 3.6 holds for higher-dimensional rings and more variables (one variable must be a power series variable), but to fit our focus in this article, we consider prime ideals of A, where $A = R[\![x, y]\!]$, $R[y][\![x]\!]$, or $R[\![x]\!][y]$ has dimension three. One case of Proposition 3.6 is given in [5, Proposition 3.8].

Proposition 3.6 ([4, Proposition 2.18]). *Assume Setting 3.1 and let Q and $\mathcal{M}$ be prime ideals of A with $x \notin Q$, $\mathrm{ht}(Q) = 1$, and $\mathrm{ht}(\mathcal{M}) = 3$. Then $Q^{\uparrow} \cap \mathcal{M}^{\downarrow}$ contains exactly $|R[x]|$ height-two prime ideals.*

3.1   *j*-Spectra of Quotients of Mixed Polynomial-Power Series Rings

We use Setting and Notation 3.7 in the remainder of this section.

Setting and Notation 3.7. Let R be a one-dimensional Noetherian domain and let x and y be indeterminates. Let A be $R[\![x]\!][y]$, $R[y][\![x]\!]$, or $R[\![x, y]\!]$ and let Q be a height-one prime ideal of A such that $x \notin Q$ and $(Q, x)A \neq A$. Set $B := A/Q$. By Remarks 2.2, A is catenary and has dimension three, and so B is a Noetherian integral domain with $\dim(B) \leq 2$. Let I be a nonzero ideal of $R[y]$ such that $(I, x)A = (Q, x)A$; that is, $I = \{$ all constant terms in $R[y]$ of power series in $Q\}$.

Note 3.8. If $I \neq R[y]$, then the ideal I from Setting and Notation 3.7 is a nonzero height-one ideal of $R[y]$; that is, every prime ideal P of $R[y]$ minimal over I has height one.

Proof. Let P be a prime ideal of $R[y]$ minimal over I. If $I = (0)$, then $(I, x) = (x) \neq (Q, x)$, since $Q \neq (0)$ and $x \notin Q$. Thus $I \neq (0)$, and so $\mathrm{ht}(P) \geq 1$. Now $(Q, x)A \neq A$ by assumption and $1 = \mathrm{ht}(Q) < \mathrm{ht}(Q, x)$ since $x \notin Q$ and A is catenary by Remarks 2.2. Also $\mathrm{ht}(Q, x) \leq 2$ by Krull's principal ideal theorem. Thus $\mathrm{ht}(Q, x) = 2$. Now $(P, x) \neq A$ since $P \in \mathrm{Spec}(R[y])$. Also (P, x) is a minimal prime ideal of $(I, x) = (Q, x)$. Thus $\mathrm{ht}(P, x) = 2$, and so $P \in \mathrm{Spec}(R[y])$ implies $\mathrm{ht}(P) = 1$.   □

We show in this subsection that the j-primes of A that contain Q also contain x. It follows that each j-prime of A corresponds to a minimal prime ideal of $R[y]/I$. We begin to demonstrate this correspondence with the following remarks.

Remarks 3.9. With Setting and Notation 3.7, consider the following canonical surjections:

$$\pi : A \longrightarrow B = A/Q \text{ with } \ker(\pi) = Q,$$

$$\pi_x : A \longrightarrow R[y] = A/xA \text{ with } \ker(\pi_x) = (x).$$

(i) The maps π and π_x yield isomorphisms:

$$\mathrm{Spec}(B) \cong \mathrm{Spec}\left(\frac{A}{Q}\right); \quad \text{and} \quad \mathrm{Spec}(R[y]) \cong \mathrm{Spec}\left(\frac{A}{xA}\right);$$

$$\mathrm{Spec}\left(\frac{B}{xB}\right) \cong \mathrm{Spec}\left(\frac{A}{(x,Q)A}\right) = \mathrm{Spec}\left(\frac{A}{(x,I)A}\right) \cong \mathrm{Spec}\left(\frac{R[y]}{I}\right).$$

(ii) Since A is catenary, the correspondences in Remark 3.9(i) above imply that for each $n \le 2$, the ht-n prime ideals of B can be identified with the ht-$(n+1)$ prime ideals of A containing Q; the ht-n prime ideals of $R[y]$ can be identified with the ht-$(n+1)$ prime ideals of A containing x; and the ht-n primes of B containing x can be identified with the ht-n prime ideals of $R[y]$ containing I.

Proposition 3.10 ([4, Proposition 3.20]). *Assume Setting 3.7.*

(1) $\mathrm{Spec}(B/xB) \cong \mathrm{Spec}(R[y]) \cap (V_{R[y]}(I)) \cong \mathrm{Spec}(R[y]/I).$
(2) The height-one prime ideals of B that contain x correspond to the height-one prime ideals of $R[y]$ that contain I.
(3) Every nonmaximal j-prime ideal of B contains x and thus corresponds to a j-prime ideal of $R[y]$ containing I.
(4) j-$\mathrm{Spec}(B) \setminus \{(0)\} \setminus \{height$-$one\ maximal\ elements\} \cong j$-$\mathrm{Spec}(B/xB) \cong j$-$\mathrm{Spec}(R[y]/I).$
(5) If $\max(R)$ is infinite, then j-$\mathrm{Spec}(B/xB) = \mathrm{Spec}(B/xB)$; that is, every prime ideal of B containing x is a j-prime, and every prime ideal of $R[y]$ containing I is a j-prime.

Proof. Items (1)–(4) follow from Remarks 3.9; see [4, Proposition 3.22]. For item (5), if $P \in \mathrm{Spec}(B)$ has height one and contains x, then P corresponds to a height-one prime ideal P' of $R[y]$ that contains I. Therefore it suffices to show that every prime ideal P' of $R[y]$ containing I is contained in an infinite number of height-two maximal ideals. If $P' = \mathbf{m}R[y]$, for some $\mathbf{m} \in \max(R)$, then

$$|(P')^{\uparrow(R[y])}| = |(\mathbf{m}R[y])^{\uparrow(R[y])}| = |R[y]/(\mathbf{m}R[y])| = |(R/\mathbf{m})[y]| = |R/\mathbf{m}| \cdot \aleph_0,$$

using Lemma 2.6; thus P'/I is a j-prime of $R[y]/I$. On the other hand, if $P' \cap R = (0)$, then every element of P' has positive degree and $P' = h(y)K[y] \cap R[y]$, where K is the field of fractions of R and $h(y) \in R[y]$, by Remarks 2.7(3). The leading coefficient h_n of $h(y)$ is contained in at most finitely many maximal ideals of R.

Claim 3.11. $P' \subsetneq (\mathbf{m}, P')R[y] \neq R[y]$, for every maximal ideal $\mathbf{m}$ such that $h_n \notin \mathbf{m}$.

Proof of Claim 3.11. Since $\mathbf{m}R[y]$ contains elements of degree 0, $(\mathbf{m}, P')$ is properly bigger than P'. For the inequality, write $h(y) = h_n y^n + h_{n-1} y^{n-1} + \cdots + h_0$, where $n \geq 1$, each $h_i \in R$ and $h_n \neq 0$. If $h_n \notin \mathbf{m}$, then $h_i/h_n \in R_\mathbf{m}$, for each i with $0 \leq i \leq n$. Thus

$$P'R_\mathbf{m}[y] = h(y)K[y] \cap R_\mathbf{m}[y] = \left(y^n + \frac{h_{n-1}}{h_n} y^{n-1} + \cdots + \frac{h_0}{h_n} \right) R_\mathbf{m}[y]$$

$$\implies \frac{P'R_\mathbf{m}[y] + \mathbf{m}R_\mathbf{m}[y]}{\mathbf{m}R_\mathbf{m}[y]} \subsetneq \frac{R_\mathbf{m}[y]}{\mathbf{m}R_\mathbf{m}[y]},$$

and so Claim 3.11 is proved.   □

For item (5), since $\max(R)$ is infinite, there are infinitely many $\mathbf{m} \in \max(R)$ such that $h_n \notin \mathbf{m}$. Each pair $(\mathbf{m}, P')$ with $\mathbf{m} \in \max(R)$ is in a distinct maximal ideal of $R[y]$; that is, a maximal ideal containing $(\mathbf{m}, P')$ cannot contain $(\mathbf{m}', P')$ if $\mathbf{m} \neq \mathbf{m}'$ and $\mathbf{m}, \mathbf{m}' \in \max(R)$. Thus $|(P')^{\uparrow(R[y])}| = |\max(R)|$, since removing finitely many $\mathbf{m} \in \max(R)$ such that $h_n \in \mathbf{m}$ from the infinite set $\max(R)$ leaves the same number. This completes the proof of item (5) and thus Proposition 3.10 is proved.   □

Remark 3.12. With Setting 3.7, assume that R is semilocal and $I \subseteq P' \in \mathrm{Spec}(R[y])$. If P' is $\mathbf{m}R[y]$, for some $\mathbf{m} \in \max(R)$, or P' is a maximal ideal of $R[y]$, then P' is a j-prime ideal. However not every prime ideal containing I is necessarily a j-prime ideal. See Example 3.13 and Theorem 7.2.

Example 3.13. Let $R = \mathbb{Z}_{(2)}$ and $I = 2y(2y - 1)(y + 2)$. Then $\mathrm{Spec}(R/I)$ is shown below:

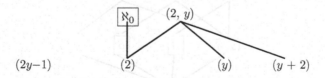

Diagram 3.13.0: $\mathrm{Spec}(\mathbb{Z}_{(2)}[y]/(2y(2y - 1)(y + 2)))$

The structure of $\mathrm{Spec}(\mathbb{Z}_{(2)}[y]/(2y(2y - 1)(y + 2)))$ is determined by the finite partially ordered subset

$$F = \{(2y - 1), (2), (y), (y + 2), (2, y)\}.$$

We see that $\text{Spec}(\mathbb{Z}_{(2)}[y]/(2y(2y-1)(y+2)))$ and its j-spec are not the same, since (y) and $(y+2)$ are prime ideals that are not j-prime ideals.

4 Two-Dimensional Prime Spectra of Form $R[\![x, y]\!]/Q$

In this short section we discuss the prime spectra of homomorphic images by a height-one prime ideal of the ring of power series in two variables over a one-dimensional Noetherian domain. These prime spectra are similar to those for images of mixed polynomial-power series rings. If both variables x and y are power series variables, however, the analysis is simplified.

Theorem 4.1. *Let R be a one-dimensional Noetherian domain, let x and y be indeterminates, and let Q be a height-one prime ideal of $R[\![x, y]\!]$. Set $B = R[\![x, y]\!]/Q$ and $\beta = |R[\![x]\!]|$. Then:*

(1) If $Q \nsubseteq (x, y)R[\![x, y]\!]$, then there exist $n \in \mathbb{N}$ and $\mathbf{m}_1, \dots, \mathbf{m}_n \in \max(R)$ such that $\text{Spec}(B)$ has the form shown below:

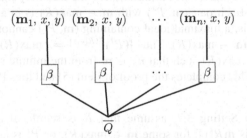

(2) If $Q \subseteq (x, y)R[\![x, y]\!]$, then $\text{Spec}(B)$ is order-isomorphic to $\text{Spec}(R[\![x]\!])$; that is, $\text{Spec}(B)$ has the form shown below:

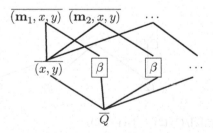

where the $\mathbf{m}_i$ range over all the elements of $\max(R)$ and $\beta = |R[\![x]\!]|$.

As the diagrams show, $\text{Spec}(B)$ is characterized by the description for each case.

Proof. For both items we use that every maximal ideal of $R[x, y]$ has the form $(\mathbf{m}, x, y)$ where $\mathbf{m} \in \max(R)$, by Proposition 3.2(3). For item (1), let Q_0 be the ideal of R generated by all the constant terms of elements of Q. Then $1 \notin Q_0$, since an element with constant term 1 is a unit of $R[x, y]$—every such element is outside every maximal ideal. Also $Q_0 \neq (0)$, since $Q \not\subseteq (x, y)$. Therefore Q_0 is contained in finitely many maximal ideals of R, say $\mathbf{m}_1, \ldots, \mathbf{m}_n$. It follows that Q is contained in n maximal ideals of $R[x, y]$, namely $(\mathbf{m}_1, x, y), \ldots, (\mathbf{m}_n, x, y)$. By Proposition 3.6, we have $|R[x]|$ height-two prime ideals between Q and $(\mathbf{m}_i, x, y)$, for each i. Every height-two prime ideal P between Q and a maximal ideal $(\mathbf{m}, x, y)$ is missing either x or y, since $Q \not\subseteq (x, y)R[x, y]$. Therefore by Proposition 3.4 each such prime ideal P is contained in a unique maximal ideal, namely $(\mathbf{m}, x, y)$. Thus $\mathrm{Spec}(R[x, y]/Q)$ has the form given in the first diagram.

For item (2), since $Q \subseteq (x, y)$, we have $Q \subseteq (\mathbf{m}, x, y)$, for every $\mathbf{m} \in \max(R)$. By Proposition 3.6 again the number of primes between Q and every maximal ideal $(\mathbf{m}, x, y)$ is $|R[x]|$, and so we have $|R[x]|$ height-two prime ideals between Q and each $(\mathbf{m}_i, x, y)$, for each i. As for item (1), every height-two prime ideal P other than (x, y) that is between Q and a maximal ideal $(\mathbf{m}, x, y)$ is missing either x or y, since $Q \not\subseteq (x, y)R[x, y]$. Therefore every such prime ideal P is in a unique maximal ideal of $R[x, y]$, and so we get the form of the second diagram in this case. $\square$

5 Spectra of Quotients of Mixed Polynomial-Power Series Rings

Let x and y be indeterminates over a one-dimensional Noetherian domain R, let $A = R[x][[y]]$ or $R[y][[x]]$, and let Q be a height-one prime ideal of A. In this section we describe $\mathrm{Spec}(A/Q)$; this is work in progress from [4]. In some cases, we determine the spectra precisely. We need not consider $A = R[x, y]$, since Theorem 4.1 contains a complete description of $\mathrm{Spec}(R[x, y]/Q)$, for Q a height-one prime ideal of $R[x, y]$.

First we consider the exceptional cases where the dimension of A/Q is 1. We need a definition:

Definition 5.1. A *fan* is a one-dimensional poset with a unique minimal element.

Theorem 5.2 ([4, Theorem 5.2]). *Let R be a one-dimensional Noetherian domain and let x and y be indeterminates over R. Let $A = R[x][[y]]$ or $R[y][[x]]$, let Q be a height-one prime ideal of A, and let $B = A/Q$. Then $\mathrm{Spec}(B)$ is a fan if one of the following two cases occur:*

(i) Every height-two prime ideal of A containing $(Q, x)A$ is maximal.
(ii) For every $\mathbf{m} \in \max(R)$, every height-two prime ideal of A containing $(Q, \mathbf{m})A$ is maximal.

Moreover, if $A = R[y][[x]]$, then $\mathrm{Spec}(B)$ is a fan with a finite number of elements, but at least two. If $A = R[x][[y]]$, then $\mathrm{Spec}(B)$ is a fan with $|R[x]|$ elements.

Proof. By Proposition 3.5, either of these conditions implies that $\operatorname{Spec}(B)$ is a fan.

For the "Moreover" statement, every maximal ideal of B is the image of a height-two maximal ideal of A that contains Q. In case $A = R[y][\![x]\!]$, every height-two maximal ideal has the form (M, x), where M is a height-one maximal ideal of $R[\![y]\!]$, by Remarks 2.14. There are just finitely many such height-two maximal ideals that contain $(Q, x)A$. For both of the rings $A = R[y][\![x]\!]$ and $A = R[\![x]\!][y]$, since A has no height-one maximal ideals by Proposition 3.3, there must be a maximal ideal containing Q that is bigger than Q and so the cardinality of the fan is at least two. For $A = R[\![x]\!][y]$, $|\operatorname{Spec}(A/Q)| = |R[\![x]\!]|$; see [4, Theorem 5.2].   □

Except for the special cases of Theorem 5.2, prime spectra of homomorphic images of mixed polynomial-power series rings $R[\![x]\!][y]$ and $R[y][\![x]\!]$ by height-one prime ideals are two dimensional. In order to describe the partially ordered sets that arise, we need a kind of *genetic code*. Definition 5.5 of this section contains such a code and a general set of axioms involving the code that are satisfied by two-dimensional images $B = R[\![x]\!][y]/Q$ and $B' = R[y][\![x]\!]/Q'$, where Q and Q' are height-one prime ideals of $R[\![x]\!][y]$ and $R[y][\![x]\!]$, respectively. Basically the code tells us, for each two-dimensional partially ordered set, how many elements are at each level and what relationships hold between elements.

We use the following setting and notation for the rest of this section.

Setting and Notation 5.3. Let x and y be indeterminates over a one-dimensional Noetherian domain R, let $A = R[\![x]\!][y]$ or $R[y][\![x]\!]$, and set $\beta := |R[\![x]\!]|$. Let Q be a height-one prime ideal of A such that $x \notin Q$ and the domain $B := A/Q$ has dimension two. Let I be the height-one ideal of $R[y]$ such that $(I, x)A = (Q, x)A$, and let $\{q_1, \ldots, q_\ell\}$, for $\ell \in \mathbb{N}$, be the minimal primes of I in $R[y]$. Define:

- $F := \{q_1, \ldots, q_\ell\} \cup \{q_i^\uparrow \cap q_j^\uparrow\}_{1 \le i < j \le \ell}$, a subset of $V_{R[y]}(I)$;
- $\gamma_i := |q_i^\uparrow \setminus (\bigcup_{j \ne i} q_j^\uparrow)|$, for each i with $1 \le i \le \ell$; that is, each γ_i is the number of height-two maximal ideals of $R[y]$ that contain q_i but none of the other q_js; and
- $\varepsilon := |\{\text{ht } 1 \text{ maximal ideals of } B\}|$.

The main theorem of [4] describes $\operatorname{Spec}(B)$ in Setting and Notation 5.3. We remark that, as might be expected from Proposition 3.10, these prime spectra are largely determined by $\operatorname{Spec}(R[y]/I)$.

Main Theorem 5.4 ([4, Theorem 6.5]). *Assume Setting and Notation 5.3. Then there exists an order-monomorphism $\varphi : F \to U$ such that U and φ have the following properties:*

(1) $|U| = \beta$, $\dim(U) = 2$, and U has a unique minimal element u_0.

(2) $|\{\mathcal{H}_1(U) \cap \max(U)\}| = \varepsilon$; $\{\varphi(q_1), \ldots, \varphi(q_\ell)\} \subseteq \mathcal{H}_1(U)$.

(3) $\mathcal{H}_2(U) = \bigcup \varphi(q_i)^\uparrow = (\varphi(F) \setminus \{\varphi(q_1), \ldots, \varphi(q_\ell)\}) \cup \bigcup_{i=1}^\ell T_i$, where each $T_i = \varphi(q_i)^\uparrow \setminus (\bigcup_{j \ne i} \varphi(q_j)^\uparrow)$ and $|T_i| = \gamma_i$.

(4) $\{\varphi(q_1), \ldots, \varphi(q_\ell)\}$ contains the set $\{u \in U \mid |u^\uparrow| = \infty,\ \operatorname{ht}(u) = 1\}$ of nonmaximal nonzero j-elements of U.

(5) For every $u \in \mathcal{H}_1(U) \setminus \varphi(F)$, there exists a unique maximal element in U that is greater than or equal to u.

(6) For every $1 \leq i < j \leq \ell$, $\varphi(q_i)^\uparrow \cap \varphi(q_j)^\uparrow = \varphi(q_i^\uparrow \cap q_j^\uparrow) \subseteq \varphi(F)$.

(7) For every finite nonempty subset $T \subseteq \mathcal{H}_2(U) \setminus F$, $L_e(T) = \emptyset$ if $|T| > 1$ and $|L_e(T)| = \beta$ if $|T| = 1$.[4]

These properties determine U as a partially ordered set. Moreover $\varepsilon \leq \beta$. If $A = R[y][\![x]\!]$, then ε is finite; if $A = R[y][\![x]\!]$ and $\max(R)$ is infinite, then $\varepsilon = 0$.

Proof. We give some notes about the proof: The map $\varphi : F \hookrightarrow U$ is given by $\varphi(P) = P/I$, for every $P \in F$, so that $\mathrm{ht}(\varphi(P)) = 1 + \mathrm{ht}(P)$. Then item (4) follows from Note 3.8 and Proposition 3.10, and items (5) and (7) follow from Proposition 3.4. The "Moreover" statement holds since every ideal of A is finitely generated, and thus the total number of prime ideals of A and of B is at most β. The remaining statements follow from Remark 2.14(1) and Lemma 2.8; if $\max(R)$ is finite, every height-one maximal ideal of B corresponds to a height-two maximal ideal of A such that $N = (M, x)$, where M is a height-one maximal ideal of $R[y]$ and $(Q, x) \subseteq N$. There are just finitely many of these. For more details, see [4]. $\square$

Definition 5.5. Let $\ell \in \mathbb{N}_0$ and let $\varepsilon, \beta, \gamma_1, \ldots, \gamma_\ell$ be cardinal numbers with $\varepsilon, \gamma_i \leq \beta$, for each γ_i. We say that a partially ordered set U is *image polynomial-power series of type* $(\varepsilon; \beta; F; \ell; (\gamma_1, \ldots, \gamma_\ell))$, if there exist a finite partially ordered set F of dimension at most one with ℓ minimal elements such that every non-minimal maximal element of F is greater than at least two minimal elements of F and an order-monomorphism φ such that U satisfies properties (1)–(7) of Theorem 5.4.

For examples of these prime spectra, see Sects. 6 and 7.

6 Prime Spectra of Simple Birational Extensions of Power Series Rings

By a *simple birational extension* of an integral domain A with field of fractions K, we mean a ring of form $A[g/f]$ between A and K, where $f, g \in A$ with $f \neq 0$, and either f, g is an A-sequence or $(f, g)A = A$. As noted in Remarks 2.11(c), the prime spectra of simple birational extensions of $\mathbb{Z}[y]$ are order-isomorphic to $\mathrm{Spec}(\mathbb{Z}[y])$; see [14]. In this section, for R a one-dimensional Noetherian domain and x an indeterminate, we present some recent work of Eubanks-Turner, Luckas, and Saydam on prime spectra of simple birational extensions of $R[\![x]\!]$; see [5]. Generally the prime spectrum of a simple birational extension of $R[\![x]\!]$ is rather more complicated than that of $R[\![x]\!]$.

---

[4] The term "$L_e(T)$" is defined in Notation 2.1.

Theorem 6.1 summarizes the possible prime spectra of simple birational extensions of a power series ring $R[\![x]\!]$ if R is a one-dimensional Noetherian domain
with infinitely many maximal ideals. The original statement of this theorem is given
incorrectly in [5]. We do not necessarily know that all the γ_i are the same, as was
assumed there.

Theorem 6.1 ([5, Theorem 4.1]). *Let R be a one-dimensional Noetherian domain
such that $\alpha = |\max(R)|$ is infinite, let x and y be indeterminates, and let f and
g be elements of $R[\![x]\!]$ with $f \neq 0$. Let a and b be the constant terms of f and g
respectively. Set $\beta = |R[\![x]\!]|$, and $v = |V_R(a,b)|$. Let $V_R(a,b) = \{\mathbf{m}_1, \ldots, \mathbf{m}_v\}$
be a numbering of the maximal ideals of R that contain a and b. For each i with
$1 \leq i \leq v$, let $\gamma_i := |R/\mathbf{m}_i| \cdot \aleph_0$. Let $B = R[\![x]\!][g/f]$.*

(1) Suppose that $(f,g)R[\![x]\!] = R[\![x]\!]$ and x divides f; equivalently $B = R[\![x]\!][1/f]$, $a = 0$, and b is a unit. Then $\operatorname{Spec}(B)$ is a fan of cardinality β.

*(2) Suppose that $g = 0$ or $(f,g)R[\![x]\!] = R[\![x]\!]$ and x does not divide f, so that
$B = R[\![x]\!]$ or $B = R[\![x]\!][1/f]$, $a \neq 0$, and $(a,b)R = R$. Then $\operatorname{Spec}(B)$ is
either order-isomorphic to $\operatorname{Spec}(R[\![x]\!])$ or to $\operatorname{Spec}(R[\![x]\!])$ with $|R[\![x]\!]|$ height-
one maximal elements adjoined.*

*(3) (a) If f, g is an $R[\![x]\!]$-sequence and x divides f, then $a = 0$ and b is a nonzero
    nonunit.*

    *(b) If $a = 0$ and b is a nonzero nonunit, then $\operatorname{Spec}(B)$ is D-birational of type
    $(\beta; \beta; (v, 0); \gamma_1, \cdots, \gamma_v)$, as defined in Definition 6.3.*

*(4) (a) If f, g is an $R[\![x]\!]$-sequence and x does not divide f, then $a \neq 0$ and
    $(a,b)R \neq R$.*

    *(b) If $a \neq 0$ and $(a,b)R \neq R$, then $\operatorname{Spec}(B)$ is N-birational of type $(\beta; \beta; v +
    1; \gamma_1, \cdots, \gamma_v, \alpha; t_1, \cdots, t_v)$, for some $t_1, \cdots, t_v$ in $\mathbb{N}_0$; the list $\mathbf{m}_1, \ldots, \mathbf{m}_v$ is
    to be reordered so that the corresponding list $t_1, \cdots, t_v$ is in increasing order.
    See Definition 6.2.*

Proof. See [5, Theorem 4.1]; it is an easy adjustment to put in the γ_i instead of γ.

□

The "type" referred to in Definitions 6.2 and 6.3 below is like a *genetic code*
that describes the numbers of prime ideals in various positions of $\operatorname{Spec}(R[\![x]\!][g/f])$
in general. These definitions are related to Definition 5.5; here we give more details
and restrictions on the partially ordered set F than in that definition.

Definition 6.2. Let $\ell, t_1, \ldots, t_{\ell-1} \in \mathbb{N}_0$ be such that $t_1 \leq t_2 \leq \cdots \leq t_{\ell-1}$, and
let $\beta, \gamma_1, \ldots, \gamma_\ell$ be infinite cardinal numbers with each $\gamma_i \leq \beta$. Let $\varepsilon = 0$ or β; if
$\ell = 0$, there are no t_i or γ_i and we require $\varepsilon = \beta \neq 0$. Then a partially ordered set
U is *N-birational of type* $(\varepsilon; \beta; \ell; \gamma_1, \ldots, \gamma_\ell; t_1, \ldots, t_{\ell-1})$ if axioms 1–6 hold:

(1) $|U| = \beta$, and U has a unique minimal element u_0.
(2) $|\{\text{height-one maximal elements of } U\}| = \varepsilon$.
(3) If $\ell \neq 0$, then $\dim(U) = 2$. If $\ell = 0$ (and so $\varepsilon = \beta \neq 0$), then $\dim(U) = 1$
    and U is a fan; see Definition 5.1.

(4) U has exactly ℓ height-one elements $u \in U$ such that $|u^\uparrow| = \infty$. Moreover, if we list these elements as $P_1, P_2, \ldots, P_\ell$, then they satisfy:

- $|P_i^\uparrow| = \gamma_i$, for each i with $1 \leq i \leq \ell$;

- $\bigcup_{i=1}^{\ell} P_i^\uparrow = \{\text{height-two maximal elements of } U\}$; and

- $1 \leq i, j < \ell$, and $i \neq j \implies |P_i^\uparrow \cap P_\ell^\uparrow| = t_i$ and $|P_i^\uparrow \cap P_j^\uparrow| = 0$.

(5) For every height-one element $u \in U \setminus \{P_1, P_2, \ldots, P_\ell\}$, there exists a unique maximal element in U that is greater than or equal to u.

(6) For every height-two element $t \in U$, $|L_e(t)| = \beta$.

If each $t_i = 0$, then every pair (P_i, P_j) with $i \neq j$ is comaximal by the condition of axiom 4; otherwise P_ℓ is usually distinguishable from the other P_i because there are t_i maximal elements bigger than both P_ℓ and P_i, for each i with $1 \leq i < \ell$. Schematically, the condition of axiom 4 yields the following j-sets, where the unique minimal element below all other elements has been removed:

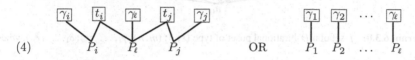

(4)

Diagram 6.2.0: Parts of N-birational posets; on the right each $t_i = 0$

Abbreviations 6.2.a. If $\gamma_1 = \cdots = \gamma_\ell = \gamma$ (as in all of our examples), then we write the type as $(\varepsilon; \beta; \ell; \gamma; t_1, \ldots, t_{\ell-1})$.

Definition 6.3 for D-*birational* is the case where every $t_i \leq 1$ of the N-birational condition. The D-birational posets correspond to prime spectra for simple birational extensions of $D[\![x]\!]$, for some Dedekind domain D; see Theorem 6.5. In this case, we group the nonmaximal j-primes into a comaximal subset and a non-comaximal subset.

Definition 6.3. Let $m, n \in \mathbb{N}_0$ and let $\beta, \gamma_1, \ldots, \gamma_m, \delta_1, \ldots, \delta_n$ be infinite cardinal numbers with each $\gamma_i, \delta_j \leq \beta$. Let $\varepsilon = 0$ or β; if $m = n = 0$, require $\varepsilon = \beta \neq 0$. Then a partially ordered set U is D-*birational of type* $(\varepsilon; \beta; (m, n); \gamma_1, \ldots, \gamma_m; \delta_1, \ldots, \delta_n)$ if axioms (1), (2), (5), (6) of Definition 6.2 hold as well as axioms (3') and (4') below:

(3') If $m \neq 0$ or $n \neq 0$, then $\dim(U) = 2$. If $m = n = 0 = \beta$, then $\varepsilon = \beta \neq 0$, $\dim(U) = 1$, and U is a fan.

(4') U has exactly $m + n$ height-one elements u such that $u^\uparrow$ is infinite: $P_1, P_2, \ldots, P_m, Q_1, \cdots, Q_n$, where for $i, j, r, i', j' \in \mathbb{N}$ with $1 \leq i, j \leq m, i \neq j, 1 \leq r \leq n, 1 \leq i', j' < n$, and $i' \neq j'$, we have:

- $|P_i^\uparrow| = \gamma_i$, $|Q_r^\uparrow| = \delta_r$;

- $|P_i^\uparrow \cap P_j^\uparrow| = 0 = |P_i^\uparrow \cap Q_r^\uparrow| = |Q_{i'}^\uparrow \cap Q_{j'}^\uparrow|$ and $|Q_{i'}^\uparrow \cap Q_n^\uparrow| = 1$;
- $\bigcup_{i=1}^{m} P_i^\uparrow \cup \bigcup_{r=1}^{n} Q_r^\uparrow = \{\text{height-two maximal elements of } U\}$.

Abbreviations 6.3.a. If $\gamma_i = \gamma_j = \delta_s = \delta_t = \gamma$, for every i, j, s, t with $1 \le i < j \le m$, $1 \le s < t \le n$, we write the type as $(\varepsilon; \beta; (m, n); \gamma)$.

Diagram 6.3.0 shows the j-set of a D-birational poset of type $(\varepsilon; \beta; (m, n); \gamma_1,$ $\ldots, \gamma_m; \delta_1, \ldots, \delta_n)$, where $n > 1$. (The complete poset U would have clumps of size β beneath each height-two maximal element by axiom 6.)

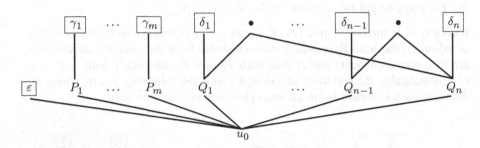

Diagram 6.3.0: j-set of a D-birational poset of type $(\varepsilon; \beta; (m, n); \gamma_1, \ldots, \gamma_m; \delta_1, \ldots, \delta_n)$, where $n > 1$

Remark 6.4. Ambiguity: If U is an N-birational poset of type $(\varepsilon; \beta; \ell; \alpha; 0, \cdots, 0)$ where $\ell = 1$, or where $\ell > 1$ and each $t_i = 0$, then U is D-birational, but there is some ambiguity about the type. We could take either $(m, n) = (\ell, 0)$ or $(m, n) = (\ell - 1, 1)$. The picture for the spectra is the same in either case. A D-birational partially ordered set of type $(\varepsilon; \beta; (1, 0); \alpha)$ is order-isomorphic to one of type $(\varepsilon; \beta; (0, 1); \alpha)$ and a D-birational partially ordered set of type $(\varepsilon; \beta; (m, 0); \alpha)$, for $m > 1$, is order-isomorphic to one of type $(\varepsilon; \beta; (m - 1, 1); \alpha)$, but *not* to one of type $(\varepsilon; \beta; (m - 2, 2); \alpha)$. We keep this ambiguity because the different types arise in different circumstances when the notation is applied to $\text{Spec}(R[\![x]\!][g/f])$.

When R is a countable Dedekind domain, the cardinalities in Theorem 6.1 can be given more explicitly, yielding a true characterization. For R countable, all the γ_i and δ_j are equal, by Lemma 2.6, and so we use the abbreviated form of the code in Abbreviations 6.3.a. Recall that $(aR :_R b) = \{c \in R \mid bc \in aR\}$, if $a, b \in R$.

Theorem 6.5 ([5, Theorem 4.3]). *Let R be a countable Dedekind domain with quotient field K such that $\max(R)$ is infinite, let x be an indeterminate, and let B be a simple birational extension of $R[\![x]\!]$, as described below for $f, g \in R[\![x]\!]$ an $R[\![x]\!]$-sequence such that f, g have constant terms a, b respectively. Set $v = |V_R(a, b)|$ and $w = |\{q \in V_R(a, b) \mid (aR : b) \not\subseteq q\}|$. Then:*

(1) If $a = 0$ and $B := R[\![x]\!][1/f]$, then $\text{Spec}(B)$ is a fan.

(2) If $a \neq 0$ and $B := R[x][1/f]$, then $\mathrm{Spec}(B)$ is order-isomorphic to $\mathrm{Spec}(\mathbb{Q}[y][x])$ or $\mathrm{Spec}(\mathbb{Q}[x][y])$.

(3) If $B = R[x][g/f]$, $a = 0$, $b \neq 0$, and $bR \neq R$, then $\mathrm{Spec}(B)$ is D-birational of type $(|\mathbb{R}|; |\mathbb{R}|; (v, 0); \aleph_0)$.

(4) If $B = R[x][g/f]$, $a \neq 0$, and $(a, b)R \neq R$, then $\mathrm{Spec}(B)$ is D-birational of type

$$(|\mathbb{R}|; |\mathbb{R}|; (v - w, w + 1); \aleph_0).$$

In order to show that Theorem 6.5 is a characterization, we show every D-birational poset occurs, for some Dedekind domain D. In fact this is true with $D = \mathbb{Z}$, as we see in Theorem 6.6.

Theorem 6.6 ([5, Theorem 4.8]). *Let R be a PID with α maximal ideals, where α is infinite, let x and y be indeterminates, set $\beta = |R[x]|$, and suppose that $\alpha = |R| \cdot \aleph_0 = |R/\mathbf{m}| \cdot \aleph_0$ is constant, for each $\mathbf{m} \in \max(R)$. Then, for each $m, n \in \mathbb{N}_0$, there exists a simple birational extension of $R[x]$ that is D-birational of type $(\beta; \beta; (m, n); \alpha)$. In particular, if R is a PID with $|R| = |\max(R)| = \aleph_0$ and $m, n \in \mathbb{N}_0$, then, for every D-birational partially ordered set U of type $(|\mathbb{R}|; |\mathbb{R}|; (m, n); \aleph_0)$ from Definition 6.3, there is a simple birational extension $B := R[x][g/f]$ of $R[x]$ so that the prime spectrum of B is order-isomorphic to U.*

We give an example adjusted from [5] to illustrate Theorems 6.5 and 6.6.

Example 6.7. Let B be the simple birational extension $B := \mathbb{Z}[x][g/f]$ of $\mathbb{Z}[x]$, where $f = x + 2079$ and $g = x + 4851$. Then, in the notation of Theorem 6.5, $a = 2079 = 3^3 \cdot 7 \cdot 11, b = 4851 = 3^2 \cdot 7^2 \cdot 11, (a\mathbb{Z} : b) = 3\mathbb{Z}$, and so $v = |\{3, 7, 11\}| = 3$, $w = 2$, and $(f, g)\mathbb{Z}[x] \neq \mathbb{Z}[x]$, since $(f, g) \subseteq (x, 3)$. Since $7f - 3g = 4x \in (f, g)$, we have $(f, x) \subseteq P$ or $(f, 2) \subseteq P$, for every prime ideal P minimal over (f, g). Therefore the ideal (f, g) has height two, and so, by [15, Theorem 17.4], f, g is a $\mathbb{Z}[x]$-sequence. Also, in $\mathbb{Z}[y]$, the ideal

$$(fy - g)\mathbb{Z}[y] = (2079y - 4851)\mathbb{Z}[y] = (3^2 \cdot 7 \cdot 11(3y - 7))\mathbb{Z}[y].$$

The height-one prime ideals in $\mathbb{Z}[y]$ containing this ideal are (3), (7), (11), and $(3y - 7)$. By Theorem 6.5, $\mathrm{Spec}(B)$ is D-birational of type $(|\mathbb{R}|; |\mathbb{R}|; (1, 3); \aleph_0)$, since the cardinality of $\mathbb{Z}[x]$ is $|\mathbb{R}|$ and the cardinality of $\max(\mathbb{Z})$ is $|\mathbb{Z}| \cdot \aleph_0 = \aleph_0$. Thus Diagram 6.7.1 shows the partially ordered set $\mathrm{Spec}(B)$, except that we cannot show the clumps of size $|\mathbb{R}|$ beneath every height-two maximal ideal. Here π is the canonical map from $\mathbb{Z}[x][y] \to B := \mathbb{Z}[x][g/f]$, and $\pi(x, 3)$ denotes the image in B under π of the ideal $(x, 3)\mathbb{Z}[x]$. There is one j-prime ideal, namely $\pi(x, 3)$, that is unrelated to the others; the other three are connected by height-two maximal ideals that contain the last j-prime ideal, $\pi(x, 3y - 7)$.

Diagram 6.7.2 is a close-up picture showing relations for elements of the set labeled $C_{\aleph_0}$, to show, for every $M \in C_{\aleph_0}$, that $|L_e(M)| = |\mathbb{R}|$.

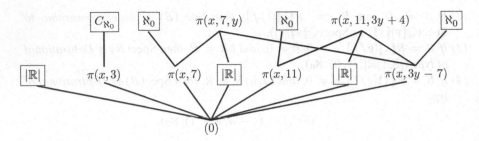

Diagram 6.7.1: Spec(B) for Example 6.7

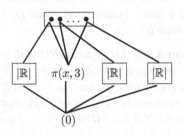

Diagram 6.7.2: Relations in $C_{\aleph_0}$ from Diagram 6.7.1

Remark 6.8. If R is a countable one-dimensional Noetherian domain R such that Spec($R[y]/(ay-b)$) is know then one can also find Spec(B); see [5]. When R is not Dedekind, however, the relations among the minimal elements of Spec($R[y]/(ay - b)$) may be more complex than they are for Dedekind domains, and we do not know what posets are realizable as Spec($R[y]/(ay - b)$). It is not clear that every form of the axioms for that situation can be realized.

The following example from [5] gives a simple birational extension B of $\mathbb{Z}[5i][x]$ that has two distinct maximal ideals containing two distinct nonmaximal height-one j-primes. Thus Spec(B) is not D-birational.

Example 6.9. For $R = \mathbb{Z}[5i]$, a non-Dedekind ring, let $B = \mathbb{Z}[5i][x][g/f]$, $f = x + 5$, $g = 5i$. Then the two nonmaximal height-one j-primes of B correspond in $\mathbb{Z}[5i][y]$ to $(5, 5i)\mathbb{Z}[5i][y]$ and to

$$\mathbf{p} := (y - i)\mathbb{Z}[i][y] \cap \mathbb{Z}[5i][y] = (y^2 - 1, 5y - 5i, 5iy + 5)\mathbb{Z}[5i][y].$$

Therefore $(5, 5i) + \mathbf{p} = (y^2 - 1, 5, 5i) = (y^2 - 4, 5, 5i) \subseteq (y - 2, 5, 5i) \cap (y + 2, 5, 5i)$. If we let $M_1 = (y - 2, 5, 5i)\mathbb{Z}[5i][y]$, $M_2 = (y + 2, 5, 5i)\mathbb{Z}[5i][y]$, and $\overline{}$ denotes the image in B of the map Spec($\mathbb{Z}[5i][y]/(5y - 5i)) \to V_B(x)$ from Remarks 3.9.1, we have j-Spec(B) in Diagram 6.9.1:

To make Diagram 6.9.1 show all of Spec(B), we would add clumps of size $|\mathbb{R}|$ beneath every height-two prime ideal but beneath no other height-two prime ideal. This partially ordered set is N-birational of type $(|\mathbb{R}|; |\mathbb{R}|; 2; \aleph_0; 2)$, since the number of height-one maximal ideals is $|\mathbb{R}|$ and $|L_e(P)| = |\mathbb{R}|$ for every height-two

Diagram 6.9.1: j-Spec(B),
for $B := \mathbb{Z}[5i][x][g/f]$,
$f = x + 5, g = 5i$

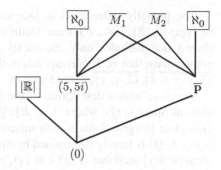

$P \in$ Spec(B); also ℓ, the number of nonmaximal j-prime ideals, is 2. Since the number t_1 of maximal ideals containing both of them is 2, we have that j-Spec(B) is not D-birational.

The following question raised in [5] is still unknown:

Question 6.10. Does every N-birational poset of type $(|\mathbb{R}|; |\mathbb{R}|; \ell; \aleph_0; t_1, \cdots, t_{\ell-1})$, for all values of ℓ, and $t_1 \leq \cdots \leq t_\ell$, occur as Spec($R[x][g/f]$), where g, f are as described in Theorem 6.1?

7 Examples of Two-Dimensional Polynomial-Power Series Prime Spectra

To illustrate Theorem 5.4, we give an example with $R = \mathbb{Z}$; this partially ordered set is the prime spectrum of $\mathbb{Z}[y][x]/Q$, for an appropriately chosen height-one prime ideal Q of $\mathbb{Z}[y][x]$:

Example 7.1. For $\alpha = (2y - 1) \cdot 3 \cdot (y + 1) \cdot y \cdot (y(y + 1) + 6) \cdot 2 \cdot (3y + 1)$, we describe Spec($\mathbb{Z}[y][x]/(x - \alpha)$), by displaying j-Spec($\mathbb{Z}[y][x]/(x - \alpha)$) in this diagram:

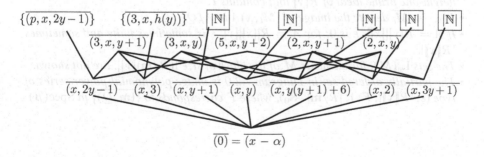

Diagram 7.1.1: j-Spec($\mathbb{Z}[y][x]/(x - \alpha)$)

The partially ordered set in Diagram 7.1.1 is image polynomial-power series of type $(0; |\mathbb{R}|; 7; \aleph_0; F)$ from Definition 5.5, where F is the part of the poset shown that includes only the ideals at the second and third levels and their connections; that is, F corresponds to the twelve ideals $\overline{(x, 2y - 1)}, \overline{(x, 3)}, \ldots$ and $\overline{(3, x, y + 1)}, \overline{(3, x, y)}, \ldots$ and with the relations (lines) connecting them.

We close with a description of the two-dimensional partially ordered sets that arise as $\mathrm{Spec}(A/Q)$, where $A = R[x][y]$ or $R[y][x]$ and R is a one-dimensional Henselian integral domain with unique maximal ideal $\mathbf{m}$. As with Example 7.1, $\mathrm{Spec}(A/Q)$ is largely determined by $\mathrm{Spec}(R[y]/I)$, where I is a height-one prime ideal of $R[y]$ such that $(I, x)A = (Q, x)A$.

Theorem 7.2 ([4, Theorem 7.2]). *Let $(R, \mathbf{m})$ be a Henselian integral domain. Let x and y be indeterminates; let $A = R[x][y]$ or $R[y][x]$. Let Q be a height-one prime ideal of A and let $B = A/Q$. Then $\mathrm{Spec}(B) \setminus \{the\ set\ of\ height-one\ maximal\ ideals\}$ is determined by $\mathrm{Spec}(R[y]/I)$, where I is a height-one prime ideal of $R[y]$ such that $(I, x)A = (Q, x)A$. If I is contained in $\mathbf{m}R[y]$, then $\mathrm{Spec}(B)$ is given in Diagram 7.2.0.*

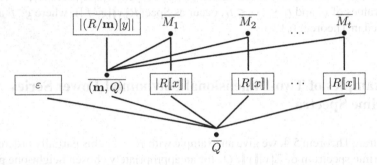

Diagram 7.2.0: $\mathrm{Spec}(B)$ if $I \subseteq \mathbf{m}R[y]$

Notes

- $M_1, M_2, \cdots, M_t$ *are the height-two ideals of $R[y]$ that contain $\mathbf{m}$ and another height-one prime ideal of $R[y]$ that contains I.*
- $\overline{M_1}, \cdots, \overline{M_t}$ *denote the image of $(M_i, x)A$ in A/Q.*
- *If $A = R[y][[x]]$, ε is 0; for $A = R[[x]][y]$, ε is sometimes finite and sometimes $|R[[x]]|$.*
- *The sets $\mathrm{L}_e(M)$, for elements M of the block of size $|(R/\mathbf{m})[y]|$, are not shown.*
- *The partially ordered set in Diagram 7.2.0 is image polynomial-power series of type $(\varepsilon; |R[[x]]|; F; 1; |R/\mathbf{m}| \cdot \aleph_0)$, where F corresponds to $\{\overline{(\mathbf{m}, Q)}\}$ in $\mathrm{Spec}(B)$.*

References

1. J.T. Arnold, Prime ideals in power series rings. In *Conference on Commutative Algebra (University of Kansas, Lawrence, Kansas 1972)*. Lecture Notes in Mathematics, vol. 311 (Springer, Berlin, 1973), pp. 17–25
2. H. Bass, *Algebraic K-Theory*. Mathematics Lecture Note Series (W.A. Benjamin Inc, New York/Amsterdam, 1968), pp. 762
3. J.W. Brewer, *Power Series Over Commutative Rings*. Lecture Notes in Pure and Applied Mathematics, vol. 64 (Marcel Dekker Inc, New York, 1981)
4. E. Celikbas, C. Eubanks-Turner, S. Wiegand, *Prime Ideals in Quotients of Mixed Polynomial-Power Series Rings*. See http://www.math.unl.edu/~swiegand1 (preprint)
5. C. Eubanks-Turner, M. Luckas, S. Saydam, Prime ideal in birational extensions of two-dimensional power series rings. Comm. Algebra **41**(2), 703–735 (2013)
6. W. Heinzer, S. Wiegand, Prime ideals in two-dimensional polynomial rings. Proc. Am. Math. Soc. **107**(3), 577–586 (1989)
7. W. Heinzer, C. Rotthaus, S. Wiegand, Mixed polynomial-power series rings and relations among their spectra. In *Multiplicative Ideal Theory in Commutative Algebra* (Springer, New York, 2006), pp. 227–242
8. R.C. Heitmann, Prime ideal posets in Noetherian rings. Rocky Mountain J. Math. **7**(4), 667–673 (1977)
9. M. Hochster, Prime ideal structure in commutative rings. Trans. Am. Math. Soc. **142**, 43–60 (1969)
10. I. Kaplansky, *Commutative Rings* (Allyn and Bacon Inc, Boston, 1970)
11. K. Kearnes, G. Oman, Cardinalities of residue fields of Noetherian integral domains. Comm. Algebra, **38**, 3580–3588 (2010)
12. W. Lewis, The spectrum of a ring as a partially ordered set. J. Algebra **25**, 419–434 (1973)
13. W. Lewis, J. Ohm, The ordering of Spec R. Can. J. Math. **28**, 820–835 (1976)
14. A. Li, S. Wiegand, Prime ideals in two-dimensional domains over the integers. J. Pure Appl. Algebra **130**, 313–324 (1998)
15. H. Matsumura, *Commutative Ring Theory*, 2nd edn. Cambridge Studies in Advanced Mathematics, vol. 8 (Cambridge University Press, Cambridge, 1989). Translated from the Japanese by M. Reid
16. S. McAdam, Saturated chains in Noetherian rings. Indiana Univ. Math. J. **23**, 719–728 (1973/1974)
17. S. McAdam, Intersections of height 2 primes. J. Algebra **49**(2), 315–321 (1977)
18. M.P. Murthy, A note on factorial rings. Archiv der Mathematik **15**(1), 418–420 (1964)
19. M. Nagata, On the chain problem of prime ideals. Nagoya Math. J. **10**, 51–64 (1956)
20. M. Nagata, *Local Rings*. Interscience Tracts in Pure and Applied Mathematics, vol. 13 (Interscience Publishers a division of John Wiley and Sons, New York-London, 1962)
21. L.J. Ratliff, *Chain Conjectures in Ring Theory: An Exposition of Conjectures on Catenary Chains*. Lecture Notes in Mathematics, vol. 647 (Springer, Berlin, 1978)
22. A.S. Saydam, S. Wiegand, Noetherian domains with the same prime ideal structure as $\mathbb{Z}_{(2)}[x]$. Arab. J. Sci. Eng. Sect. C Theme Issues (Comm. Algebra) **26**(1), 187–198, (2001)
23. C. Shah, Affine and projective lines over one-dimensional semilocal domains. Proc. Am. Math. Soc. **124**(3), 697–705 (1996)
24. S. Wiegand, Locally maximal Bezout domains. Am. Math. Soc. **47**, 10–14 (1975)
25. S. Wiegand, Intersections of prime ideals in Noetherian rings. Comm. Algebra **11**, 1853–1876 (1983)
26. R. Wiegand, The prime spectrum of a two-dimensional affine domain. J. Pure Appl. Algebra **40**(2), 209–214 (1986)
27. S. Wiegand, R. Wiegand, The maximal ideal space of a Noetherian ring. J. Pure Appl. Algebra **08**, 129–141 (1976)

28. R. Wiegand, S. Wiegand, Prime ideals and decompositions of modules. In *Non-Noetherian Commutative Ring Theory*, ed. by S. Chapman, S. Glaz. Mathematics and its Applications, vol. 520 (Kluwer Academic Publishers, Dordrecht, 2000), pp. 403–428
29. R. Wiegand, S. Wiegand, Prime ideals in Noetherian rings: a survey. In *Ring and Module Theory*, ed. by T. Albu, G.F. Birkenmeier, A. Erdo **u** gan, A. Tercan (Birkhäuser, Boston, 2010), pp. 175–193

Integer-Valued Polynomials: Looking for Regular Bases (A Survey)

Jean-Luc Chabert

Abstract This paper reviews recent results about the additive structure of algebras of integer-valued polynomials and, particularly, the question of the existence and the construction of regular bases. Doing this, we will be led to consider questions of combinatorial, arithmetical, algebraic, ultrametric, or dynamical nature.

Keywords Integer-valued polynomials • Generalized factorials • v-Orderings • Kempner's formula • Regular basis • Pólya fields • Divided differences • Mahler's theorem

2010 MSC. Primary 13F20; Secondary 11S05, 11R21, 11B65

1 Introduction

The $\mathbb{Z}$-algebra

$$\text{Int}(\mathbb{Z}) = \{f(X) \in \mathbb{Q}[X] \mid f(\mathbb{Z}) \subseteq \mathbb{Z}\}$$

is a paradigmatic example because several of its properties still hold for some general algebras of integer-valued polynomials. As a ring, $\text{Int}(\mathbb{Z})$ has a lot of interesting properties, but here we focus our survey on its additive structure since it is a cornerstone for the study of quite all other properties. The $\mathbb{Z}$-module $\text{Int}(\mathbb{Z})$ is

J.-L. Chabert (✉)
LAMFA CNRS-UMR 7352, Université de Picardie, 33 rue Saint Leu, 80039 Amiens, France
e-mail: jean-luc.chabert@u-picardie.fr

M. Fontana et al. (eds.), *Commutative Algebra: Recent Advances in Commutative Rings, Integer-Valued Polynomials, and Polynomial Functions*, DOI 10.1007/978-1-4939-0925-4_5, © Springer Science+Business Media New York 2014

free: it admits a basis formed by the binomial polynomials, and this basis turns out to be also an orthonormal basis of the ultrametric Banach space $\mathscr{C}(\mathbb{Z}_p, \mathbb{Q}_p)$ whatever the prime p.

Now replace $\mathbb{Z}$ by the ring of integers of a number field as Pólya [43] and Ostrowski [41] did, or more generally, on the one hand, replace $\mathbb{Z}$ by any Dedekind domain D with quotient field K and, on the other hand, consider a subset S of D and the D-algebra of polynomials whose values on S are in D:

$$\text{Int}(S, D) = \{f \in K[X] \mid f(S) \subseteq D\}.$$

During the last decades of the last century, there were many works about this D-algebra from the point of view of commutative algebra. As said above, our aim here is to characterize the cases where the D-module $\text{Int}(S, D)$ admits bases and, more precisely, regular bases, that is, with one and only one polynomial of each degree and, when there are such bases, we want to describe them. This was already the subject of [17, Chap. II] but, during the last 15 years, a lot of new results were obtained, especially thanks to several notions of v-ordering introduced by Bhargava from 1998 to 2009 [9–12].

In Sect. 2, we recall a few properties of $\text{Int}(\mathbb{Z})$ that will be generalized, in particular those concerning factorials. Then, in Sect. 3, we consider the factorial ideals and, in Sect. 4, the notion of v-ordering and its links with integer-valued polynomials. In Sect. 5 we study the existence and the construction of regular bases while Sect. 6 is devoted to effective computations. Then, in Sect. 7, we consider some particular sub-algebras of $\text{Int}(S, D)$ and, in Sect. 8, we apply our knowledge of regular bases to obtain orthonormal bases of ultrametric Banach spaces. We end in Sect. 9 with the case of several indeterminates.

2   The Paradigmatic Example: $\text{Int}(\mathbb{Z})$

In the ring of *integer-valued polynomial* $\text{Int}(\mathbb{Z}) = \{f(X) \in \mathbb{Q}[X] \mid f(\mathbb{Z}) \subseteq \mathbb{Z}\}$, there are polynomials without any integral coefficients, for instance,

$$\binom{X}{n} = \frac{X(X-1)\cdots(X-n+1)}{n!} \quad (n \geq 2) \quad \text{or} \quad F_p(X) = \frac{X^p - X}{p} \quad (p \in \mathbb{P}).$$

2.1   Some Algebraic Structures

As a subset of $\mathbb{Q}[X]$, $\text{Int}(\mathbb{Z})$ is stable by addition, multiplication, and composition.

Proposition 1. *The binomial polynomials form a basis of the $\mathbb{Z}$-module $\text{Int}(\mathbb{Z})$: every $g(X) \in \text{Int}(\mathbb{Z})$ may be uniquely written as*

$$g(X) = \sum_{k=0}^{\deg(g)} c_k \binom{X}{k} \quad \text{with } c_k \in \mathbb{Z} \qquad \text{(Pólya [42], 1915).} \qquad (1)$$

Proposition 2 ([17, Sect. 2.2]). *The set $\{1, X\} \cup \{(X^p - X)/p \mid p \in \mathbb{P}\}$ is a minimal system of polynomials with which every element of $\mathrm{Int}(\mathbb{Z})$ may be constructed by means of sums, products, and composition (see Example 3 below).*

Proposition 3 ([17]). $\mathrm{Int}(\mathbb{Z})$ *is a two-dimensional non-Noetherian Prüfer domain.*

For instance, the ideal $\mathfrak{I} = \{f(X) \in \mathrm{Int}(\mathbb{Z}) \mid f(0) \text{ is even}\}$ is not finitely generated.

2.2 A Polynomial Approximation in Ultrametric Analysis

Let p be a fixed prime number. Recall that $|x| = e^{-v_p(x)}$ is an absolute value on $\mathbb{Q}_p$ where v_p denotes the p-adic valuation. For every compact subset S, the norm of a continuous function $\varphi : S \to \mathbb{Q}_p$ is then $\|\varphi\|_S = \sup_{x \in S} |\varphi(x)|$.

Proposition 4 (Mahler [39], 1958). *Every function $\varphi \in \mathscr{C}(\mathbb{Z}_p, \mathbb{Q}_p)$ may be written*

$$\varphi(x) = \sum_{n=0}^{\infty} c_n \binom{x}{n} \quad \text{with } c_n \in \mathbb{Q}_p \text{ and } \lim_{n \to +\infty} v_p(c_n) = +\infty. \qquad (2)$$

Moreover, $\|\varphi\|_{\mathbb{Z}_p} = \|\{c_n\}_{n \in \mathbb{N}}\|$, *that is,* $\inf_{x \in \mathbb{Z}_p} v_p(\varphi(x)) = \inf_{n \in \mathbb{N}} v_p(c_n)$.

One says that the binomial functions $\binom{x}{n}$ form an *orthonormal basis* of the Banach space $\mathscr{C}(\mathbb{Z}_p, \mathbb{Q}_p)$. The coefficients c_n are unique and may be computed recursively:

$$c_n = \varphi(n) - \sum_{k=0}^{n-1} c_k \binom{n}{k}.$$

2.3 Some Properties of the Factorials

As denominators of the binomial polynomials, the factorials and their generalizations will play an important rôle in the description of bases. We recall some properties which will be preserved in a more general context.

Property A. For all $k, l \in \mathbb{N}$, $\binom{k+l}{k} = \frac{(k+l)!}{k! \times l!} \in \mathbb{N}$.

Equivalently, the product of l consecutive integers is divisible by $l!$:

$$\frac{(k+1)(k+2)\cdots(k+l)}{l!} \in \mathbb{N}. \qquad (3)$$

When considering integers which are not consecutive, we still have the following:

Property B. For every sequence $x_0, x_1, \ldots, x_n$ of $n + 1$ integers, the product

$$\prod_{0 \le i < j \le n} (x_j - x_i) \quad \text{is divisible by} \quad 1! \times 2! \times \cdots \times n!. \tag{4}$$

We now consider links between factorials and polynomials.

Property C. For every monic polynomial $f \in \mathbb{Z}[X]$ of degree n,

$$d(f) = \gcd\{f(k) \mid k \in \mathbb{Z}\} \quad \text{divides} \quad n! \qquad \text{(Pólya [42], 1915).} \tag{5}$$

Property D. For every integer-valued polynomial g of degree n

$$n! \times g(X) \in \mathbb{Z}[X] . \tag{6}$$

Property E. The subset formed by the leading coefficients of the integer-valued polynomials of degree $\le n$ is $\frac{1}{n!}\mathbb{Z}$.

Property F. The number of polynomial functions from $\mathbb{Z}/n\mathbb{Z}$ to $\mathbb{Z}/n\mathbb{Z}$ is

$$\prod_{k=0}^{n-1} \frac{n}{\gcd(n, k!)} \qquad \text{(Kempner [36], 1921),} \tag{7}$$

where functions are induced by polynomials of $\mathbb{Z}[X]$.

Finally, recall Legendre's formula:

$$n! = \prod_{p \in \mathbb{P}} p^{w_p(n)}, \quad \text{where } w_p(n) = \sum_{k \ge 1} \left[\frac{n}{p^k} \right] \qquad \text{(Legendre, 1808).} \tag{8}$$

3 General Integer-Valued Polynomials and Generalized Factorials

Notation. *In the sequel, D always denotes a Dedekind domain with quotient field K and S a subset of D.*

The D-algebra of *integer-valued polynomials on D* is

$$\text{Int}(D) = \{f(X) \in K[X] \mid f(D) \subseteq D\} \tag{9}$$

and the D-algebra of *integer-valued polynomials on S* (with respect to D) is

$$\text{Int}(S, D) = \{f(X) \in K[X] \mid f(S) \subseteq D\}. \tag{10}$$

As $S \subseteq D$, we have $D[X] \subseteq \text{Int}(D) \subseteq \text{Int}(S, D) \subseteq K[X]$. Let us consider a more general situation by introducing a D-algebra $\mathbb{B}$ such that $D[X] \subseteq \mathbb{B} \subseteq K[X]$.

Definition 1 (Pólya [43]). A basis of the D-module $\mathbb{B}$ is said to be a *regular basis* if it is formed by one and only one polynomial of each degree.

3.1　Characteristic Ideals and Regular Bases

Definition 2. The *characteristic ideal* of index n of the D-algebra $\mathbb{B}$ is the set $\mathfrak{I}_n(\mathbb{B})$ formed by 0 and the leading coefficients of the polynomials in $\mathbb{B}$ of degree n.

If $\mathbb{B} = \text{Int}(S, D)$, we write $\mathfrak{I}_n(S, B)$ instead of $\mathfrak{I}_n(\text{Int}(S, D))$.
Clearly, $\{\mathfrak{I}_n(\mathbb{B})\}_{n \in \mathbb{N}}$ is an increasing sequence of D-modules such that

$$\forall k, l \in \mathbb{N} \quad D \subseteq \mathfrak{I}_k(\mathbb{B}) \subseteq K \text{ and } \mathfrak{I}_k(\mathbb{B}) \cdot \mathfrak{I}_l(\mathbb{B}) \subseteq \mathfrak{I}_{k+l}(\mathbb{B}). \tag{11}$$

Recall that a *fractional ideal* of D is a sub-D-module $\mathfrak{J}$ of K for which there exists a nonzero element $d \in D$ such that $d\mathfrak{J} = \{dj \mid j \in \mathfrak{J}\} \subseteq D$. A very simple argument using Vandermonde's determinant leads to:

Lemma 1 ([17, Proposition I.3.1]). *Let f be a polynomial of $K[X]$ with degree n. Assume that $x_0, x_1, \ldots, x_n$ are distinct elements of K such that $f(x_i) \in D$ for $0 \leq i \leq n$, then df belongs to $D[X]$ where $d = \prod_{0 \leq i < j \leq n}(x_j - x_i)$.*

As a consequence,

– if $n < \text{Card}(S)$, then $\mathfrak{I}_n(S, D)$ is a fractional ideal of D,
– if $n \geq \text{Card}(S)$, then $\mathfrak{I}_n(S, D) = K$ because $\left(\prod_{s \in S}(X - s)\right) K[X] \subseteq \text{Int}(S, D)$.

In particular, if $\text{Card}(S)$ is infinite, all the $\mathfrak{I}_n(S, D)$'s are fractional ideals and, more generally, so are all the $\mathfrak{I}_n(\mathbb{B})$'s if $\mathbb{B} \subseteq \text{Int}(S, D)$.
Clearly, by definition of the characteristic ideals, we have:

Proposition 5 ([17, Proposition II.1.4]). *A sequence of polynomials $\{f_n\}_{n \geq 0}$ where $\deg(f_n) = n$ is a regular basis of $\mathbb{B}$ if and only if, for every $n \geq 0$, the leading coefficient of f_n generates the ideal $\mathfrak{I}_n(\mathbb{B})$. In particular, the D-algebra $\text{Int}(\mathbb{B})$ admits a regular basis as a D-module if and only if all the $\mathfrak{I}_n(\mathbb{B})$'s are principal.*

Thus, if S is finite, the D-module $\text{Int}(S, D)$ cannot admit any regular basis since $\mathfrak{I}_n(S, D) = K$ for $n \geq \text{Card}(S)$.

3.2   The Factorial Ideals of a Subset S

When $S = D = \mathbb{Z}$, $\mathfrak{I}_n(\mathbb{Z}, \mathbb{Z}) = \frac{1}{n!}\mathbb{Z}$. Thus, it is natural to define new factorials as the inverses of the characteristic ideals. In general, D is not a principal ideal domain and we cannot define these new factorials as numbers, but as ideals.

Recall that the inverse of a nonzero fractional ideal $\mathfrak{I}$ of D is the fractional ideal $\mathfrak{I}^{-1} = \{x \in K \mid x\mathfrak{I} \subseteq D\}$. If $\mathfrak{I}$ contains 1, then $\mathfrak{I}^{-1}$ is an integral ideal. Since D is assumed to be a Dedekind domain, the nonzero fractional ideals of D form a multiplicative group (and $\mathfrak{I} \cdot \mathfrak{I}^{-1} = D$) and an ideal $\mathfrak{a}$ divides an ideal $\mathfrak{b}$ if and only if $\mathfrak{b} \subseteq \mathfrak{a}$. By convention, we write $K^{-1} = (0)$ and $(0)^{-1} = K$.

Definition 3. The *factorial ideal* $(n!)_S^D$ of index n of the subset S with respect to the domain D is the inverse of the fractional ideal $\mathfrak{I}_n(S, D)$:

$$n!_S^D = \mathfrak{I}_n(S, D)^{-1}. \tag{12}$$

The sequence $\{n!_S^D\}_{n \in \mathbb{N}}$ is a decreasing sequence of integral ideals of D and

$$\forall n \quad n!_S^D \mid (n+1)!_S^D, \quad 0!_S^D = D, \quad [n!_S^D = (0) \Leftrightarrow n \geq \mathrm{Card}(S)], \tag{13}$$

3.3   First Generalized Properties of the Factorial Ideals

Proposition 6 (Generalized Property A). *Let k, $l \in \mathbb{N}$ be any integers then*

$$k!_S^D \times l!_S^D \text{ divides } (k+l)!_S^D. \tag{14}$$

This is a straightforward consequence of (11).

Proposition 7 (Generalized Property D [17, Proposition II.1.7]). *For every polynomial $g(X) \in \mathrm{Int}(S, D)$ of degree n, we have*

$$n!_S^D \times g(X) \subseteq D[X]. \tag{15}$$

Recall that for every polynomial $f \in K[X]$:

- the *content* of f is the ideal $c(f)$ of D generated by the coefficients of f,
- the *fixed divisor* of f over S is the ideal $d(S, f)$ of D generated by the values of f on S.

Proposition 8 (Generalized Property C [10, Theorem 2]). *With the previous notation, for every $f \in K[X]$ of degree n,*

$$d(S, f) \text{ divides } c(f) \times n!_S^D. \tag{16}$$

Proof. By definition, $d(S, f)^{-1} \times f \subseteq \text{Int}(S, D)$. Then, $n!_S^D \times d(S, f)^{-1} \times f \subseteq D[X]$ by Proposition 7. Finally, $n!_S^D \times d(S, f)^{-1} \times c(f) \subseteq D$.   □

Proposition 9. *1. For every* $b \in D \setminus \{0\}$ *and every* $c \in D$, $n!_{bS+c}^D = b^n n!_S^D$.

2. For every $T \subseteq S$, $n!_S^D$ *divides* $n!_T^D$. *In particular,* $n!$ *divides* $n!_S^{\mathbb{Z}}$.

Proof. 1. The equality follows from the isomorphism of D-algebras:

$$f(X) \in \text{Int}(S, D) \mapsto f\left(\frac{X - c}{b}\right) \in \text{Int}(bS + c, D).$$

2. The divisibility relation follows from the containment $\text{Int}(S, D) \subseteq \text{Int}(T, D)$.   □

We generalized Properties A, C, and D and, by definition, Property E is satisfied by the ideals $n!_S^D$. We will study Properties B and F in the next section by means of localization.

3.4 Localization

Clearly,

$$\text{Int}(S, D) = \cap_{\mathfrak{m} \in \text{Max}(D)} \text{Int}(S, D_{\mathfrak{m}}) \quad \text{and} \quad \forall \mathfrak{p} \in \text{Spec}(D) \quad \text{Int}(S, D)_{\mathfrak{p}} \subseteq \text{Int}(S, D_{\mathfrak{p}}).$$

Since D is Noetherian, we have the reverse containment [17, Proposition I.2.7]:

$$\forall \mathfrak{p} \in \text{Spec}(D) \qquad \text{Int}(S, D)_{\mathfrak{p}} = \text{Int}(S, D_{\mathfrak{p}}). \tag{17}$$

We deduce localization formulas for the characteristic ideals and the factorial ideals:

$$\mathfrak{I}_n(S, D) = \cap_{\mathfrak{m} \in \text{Max}(D)} \mathfrak{I}_n(S, D_{\mathfrak{m}}), \quad n!_S^D = \cap_{\mathfrak{m} \in \text{Max}(D)} n!_S^{D_{\mathfrak{m}}}, \tag{18}$$

$$\mathfrak{I}_n(S, D)_{\mathfrak{m}} = \mathfrak{I}_n(S, D_{\mathfrak{m}}) \quad \text{and} \quad (n!_S^D)_{\mathfrak{m}} = n!_S^{D_{\mathfrak{m}}}. \tag{19}$$

4 Local Studies (Bhargava's v-Orderings) and Globalizations

Is there an easy way to compute these factorials? Yes, by means of the notion of v-ordering introduced by Bhargava [9]. This is a local notion; thus, we consider localizations, which are discrete valuation domains.

Notation for Sects. 4.1 and 4.2. *We denote by v a discrete valuation on K, by V the corresponding valuation domain and by S any subset of V.*

4.1 Definitions and Examples

Before v-orderings, we used sequences introduced by Helsmoortel to study the case where $S = V$. The sequences called VWDWO sequences are the sequences which satisfy the equivalent statements of the following proposition.

Proposition 10 ([17, Sect. II.2]). *Assume that the residue field of V is finite with cardinality q and denote by $\mathfrak{m}$ the maximal ideal. For a sequence $\{a_n\}_{n \geq 0}$ of elements of V, the three following assertions are equivalent:*

1. Denoting by $v_q(x)$ the largest exponent k such that q^k divides x,

$$\forall m, n \in \mathbb{N} \quad v(a_n - a_m) = v_q(m - n).$$

2. For all $r, s \in \mathbb{N}$, $\{a_{r+1}, a_{r+2}, \ldots, a_{r+q^s}\}$ is a complete system of representatives of V (mod $\mathfrak{m}^s$).

3. For all $n > m \geq 0$, $v\left(\prod_{k=m}^{n-1}(a_n - a_k)\right) = \sum_{k \geq 1}\left[\frac{n-m}{q^k}\right]$.

Example 1. The following sequence $\{a_n\}_{n \geq 0}$ is a VWDWO sequence of V. Choose a generator π of $\mathfrak{m}$ and a system of representatives $\{a_0 = 0, a_1, \ldots, a_{q-1}\}$ of V modulo $\mathfrak{m}$.

$$\text{For } n = n_0 + n_1 q + n_2 q^2 + \cdots + n_r q^r \text{ with } 0 \leq a_i < q, \tag{20}$$

$$\text{let } a_n = a_{n_0} + a_{n_1}\pi + a_{n_2}\pi^2 + \cdots + a_{n_r}\pi^r. \tag{21}$$

We also had the VWD sequences introduced by Amice [8] for her regular compact subsets of local fields. But Bhargava's v-orderings are more general.

Definition 4. A v-*ordering* of a subset S of V is a sequence $\{a_n\}_{n \geq 0}$ of elements of S such that, for every $n \geq 1$,

$$v\left(\prod_{k=0}^{n-1}(a_n - a_k)\right) = \min_{x \in S} v\left(\prod_{k=0}^{n-1}(x - a_k)\right). \tag{22}$$

Since v is discrete, there always exist v-orderings. Such sequences may be constructed inductively on n choosing any element of S for a_0.

Example 2. 1. For every prime p, the sequence $\{n\}_{n \geq 0}$ is a p-ordering of $\mathbb{Z}$.

2. For every integer $q \geq 2$ and every prime p, the sequence $\{q^n\}_{n \geq 0}$ is a p-ordering of the set $S_q = \{q^n \mid n \in \mathbb{N}\}$ (cf. [17, Exercise II.15]).

3. If $\text{Card}(S) = s < +\infty$ and $\{a_n\}_{n \geq 0}$ is a v-ordering, then $S = \{a_i \mid 0 \leq i < s\}$.
4. If $\{a_n\}_{n \geq 0}$ is a VWDWO sequence of V as defined in Proposition 10 then, for every $k \geq 0$, the sequence $\{a_n\}_{n \geq k}$ is a v-ordering of V.

4.2 v-Orderings and Integer-Valued Polynomials

There are strong links between v-orderings and integer-valued polynomials:

Proposition 11 ([9]). *Let $\{a_n\}_{n \geq 0}$ be a sequence of distinct elements of S. Consider the associated sequence of polynomials:*

$$f_0(X) = 1 \quad \text{and} \quad f_n(X) = \prod_{k=0}^{n-1} \frac{X - a_k}{a_n - a_k} \quad \text{(for } n \geq 1). \tag{23}$$

Then the following assertions are equivalent:

1. *The sequence $\{a_n\}_{n \geq 0}$ is a v-ordering of S.*
2. *For every $n \geq 0$, $f_n(S) \subseteq V$.*
3. *The sequence $\{f_n \mid n \in \mathbb{N}\}$ is a basis of the V-module $\text{Int}(S, V)$.*
4. *For every $d \in \mathbb{N}$, for every $g \in K[X]$ with degree d, one has*

$$g(a_0), g(a_1), \ldots, g(a_d) \in V \iff g(S) \subseteq V. \tag{24}$$

Corollary 1. *If $\{a_n\}_{n \in \mathbb{N}}$ is a v-ordering of S, then the following numbers do not depend on the choice of the v-ordering $\{a_n\}_{n \in \mathbb{N}}$ of S:*

$$w_S(n) = v\left(\prod_{k=0}^{n-1} (a_n - a_k) \right). \tag{25}$$

Proof.

$$w_S(n) = \begin{cases} -v(\mathfrak{I}_n(S, V)) & \text{for } 0 \leq n < \text{Card}(S), \\ +\infty & \text{for } n \geq \text{Card}(S). \end{cases} \tag{26}$$

$\square$

For instance, by Proposition 10 (Pólya [43]),

$$w_V(n) = \begin{cases} w_q(n) = \sum_{k \geq 1} \left[\frac{n}{q^k} \right] & \text{if } q = \text{Card}(V/\mathfrak{m}) < +\infty, \\ w_q(n) = 0 & \text{if } q = \text{Card}(V/\mathfrak{m}) = +\infty. \end{cases} \tag{27}$$

4.3   Globalization: v-Orderings and Factorials

Notations. Consider again a Dedekind domain D. For every maximal ideal $\mathfrak{m}$ of D, we denote by $v_{\mathfrak{m}}$ the corresponding valuation of K and by $w_{S,\mathfrak{m}}(n)$ the integers defined in Formula (25) (for the valuation $v = v_{\mathfrak{m}}$); finally we speak of $\mathfrak{m}$-orderings instead of $v_{\mathfrak{m}}$-orderings.

Proposition 12. *For every $n \in \mathbb{N}$ such that $n < \mathrm{Card}(S)$, we have*

$$n!_S = \prod_{\mathfrak{m} \in \mathrm{Max}(D)} \mathfrak{m}^{w_{S,\mathfrak{m}}(n)}. \tag{28}$$

Proof. It follows from (19) and (26) that $\left(n!_S^D\right)_{\mathfrak{m}} = n!_S^{D_{\mathfrak{m}}} = \mathfrak{m}^{w_{S,\mathfrak{m}}(n)} D_{\mathfrak{m}}$. $\square$

Proposition 13 ([40, Lemma 8]). *Whatever the infinite subset S of $\mathbb{Z}$, there are no three equal consecutive terms in the sequence $\{n!_S^{\mathbb{Z}}\}_{n \geq 0}$.*

Question 1 ([40]). Does there exist an infinite subset S of $\mathbb{Z}$ such that there are infinitely many two equal consecutive terms in the sequence $\{n!_S^{\mathbb{Z}}\}_{n \geq 0}$?

Proposition 14 (Generalized Property B). *For all $x_0, x_1, \ldots, x_n \in S$:*

$$\prod_{0 \leq i < j \leq n} (x_j - x_i)D \quad \text{is divisible by} \quad 1!_S^D \times 2!_S^D \times \cdots \times n!_S^D. \tag{29}$$

Bhargava's proof [11] is given for $\mathbb{Z}$, but it also works for D and it really deserves to be read because it shows how powerful is the notion of v-ordering. There are many generalizations of Kempner's formula (7) (see for instance [29]), but Bhargava seems to be the first who considered functions defined on subsets.

Proposition 15 (Property F [9, Theorem 5]). *Let D be a Dedekind domain, let $\mathfrak{J}$ be a proper ideal of D with finite norm $N = \mathrm{Card}(D/\mathfrak{J})$, and let S be a subset of D whose elements are noncongruent modulo $\mathfrak{J}$. Then, the number of polynomial functions from S to $D/\mathfrak{J}$ (induced by a polynomial of $D[X]$) is equal to*

$$\prod_{k=0}^{N-1} \frac{N}{\mathrm{Card}(D/(\mathfrak{J}, k!_S))}, \tag{30}$$

where $(\mathfrak{J}, k!_S)$ denotes the ideal of D generated by $\mathfrak{J}$ and $k!_S$.

Note that $k!_S = (0)$ for k greater than the number of classes of S modulo $\mathfrak{J}$, and then, $(\mathfrak{J}, k!_S) = \mathfrak{J}$.

5 Regular Bases

From now on, S is suppose to be infinite. Recall that a regular basis is a basis with one and only one polynomial of each degree and that $Int(S, D)$ admits a regular basis if and only if the factorial ideals $n!_S^D$ are (nonzero) principal ideals. Yet, even when this is the case, it may be difficult to describe a regular basis.

5.1 The Local Case

In the local case, that is, in D_m, there always exist regular bases (as we suppose S to be infinite) constructed by means of an m-ordering of S as shown by Proposition 11. We recall here an example of regular basis constructed in a different way.

Example 3 ([17, Sect. II.2]). Given a polynomial $g(X)$, we denote by g^{*k} the kth iterate of g by composition and we let $g^{*0}(X) = X$. In particular, for a fixed prime number p, starting with $F_p(X) = \frac{X^p - X}{p}$, $F_p^{*0}(X) = X$, $F_p^*(X) = F_p(X)$, $F_p^{*2}(X) = F_p(F_p(X))$, and by iteration, $F_p^{*k}(X) = F_p(F_p^{*k-1}(X))$. Finally, for every integer $n = n_0 + n_1 p + \cdots + n_s p^s$ where $0 \le n_j < p$, we let $F_{p,n} = \prod_{k=0}^s (F_p^{*k})^{n_k}$. Note that $F_{p,0} = 1$, $F_{p,1} = X$, and $F_{p,p^k} = F_p^{*k}$.

Then, the polynomials $\{F_{p,n}(X)\}_{n \ge 0}$ form a basis of the $\mathbb{Z}_{(p)}$-module $Int(\mathbb{Z}_{(p)})$.

Moreover, as a $\mathbb{Z}_{(p)}$-algebra, $Int(\mathbb{Z}_{(p)})$ is generated by the set $\left\{ F_p^{\circ k}(X) \mid k \ge 0 \right\}$ and this is a minimal set of generators. Finally, every polynomial of $Int(\mathbb{Z}_{(p)})$ is obtained from $1, X$ and $\frac{X^p - X}{p}$ by means of sums, products, and composition.

Remark 1. Analogously to Example 3, we may obtain in the local case minimal sets of generators of the V-algebra $Int(S, V)$ by means of the sequence $\{w_S(n) = -v(\mathfrak{I}_n(S, V))\}_{n \ge 0}$. If g_n is a regular basis, we obtain a set of generators by considering only the g_n's where n satisfies the following:

$$w_S(n) > w_S(i) + w_S(j) \text{ for all } i, j > 0 \text{ such that } i + j = n \quad [33].$$

5.2 Simultaneous Orderings

Computation of factorials and description of bases of the ring of integer-valued polynomials is easy when there exist simultaneous orderings as for $\mathbb{Z}$.

Definition 5. A sequence $\{a_n\}_{n \in \mathbb{N}}$ of elements of S which is an m-ordering of S for every maximal ideal m of D is called a *simultaneous ordering* of S.

The following proposition is the global version of Proposition 11.

Proposition 16. *Let $\{a_n\}_{n\geq 0}$ be a sequence of distinct elements of S. Consider the associated sequence of polynomials $f_n(X) = \prod_{k=0}^{n-1} \frac{X-a_k}{a_n-a_k}$ $(n \geq 0)$. Then the following assertions are equivalent:*

1. *The sequence $\{a_n\}_{n\geq 0}$ is a simultaneous ordering of S.*
2. *For every $n \geq 0$, $f_n(S) \subseteq D$.*
3. *The polynomials $\{f_n \mid n \in \mathbb{N}\}$ form a basis of the D-module $\mathrm{Int}(S, D)$.*
4. *For each $g \in K[X]$ with $\deg(g) = d$, $g(S) \subseteq D$ if and only if $g(a_0), g(a_1), \ldots, g(a_d) \in D$.*
5. *For every $n \geq 1$, we have $n!_S^D = \prod_{k=0}^{n-1}(a_n - a_k)D$.*

Are there simultaneous orderings? In particular, which Dedekind domains admit simultaneous orderings [9, Question 3]?

Example 4. 1. $\mathbb{Z}$ admits the simultaneous ordering $\{n\}_{n\geq 0}$.
2. Every semi-local Dedekind domain admits simultaneous orderings (obtained by the Chinese remainder theorem, see Proposition 18 below).
3. $\mathbb{F}_q[T]$, the analog of $\mathbb{Z}$ for function fields, admits a simultaneous ordering $\{a_n\}_{n\geq 0}$, given by Formulas (20) and (21) where $\mathbb{F}_q^* = \{a_1, \ldots, a_{q-1}\}$ and π is replaced by T [11, Sect. 10], leading to *Carlitz factorials* [20].
4. Let K be a number field with ring of integers $\mathscr{O}_K$ and T be a multiplicative subset of $\mathscr{O}_K$. Then $\{n\}_{n\geq 0}$ is a simultaneous ordering of $D = T^{-1}\mathscr{O}_K$ if and only if every prime p is either invertible or completely split in D [17, Theorem IV.3.1].

Conjecture. *If K is a number field distinct from $\mathbb{Q}$, then its ring of integers $\mathscr{O}_K$ does not admit any simultaneous ordering.*

In 2003, Wood [46] proved this conjecture for imaginary quadratic number fields, while Adam [1] did an analogous study for "imaginary" quadratic function fields in 2005. Adam and Cahen [5] proved in 2010 that there are at most finitely many real quadratic number fields whose ring of integers admits a simultaneous ordering.

Let us restrict our question on the existence of simultaneous orderings of subsets of $\mathbb{Z}$ [11, Question 30]. Let us say that a sequence $\{a_n\}_{n\geq 0}$ is *self-simultaneously ordered* if it is a simultaneous ordering of the subset $S = \{a_n \mid n \in \mathbb{N}\}$ (formed by its own terms). We then have the following examples:

Example 5 ([7, 11]). 1. The sequence $\{q^n\}_{n\geq 0}$ where $|q| \geq 2$ is self-simultaneously ordered. Denoting by S_q the subset $S_q = \{q^n \mid n \in \mathbb{N}\}$, it follows that

$$n!_{S_q} = q^{\frac{n(n-1)}{2}}(q^n - 1)(q^{n-1} - 1) \cdots (q - 1) \text{ (Jackson's factorials)}.$$

2. The sequence $\{n^2\}_{n\geq 0}$ is self-simultaneously ordered. Denoting by $\mathbb{N}^{(2)}$ the subset $\mathbb{N}^{(2)} = \{n^2 \mid n \in \mathbb{N}\}$, it follows that $n!_{\mathbb{N}^{(2)}} = \frac{(2n)!}{2}$. Moreover, the subset

$\mathbb{N}^{(k)} = \{n^k \mid n \in \mathbb{N}\}$ admits a simultaneous ordering if and only if $k = 1$ or 2 (recall that the sequence $\{n\}_{n \geq 0}$ of natural numbers is a simultaneous ordering of $\mathbb{N}$).

3. The sequence $\left\{\frac{n(n+1)}{2}\right\}_{n \geq 0}$ of triangular numbers is self-simultaneously ordered.

Denoting by S the subset $S = \left\{\frac{n(n+1)}{2} \mid n \geq 0\right\}$, it follows that $n!_S = \frac{(2n)!}{2^n}$.

Noticing that if S admits a simultaneous ordering $\{a_n\}_{n \geq 0}$, then $T = bS + c$ where $b, c \in \mathbb{Z}, b \neq 0$, admits also a simultaneous ordering, namely $\{ba_n + c\}_{n \geq 0}$, we thus have many other simultaneously ordered subsets of $\mathbb{Z}$.

We also have the following for discrete dynamical systems.

Proposition 17 ([7, Proposition 18]). *Consider the dynamical system $(\mathbb{Z}, f)$ formed by the set $\mathbb{Z}$ and a nonconstant polynomial $f \in \mathbb{Z}[X]$ distinct from $\pm X$. Then, for every $x \in \mathbb{Z}$, the sequence $\{f^n(x)\}_{n \geq 0}$, where f^n denotes the nth iterate of f, is self-simultaneously ordered. In other words, each orbit admits a simultaneous ordering.*

Equivalently, for every $x \in \mathbb{Z}$ and for all $m, n \in \mathbb{N}$ with $m \geq n \geq 1$:

$$\prod_{j=0}^{n-1}(f^n(x) - f^j(x)) \text{ divides } \prod_{j=0}^{n-1}(f^m(x) - f^j(x)).$$

Note that the sequence $\{q^n\}_{n \geq 0}$ in Example 5(1) stems from a dynamical system with $f(X) = qX$ and $x = 1$. By considering the orbit of 3 under the iteration of $X^2 - 2X + 2$, we obtain:

Corollary 2. *The sequence formed by the Fermat numbers $\{F_n = 2^{2^n} + 1\}_{n \geq 0}$ is self-simultaneously ordered.*

Question 2. Are there other natural examples of subsets of $\mathbb{Z}$ admitting simultaneous orderings?

5.3 The General Case

To obtain regular bases, if any, we use the Chinese remainder theorem. Analogously to [17, Lemma II.3.4] or following [9, Theorem 11], we have:

Proposition 18. *For each $\mathfrak{m} \in \mathrm{Max}(D)$, let $\{a_{\mathfrak{m},n}\}_{n \geq 0}$ be an $\mathfrak{m}$-ordering of S. For $n > 0$, let $\{b_{n,k}\}_{0 \leq k < n}$ be elements of D such that*

$$v_{\mathfrak{m}}(b_{n,k} - a_{\mathfrak{m},k}) > w_{S,\mathfrak{m}}(n) \text{ for all } \mathfrak{m} \text{ such that } w_{S,\mathfrak{m}}(n) \neq 0. \tag{31}$$

Finally, let $g_n(X) = \prod_{k=0}^{n-1}(X - b_{n,k})$. Then, the fixed divisor $d(g_n, S)$ of g on S (as defined before Proposition 8) is equal to the nth factorial $n!_S^D$.

We derive a few corollaries with the g_n's as defined in Proposition 18.

Corollary 3. *We have the following isomorphism of D-modules:*

$$\text{Int}(S, D) = \oplus_{n \geq 0} \mathfrak{I}_n(S, D) \, g_n(X). \tag{32}$$

We just have to verify that $c_0 g_0(X) + c_1 g_1(X) + \cdots + c_n g_n(X)$ with $c_j \in K$ belongs to $\text{Int}(S, D)$ if and only if for $0 \leq j \leq n$, $c_j \in \mathfrak{I}_j(S, D)$.

Corollary 4. *If the factorial ideals are principal, writing $n!_S^D = d_n D$, the polynomials $\frac{1}{d_n} g_n(X)$ then form a regular basis of the D-module $\text{Int}(S, D)$.*

But at any rate, since a nonfinitely generated projective module over a Dedekind domain is free:

Corollary 5. *The D-module $\text{Int}(S, D)$ is free.*

Yet, if there is no regular basis, that is, if the factorial ideals are not principal, it may be difficult to describe a basis.

5.4   Pólya Groups and Pólya Fields

In this paragraph, K denotes an algebraic number field and $\mathscr{O}_K$ its ring of integers. We restrict here our study to the case where $S = D = \mathscr{O}_K$. The following group is a measure of the obstruction for $\text{Int}(\mathscr{O}_K)$ to have a regular basis.

Definition 6 ([17, Definition II.3.8]). The *Pólya group* of K is the subgroup $\mathscr{P}o(K)$ of the class group of K generated by the classes of the factorial ideals of $\mathscr{O}_K$.

One knows [17, II.3.9] that $\mathscr{P}o(D)$ is also generated by the classes of the ideals:

$$\Pi_q(D) = \prod_{\mathfrak{m} \in \text{Max}(D), \, N(\mathfrak{m}) = q} \mathfrak{m} \quad (q \geq 2).$$

Proposition 19. *A Pólya field is a number field K which satisfies the following equivalent assertions:*

1. *$\text{Int}(\mathscr{O}_K)$ admits a regular basis.*
2. *The fractional ideals $\mathfrak{I}_n(\mathscr{O}_K)$ are principal.*
3. *The integral ideals $(n!)_{\mathscr{O}_K}$ are principal.*
4. *The ideals $\Pi_q(\mathscr{O}_K)$ are principal.*
5. *$Po(K) \simeq \{1\}$.*

If $K/\mathbb{Q}$ is a galoisian extension, for every prime p, one has $p\mathscr{O}_K = \Pi_{p^f}(\mathscr{O}_K)^e$, and hence, to know whether K is a Pólya field, we just have to consider the ideals

$\Pi_q(\mathscr{O}_K)$ such that the maximal ideals $\mathfrak{m}$ with $N(\mathfrak{m}) = q$ lye over the primes p that are ramified in K [41]. We also have:

Proposition 20 ([22, Proposition 3.6]). *If $K_1/\mathbb{Q}$ and $K_2/\mathbb{Q}$ are two galoisian extensions whose degrees are relatively prime, then*

$$Po(K_1 K_2) \simeq Po(K_1) \times Po(K_2). \tag{33}$$

For quadratic fields, the Pólya group corresponds to the group of ambiguous classes whose description was done by Hilbert (see [17, Proposition II.4.4]) and the characterization of the quadratic Pólya fields was done by Zantema [47]. Every cyclotomic field is a Pólya field [47]. A systematic study of the galoisian Pólya fields of degree ≤ 6 was recently undertaken by Leriche. For instance,

Proposition 21 ([37, Proposition 3.2]). *The cyclic cubic Pólya fields are the fields $\mathbb{Q}[t]$ where t is a root of $X^3 - 3X + 1$ or of $X^3 - 3pX - pu$ where p is a prime of the form $\frac{1}{4}(u^2 + 27w^2)$, $u \equiv 2 \pmod 3$ and $w \neq 0$.*

She also characterized the galoisian Pólya fields in the cases of cyclic quartic or sextic fields (the latter, compositum of a cyclic cubic Pólya field and a quadratic Pólya field) as well as in the cases of cyclic fields of the form $\mathbb{Q}[j, \sqrt[3]{m}]$, and of biquadratic fields [37].

Adam [3] undertook a similar study for functions fields. He proved analogously that every cyclotomic function field in the sense of Carlitz is a Pólya field and characterized the Kummer extensions and the "totally imaginary" Artin-Schreier extensions of $\mathbb{F}_q(T)$ which are Pólya function fields.

Another interesting notion is the notion of Pólya extension: L/K is a *Pólya extension* if the $\mathscr{O}_L$-module $\mathrm{Int}(\mathscr{O}_K, \mathscr{O}_L)$ admits a regular basis. By the capitulation theorem, the Hilbert class field H_K of every number field K is a Pólya extension of K. Moreover, H_K gives an answer to the following embedding problem: is every number field contained in a Pólya number field? The answer is yes because it turns out that every Hilbert class field is a Pólya field [38, Corollary 3.2]. An open question is to determine the minimal degree of a Pólya field containing K.

6   Computation and Explicit Formulas

In this section, we restrict our study to the local case and consider a slightly more general situation because the notion of v-ordering may be defined for rank-one valuations v of K, that is, valuations v such that $v(K^*) \subseteq \mathbb{R}$. But since there do not always exist v-orderings, we have to assume conditions on S, for instance, that S is *precompact*, that is, that its completion is compact (cf. [19, Corollary 1.6]).

Notation for Sect. 6. *K is a valued field endowed with a rank-one valuation v, the valuation domain is denoted by V, its maximal ideal by $\mathfrak{m}$, and S is an infinite precompact subset of V.*

In this general framework, all results of Sect. 4.2 hold.

For $a \in V$ and $\gamma \in \mathbb{R}$, we denote by $B(a, \gamma)$ the ball of center a and radius $e^{-\gamma}$:

$$B(a, \gamma) = \{x \in V \mid v(a - x) \geq \gamma\}.$$

6.1   How Can We Compute the Function $w_S(n)$?

We are interested here in the function $w_S(n) = -v(\mathfrak{I}_n(S, D)) = v(n!_S^D)$. The sequence $\{w_S(n)\}_{n \geq 0}$ is called the *characteristic sequence* of S.

Lemma 2. *If $\{a_n\}_{n \geq 0}$ is an $\mathfrak{m}$-ordering of S, then, for every nonzero $b \in V$ and every $c \in V$, $\{ba_n + c\}_{n \geq 0}$ is an $\mathfrak{m}$-ordering of $bS + c = \{bs + c \mid s \in S\}$. Thus,*

$$\forall b, c \in V \quad w_{bS+c}(n) = nv(b) + w_S(n).$$

Lemma 3 ([16, Lemma 3.4]). *If $\{a_n\}_{n \geq 0}$ is a v-ordering of S, then the subsequence formed by the a_n's which are in a ball $B(a, \gamma)$ is a v-ordering of $S \cap B(a, \gamma)$.*

Proposition 22 ([31, Lemma 2]). *Let $\{s_i \mid 1 \leq i \leq r\}$ be a system of representatives of S modulo $\mathfrak{m}$, that is, $S = \cup_{i=1}^{r}(S \cap (s_i + \mathfrak{m}))$ (where $s_i \not\equiv s_j$ (mod $\mathfrak{m}$) $\forall i \neq j$). If, for each i, $\{a_{i,n}\}_{n \geq 0}$ is a v-ordering of $S \cap (s_i + \mathfrak{m})$, we obtain a v-ordering of S by shuffling these v-orderings in such a way that the shuffling of the corresponding characteristic sequences leads to a nondecreasing sequence of integers.*

In particular, the characteristic sequence of S is the disjoint union of the characteristic sequences of the $S \cap (s_i + \mathfrak{m})$ sorted into a nondecreasing order.

Example 6. Assume that $\mathfrak{m} = \pi V$ and $\mathrm{Card}(V/\mathfrak{m}) = q < +\infty$. Consider

$$S = V \setminus \mathfrak{m} = \cup_{i=1}^{q-1}(a_i + \mathfrak{m}). \tag{34}$$

Let $\{a_n\}_{n \geq 0}$ be the v-ordering of V given by (21). For $1 \leq i \leq q - 1$, $\{a_i + \pi a_n\}_{n \geq 0}$ is a v-ordering of $a_i + \mathfrak{m}$, and $w_{a_i+\mathfrak{m}}(n) = w_{\pi V}(n) = n + w_V(n) = n + w_q(n)$. We construct a v-ordering of S by taking successively one element of each of the $q - 1$ partial orderings since the characteristic sequences of the sets $a_i + \mathfrak{m}$ are equal, we obtain the subsequence formed by the a_n's such that $v(a_n) = 0$, and we have

$$w_{V \setminus \mathfrak{m}} = \left[\frac{n}{q-1}\right] + w_q\left(\left[\frac{n}{q-1}\right]\right) = \sum_{k \geq 0}\left[\frac{n}{(q-1)q^k}\right]. \tag{35}$$

6.2 Toward Symmetry: Homogeneous Subsets

As seen in Example 6, symmetry may help for the shuffling. We have a kind of symmetry when we consider *homogeneous subsets*, that is, subsets S for which there exists some $\gamma \in \mathbb{R}$ such that $S = \cup_{s \in S} B(s, \gamma)$.

Proposition 23 ([16, Theorem 3.6]). *Assume that*

$$S = \cup_{i=1}^{r} B(b_i, \gamma) \text{ where } v(b_i - b_j) < \gamma (1 \leq i \neq j \leq r). \text{ Then,}$$

$$w_S(n) = \max_{\delta_1 + \cdots + \delta_r = n} \left(\min_{1 \leq i \leq r} w_S^i(\delta_1, \ldots, \delta_r) \right) \quad (\delta_1, \ldots, \delta_r \in \mathbb{N}) \text{ where}$$

$$w_S^i(\delta_1, \ldots, \delta_r) = w_q(\delta_i) + \gamma \delta_i + \sum_{j \neq i} v(b_j - b_i) \delta_j \text{ and } w_q(m) = \sum_{k \geq 1} \left[\frac{m}{q^k} \right].$$

We may introduce more symmetry by assuming that $q = +\infty$ ($w_q(\delta_i) = 0$) :

Proposition 24 ([15, Theorem 4.4]). *Assume that* $\text{Card}(V/\mathfrak{m}) = +\infty$ *and that*

$$S = \cup_{i=1}^{r} B(b_i, \gamma) \quad \text{where} \quad v(b_i - b_j) < \gamma \; (1 \leq i \neq j \leq r). \tag{36}$$

Consider the symmetric matrix $B = (\beta_{i,j}) \in \mathscr{M}_r(\mathbb{R})$ *defined by*

$$\beta_{i,j} = v(b_i - b_j) \text{ for } 1 \leq i \neq j \leq r, \text{ and } \beta_{i,i} = \gamma \text{ for } 1 \leq i \leq r. \tag{37}$$

Denote by B_i *the matrix deduced from* B *by replacing every coefficients in ith column by 1 and let* $v(B) = \sum_{i=1}^{r} \det(B_i)$. *If* $n = mv(B) + n_0$ *where* $0 \leq n_0 < v(B)$, *then* $w_S(n) = m \det(B) + w_S(n_0)$.

6.3 Preregular Subsets: Generalized Legendre's Formula

The best way to obtain symmetry is by considering the notion of preregular subset which extends Amice's notion of regular compact subset of a local field [8].

To explain this notion, we introduce the following equivalence relations on V :

$$\forall \gamma \in \mathbb{R} \; \forall x, y \in V \quad x \equiv y \pmod{\gamma} := v(x - y) \geq \gamma.$$

We denote by $S \bmod \gamma$ the set of equivalence classes of the elements of S, and let

$$q_\gamma = \text{Card}(S \bmod \gamma)$$

The fact that S is an infinite precompact subset of V is equivalent to

$$\text{all the } q_\gamma\text{'s are finite and} \lim_{\gamma \to +\infty} q_\gamma = +\infty.$$

Definition 7. The precompact subset S is *preregular* if, for all $\gamma < \delta$, for every $x \in S$, $S \cap B(x, \gamma)$ contains exactly $\frac{q_\delta}{q_\gamma}$ nonempty subsets of the form $S \cap B(y, \delta)$.

This notion will allow us to generalize the following well-known formulas obtained for v discrete, $\text{Card}(V/\mathfrak{m}) = q$, and S regular:

$$v_p(n!) = \sum_{k \geq 1} \left[\frac{n}{p^k}\right] \begin{matrix} \text{Legendre} \\ 1808 \end{matrix} \quad v(n!_V) = \sum_{k \geq 1} \left[\frac{n}{q^k}\right] \begin{matrix} \text{Pólya} \\ 1909 \end{matrix} \quad v(n!_S) = \sum_{k \geq 1} \left[\frac{n}{q_k}\right] \begin{matrix} \text{Amice} \\ 1964 \end{matrix}.$$

Proposition 25 ([26, Theorem 1.5]). *The precompact subset S is preregular if and only if, denoting by γ_k the critical valuations of S,*

$$v(n!_S) = n\gamma_0 + \sum_{k \geq 1} \left[\frac{n}{q_{\gamma_k}}\right] (\gamma_k - \gamma_{k-1}) \quad \begin{matrix} [26] \\ 2013. \end{matrix} \qquad (38)$$

Recall that the sequence $\{\gamma_k\}_{k \geq 0}$ of *critical valuation* of S is characterized by [23, Proposition 5.1] $\gamma_0 = \inf_{x \in S} v(x)$ and for $k \geq 1 : \gamma_{k-1} < \gamma \leq \gamma_k \Leftrightarrow q_\gamma = q_{\gamma_k}$.

Let us mention an application of regularity to dynamical systems:

Proposition 26 ([25, Corollary 4]). *Assume that S is a regular compact subset of V and let $\varphi : S \to S$ be an isometry. Then, the discrete dynamical system (S, φ) is minimal (i.e., for every $x \in S$, the orbit $\Omega(x) = \{\varphi^n(x) \mid n \in \mathbb{N}\}$ is dense in S) if and only if, for every $x \in S$, the sequence $\{\varphi^n(x)\}_{n \geq 0}$ is a v-ordering of S.*

6.4 Valuative Capacity

Since the function w_S is *super-additive* (i.e., $w_S(n + m) \geq w_S(n) + w_S(m)$), the following limit, finite or infinite, called the *valuative capacity* of S, exists:

$$\delta_S = \lim_{n \to +\infty} \frac{w_S(n)}{n}. \qquad (39)$$

The larger S, the smaller is δ_S. It is also equal to the limit [21, Theorem 4.2]:

$$\lim_{n \to +\infty} \frac{2}{n(n+1)} \min_{x_0, \dots, x_n \in S} v\left(\prod_{0 \leq i < j \leq n} (x_j - x_i)\right) = \delta_S. \qquad (40)$$

In some sense its generalizes the notion of transfinite diameter in archimedean metric. For instance, $\delta_{\mathbb{Z}_{(p)}} = \frac{1}{p-1}$, $\delta_{p\mathbb{Z}_{(p)}} = \frac{p}{p-1}$, $\delta_{\mathbb{Z}_{(p)}\setminus p\mathbb{Z}_{(p)}} = \frac{p}{(p-1)^2}$.

Proposition 27 ([31, Corollary 10]). *Assume that* $S = \bigcup_{i=1}^{r}(S \cap (s_i + \mathrm{m}))$ *(where* $s_i \not\equiv s_j$ *(mod m)* $\forall i \neq j$). *If, for* $i = 1,\dots,r$, $\delta_{S\cap(s_i+\mathrm{m})} \neq 0$, *then* $\frac{1}{\delta_S} = \sum_{i=1}^{r}\frac{1}{\delta_{S\cap(s_i+\mathrm{m})}}$.

This last result allows to compute easily some valuative capacities. For instance,

Corollary 6 ([31, Proposition 11]). *Assume* v *is discrete and* $\mathrm{Card}(V/\mathrm{m}) = q < +\infty$:

$$\delta_{V\setminus\mathrm{m}^k} = \frac{1}{(q-1)^2}\left(q - \frac{q^{2k}-q^2}{q^{2k}-1}\right).$$

More generally, in the spirit of Proposition 24:

Proposition 28 ([15, Theorem 5.3]). *Without particular hypothesis, denote by* q *the cardinality, finite or infinite, of the residue field. Assume that*

$$S = \bigcup_{i=1}^{r} B(b_i, \gamma) \quad \text{where } v(b_i - b_j) < \gamma \ (1 \leq i \neq j \leq r).$$

Consider the matrix $B^* = B + \frac{1}{q-1}I_r$ *and the number* $v(B^*)$ *with* B *and* $v(B^*)$ *as defined in Proposition 24. Then,* $\delta_S = \frac{\det(B^*)}{v(B^*)}$.

6.5 A Generalized Exponential Function

Returning to $\mathbb{Z}$, by analogy with the classical factorials and following [11, Question 33], we introduce an *exponential function* associated to any subset S of $\mathbb{Z}$:

$$\exp_S(x) = \sum_{n\geq 0} \frac{x^n}{n!_S}, \tag{41}$$

where $n!_S$ denotes here the positive generator of the corresponding factorial ideal. By Proposition 9, $n!$ divides $n!_S$, and hence, the power series converges for all x.
    We have the obvious formula

$$\exp_{bS+c}(x) = \sum_{n\geq 0} \frac{x^n}{b^n n!_S} = \exp_S\left(\frac{x}{b}\right). \tag{42}$$

When there exists a simultaneous ordering, it is sometimes easy to compute this exponential function. For instance, Example 4 leads to

$$\exp_{\mathbb{N}^{(2)}}(x) = \sum_{n\geq 0} \frac{x^n}{\frac{(2n)!}{2}} = 2\cosh\sqrt{|x|}, \ \exp_{\left\{\frac{n(n+1)}{2}|n\geq 0\right\}}(x) = \sum_{n\geq 0} \frac{x^n}{\frac{(2n)!}{2^n}} = \cosh\sqrt{2|x|}.$$

In particular, let us consider the value for $x = 1$ and introduce the number

$$e_S = \sum_{n \geq 0} \frac{1}{n! s}. \tag{43}$$

For instance, for $S = \mathbb{N}^{(2)}$, $e_{\mathbb{N}^{(2)}} = e + \frac{1}{e}$.

Proposition 29 (Mingarelli [40, Theorem 28]). *The number e_S is irrational.*

Question 3 ([40]). For which subsets S is e_S a transcendental number?

7 Sub-algebras of Int(S, D)

7.1 Derivatives and Finite Differences

Among the interesting sub-D-algebras of Int(S, D), one may consider the algebras Int$^{(k)}(S, D)$ formed by the polynomials that are integer valued on S together with their *derivatives* up to the order k :

$$\text{Int}^{(k)}(S, D) = \{f(X) \in K[X] \mid f^{(h)} \in \text{Int}(S, D) \ 0 \leq h \leq r\}, \tag{44}$$

and the algebras Int$^{[k]}(D)$ formed by the polynomials that are integer valued on D together with their *finite differences* up to the order k defined inductively by Int$^{[0]}(D) = \text{Int}(D)$ and, for $k \geq 1$,

$$\text{Int}^{[k]}(D) = \{f \in K[X] \mid \forall h \in D, h \neq 0 \ (f(X + h) - f(X))/h \in \text{Int}^{[k-1]}(D)\}.$$

A review on these algebras is given in [17, Chap. IX]. Yet, new results appeared in characteristic $p > 0$. For instance:

Proposition 30 ([4, Theorem 2.11]). *Let* Int$^{(\infty)}(S, D) = \cap_{k \geq 0}\text{Int}^{(k)}(S, D)$. *If* char$(D) = p > 0$, *then*

$$\mathfrak{I}_n(\text{Int}^{(\infty)}(S, D))^{-1} = \prod_{\mathfrak{m} \in \text{Max}(D)} \mathfrak{m}^{w_{S,\mathfrak{m}}([\frac{n}{p}])}.$$

Moreover, if $\{f_n(X)\}_{n \geq 0}$ *is a regular basis of* Int(S, D), *then the polynomials* $F_{m,j} = f_m(X)^p X^j$ $(m \in \mathbb{N}, j \in \{0, \ldots, p - 1\})$ *form a regular basis of* Int$^{(\infty)}(S, D)$.

In particular, we have an explicit basis for Int$^{(\infty)}(\mathbb{F}_q[T])$, thanks to Example 4(3).

Proposition 31 ([2, Theorem 16]). *If* $\mathrm{char}(D) = p > 0$, *then*

$$\mathfrak{I}_n(\mathrm{Int}^{[k]}(D))^{-1} = \prod_{q \leq n} \left(\prod_{\mathfrak{m} \in \mathrm{Max}(D), N(\mathfrak{m})=q} \mathfrak{m} \right)^{\delta_q^{[k]}(n)}$$

where $\delta_q^{[k]}(n) = w_q(n) - \lambda_q^{[k]}(n)$ *and*

$$\lambda_q^{[k]}(n) = \sup \left\{ v_q(j_1) + \cdots + v_q(j_r) \mid r \leq k, j_1 + \cdots + j_r \leq n, j_i \geq 1, p \nmid \binom{n}{j_1, \ldots, j_r} \right\}.$$

There are also results about a multiplicative analog of finite differences, namely, the *Euler-Jackson differences* (see [6]): let $S_q = \{q^n \mid n \geq 0\}$ where q denotes a nonzero element of D which is not a root of unity, and, for $h \in \mathbb{N}^*$, let

$$\delta_{q^h} f(X) = \frac{f(q^h X) - f(X)}{(q^h - 1)X}. \tag{45}$$

Then, $\mathrm{Int}_J^{[k]}(S_q, D)$ is defined inductively by

$$\mathrm{Int}_J^{[k]}(S_q, D) = \{ f(X) \in K[X] \mid \forall h \in \mathbb{N}^* \; \delta_{q^h} f(X) \in \mathrm{Int}_J^{[k-1]}(S, D) \}. \tag{46}$$

7.2 Divided Differences

Contrarily to finite differences, divided differences make sense on subsets.

Definition 8. The *divided difference of order* k of a function $f : S \to K$ of one variable is defined inductively on k by $\Phi^0(f) = f$ and for $k \geq 1$,

$$\Phi^k(f)(x_0, \ldots, x_{k-1}, x_k) \mapsto \frac{\Phi^{k-1}(f)(\ldots, x_{k-2}, x_{k-1}) - \Phi^{k-1}(f)(\ldots, x_{k-2}, x_k)}{x_{k-1} - x_k},$$

defined on $S^{k+1} \setminus \Delta_k$ where $\Delta_k = \{ (x_0, \ldots, x_k) \in S^{k+1} \mid x_i = x_j \text{ for some } i \neq j \}$.

The function $\Phi^k(f)$ is symmetric with respect to the $k + 1$ variables $x_0, \ldots, x_k$.

Definition 9. The ring $\mathrm{Int}^{\{k\}}(S, D)$ of polynomials *integer valued on* S *together with their divided differences* up to the order k is

$$\mathrm{Int}^{\{k\}}(S, D) = \{ f \in K[X] \mid \Phi^h(f)(S^{h+1}) \subseteq D \; 0 \leq h \leq k \}. \tag{47}$$

The algebraic properties of this ring are studied in [14]. One has the containments:

$$\mathrm{Int}^{\{k\}}(D) \subseteq \mathrm{Int}^{[k]}(D) \subseteq \mathrm{Int}^{(k)}(D) \tag{48}$$

with the equality $\text{Int}^{\{1\}}(D) = \text{Int}^{[1]}(D)$ and, for every subset S, $\text{Int}^{\{k\}}(S, D) \subseteq \text{Int}^{(k)}(S, D)$. The construction of bases of $\text{Int}^{\{k\}}(S, D)$ is described in [12].

Let us focus on the local case. As in Sect. 6, V denotes a rank-one valuation domain and S a precompact subset of V.

Definition 10 (Bhargava [12]). A *r-removed v-ordering* of S is a sequence $\{a_n\}_{n \geq 0}$ of elements of S where $a_0, a_1, \ldots, a_r$ are chosen arbitrarily and, for $n > r$, there exist r distinct integers $i_1, \ldots, i_r \in \{0, 1, \ldots, n-1\}$ such that

$$v \left(\prod_{\substack{0 \leq k < n \\ k \neq i_1, \ldots, i_r}} (a_n - a_k) \right) = \inf_{\substack{x \in S \\ 0 \leq j_1 < \cdots < j_r < n}} v \left(\prod_{\substack{0 \leq k < n \\ k \neq j_1, \ldots, j_r}} (x - a_k) \right).$$

Let $\alpha_n = \{1, \ldots, n\} \setminus \{i_1, \ldots, i_r\}$ be the set formed by the remaining indices.

Proposition 32 (Bhargava [12]). *If $\{a_n\}_{n \in \mathbb{N}}$ is a r-removed v-ordering of S, then the following polynomials form a basis of the V-module $\text{Int}^{\{r\}}(S, V)$:*

$$\binom{x}{n}_{\{a_k\}}^{\{r\}} = \frac{(x - a_0)(x - a_1) \cdots (x - a_{n-1})}{\prod_{k \in \alpha_n} (a_n - a_k)}. \tag{49}$$

In general, it is not so easy to construct a *r*-removed *v*-ordering of S, nor to compute the valuation of the denominator, that is, the number:

$$w_S^{\{r\}}(n) = v \left(\prod_{k \in \alpha_n} (a_n - a_k) \right). \tag{50}$$

Johnson [32] was the first one to give an explicit formula for $w_S^{\{r\}}(n)$ in case $S = V = \mathbb{Z}_{(p)}$. Given n, there is a unique integer l such that $rp^l \leq n < rp^{l+1}$, and with this l, one has the formula

$$w_{\mathbb{Z}_{(p)}}^{\{r\}}(n) = \sum_{k=1}^{l} \left[\frac{n}{p^k} \right] - r \times l. \tag{51}$$

This formula can be generalized to preregular precompact subsets as defined in Sect. 6.3. For such a subset, the q_γ's (for $\gamma \in \mathbb{R}$) and the critical valuations γ_k's are also defined in Sect. 6.3. Note that, given n, there is a unique integer l such that $rq_{\gamma_l} \leq n < rq_{\gamma_{l+1}}$ and one has:

Proposition 33 ([26]). *If S is a preregular precompact subset of a rank-one valuation domain, then, with the previous notation, for* $r\,q_{\gamma_l} \leq n < r\,q_{\gamma_l+1}$ *we have*

$$w_S^{\{r\}}(n) = n\gamma_0 + \sum_{k=1}^{l} \left\lceil \frac{n}{q_{\gamma_k}} \right\rceil (\gamma_k - \gamma_{k-1}) - r \times \gamma_l . \tag{52}$$

Proposition 34 ([26]). *If S is a preregular precompact subset, then every VWDWO sequence of S is a r-removed v-ordering of S whatever r.*

Recall that a VWDWO sequence of a preregular subset generalizes the VWDWO sequences defined in Proposition 10 and is characterized by

$$\forall n \neq m \; [v(a_n - a_m) > \gamma \; \Leftrightarrow \; q_\gamma | n - m]. \tag{53}$$

Such sequences are easy to describe: they are the "most regular sequences" in S!

7.3 Integer-Valued Polynomials of a Given Modulus

Definition 11. For every nonzero element a of D, the ring of *integer-valued polynomials on S of modulus a* is the ring

$$\mathrm{Int}_a(S, D) = \{f(X) \in K[X] \mid \forall s \in S \;\; f(aX + s) \in D[X]\}. \tag{54}$$

The algebraic properties of this ring are studied in [14]. We have the following containments: for all a, $\mathrm{Int}_a(S, D) \subseteq \mathrm{Int}(S, D)$, and if a divides b in D, then $\mathrm{Int}_a(S, D) \subseteq \mathrm{Int}_b(S, D)$. On the other hand, $\mathrm{Int}(S, D) = \cup_{a \in D \setminus \{0\}} \mathrm{Int}_a(S, D)$.

Once more, let us focus on a local study: D is assumed to be a rank-one valuation domain V.

Definition 12 (Bhargava [12]). Let $\alpha \in \mathbb{R}_+$. A *v-ordering of order* α of S is a sequence $\{a_n\}_{n \geq 0}$ of elements of S where a_0 is arbitrarily chosen and, for $n \geq 1$, a_n is chosen such that

$$\sum_{k=0}^{n-1} \inf(\alpha, v(a_n - a_k)) = \inf_{s \in S} \left(\sum_{k=0}^{n-1} \inf(\alpha, v(s - a_k)) \right). \tag{55}$$

For such a v-ordering of order α, let

$$w_S^{(\alpha)}(n) = \sum_{k=0}^{n-1} \inf(\alpha, v(a_n - a_k)). \tag{56}$$

Proposition 35 (Bhargava [12]). *If $v(a) = \alpha$, if $\{a_n\}_{n \in \mathbb{N}}$ is a v-ordering of order α of S, and if $v(t_n) = w_S^{(\alpha)}(n)$, then the following polynomials form a basis of the V-module $\mathrm{Int}_a(S, V)$:*

$$\binom{x}{n}^{(\alpha)}_{\{a_k\}} = \frac{(x - a_0)(x - a_1) \cdots (x - a_{n-1})}{t_n}. \tag{57}$$

For instance, for $\mathrm{Int}_{p^h}(\mathbb{Z}_{(p)})$, we have $w_{\mathbb{Z}_{(p)}}^{(h)}(n) = \sum_{k=1}^{h} \left[\frac{n}{p^k} \right]$ (Johnson [32]). This formula may generalized:

Proposition 36. *If S is a preregular precompact subset, then*

$$w_S^{(\alpha)}(n) = n\gamma_0 + \sum_{k=1}^{l} \left[\frac{n}{q_{\gamma_k}} \right] (\gamma_k - \gamma_{k-1}) \quad \text{where } \gamma_l \leq \alpha < \gamma_{l+1}, \tag{58}$$

where the q_γ's and the γ_k's are defined in Sect. 6.3.

As for r-removed v-orderings [Proposition 34], we still have:

Proposition 37. *If S is a preregular precompact subset, then every VWDWO sequence of S is a v-ordering of S of order α whatever α.*

8 Ultrametric Analysis: Extensions of Mahler's Theorem

Mahler's approximation theorem for the Banach ultrametric space $\mathscr{C}(\mathbb{Z}_p, \mathbb{Q}_p)$ (Proposition 4 above) may be generalized by replacing $\mathbb{Q}_p$ by a complete valued field and $\mathbb{Z}_p$ by a precompact subset.

Hypotheses: *K is a valued field endowed with a rank-one valuation v (V denotes the valuation domain) and S an infinite precompact subset of K.*

8.1 Polynomial Approximation in $\mathscr{C}(\hat{S}, \hat{K})$

We denote by $\mathscr{C}(\hat{S}, \hat{K})$ the ultrametric Banach space of continuous functions from the completion $\hat{S}$ of S to the completion $\hat{K}$ of K endowed with the uniform convergence topology.

Proposition 38 ([13] and [18, Theorem 2.4]). *Let K be a valued field and S be a precompact subset of K. Let $\{a_n\}_{n \geq n}$ be a v-ordering of S. Then, every function $\varphi \in \mathscr{C}(\hat{S}, \hat{K})$ can be developed in series as follows:*

$$\varphi(x) = \sum_{n=0}^{\infty} c_n \prod_{k=0}^{n-1} \frac{X - a_k}{a_n - a_k} \quad \text{with } c_n \in \hat{K} \text{ and } \lim_{n \to +\infty} v(c_n) = +\infty. \quad (59)$$

Moreover, $\inf_{x \in S} v(\varphi(x)) = \inf_{n \in \mathbb{N}} v(c_n)$.

The generalized binomial polynomials $\prod_{k=0}^{n-1} \frac{X-a_k}{a_n-a_k}$ $(n \geq 0)$ form an orthonormal basis of the Banach space $\mathscr{C}(\hat{S}, \hat{K})$. The coefficients c_n are unique and may be obtained inductively by

$$c_n = \varphi(a_n) - \sum_{k=0}^{n-1} c_k \prod_{h=0}^{k-1} \frac{a_n - a_h}{a_k - a_h}. \quad (60)$$

Once one knows Formula (59), it is easy to prove that:

Proposition 39 ([18, Theorem 2.7]). *Every basis of the* V*-module* $\text{Int}(S, V)$ $= \mathscr{C}(\hat{S}, \hat{V}) \cap K[X]$ *is an orthonormal basis of the Banach space* $\mathscr{C}(\hat{S}, \hat{K})$.

8.2 Polynomial Approximation in $\mathscr{C}^r(\hat{S}, \hat{K})$ and $LA_\alpha(\hat{S}, \hat{K})$

Recall that, in ultrametric analysis, the Banach space $\mathscr{C}^r(\hat{S}, \hat{K})$ of *functions of class* $\mathscr{C}^r$ is formed by the function $f : \hat{S} \to \hat{K}$ such that $\Phi^k(f)$ may be extended continuously to $\hat{S}^{k+1}$. Proposition 38 may be generalized in the following way:

Proposition 40 ([12, Theorem 21]). *Assume that the precompact subset S has no isolated points. Then, every basis of the V-module* $\text{Int}^{\{r\}}(S, V) = \mathscr{C}^r(\hat{S}, \hat{V}) \cap K[X]$ *is an orthonormal basis of the Banach space* $\mathscr{C}^r(\hat{S}, \hat{K})$.

Example 7. 1. The following polynomials form a basis of the $\mathbb{Z}$-module $\text{Int}^{\{1\}}(\mathbb{Z})$ and thus an orthonormal basis of $\mathscr{C}^1(\mathbb{Z}_p, \mathbb{Q}_p)$ for all p : $\prod_{p \in \mathbb{P}} p^{\left\lfloor \frac{\ln n}{\ln p} \right\rfloor} \binom{x}{n}$ (Johnson [32]).

2. The following polynomials are the first terms of a basis of $\text{Int}^{\{1\}}(\mathbb{P}, \mathbb{Z})$:

$$1, \ X - 1, \ (X - 1)(X - 2), \ \frac{1}{2}(X - 1)(X - 2)(X - 3),$$

$$\frac{1}{4}(X-1)(X-2)(X-3)(X-5), \ \frac{1}{48}(X-1)(X-2)(X-3)(X-5)(X-7), \dots [24].$$

Let $\alpha \in \mathbb{R}^+$. The Banach space $LA_\alpha(\hat{S}, \hat{K})$ of *locally analytic functions of order α* from $\hat{S}$ to $\hat{K}$ is formed by the function $f : \hat{S} \to \hat{K}$ such that, for each $s \in S$, the restriction of f to $S \cap B(s, \alpha)$ is extendable to an analytic function on the whole ball $B(s, \alpha)$. Note that if $a \in V$ is such that $v(a) = \alpha$, the polynomials of $K[X]$ which are in $LA_\alpha(\hat{S}, \hat{V})$ are the elements of $\text{Int}_a(S, V)$.

Proposition 41 ([12, Theorem 28]). *Assume that the precompact subset S has no isolated points and let $a \in V$ and $\alpha = v(a)$. Then, every basis of the V-module $\mathrm{Int}_a(S, V)$ is an orthonormal basis of the Banach space $LA_\alpha(\hat{S}, \hat{K})$.*

9 Generalizations

Let again D be a Dedekind domain with quotient field K.

9.1 Integer-Valued Polynomials in Several Indeterminates

Let m be a positive integer, let $\underline{S}$ be a subset of D^m, and consider the D-algebra

$$\mathrm{Int}(\underline{S}, D) = \{ f(X_1, \dots, X_m) \in K[X_1, \dots, X_m] \mid f(\underline{S}) \subseteq D \}. \tag{61}$$

Most of the results about $\mathrm{Int}(\underline{S}, D)$ which are gathered in [17, Chap. XI] concern subsets $\underline{S}$ of the form $S = \prod_{i=1}^m S_i$. In 2000, Bhargava [11, Sect. 12] suggested some ways to define generalized factorials for all subsets $\underline{S}$ and, only recently, several interesting results were published (Evrard [27]).

Following [27], let $\mathfrak{I}_n(\underline{S}, D)$ be the D-module generated by *all* the coefficients of the polynomials of total degree n in $\mathrm{Int}(\underline{S}, D)$ and let

$$n!_{\underline{S}}^D = \mathfrak{I}_n(\underline{S}, D)^{-1} = \{ x \in D \mid xf \in D[X_1, \dots, X_m] \ \forall f \in \mathrm{Int}(\underline{S}, D), \deg(f) \leq n \}.$$

One may compute these factorials by means of a generalized notion of v-ordering. For this, we must first assume that no nonzero polynomial $f \in K[X_1, \dots, X_m]$ is such that $f(\underline{S}) = 0$ (the analog of $\mathrm{Card}(S) = \infty$ for one variable).

Then, write all the monomials in a sequence $(m_l)_{l \geq 0}$ in a way compatible with the total degree, that is, such that $\deg(m_l) < \deg(m_{l'}) \Rightarrow l < l'$.

Finally, for $l \geq 1$ and any sequence $(\underline{x}_0, \dots, \underline{x}_{l-1})$ of elements of D^m, let

$$\Delta(\underline{x}_0, \dots, \underline{x}_{l-1}) = \det \left(m_j(\underline{x}_i) \right)_{0 \leq i, j < l}. \tag{62}$$

Definition 13. A v-ordering of $\underline{S}$ is a sequence $\{\underline{a}_k\}_{k \geq 0}$ of elements of $\underline{S}$ such that for every $k \geq 1 : v(\Delta(\underline{a}_0, \dots, \underline{a}_k)) = \inf_{\underline{x} \in S} v\left(\Delta(\underline{a}_0, \dots, \underline{a}_{k-1}, \underline{x})\right)$.

Proposition 42 ([27]). *Let $\{\underline{a}_k\}_{k \geq 0}$ be a sequence of elements of $\underline{S}$ such that, for every $k \geq 0$, $\Delta(\underline{a}_0, \dots, \underline{a}_k) \neq 0$ and consider the associated sequence of polynomials:*

$$F_k(\underline{X}) = \frac{\Delta(\underline{a}_0, \dots, \underline{a}_{k-1}, \underline{X})}{\Delta(\underline{a}_0, \dots, \underline{a}_k)}.$$

Then the following assertions are equivalent:

1. $\{\underline{a_k}\}_{k \geq 0}$ *is a v-ordering of* $\underline{S}$.
2. *For every* $k \geq 0$, $F_k \in Int(\underline{S}, D)$.
3. $\{F_k(\underline{X})\}_{k \geq 0}$ *is a basis of the V-module* $Int(\underline{S}, V)$.
4. *For every* $f(\underline{X}) \in K[\underline{X}]$, *if the indices of the monomials of* f *are* $< k$, *then*

$$f \in Int(\underline{S}, V) \Leftrightarrow f(\underline{a_0}), \ldots, f(\underline{a_{k-1}}) \in V.$$

Then, we can compute the factorials of $\underline{S}$ by globalization. Properties A, B, C, and D still hold for these factorials [27]. We do not know whether Property E is still true and if we have a generalized Property F (Kempner's formula).

9.2 Other Generalizations

9.2.1 Homogeneous Integer-Valued Polynomials

Johnson and Patterson [34] introduced a notion of *projective v-ordering* to construct bases of homogeneous polynomials in (only) two variables. For instance, they considered the $\mathbb{Z}_2$-module:

$$\{f \in \mathbb{Q}_2[X, Y] \mid f \text{ homogeneous, } \deg(f) = 3, f(\mathbb{Z}_2 \times \mathbb{Z}_2) \subseteq \mathbb{Z}_2\}$$

and obtained the following basis:

$$Y^3, XY^2, X^2(X - Y), XY(X - Y)/2.$$

9.2.2 Integer-Valued Polynomials on Noncommutative Algebras

There are several works about algebras of *integer-valued polynomials on quaternions* (Werner [44], Johnson and Pavlovski [35]) but only partial results about the additive structure.

There are also several works about algebras of *integer-valued polynomials on matrices* (Frisch [30], Werner [45]). Let us recall the only case where we know a basis, that is, the case of integer-valued polynomials on triangular matrices.

Let $\mathcal{M}_n(D)$ denote the ring of $n \times n$ matrices with coefficients in D and, for every subset S of $\mathcal{M}_n(D)$, let

$$Int(S, \mathcal{M}_n(D)) = \{f(X) \in K[X] \mid f(S) \subseteq \mathcal{M}_n(D)\}. \tag{63}$$

Denoting by $T_n(D)$ the subring of $\mathcal{M}_n(D)$ formed by triangular matrices, Evrard, Fares, and Johnson [28] obtained the equality:

$$\mathrm{Int}(T_n(D), \mathcal{M}_n(D)) = \{f \in K[X] \mid f(T_n(D) \subseteq \mathcal{M}_n(D)\} = \mathrm{Int}^{\{n-1\}}(D). \quad (64)$$

It seems that today there is no other published construction of bases of such integer-valued polynomials on noncommutative algebras. . ..

Acknowledgements The author thanks the anonymous referee for many valuable suggestions.

References

1. D. Adam, Simultaneous orderings in function fields. J. Number Theory **112**, 287–297 (2005)
2. D. Adam, Finite differences in finite characteristic. J. Algebra **296**, 285–300 (2006)
3. D. Adam, Pólya and Newtonian function fields. Manuscripta Math. **126**, 231–246 (2008)
4. D. Adam, Polynômes à valeurs entières ainsi que leurs dérivées en caractéristique p. Acta Arith. **148**, 351–365 (2011)
5. D. Adam, P.-J. Cahen, Newtonian and Schinzel quadratic fields. J. Pure Appl. Algebra **215**, 1902–1918 (2011)
6. D. Adam, Y. Fares, Integer-valued Euler-Jackson's finite differences. Monatsh. Math. **161**, 15–32 (2010)
7. D. Adam, J.-L. Chabert, Y. Fares, Subsets of $\mathbb{Z}$ with simultaneous orderings. Integers **10**, 437–451 (2010)
8. Y. Amice, Interpolation p-adique. Bull. Soc. Math. France **92**, 117–180 (1964)
9. M. Bhargava, P-orderings and polynomial functions on arbitrary subsets of Dedekind rings. J. Reine Angew. Math. **490**, 101–127 (1997)
10. M. Bhargava, Generalized factorials and fixed divisors over subsets of a Dedekind domain. J. Number Theory **72**, 67–75 (1998)
11. M. Bhargava, The factorial function and generalizations. Am Math. Monthly **107**, 783–799 (2000)
12. M. Bhargava, On p-orderings, integer-valued polynomials, and ultrametric analysis. J. Am. Math. Soc. **22**, 963–993 (2009)
13. M. Bhagava, K. Kedlaya, Continuous functions on compact subsets of local fields. Acta Arith. **91**, 191–198 (1999)
14. M. Bhargava, P.-J. Cahen, J. Yeramian, Finite generation properties for various rings of integer-valued polynomials. J. Algebra **322**, 1129–1150 (2009)
15. J. Boulanger, J.-L. Chabert, Asymptotic behavior of characteristic sequences of integer-valued polynomials. J. Number Theory **80**, 238–259 (2000)
16. J. Boulanger, J.-L. Chabert, S. Evrard, G. Gerboud, The characteristic sequence of integer-valued polynomials on a subset. In *Advances in Commutative Ring Theory*. Lecture Notes in Pure and Applied Mathematics, vol. 205 (Dekker, New York, 1999), pp. 161–174
17. P.-J. Cahen, J.-L. Chabert, *Integer-Valued Polynomials*. American Mathematical Society Surveys and Monographs, vol. 48 (American Mathematical Society, Providence, 1997)
18. P.-J. Cahen, J.-L. Chabert, On the ultrametric Stone-Weierstrass theorem and Mahler's expansion. J. Théor. Nombres Bordeaux **14**, 43–57 (2002)
19. P.-J. Cahen, J.-L. Chabert, K.A. Loper, High dimension Prüfer domains of integer-valued polynomials. J. Korean Math. Soc. **38**, 915–935 (2001)
20. L. Carlitz, A class of polynomials. Trans. Am. Math. Soc. **43**, 167–182 (1938)
21. J.-L. Chabert, Generalized factorial ideals. Arab. J. Sci. Eng. Sect. C **26**, 51–68 (2001)
22. J.-L. Chabert, Factorial groups and Pólya groups in Galoisian extensions of $\mathbb{Q}$. In *Commutative Ring Theory and Applications*. Lecture Notes in Pure and Applied Mathematics, vol. **231** (Marcel Dekker, New York, 2003), pp. 77–86

23. J.-L. Chabert, Integer-valued polynomials in valued fields with an application to discrete dynamical systems. In *Commutative Algebra and Applications* (de Gruyter, Berlin, 2009), pp. 103–134
24. J.-L. Chabert, About polynomials whose divided differences are integer valued on prime numbers. Comm. Algebra, (to appear)
25. J.-L. Chabert, A.-H. Fan, Y. Fares, Minimal dynamical systems on a discrete valuation domain. Discrete Contin. Dyn. Syst. **35**, 777–795 (2009)
26. J.-L. Chabert, S. Evrard, Y. Fares, Regular subsets of valued fields and Bhargava's v-orderings. Math. Zeitschrift **274**, 263–290 (2013)
27. S. Evrard, Bhargava's factorials in several variables. J. Algebra **372**, 134–148 (2012)
28. S. Evrard, Y. Fares, K. Johnson, Integer valued polynomials on lower triangular integer matrices. Monatsh. Math. **170**, 147–160 (2013)
29. S. Frisch, Polynomial functions on finite commutative rings. In *Advances in Commutative Ring Theory*. Lecture Notes in Pure and Applied Mathematics, vol. 205 (Dekker, New York, 1999), pp. 323–336
30. S. Frisch, Integer-valued polynomials on algebras. J. Algebra **373**, 414–425 (2013)
31. K. Johnson, Limits of characteristic sequences of integer-valued polynomials on homogeneous sets. J. Number Theory **129**, 2933–2942 (2009)
32. K. Johnson, Computing r-removed P-orderings and P-orderings of order h. Actes des rencontres du CIRM **2**(2), 147–160 (2010)
33. K. Johnson, Super-additive sequences and algebras of polynomials. Proc. Am. Math. Soc. **139**, 3431–3443 (2011)
34. K. Johnson, D. Patterson, Projective p-orderings and homogeneous integer-valued polynomials. Integers **11**, 597–604 (2011)
35. K. Johnson, M. Pavlovski, Integer-valued polynomials on the Hurwitz ring of integral quaternions. Comm. Algebra **40**, 4171–4176 (2012)
36. A.J. Kempner, Polynomials and their residue systems. Trans. Am. Math. Soc. **22**, 240–288 (1921)
37. A. Leriche, Cubic, quatric and sextic Pólya fields. J. Number Theory **133**, 59–71 (2013)
38. A. Leriche, About the embedding of a number field in a Pólya field. J. Number Theory, (to appear)
39. K. Mahler, An interpolation series for continuous functions of a p-adic variable. J. Reine Angew. Math. **199**, 23–34 (1958); **208**, 70–72 (1961)
40. A. Mingarelli, Abstract factorials. arXiv:00705.4299v3 [math.NT]. Accessed 10 Jul 2012
41. A. Ostrowski, Über ganzwertige Polynome in algebraischen Zahlkörpern. J. Reine Angew. Math. **149**, 117–124 (1919)
42. G. Pólya, Über ganzwertige ganze Funktionen. Rend. Circ. Mat. Palermo **40**, 1–16 (1915)
43. G. Pólya, Über ganzwertige Polynome in algebraischen Zahlkörpern. J. Reine. Angew. Math. **149**, 97–116 (1919)
44. N. Werner, Integer-valued polynomials over quaternions rings. J. Algebra **324**, 1754–1769 (2010)
45. N. Werner, Integer-valued polynomials over matrix rings. Comm. Algebra **40**, 4717–4726 (2012)
46. M. Wood, P-orderings: a metric viewpoint and the non-existence of simultaneous orderings. J. Number Theory **99**, 36–56 (2003)
47. H. Zantema, Integer valued polynomials over a number field. Manuscr. Math. **40**, 155–203 (1982)

On Boolean Subrings of Rings

Ivan Chajda and Günther Eigenthaler

Abstract We determine Boolean subrings of commutative unitary rings satisfying the identity $x^{p+k} = x^p$ for some integer $p \geq 1$ where $k = 2^s$ or $k = 2^s - 1$.

Keywords Boolean ring • Commutative unitary ring • Characteristic 2 • Subring

MS Classification: 06E20, 16R50, 16B70

It is well known (and easy to prove) that every finite ring (in fact, even every finite semigroup) satisfies an identity

$$x^{p+k} = x^p \text{ for some integers } p, k \geq 1. \qquad (*)$$

This motivates us to study not necessarily finite rings satisfying such an identity and to check which of them are Boolean. A ring is *Boolean* (see [1]) if it satisfies the identity $x^2 = x$. In fact every Boolean ring is commutative and of *characteristic* 2, i.e., it satisfies $x + x = 0$.

It was already proved in [2] that every unitary ring satisfying an identity $x^{p+1} = x^p$ is Boolean. In [3], results are given for rings satisfying an identity of the form $x^{1+k} = x$. The question is how it works in the general case $x^{p+k} = x^p$. In Example 5 we describe a ring satisfying $x^{p+2} = x^p$ which is not Boolean although it is unitary, commutative, and of characteristic 2.

I. Chajda
Department of Algebra and Geometry, Palacký University Olomouc, 17. listopadu 12, 77146 Olomouc, Czech Republic
e-mail: ivan.chajda@upol.cz

G. Eigenthaler (✉)
Institut für Diskrete Mathematik und Geometrie, Technische Universität Wien, Wiedner Hauptstraße 8-10/104, 1040 Wien, Austria
e-mail: guenther.eigenthaler@tuwien.ac.at

M. Fontana et al. (eds.), *Commutative Algebra: Recent Advances in Commutative Rings,*   113
Integer-Valued Polynomials, and Polynomial Functions, DOI 10.1007/978-1-4939-0925-4_6,
© Springer Science+Business Media New York 2014

The aim of this paper is to describe the largest Boolean subring for some particular values of k. The set of all positive integers is denoted by $\mathbb{N}$. For a ring $\mathcal{R}$ we put

$$B(\mathcal{R}) = \{x \in R \mid x^2 = x \text{ and } x + x = 0\}.$$

Since $x^2 = x$ implies $x^k = x$, we have

$$B(\mathcal{R}) \subseteq \{x^k \mid x \in R\} \text{ for all } k \in \mathbb{N}. \tag{**}$$

Note that for any ring $\mathcal{R} = (R; +, \cdot, 0)$ we have $0 \in B(\mathcal{R})$. In case that $\mathcal{R} = (R; +, \cdot, 0, 1)$ is a unitary ring, we have $1 \in B(\mathcal{R})$ if and only if char $\mathcal{R} = 2$.

Proposition 1. *Let $\mathcal{R}$ be a ring, then the following are equivalent:*

(i) $B(\mathcal{R})$ *is a subring of $\mathcal{R}$.*
(ii) $B(\mathcal{R})$ *is the largest Boolean subring of $\mathcal{R}$.*
(iii) *There exists a largest Boolean subring of $\mathcal{R}$.*
(iv) $xy = yx$ *for all $x, y \in B(\mathcal{R})$.*

Proof. (i) $\Rightarrow$ (ii): Clearly, $B(\mathcal{R})$ is a Boolean subring of $\mathcal{R}$, and any Boolean subring of $\mathcal{R}$ is contained in $B(\mathcal{R})$.

(ii) $\Rightarrow$ (iii): Obvious.

(iii) $\Rightarrow$ (iv): Suppose that there exist elements $x, y \in B(\mathcal{R})$ with $xy \neq yx$. $\{0, x\}$ and $\{0, y\}$ are Boolean subrings of $\mathcal{R}$. Since every Boolean ring is commutative, there does not exist a Boolean subring of $\mathcal{R}$ containing $\{0, x\}$ and $\{0, y\}$. Hence there does not exist a largest Boolean subring of $\mathcal{R}$.

(iv) $\Rightarrow$ (i): Note that $0 \in B(\mathcal{R})$. Suppose that $x, y \in B(\mathcal{R})$. Then we have $(xy)^2 = (xy)(xy) = x^2 y^2 = xy$ and $xy + xy = x(y + y) = x \cdot 0 = 0$; thus $xy \in B(\mathcal{R})$. Furthermore, $(x + y)^2 = x^2 + xy + xy + y^2 = x^2 + y^2 = x + y$ and $(x+y)+(x+y) = (x+x)+(y+y) = 0+0 = 0$; thus also $x+y \in B(\mathcal{R})$. Therefore, $B(\mathcal{R})$ is a subring of $\mathcal{R}$. $\square$

Corollary 1. *If $\mathcal{R}$ is a commutative ring, then $B(\mathcal{R})$ is the largest Boolean subring of $\mathcal{R}$.*

Example 1. Let $\mathcal{R}$ be an integral domain, then $B(\mathcal{R}) = \{0, 1\}$ if $\mathcal{R}$ is of characteristic 2, otherwise $B(\mathcal{R}) = \{0\}$.

Example 2. Let $\mathcal{R}$ be the factor ring $\mathbb{Z}/I$, where $\mathbb{Z}$ is the ring of integers and $I = (n)$ is the ideal generated by $n \in \mathbb{N}$. Then $B(\mathcal{R}) = \{I, m + I\}$ in the case that $n = 2m$ where m is odd, and $B(\mathcal{R}) = \{I\}$ in all other cases. The proof is an easy exercise.

Example 3. Let $\mathcal{R}$ be the ring of (2×2)-matrices over the 2-element field GF(2). Then $B(\mathcal{R}) = \left\{ \begin{pmatrix} 0 & 0 \\ 0 & 0 \end{pmatrix}, \begin{pmatrix} 0 & 0 \\ 0 & 1 \end{pmatrix}, \begin{pmatrix} 1 & 0 \\ 0 & 0 \end{pmatrix}, \begin{pmatrix} 0 & 0 \\ 1 & 1 \end{pmatrix}, \begin{pmatrix} 0 & 1 \\ 0 & 1 \end{pmatrix}, \begin{pmatrix} 1 & 1 \\ 0 & 0 \end{pmatrix}, \right.$ $\left. \begin{pmatrix} 1 & 0 \\ 1 & 0 \end{pmatrix}, \begin{pmatrix} 1 & 0 \\ 0 & 1 \end{pmatrix} \right\}$ which is not a subring of $\mathcal{R}$.

Example 4. Let $\mathcal{R}$ be a direct product, say $\mathcal{R} = \prod_{\alpha \in A} \mathcal{R}_\alpha$. Then $B(\mathcal{R}) = \prod_{\alpha \in A} B(\mathcal{R}_\alpha)$.

Theorem 1. *Let* $\mathcal{R} = (R; +, \cdot, 0, 1)$ *be a unitary ring of characteristic 2 satisfying (*) with* $k = 2^s, s \in \mathbb{N}$. *Then* $B(\mathcal{R}) = \{x^k \mid x \in R\}$.

Proof. First recall that for $c = 2^n$ we have $(x + 1)^c = x^c + 1$ and every element of R is of the form $y = x + 1$ due to the fact that $y = (y + 1) + 1$. Moreover, if $\mathcal{R}$ satisfies (*) then it satisfies also $x^{q+k} = x^q$ for every integer $q \geq p$.

Choose $m \in \mathbb{N}$ such that $p \leq q = 2^m$. Then

$$(x + 1)^{q+k} = (x+1)^q \cdot (x + 1)^k = (x^q + 1)(x^k + 1) = x^{q+k} + x^q + x^k + 1$$
$$= x^q + x^q + x^k + 1 = x^k + 1 = (x + 1)^k,$$

thus $\mathcal{R}$ satisfies the identity

$$x^{q+k} = x^k$$

and, as $x^{q+k} = x^q$, also

$$x^q = x^k.$$

Therefore,

$$x^k \cdot x^k = x^q \cdot x^k = x^{q+k} = x^q = x^k.$$

Since char $\mathcal{R} = 2$, we have $\{x^k \mid x \in R\} \subseteq B(\mathcal{R})$ and (**) implies that $B(\mathcal{R}) = \{x^k \mid x \in R\}$. □

In case that $\mathcal{R}$ is commutative, using Corollary 1, we obtain

Corollary 2. *If* $\mathcal{R} = (R; +, \cdot, 0, 1)$ *is a commutative unitary ring of characteristic 2 satisfying (*) with* $k = 2^s, s \in \mathbb{N}$, *then* $\{x^k \mid x \in R\}$ *is the largest Boolean subring of* $\mathcal{R}$. *In particular, if* $\mathcal{R}$ *satisfies* $x^{p+2} = x^p$, *then* $\{x^2 \mid x \in R\}$ *is the largest Boolean subring of* $\mathcal{R}$.

Example 5. Let GF(2)$[X]$ be the polynomial ring in X over the field GF(2) and $I = (X^2)$ the ideal of GF(2)$[X]$ generated by the polynomial X^2. We consider the factor ring $\mathcal{R} = $ GF(2)$[X]/I$. Its elements are $0 + I = I, 1 + I, X + I$, and $(1 + X) + I$. It is easily seen that char $\mathcal{R} = 2$ and $\mathcal{R}$ is a commutative unitary ring satisfying $x^{p+2} = x^p$ for each integer $p \geq 2$. However, $\mathcal{R}$ is not Boolean. Its largest Boolean subring is $\{I, 1 + I\}$.

We turn to studying rings satisfying (*) with $k = 2^s - 1$, but first introduce a more general result which will be useful in the case where $k = 2^s - 1$.

Lemma 1. *Let* $\mathcal{R} = (R; +, \cdot, 0, 1)$ *be a unitary ring satisfying (*) with an odd* $k \in \mathbb{N}$. *Then* $\operatorname{char} \mathcal{R} = 2$.

Proof. Since $\mathcal{R}$ is unitary, it contains 1 and hence also -1. Since k is odd, we have $(-1)^k = -1$. The identity $x^{p+k} = x^p$ yields $x^p(x^k - 1) = 0$. For $x = -1$ we get $(-1)^p(-1 - 1) = 0$ whence $1 + 1 = 0$. Thus $\operatorname{char} \mathcal{R} = 2$. $\square$

Theorem 2. *Let* $\mathcal{R} = (R; +, \cdot, 0, 1)$ *be a unitary ring satisfying (*) with* $k = 2^s - 1, s \in \mathbb{N}$. *Then* $\mathcal{R}$ *satisfies the identity* $x^{1+k} = x$. *In particular,* $\mathcal{R}$ *is Boolean in case that* $k = 1$. *(Cf. [2], Theorem 4.)*

Proof. Since $k = 2^s - 1$ is odd, we have $\operatorname{char} \mathcal{R} = 2$. Pick $m \in \mathbb{N}$ such that $p \leq q = 2^m + 1$, and let us note that (as in the proof of Theorem 1) $x^{p+k} = x^p$ implies $x^{q+k} = x^q$. Then $q - 1 = 2^m, k + 1 = 2^s$ and hence

$$
\begin{aligned}
(x + 1)^q &= (x + 1)^{q+k} = (x + 1)^{q-1} \cdot (x + 1)^{k+1} = (x^{q-1} + 1)(x^{k+1} + 1) \\
&= x^{(q-1)+(k+1)} + x^{q-1} + x^{k+1} + 1 = x^{q+k} + x^{q-1} + x^{k+1} + 1 \\
&= x^{q-1}(x + 1) + (x + 1)^{k+1}.
\end{aligned}
$$

For $y = x + 1$ we obtain

$$
y^q = (y + 1)^{q-1} \cdot y + y^{k+1} = (y^{q-1} + 1) \cdot y + y^{k+1} = y^q + y + y^{k+1}
$$

and hence $0 = y + y^{k+1}$. This yields $y^{k+1} = y$ for each $y \in R$. $\square$

Since $x^{1+k} = x$ evidently implies $x^{p+k} = x^p$ for all $p \in \mathbb{N}$, we have the following:

Corollary 3. *Let* $\mathcal{R} = (R; +, \cdot, 0, 1)$ *be a unitary ring,* $p \in \mathbb{N}$, *and* $k = 2^s - 1$, $s \in \mathbb{N}$. *Then* $\mathcal{R}$ *satisfies*

$$
x^{1+k} = x
$$

if and only if it satisfies

$$
x^{p+k} = x^p.
$$

The following result was proved by Jedlička [4]:

Lemma 2. *Let* $k \in \mathbb{N}$. *There exists a non-Boolean unitary ring of characteristic 2 satisfying* $x^{1+k} = x$ *if and only if* $k = l \cdot (2^s - 1)$ *for some* $l \in \mathbb{N}$ *and some integer* $s \geq 2$.

The rings of Theorem 2 satisfy $x^{1+k} = x$ for $k = 2^s - 1$, thus, by Lemma 2, we cannot expect those rings to be Boolean if $s \geq 2$. However, we can describe—even for arbitrary odd k—their largest Boolean subrings similarly as in Theorem 1.

Theorem 3. *Let* $\mathcal{R} = (R; +, \cdot, 0)$ *be a unitary ring satisfying* $x^{1+k} = x$ *for some odd* $k \in \mathbb{N}$, *then* $B(\mathcal{R}) = \{x^k \mid x \in R\}$.

Proof. We have $x^k \cdot x^k = x^{k+k} = x^{k+1} \cdot x^{k-1} = x \cdot x^{k-1} = x^k$ for each $x \in R$. By Lemma 1, char $\mathcal{R} = 2$ and therefore $\{x^k \mid x \in R\} \subseteq B(\mathcal{R})$, thus (**) implies that $B(\mathcal{R}) = \{x^k \mid x \in R\}$. $\qquad\square$

Using Corollary 1 and Corollary 3 we obtain

Corollary 4. *If* $\mathcal{R} = (R; +, \cdot, 0, 1)$ *is a commutative unitary ring satisfying* (*) *with* $k = 2^s - 1, s \in \mathbb{N}$, *then* $\{x^k \mid x \in R\}$ *is the largest Boolean subring of* $\mathcal{R}$. *In particular, if* $\mathcal{R}$ *satisfies* $x^{p+3} = x^p$, *then* $\{x^3 \mid x \in R\}$ *is the largest Boolean subring of* $\mathcal{R}$.

Example 6. The finite field $GF(2^s)$ satisfies (*) with $k = 2^s - 1$ and any $p \in \mathbb{N}$, and $B(GF(2^s)) = \{0, 1\}$. Using Example 4, we have $B((GF(2^s))^m) = \{0, 1\}^m$ for all $m \in \mathbb{N}$.

Remark. As Martin Goldstern pointed out, the following generalization of Theorem 3 holds:

Let $\mathcal{R}$ *be a ring of characteristic 2 satisfying* (*), *and let* r *be any multiple of* k *with* $r \geq p$. *Then* $B(\mathcal{R}) = \{x^r \mid x \in R\}$.

Proof. By (*), $x^{a+p+lk} = x^{a+p}$ for all integers $a \geq 0, l \geq 1$. Let $a = r - p, l = \frac{r}{k}$, then $x^r \cdot x^r = x^{r+r} = x^r$. Hence $x^r \in B(\mathcal{R})$, and (**) implies that

$$B(\mathcal{R}) = \{x^r \mid x \in R\}.$$

$\qquad\square$

Acknowledgements This work is supported by ÖAD, Cooperation between Austria and Czech Republic in Science and Technology, Grant Number CZ 03/2013, and by the Project CZ1.07/2.3.00/20.0051 Algebraic Methods in Quantum Logics.

References

1. G. Birkhoff, *Lattice Theory*, 3rd edn. Colloquium Publications, vol. 25 (American Mathematical Society, Providence, 1967)
2. I. Chajda, F. Švrček, Lattice-like structures derived from rings. In *Contributions to General Algebra*, vol. 20 (Verlag Johannes Heyn, Klagenfurt, 2012), pp. 11–18
3. I. Chajda, F. Švrček, The rings which are Boolean. Discuss. Mathem. General Algebra Appl. **31**, 175–184 (2011)
4. P. Jedlička, The rings which are Boolean II. Acta Univ. Carolinae (to appear)

On a New Class of Integral Domains
with the Portable Property

David E. Dobbs, Gabriel Picavet, and Martine Picavet-L'Hermitte

Abstract A (commutative integral) domain R is said to be a pseudo-almost divided domain if for all $P \in \mathrm{Spec}(R)$ and $u \in PR_P$, there exists a positive integer n such that $u^n \in P$. Such domains are related to several known kinds of domains, such as divided domains and straight domains. It is shown that "locally pseudo-almost divided" is a portable property of domains. Hence, if T is a domain with a maximal ideal Q and D is a subring of T/Q, then the pullback $R := T \times_{T/Q} D$ is locally pseudo-almost divided if and only if both T and D are locally pseudo-almost divided. A similar pullback transfer result is given for the "straight domain" property (which is not known to be portable) by imposing additional restrictions on the data T, Q, D.

Keywords Integral domain • Pullback • Portable property • Straight domain • Pseudo-almost divided domain • PAVD • APVD • Almost Prüfer domain • Divided domain • Root closed

Subject Classifications:[2010] Primary: 13G05; Secondary: 13A15, 13F05, 13B21

D.E. Dobbs
Department of Mathematics, University of Tennessee, Knoxville, TN 37996-1320, USA
e-mail: ddobbs1@utk.edu

G. Picavet (✉) • M. Picavet-L'Hermitte
Laboratoire de Mathématiques, Université Blaise Pascal, UMR6620 CNRS, Les Cézeaux, 24 avenue des Landais, BP 80026, 63177 Aubière Cedex, France
e-mail: Gabriel.Picavet@math.univ-bpclermont.fr; picavet.gm@wanadoo.fr;
Martine.Picavet@math.univ-bpclermont.fr

M. Fontana et al. (eds.), *Commutative Algebra: Recent Advances in Commutative Rings,*
Integer-Valued Polynomials, and Polynomial Functions, DOI 10.1007/978-1-4939-0925-4_7,
© Springer Science+Business Media New York 2014

1 Introduction and Notation

We assume throughout that R denotes a (commutative integral) domain with quotient field K and integral closure R' (in K). As usual, if D is a domain, then $\mathrm{Spec}(D)$ denotes the set of prime ideals of D; $\mathrm{Max}(D)$ the set of maximal ideals of D; and $\sqrt{I}$ the radical of an ideal I of D.

One purpose of the present work is to contribute to the list of the classes of domains with the portable property. If $\mathfrak{P}$ is a property of (some) domains, then as in [16] (cf. also [19]), we say that $\mathfrak{P}$ is a *portable property* (*of domains*) if the following three conditions hold:

- (•) If $A \subset B$ are domains such that $\mathrm{Spec}(A) = \mathrm{Spec}(B)$ (as sets), then A has $\mathfrak{P}$ if and only if B has $\mathfrak{P}$.
- (••) If A is a domain and $P \in \mathrm{Spec}(A)$, then $A + PA_P$ has $\mathfrak{P}$ if and only if both A_P and A/P have $\mathfrak{P}$.
- (• • •) $\mathfrak{P}$ is a local property (i.e., a domain A has $\mathfrak{P}$ if and only if A_M has $\mathfrak{P}$ for all $M \in \mathrm{Max}(A)$).

The main known benefit of portability is the following pullback transfer result. Let $\mathfrak{P}$ be a portable property, T a domain, $Q \in \mathrm{Max}(T)$, $\pi : T \to T/Q$ the canonical surjection, D a subring of T/Q, and S the pullback $S := \pi^{-1}(D) = T \times_{T/Q} D$; then S has $\mathfrak{P}$ if and only if both T and D have $\mathfrak{P}$ [16, Theorem 2.4]. In fact, if $\mathfrak{P}$ is only known to satisfy parts (•) and (••) in the definition of a portable property, then "locally $\mathfrak{P}$" is a portable property and so the above conclusion holds with "locally $\mathfrak{P}$" replacing "$\mathfrak{P}$" [19, Corollary 2]. (Note that in this paper, if $\mathfrak{P}$ is a property of some domains, then a domain A is said to satisfy locally $\mathfrak{P}$ if A_M satisfies $\mathfrak{P}$ for each $M \in \mathrm{Max}(A)$.)

In [16, Lemma 2.1], it was shown that "locally divided domain" and "going-down domain" (as defined in [9, 10]) are examples of portable properties. Subsequently, it was shown in [19, Theorem 2] that "locally almost-divided domain" and (for each n) "locally n-divided domain" are also examples of portable properties. For motivation, it may be helpful to recall the following implications: divided domain $\Leftrightarrow$ 1-divided domain $\Rightarrow$ n-divided domain $\Rightarrow$ almost-divided domain $\Rightarrow$ quasi-divided domain $\Rightarrow$ quasi-local going-down domain (cf. [18, Proposition 5.2]).

In Sect. 3, we introduce and investigate a property of domains called "pseudo-almost divided." In Sect. 4, we show that this property satisfies enough of the definition of a portable property that it supports the above kind of pullback transfer result in case (T, Q) is quasi-local: see Corollary 6. In addition, we show that "locally pseudo-almost divided" is a portable property and, hence, supports exactly the above kind of pullback transfer result: see Corollary 7. Let us note here that all quasi-divided domains and all PAVDs (as defined by Badawi [6]) are examples of pseudo-almost divided domains.

Another purpose of this paper is to show that some known classes of domains support pullback transfer results with some of the above flavor, even though these classes are not known to have (or, indeed, may not have) the portable property.

We do this, in particular, for the class of straight domains (which was defined in [17] and studied further in [18]): see Theorem 1. In the same vein, Theorem 2 and Corollary 5 involve almost Prüfer domains (or AP-domains, in the sense of [2]) and locally PAVDs. It may be helpful to note that AV-domain (in the sense of [2]) $\Rightarrow$ PAVD $\Rightarrow$ pseudo-almost divided domain and APVD (as defined by Badawi and Houston in [8]) $\Rightarrow$ divided domain and PAVD. The abovementioned kinds of domains are relevant to the above hierarchy in part because locally divided domain $\Rightarrow$ straight domain $\Rightarrow$ going-down domain [17, Corollary 3.8 and p. 764] and because all PAVDs are quasi-local going-down domains [15, Theorem 1].

The definitions and relevant background for all the various kinds of domains being considered will be given below as needed.

2 Some Straight Domains Obtained as Pullbacks Issuing from Prüfer Domains

It is convenient to begin by recalling the following material from [2]. A domain R is called an *almost valuation domain* (or AV-*domain*) if, for all nonzero elements a, b of R, there exists a positive integer n such that either $a^n \mid b^n$ or $b^n \mid a^n$; R is called an *almost Prüfer domain* (or AP-*domain*) if, for all nonzero elements a, b of R, there exists a positive integer n such that (a^n, b^n) is an invertible ideal. Each AV-domain is a quasi-local AP-domain. Moreover, R is an AP-domain if and only if R_M is an AV-domain for each $M \in \text{Max}(R)$ [2, Theorem 5.8]. Now, for motivation, we can state the following result of Mimouni [25, Theorem 2.2]. Let T be a domain with $Q \in \text{Max}(T)$, let $\varphi : T \to T/Q$ be the canonical surjection, let D be a domain with quotient field k subfield of T/Q and consider the pullback $S := \varphi^{-1}(D) = T \times_{T/Q} D$; then S is an AV-domain (resp., an AP-domain) if and only if T and D are each AV-domains (resp., each AP-domains) and $k \subseteq T/Q$ is a root extension. It seems natural to ask if some other properties (besides "AV-domain" and "AP-domain") admit similar pullback results, possibly without a need for the above "root extension" condition. We will give such a result for the "straight domain" property in Theorem 1. First, we recall some material from [17] (specialized from rings to domains to fit our current context).

If $D \subseteq E$ are domains, then D is said to be *straight in* E if the inclusion map $D \hookrightarrow S$ is a prime morphism for each ring S such that $D \subseteq S \subseteq E$ (in the sense that S/PS is torsion-free over D/P for each $P \in \text{Spec}(D)$). The ambient domain R is called a *straight domain* if R is straight in K. In [17, Corollary 4.13], we gave a pullback result for the "straight domain" property which was somewhat in the spirit of the abovementioned result of Mimouni but which lacked generality, in part because its pullback issued from a valuation domain which is not a field. Theorem 1 will improve this situation by generalizing [17, Corollary 4.13] to pullbacks that issue from a Prüfer domain which is not a field.

Theorem 1. *Let T be a Prüfer domain with Q a nonzero maximal ideal of T, let $\varphi : T \to T/Q$ be the canonical surjection, let D be a subring of T/Q, and consider the pullback $S := \varphi^{-1}(D) = T \times_{T/Q} D$. Then S is a straight domain if and only if D is straight in T/Q. In particular, if D is a field, then S is a straight domain.*

Proof. We first explain why the "In particular" assertion is a consequence of the main assertion. Suppose that D is a field. Then D is straight in T/Q, for if A is a ring such that $D \subseteq A \subseteq T/Q$, then $D \hookrightarrow A$ is a prime morphism since A is torsion-free over (the field) D (or, alternately, by [24, Proposition 2] since A is D-flat.)

We turn now to the main assertion. Note that [17, Corollary 4.13] handles the case where T is quasi-local (i.e., where T is a valuation domain which is not a field). For the general case, note that the proof of [17, Proposition 3.5(b)] shows that "straight domain" satisfies part ($\bullet \bullet \bullet$) in the definition of a portable property. In particular, S is a straight domain if and only if S_N is a straight domain for each $N \in \mathrm{Max}(S)$. For the moment, fix such an ideal N. If $N \not\supseteq Q$, then there exists a unique prime ideal $\mathfrak{N}$ of T such that $\mathfrak{N} \cap S = N$ and $S_N = T_{\mathfrak{N}}$ (cf. [22, Theorem 1.4(c)]). In this case, S_N is a valuation domain and hence, by [17, p. 764], a straight domain. Therefore, S is a straight domain if and only if S_N is a straight domain for each maximal ideal N of S such that $N \supseteq Q$. Now, fix *this* kind of N. Identifying D with S/Q, we can consider $\mathcal{N} := N/Q \in \mathrm{Max}(D)$. Then, by reasoning as in the penultimate paragraph of the proof of [16, Theorem 2.4], we get that $S_N = T_Q \times_{T/Q} D_{\mathcal{N}}$. As we already know the result for a pullback issuing from a valuation domain which is not a field, we can conclude that S_N is a straight domain (for the specific N that has been fixed) if and only if $D_{\mathcal{N}}$ is straight in T/Q. Hence, S is a straight domain if and only if $D_{N/Q}$ is straight in T/Q for all maximal ideals N of S such that $N \supseteq Q$. However, the latter condition is, as an easy consequence of [17, Proposition 2.3], equivalent to D being straight in T/Q. The proof is complete. $\square$

Perhaps the analogy between Theorem 1 and [25, Theorem 2.2] becomes clearer by noting the following special case of Theorem 1.

Corollary 1. *Let T be a Prüfer domain with Q a nonzero maximal ideal of T, let $\varphi : T \to T/Q$ be the canonical surjection, let D be a domain with quotient field T/Q, and consider the pullback $S := \varphi^{-1}(D) = T \times_{T/Q} D$. Then S is a straight domain if and only if D is a straight domain.*

We observed in the proof of Theorem 1 that "straight domain" satisfies part ($\bullet \bullet \bullet$) in the definition of a portable property. The next result pursues this theme.

Proposition 1. *Let $A \subset B$ be domains such that $\mathrm{Spec}(A) = \mathrm{Spec}(B)$ (as sets). Then A is a straight domain if and only if B is a straight domain. In other words, "straight domain" satisfies part ($\bullet$) in the definition of a portable property of domains.*

Proof. Since any field is trivially a straight domain, we may assume, without loss of generality, that neither A nor B is a field. Hence, by [3, Propositions 3.3 and 3.5], A

and B are quasi-local domains that have the same maximal ideal (say, M) and also have the same quotient field. Suppose first that A is straight. We must show that if D is an overring of B and $P \in \text{Spec}(B)$, then D/PD is torsion-free over B/P. Without loss of generality, $P \neq M$, since all modules over fields are torsion-free. Pick $c \in M \setminus P$. If the assertion fails, we have $b \in B \setminus P$ and $d \in D \setminus PD$ such that $bd \in PD$. But then $bcd \in PD$ and $bc \in M \setminus P \subseteq A \setminus P$, a contradiction since D/PD is torsion-free over A/P.

For the converse, suppose that B is straight. We must show that if E is an overring of A and $Q \in \text{Spec}(A)$, then E/QE is torsion-free over A/Q. Suppose the assertion fails, so that $\alpha\delta \in QE$ for some $\alpha \in A \setminus Q$ and $\delta \in E \setminus QE$. But since BE is an overring of B and BE/QE ($= BE/QBE$) is torsion-free over B/Q, we obtain the desired contradiction because $\alpha \in B \setminus Q$ and $\delta \in BE$. $\quad\square$

In view of the above material, one cannot resist asking if "straight domain" is a portable property. The answer is affirmative if and only if "straight domain" satisfies part ($\bullet\bullet$) in the definition of a portable property. An affirmative answer may seem desirable because, in tandem with [16, Theorem 2.4], it would lead to a generalization of Theorem 1 (where the "Prüfer" hypothesis could be deleted). Some evidence for an affirmative answer is available, as the class of straight domains is closed under localization and homomorphic images [17, Proposition 3.5(b), Theorem 4.6]. However, we do not know the answer, as it is an open question whether R_P and R/P being straight domains implies that $R + PR_P$ is a straight domain. In view of [14, 16], we note that a negative answer may shed light on a question that was left open in [17, Corollary 3.8, Example 4.5]: does there exist a straight domain which is not locally divided?

3 Pseudo-Almost Divided Domains

We begin with some background on some relevant classes of domains. As in [6], we say that (the ambient domain) R is a *pseudo-almost valuation domain* (in short, a PAVD) if each $P \in \text{Spec}(R)$ is a pseudo-strongly prime ideal of R; that is, whenever $u, v \in K$ with $uvP \subseteq P$, then there exists a positive integer n such that either $u^n \in R$ or $v^n P \subseteq P$. (By [6, Theorem 2.8], it is equivalent to require that for all $u \in K$, either there exists a positive integer n such that $u^n \in R$ or there exists a positive integer m such that $vu^{-m} \in R$ for each nonunit v of R.) It is easy to see that each almost valuation domain (in the sense of [2, Definition 5.5]) is a PAVD. Note that any such domain is quasi-local and treed. In that regard, recall from [8] that R is an *almost-pseudo-valuation domain* (in short, an APVD) if each $P \in \text{Spec}(R)$ is a strongly primary ideal of R; that is, $u, v \in K$ with $uv \in P$ implies that either $u^n \in P$ for some positive integer n or $v \in P$. Note that each PVD (i.e., each pseudo-valuation domain, in the sense of [23]) is an APVD and each APVD is a PAVD. Note that any such domain is quasi-local and treed. In fact, recall from [8,

Proposition 3.2] that each APVD is a divided domain. Since the APVDs form a natural family of PAVDs which are divided domains, this leads us to the following definition.

We say that a prime ideal P of the ambient domain R is *pseudo-almost divided* (*in R*) if for each $u \in PR_P$, there exists a positive integer n such that $u^n \in P$ (equivalently, $u^n \in R$); also R is said to be (*a*) *pseudo-almost divided* (*domain*) if each prime ideal of R is pseudo-almost divided in R. Examples of pseudo-almost divided domains are easily at hand. For instance, by [6, Proposition 2.3], each PAVD is pseudo-almost divided. Also, it is also easy to check that all almost-divided domains and all quasi-divided domains (in the sense of [18, Definition 5.1]) are pseudo-almost divided domains.

Next, recall from [11, Proposition 2.1] that quasi-local going-down domains are characterized as the domains R for which $R + PR_P$ is integral over R for each $P \in \mathrm{Spec}(R)$ (cf. also [10, Lemma 2.4(a)]). We will see that this property also holds if R is a PAVD.

Proposition 2. *The following three conditions are equivalent for a domain R:*

(1) R is a quasi-local going-down domain.
(2) $PR_P \subseteq \sqrt{PR'}$ for each $P \in \mathrm{Spec}(R)$.
(3) $R \subseteq R + PR_P$ is an integral extension for each $P \in \mathrm{Spec}(R)$.

Moreover, if any of the above equivalent conditions holds and $P \in \mathrm{Spec}(R)$, then $PR_P \subseteq \cap\{Q \in \mathrm{Spec}(R') \mid Q \cap R = P\}$.

Proof. (1) $\Leftrightarrow$ (3) by [11, Proposition 2.1], and it is clear that (2) $\Rightarrow$ (3) since integrality is transitive. It therefore suffices to show that (1) $\Rightarrow$ (2). Assume (1) and let $P \in \mathrm{Spec}(R)$. Our task is to show that if $u \in PR_P$, then there exists a positive integer n such that $u^n \in PR'$. In fact, [10, Corollary 2.6] shows more, namely, that there exists a positive integer n such that $u^m \in P$ ($\subseteq PR'$) for all integers $m \geq n$. This reasoning also establishes the "Moreover" assertion, since any $Q \in \mathrm{Spec}(R')$ that lies over P must contain PR'. $\square$

For $P \in \mathrm{Spec}(R)$, we let P' denote one of the notions of the integral closure of P that can be found in the literature, as follows: $P' = \{x \in K \mid$ there exists $p(X) = a_0 + \cdots + a_{n-1}X^{n-1} + X^n \in R[X]$ with each $a_i \in P$ such that $p(x) = 0\}$. It is instructive to compare P' with the notion of the integral closure $\bar{P}$ of P with respect to an overring of R, as defined in [5, p. 63]; it follows immediately that by taking K as the overring in question, we get $\bar{P} := \{x \in K \mid p(x) = 0$ for some $p(X) \in P[X]\} = \sqrt{PR'}$ by [5, Lemma 5.14]. We can now observe, by Proposition 2, that if R is such that $PR_P \subseteq P'$ or $PR_P \subseteq \bar{P}$ for each $P \in \mathrm{Spec}(R)$, then R is a quasi-local going-down domain, since each prime ideal P satisfies $P' \subseteq \bar{P} \subseteq \sqrt{PR'}$.

The next result follows from (the implication (2) $\Rightarrow$ (1) in) Proposition 2 and the earlier comments.

Corollary 2. *A pseudo-almost divided domain is a quasi-local going-down domain. Hence all PAVDs, all almost-divided domains, and all quasi-divided domains are quasi-local going-down domains.*

Recall that a ring extension $A \subseteq B$ is called a *root extension* if for each $b \in B$, there exists a positive integer n such that $b^n \in A$.

Corollary 3. *Let R be a domain such that $R \subseteq R'$ is a root extension. Then $R \subseteq R + PR_P$ is an integral extension for each $P \in \mathrm{Spec}(R) \Leftrightarrow R$ is a pseudo-almost divided domain $\Leftrightarrow R$ is a quasi-local going-down domain. If these equivalent conditions hold (and $R \subseteq R'$ is a root extension), then $PR_P \subseteq \sqrt{PR'}$ for each $P \in \mathrm{Spec}(R)$.*

Proof. The first implication follows easily from the "root extension" hypothesis; the rest of the first assertion follows from Proposition 2 and Corollary 2. The second assertion follows by combining the first assertion with Proposition 2.   □

Proposition 3(b) will get a partial converse of the final assertion in Corollary 3 by using the root closure $^c R \subseteq R'$ of a domain R, which is defined as follows. The root closure $^c R$ of a domain R (with quotient field K) is the smallest R-subalgebra S of K such that S is root closed in K. It is clear that $^c R = \cap\{S \mid S$ is root closed in $K\}$. An easy inductive argument shows that $^c R$ is the directed union of all the (finite-type) R-subalgebras $R[x_1, \ldots, x_n]$ of K such that for each i, some power of x_i belongs to $R[x_1, \ldots, x_{i-1}]$.

Proposition 3. *(a) If R is a pseudo-almost divided domain and $P \in \mathrm{Spec}(R)$, then $R + PR_P \subseteq {}^c R$.*
(b) Let R be a root-closed domain. Then R is a divided domain if (and only if) R is a pseudo-almost divided domain.

Proof. (a) If $x \in PR_P$, then some power of x is in R (since R is pseudo-almost divided), and so $x \in {}^c R$. Hence, $R + PR_P \subseteq \cup_x R[x] \subseteq {}^c R$.
(b) The "only if" assertion is trivial because any divided domain is pseudo-almost divided. For the converse, combine (a) with the fact that a domain D is root closed only if $^c D = D$.   □

Remark 1. Let R be a quasi-local going-down domain and S an overring of R such that R is integrally closed in S. Then, by condition (3) in Proposition 2, $PR_P \cap S = P$ for each $P \in \mathrm{Spec}(R)$. It follows, under the given hypotheses on R and S, that if a prime ideal P of R satisfies $R_P \subseteq S$, then P is divided (in R, in the sense that $PR_P = P$).

Corollary 4 will give a result in the spirit of Proposition 3(b). First, recall from [20] that a domain R is said to be *almost integrally closed* if R is integrally closed in R_P for each nonzero prime ideal P of R. Corollary 4 generalizes [20, Proposition 2.8].

Corollary 4. *Let R be an almost integrally closed domain. Then R is a divided domain $\Leftrightarrow R$ is a pseudo-almost divided domain $\Leftrightarrow R$ is a quasi-local going-down domain.*

Proof. In view of the above material, the only nontrivial implication involves showing that a quasi-local almost integrally closed going-down domain must be divided. For this, apply Remark 1, with $S := R_P$.   □

Recall that any locally divided domain is a straight domain, and it is not known if the converse holds. The new two propositions will establish special cases of that converse for certain kinds of domains related to the pseudo-almost divided domains. For additional motivation, recall that divided domains can be characterized as the straight domains that are also quasi-divided domains [18, Proposition 5.3].

Let P be a prime ideal of a domain R and let n be a positive integer. Consider the ideal $P_n := \sum_{a \in P} Ra^n$. Clearly $P_n R_P = (P R_P)_n$, the R-submodule of R_P that is generated by $\{b^n \mid b \in P R_P\}$. We say that a domain R is *spectrally pseudo-almost divided* if, for each $P \in \mathrm{Spec}(R)$, there exists a positive integer n_P such that $x^{n_P} \in P$ for each $x \in P R_P$. (Note that in that case, we have $P_{n_P} R_P \subseteq P$.) Moreover, we say that R is *n-pseudo-almost divided* if R is spectrally pseudo-almost divided and the above integer(s) n_P can be chosen independently of $P \in \mathrm{Spec}(R)$. Clearly, each n-pseudo-almost divided domain is a spectrally pseudo-almost divided domain.

Proposition 4. *Let R be a spectrally pseudo-almost divided domain (for instance, an n-pseudo-almost divided domain). Then R is a divided domain if and only if R is a straight domain.*

Proof. Only the "if" assertion needs attention. Let R be a straight domain. By the proof of [18, Corollary 5.6], if $P \in \mathrm{Spec}(R)$ and n is a positive integer, then $P R_P \subseteq P + P_n R_P$. Taking $n := n_P$, we see from the above comments that $P R_P = P$ for each $P \in \mathrm{Spec}(R)$; that is, R is divided.   □

Note that the above proposition generalizes [18, Corollary 5.6], where we have now weakened the earlier "n-AVD" hypothesis to "spectrally pseudo-almost divided." We may also define n-PAVDs as follows. We say that a domain R (with quotient field K) is an *n-PAVD* if there exists some positive integer n such that, for each $u \in K$, either $u^n \in R$ or $vu^{-n} \in R$ for each nonunit v of R. Of course, each n-PAVD is a PAVD.

Proposition 5. (a) *Each n-PAVD is an n-pseudo-almost divided domain.*
(b) *Let n be a positive integer and R an n-PAVD. Then R is a divided domain if and only if R is a straight domain.*

Proof. (a) Suppose that (R, M) is an n-PAVD. Our task is to show that if $P \in \mathrm{Spec}(R)$ and $x \in P R_P$, then $x^n \in P$. If P is M (namely, the maximal ideal of R), then $x \in M R_M = M = P \subseteq R$ and the assertion is clear. Thus, without loss of generality, $P \neq M$. Then either $x^n \in R$ or $a := vx^{-n} \in R$ for each nonunit $v \in R$. If $x^n \in R$, then $x^n \in P$, as desired. In the remaining case, pick $u \in M \setminus P$, write $x := p/s$ where $p \in P$ and $s \in R \setminus P$, and get $us^n = ap^n \in P$, the desired contradiction.
(b) Combine (a) with Proposition 4.   □

The next main goal of this section is Proposition 7, which will give a class of PAVDs that are divided domains. This paragraph will identify another such class. Let (R, M) be a quasi-local domain and consider the conductor $V = (M :_K M)$. By reworking a proof of Badawi [6, Theorem 2.15], we get that R is an n-PAVD if and only if V is an n-AVD (i.e., for each nonzero $x \in K$, either $x^n \in V$ or $x^{-n} \in V$) whose maximal ideal is $\sqrt{MV}$. Suppose that R is an n-PAVD such that $\mathbb{Q} \subseteq R$. Then V is a valuation domain by [18, Proposition 5.8], and so it follows from [8, Theorem 3.4(4)] that R is an APVD and hence, by [8, Proposition 3.2], a divided domain.

We recall two more definitions. As in [1, p. 4], an ideal I of a domain R is said to be *strongly radical* if $x \in K$ with $x^n \in I$ for some positive integer n implies $x \in I$. Also (cf. [4]), a *rooty* domain is a domain each of whose radical ideals is strongly radical.

Proposition 6. *The following conditions are equivalent for a domain R:*

(1) Each maximal ideal of R is strongly radical.

(2) Each prime ideal of R is strongly radical.

(3) Each proper radical ideal of R is strongly radical.

(4) If $M \in \mathrm{Max}(R)$, then $(M :_K M)$ is root closed and M is a radical ideal of $(M :_K M)$.

(5) If $M \in \mathrm{Max}(R)$, then M is a radical ideal of some root-closed overring of R.

(6) R is a rooty domain.

Proof. It is clear that (1), (2), and (3) are equivalent. Then (1) is equivalent to each of (4) and (5), by [1, Proposition 1.4]. Now, (2) is equivalent to (6), by [4, Theorem 1.8].   □

Proposition 7. *(a) Let R be a pseudo-almost divided domain. If R is rooty, then R is a divided domain.*

(b) Each root-closed pseudo-almost divided domain is a divided domain.

(c) A PAVD which is either rooty or root closed must be a divided domain.

Proof. (a) We must show that if $P \in \mathrm{Spec}(R)$ and $x \in PR_P$, then $x \in P$. Since R is pseudo-almost divided, there exists a positive integer n such that $x^n \in P$. Since P is a strongly radical ideal of R, it follows that $x \in P$.

Finally, (b) follows from (a), since any root-closed domain is rooty by [4, Remark 1.9], and (c) follows at once by combining (a) and (b).   □

We close the section by sketching a path to other pullback results, this time combining the notions of "AV-domain," "AP-domain," and "PAVD." Because of space limitations, some routine details are left to the reader in our sketch. For the following results, we say that a domain D is a *locally* PAVD if D_M is a PAVD for each $M \in \mathrm{Max}(D)$.

Lemma 1. *(a) Let R be a PAVD and $P \in \mathrm{Spec}(R)$. Then R/P is a PAVD; and if also $PR_P = P$, then R_P is a PAVD.*

(b) Let R be a locally PAVD and $P \in \mathrm{Spec}(R)$. Then R/P is a locally PAVD.

(c) *Let* R *be an* AP-*domain* (*hence, a locally* PAVD) *and* $P \in \mathrm{Spec}(R)$. *Then* $R + PR_P$ *is an* AP-*domain.*

(d) *Let* $A \subset B$ *be domains such that* $\mathrm{Spec}(A) = \mathrm{Spec}(B)$ (*as sets*). *Then* A *is a* PAVD *if and only if* B *is a* PAVD.

Proof. (a) For the first assertion, see [6, Proposition 2.14]. The second assertion can be proved by a straightforward computation that uses [6, Theorem 2.5].

(b) If $P \subseteq M \in \mathrm{Max}(R)$, then R_M is a PAVD, and hence by (a), so is R_M/PR_M ($\cong (R/P)_{M/P}$).

(c) By [12, Lemma 2.2(b)], if $P \subseteq M \in \mathrm{Max}(R)$, then $(R + PR_P)_{M + PR_P} = R_M + PR_P$, which is an AV-domain since it is an overring of the AV-domain R_M.

(d) We can assume that A and B are quasi-local non-fields with the same quotient field and the same unique maximal ideal. The "if" assertion is easily proved, and the "only if" assertion follows from two uses of [6, Corollary 2.21] and a suitable juxtaposition of pullback diagrams. $\quad\square$

Theorem 2. *Let* T *be an* AV-*domain with* (*unique*) *maximal ideal* Q, *let* $\varphi : T \to T/Q$ *be the canonical surjection, let* D *be a subring of* T/Q, *and consider the pullback* $S := \varphi^{-1}(D) = T \times_{T/Q} D$. *Then* S *is a PAVD* (*resp., a locally* PAVD) *if and only if* D *is a PAVD* (*resp., a locally* PAVD).

Proof. As T is quasi-local, we have $T = S_Q$ (cf. [21, Lemma 2.5 (iv)]). It then follows from Lemma 1 (a) that if S is a PAVD, then so is D. For the converse, suppose that D is a PAVD. By [6, Theorem 2.5 and Lemma 2.1], we need only show that if a nonzero element z of the quotient field of S is such that none of $z, z^2, z^3, \dots$ is in S, then $z^{-n}Q \subseteq Q$ for some positive integer n. As T is an AV-domain, some n satisfies $z^n \in T$ or $z^{-n} \in T$. The assertion is clear if $z^{-n} \in T$ and so, without loss of generality, z^n is a nonunit of T, whence $z^n \in Q \subseteq S$, a contradiction.

We turn now to the corresponding "locally PAVD" assertions. In view of Lemma 1 (b), we need only show that if D is a locally PAVD, then so is S, that is, S_N is a PAVD for each $N \in \mathrm{Max}(S)$. Without loss of generality, $N \supseteq Q$ (for we see, as in the proof of [16, Theorem 2.4], that if $N \not\supseteq Q$, then S_N is isomorphic to a ring of fractions of T and thus is an AP-domain, hence a locally PAVD). Then, by combining the fourth paragraph of the proof of [16, Theorem 2.4] with the above "PAVD" assertion, we obtain the "locally PAVD" assertions. $\quad\square$

Corollary 5. *Let* T *be an* AP-*domain with* $Q \in \mathrm{Max}(T)$, *let* $\varphi : T \to T/Q$ *be the canonical surjection, let* D *be a subring of* T/Q, *and consider the pullback* $S := \varphi^{-1}(D) = T \times_{T/Q} D$. *Then* S *is a locally PAVD if and only if* D *is a locally* PAVD.

Proof. The "only if" assertion is immediate from Lemma 1(b). As for the "if" assertion, one can reduce to the case where T is quasi-local by reasoning as in the proof of [16, Theorem 2.4], and that case is then handled by Theorem 2. $\quad\square$

4 A New Class of Domains with the Portable Property

In view of Propositions 4 and 5, it seems natural to ask if pseudo-almost divided domains admit an analogue of the main result of Sect. 2. In this section, we answer that question in the affirmative, also obtaining an affirmative answer to the analogous question about locally pseudo-almost divided domains. (Of course, a domain D is said to be *locally pseudo-almost divided* if D_M is pseudo-almost divided for each $M \in \text{Max}(D)$.) These answers are obtained by means of the machinery from [16], which will show that "locally pseudo-almost divided" is a portable property. The path to those facts begins with the following lemma.

Lemma 2. *Let R be a pseudo-almost divided domain. If $P \in \text{Spec}(R)$, then both R_P and R/P are pseudo-almost divided domains.*

Proof. To prove that R_P is pseudo-almost divided, we must show that if $Q \subseteq P$ in $\text{Spec}(R)$ and $x \in Q_P(R_P)_{Q_P}$ $(= QR_P R_Q = QR_Q)$, then there exists a positive integer n such that $x^n \in QR_Q$. Since Q is pseudo-almost divided in R, we have a positive integer n such that $x^n \in Q$ $(\subseteq QR_Q)$.

Next, to show that R/P is pseudo-almost divided, consider $P \subseteq Q$ in $\text{Spec}(R)$ and an element x of $(Q/P)(R/P)_{Q/P}$, canonically identified with QR_Q/PR_Q. Our task is to find a positive integer n such that x^n is in R/P (when R/P is viewed canonically inside R_Q/PR_Q). Write $x = r/s + PR_Q$ with $r \in Q$ and $s \in R \setminus Q$. Since $u := r/s \in QR_Q$ and Q is pseudo-almost divided in R, we have a positive integer n such that $u^n \in Q$, whence x^n is canonically identified with $u^n + P \in Q/P \subseteq R/P$, to complete the proof. □

The next result is an analogue of [7, Theorem 2.7].

Proposition 8. *Let R be a domain and $P \in \text{Spec}(R)$. Then $R + PR_P$ is a pseudo-almost divided domain if and only if both R_P and R/P are pseudo-almost divided. In other words, "pseudo-almost divided domain" satisfies part $(\bullet\bullet)$ in the definition of a portable property of domains.*

Proof. We use some facts about the CPI-extension $T := R + PR_P$ from [12, Lemmas 2.1 and 2.2]. There is an isomorphism of partially ordered sets between $\text{Spec}(T)$ (under inclusion) and $\{Q \in \text{Spec}(R) \mid Q$ is comparable to P under inclusion$\}$. Under this isomorphism, if $Q \subseteq P$ in $\text{Spec}(R)$, then Q corresponds to QR_P in $\text{Spec}(T)$, with $T_{QR_P} = R_Q$; on the other hand, if $P \subseteq Q$ in $\text{Spec}(R)$, then Q corresponds to the $Q + PR_P \in \text{Spec}(T)$, with $T_{Q+PR_P} = R_Q + PR_P$. As $R_P = T_{PR_P}$, it follows from the first assertion in Lemma 2 that if T is pseudo-almost divided, then so is R_P. Also, since $T/PR_P \cong R/P$, it follows from the second assertion in Lemma 2 that if T is pseudo-almost divided, then so is R/P. This completes the proof of the "only if" assertion.

Conversely, assume that both R_P $(= T_{PR_P})$ and R/P $(\cong T/PR_P)$ are pseudo-almost divided. Our task is to prove that T is pseudo-almost divided. Since PR_P is a divided prime ideal of T, there is no harm in replacing R with T (and P with PR_P).

Thus, we can assume that P is a divided prime ideal of R and that each prime ideal of R is comparable to P, with our task now being to show that R is pseudo-almost divided. Fix $Q \in \mathrm{Spec}(R)$ and $u \in QR_Q$. We will show that there exists a positive integer n such that $u^n \in Q$ (equivalently, $u^n \in R$). We will need to consider two cases.

Suppose first that $Q \subseteq P$. We have $u \in QR_Q = (QR_P)(R_P)_{QR_P}$. As R_P is pseudo-almost divided, it follows that there exists a positive integer n such that $u^n \in QR_P$. As P is a divided prime ideal of R, we get $u^n \in PR_P = P \subseteq R$, as required.

In the remaining case, $P \subset Q$. As in the proof of Lemma 2, identify $(Q/P)(R/P)_{Q/P}$ with QR_Q/PR_Q. Consider $v := u + PR_Q$. Since Q/P is pseudo-almost divided in R/P by hypothesis, there exists a positive integer n such that v^n is in the canonical image of Q/P. Hence, there exists $y \in Q$ such that $u^n + PR_Q = y + PR_Q$. Then $u^n - y \in PR_Q \subseteq PR_P = P \subseteq Q$, and so $u^n \in Q$, as required. $\square$

The above proof somewhat mimics that of [14, Proposition 2.12]. By reasoning in the cited proof, we see that Proposition 8 carries over if we replace "pseudo-almost divided" with "locally pseudo-almost divided." Also, by mimicking the proof of [13, Theorem 2.2], we can see that if R is a strong (in the sense of [13]) locally pseudo-almost divided domain, then so is R/P for each $P \in \mathrm{Spec}(R)$. It is clear that the analogous assertion holds for R_P.

We next take another step toward showing that "locally pseudo-almost divided domain" is a portable property.

Proposition 9. *Let $A \subset B$ be an extension of domains such that $\mathrm{Spec}(A) = \mathrm{Spec}(B)$ as sets. Then A is a pseudo-almost divided domain if and only if B is a pseudo-almost divided domain. In other words, "pseudo-almost divided domain" satisfies part ($\bullet$) in the definition of a portable property of domains.*

Proof. Since any field is trivially a pseudo-almost divided domain, we may assume, without loss of generality, that neither A nor B is a field. Hence, by [3, Propositions 3.3 and 3.5], A and B are quasi-local domains that have the same maximal ideal (say, M), have the same quotient field, and satisfy $A_P = B_{B \setminus P}$ for each nonmaximal prime ideal P of A. It follows easily that each nonmaximal prime ideal of A is pseudo-almost divided in A if and only if each nonmaximal prime ideal of B is pseudo-almost divided in B. Since the unique maximal ideal M is divided (hence pseudo-almost divided) in both A and B, the proof is complete. $\square$

We can now close by giving the two promised pullback results that are in the spirit of Theorem 1.

Corollary 6. *Let (T, Q) be a quasi-local domain, $\pi : T \to T/Q$ the canonical surjection, D a subring of T/Q, and $R := \pi^{-1}(D)$. Then $R = T \times_{T/Q} D$ is a pseudo-almost divided domain if and only if both T and D are pseudo-almost divided domains.*

Proof. Combine Propositions 8 and 9 with [16, Lemma 2.3]. □

Corollary 7. (*a*) *"Locally pseudo-almost divided domain" is a portable property of domains.*

(*b*) *Let* T *be a domain with maximal ideal* Q, $\pi : T \to T/Q$ *the canonical surjection,* D *a subring of* T/Q, *and* $R := \pi^{-1}(D)$. *Then* $R = T \times_{T/Q} D$ *is a locally pseudo-almost divided domain if and only if both* T *and* D *are locally pseudo-almost divided domains.*

Proof. (a) Combine Propositions 8 and 9 with [19, Corollary 2].
(b) Combine (a) and [16, Theorem 2.4].

References

1. D.D. Anderson, D.F. Anderson, Multiplicatively closed subsets of fields. Houston J. Math. **13**(1), 1–11 (1987)
2. D.D. Anderson, M. Zafrullah, Almost Bezout domains. J. Algebra **142**(2), 285–309 (1991)
3. D.F. Anderson, D.E. Dobbs, Pairs of rings with the same prime ideals. Can. J. Math. **32**(2), 362–384 (1980)
4. D.F. Anderson, J. Park, Rooty and root closed domains. In: *Advances in Commutative Ring Theory*. Lecture Notes Pure Applied Mathematics, vol. 205 (Dekker, New York, 1999), pp. 87–99
5. M.F. Atiyah, I.G. Macdonald, *Introduction to Commutative Algebra* (Addison-Wesley, Reading, 1969)
6. A. Badawi, On pseudo-almost valuation domains. Comm. Algebra **35**(4), 1167–1181 (2007)
7. A. Badawi, D.E. Dobbs, On locally divided rings and going down rings. Comm. Algebra **29**(7), 2805–2825 (2001)
8. A. Badawi, E. Houston, Powerful ideals, strongly primary ideals, almost pseudo-valuation domains, and conducive domains. Comm. Algebra **30**(4), 1591–1606 (2002)
9. D.E. Dobbs, On going-down for simple overrings, II. Comm. Algebra **1**, 439–458 (1974)
10. D.E. Dobbs, Divided rings and going-down. Pacific J. Math. **67**(2), 353–363 (1976)
11. D.E. Dobbs, Coherence, ascent of going-down, and pseudo-valuation domains. Houston J. Math. **4**(4), 551–567 (1978)
12. D.E. Dobbs, On locally divided integral domains and CPI-overrings. Int J. Math. Math. Sci. **4**(1), 119–135 (1981)
13. D.E. Dobbs, A note on strong locally divided domains. Tsukuba J. Math. **15**(1), 215–217 (1991)
14. D.E. Dobbs, On Henselian pullbacks. In: *Factorization in Integral Domains*. Lecture Notes Pure Applied Mathematics, vol. 189 (Dekker, New York, 1997), pp. 317–326
15. D.E. Dobbs, Pseudo-almost valuation domains are quasi-local going-down domains, but not conversely. Rend. Circ. Mat. Palermo **57**(1), 119–124 (2008)
16. D.E. Dobbs, When is a pullback a locally divided domain? Houston J. Math. **35**(2), 341–351 (2009)
17. D.E. Dobbs, G. Picavet, Straight rings. Comm. Algebra **37**(3), 757–793 (2009)
18. D.E. Dobbs, G. Picavet, Straight rings, II. In: *Commutative Algebra and Applications* (De Gruyter, Berlin, 2009), pp. 183–205
19. D.E. Dobbs, G. Picavet, On almost-divided domains. Rend. Circ. Mat. Palermo **58**(2), 199–210 (2009)
20. D.E. Dobbs, J. Shapiro, Almost integrally closed domains. Comm. Algebra **32**(9), 3627–3639 (2004)

21. D.E. Dobbs, M. Fontana, J.A. Huckaba, I.J. Papick, Strong ring extensions and pseudo-valuation domains. Houston J. Math. **8**(2), 167–184 (1982)
22. M. Fontana, Topologically defined classes of commutative rings. Ann. Mat. Pura Appl. **123**, 331–355 (1980)
23. J.R. Hedstrom, E.G. Houston, Pseudo-valuation domains. Pacific J. Math. **75**(1), 137–147 (1978)
24. D.L. McQuillan, On prime ideals in ring extensions. Arch. Math. **33**(2), 121–126 (1979/1980)
25. A. Mimouni, Prüfer-like conditions and pullbacks. J. Algebra **279**(2), 685–693 (2004)

The Probability That $\text{Int}_n(D)$ Is Free

Jesse Elliott

Abstract Let D be a Dedekind domain with quotient field K. The ring of *integer-valued polynomials on D* is the subring $\text{Int}(D) = \{f \in K[X] : f(D) \subseteq D\}$ of the polynomial ring $K[X]$. The *Pólya-Ostrowski group* $\text{PO}(D)$ of D is a subgroup of the class group of D generated by the well-known factorial ideals $n!_D$ of D. A *regular basis* of $\text{Int}(D)$ is a D-module basis consisting of one polynomial of each degree. It is well known that $\text{Int}(D)$ has a regular basis if and only if the group $\text{PO}(D)$ is trivial, if and only if the D-module $\text{Int}_n(D) = \{f \in \text{Int}(D) : \deg f \leq n\}$ is free for all n. In this paper we provide evidence for and prove special cases of the conjecture that, if $\text{PO}(D)$ is finite, then the natural density of the set of nonnegative integers n such that $\text{Int}_n(D)$ is free exists, is rational, and is at least $1/|\text{PO}(D)|$. Moreover, we compute this density or determine a conjectural value for several examples of Galois number fields of degrees 2, 3, 4, 5, and 6 over $\mathbb{Q}$.

Keywords Integer-valued polynomial • Integral domain • Pólya-Ostrowski group • Class group • Number field

MSC: 13F20, 13G05, 11R04, 11R29

1 Introduction

Let D be an integral domain with quotient field K. The ring of *integer-valued polynomials on D* is the subring

$$\text{Int}(D) = \{f \in K[X] : f(D) \subseteq D\}$$

J. Elliott (✉)
Department of Mathematics, California State University,
Channel Islands, Camarillo, CA 93012, USA
e-mail: jesse.elliott@csuci.edu

M. Fontana et al. (eds.), *Commutative Algebra: Recent Advances in Commutative Rings,* 133
Integer-Valued Polynomials, and Polynomial Functions, DOI 10.1007/978-1-4939-0925-4_8,
© Springer Science+Business Media New York 2014

of the polynomial ring $K[X]$. The study of integer-valued polynomial rings began with Pólya and Ostrowski circa 1919 [1, p. xiv]. They showed that, for any number ring D, the D-module $\text{Int}(D)$ has a *regular basis*, that is, a D-module basis consisting of exactly one polynomial of each degree, if and only if the product Π_q of the prime ideals of D of norm q is a principal ideal for every q. In fact this equivalence holds for any Dedekind domain D. More generally, if D is a Krull domain, then $\text{Int}(D)$ has a regular basis if and only if the product Π_q of the height one prime ideals of norm q is principal for every q [2, Corollary 2.5]. In particular, if D is a unique factorization domain, then $\text{Int}(D)$ has a regular basis.

If $\text{Int}(D)$ has a regular basis, then the D-module

$$\text{Int}_n(D) = \{ f \in \text{Int}_n(D) : \deg f \leq n \}$$

is free of rank $n + 1$ for any nonnegative integer n, having as a basis the polynomials of degree at most n in any regular basis of $\text{Int}(D)$. The converse is true if D is Dedekind, and in that case there is an explicit algorithm to construct a basis of $\text{Int}_n(D)$ for any n [1, Proposition II.3.14]. Although $\text{Int}(D)$ is free as a D-module for any Dedekind domain D [1, Remark II.3.7(iii)], it is more common than not for $\text{Int}(D)$ to lack a regular basis.

For any Dedekind domain D, the *Pólya-Ostrowski group* of D, denoted $\text{PO}(D)$, is the subgroup of the class group $\text{Cl}(D)$ of D generated by the images in $\text{Cl}(D)$ of the ideals Π_q. This group can be considered as a "measure of the obstruction for $\text{Int}(D)$ to have a regular basis" [2, p. 37]. In particular, $\text{Int}(D)$ has a regular basis if and only if the group $\text{PO}(D)$ is trivial. More generally, for any n the D-module $\text{Int}_n(D)$ is free if and only if $\prod_q \Pi_q^{u_q(n)}$ has trivial image in $\text{PO}(D)$ (or equivalently is principal), where the $u_q(n)$ are nonnegative integers depending only on q and n. If the $u_q(n)$ are "uniformly distributed" relative to $\text{PO}(D)$, in an appropriate sense, then one might expect that the "probability" that $\text{Int}_n(D)$ is free is $1/|\text{PO}(D)|$. In this paper we provide some evidence for and prove special cases of the conjecture that the density of the set of all $n \in \mathbb{Z}_{\geq 0}$ for which the D-module $\text{Int}_n(D)$ is free exists, is rational, and is at least $1/|\text{PO}(D)|$ if D is a Dedekind domain. We also explore further connections between the ring $\text{Int}(D)$ and the group $\text{PO}(D)$.

For any $T \subseteq \mathbb{Z}_{\geq 0}$, one defines the *(natural) density* $\delta(T)$ of T to be

$$\delta(T) = \lim_{n \to \infty} \frac{|T \cap \{0, 1, 2, \ldots, n - 1\}|}{n}.$$

One also defines the *upper density* of T by

$$\overline{\delta}(T) = \limsup_{n \to \infty} \frac{|T \cap \{0, 1, 2, \ldots, n - 1\}|}{n},$$

and the *lower density* $\underline{\delta}(T)$ of T as the corresponding lim inf. Both exist and satisfy $0 \leq \underline{\delta}(T) \leq \overline{\delta}(T) \leq 1$. Moreover, $\delta(T)$ exists if and only if $\underline{\delta}(T) = \overline{\delta}(T)$, in which case $\delta(T)$ equals that common value. Informally one can think of $\delta(T)$ as the probability that a random nonnegative integer n lies in T.

We are interested for any integral domain D in the density

$$\delta(\mathrm{Int}(D)) = \delta(\{n \in \mathbb{Z}_{\geq 0} : \text{the } D\text{-module } \mathrm{Int}_n(D) \text{ is free}\}).$$

One can think of $\delta(\mathrm{Int}(D))$ as the probability that the D-module $\mathrm{Int}_n(D)$ is free for a random nonnegative integer n. Our main conjecture is as follows.

Conjecture 1.1. Let D be a Dedekind domain such that $\mathrm{PO}(D)$ is finite. Then $\delta(\mathrm{Int}(D))$ exists, is rational, and is at least $1/|\mathrm{PO}(D)|$.

A result of this type would be a first step towards an arithmetical interpretation of the group $\mathrm{PO}(D)$ when it is not trivial. We also consider the following weaker version of the conjecture.

Conjecture 1.2. Let K be finite Galois extension of $\mathbb{Q}$. Then $\delta(\mathrm{Int}(\mathcal{O}_K))$ exists, is rational, and is at least $1/|\mathrm{PO}(\mathcal{O}_K)|$.

A primary focus in literature on the Pólya-Ostrowski group has been on determining or characterizing number fields K such that $\mathrm{Int}(\mathcal{O}_K)$ has a regular basis. For example, [2, Corollary 3.11] provides a characterization of all cyclic number fields K such that $\mathrm{Int}(\mathcal{O}_K)$ has a regular basis, and [1, Corollary II.4.5] and [7, Propositions 3.4, 3.6, and 3.19] explicitly construct all such K of degrees 2, 3, 4, and 6 over $\mathbb{Q}$. The number field $K = \mathbb{Q}(\sqrt{-5})$ is an example where $\mathrm{PO}(\mathcal{O}_K)$ has order 2, and in this case we show that $\delta(\mathrm{Int}(\mathcal{O}_K)) = 3/4$. The group $\mathrm{PO}(\mathcal{O}_K)$ also has order two for $K = \mathbb{Q}(\sqrt{-6})$, but in this case numerical data suggests that $\delta(\mathrm{Int}(\mathcal{O}_K)) = 1/2$. For certain Galois number fields K of degree at most 6, we find alternative expressions for, and in some cases compute, $\delta(\mathrm{Int}(\mathcal{O}_K))$.

This paper is organized as follows. In Sect. 2 we state several density conjectures for certain arithmetic functions and explain how these conjectures were obtained from numerical data. These conjectures form the basis for our conjectures regarding $\delta(\mathrm{Int}(D))$. In Sect. 3 we prove some simple special cases of these conjectures. Then in Sects. 4 and 5 we reduce Conjectures 1.1 and 1.2 above to the density conjectures stated in Sect. 2. Corollary 5.4 of Sect. 5 provides a verification of a very special case of Conjecture 1.2.

In Sects. 6 and 7 we characterize the quadratic number fields K such that $\mathrm{PO}(\mathcal{O}_K)$ has order 2, along with the cyclic cubic number fields K such that $\mathrm{PO}(\mathcal{O}_K)$ has order 3, and we determine all n such that $\mathrm{Int}_n(\mathcal{O}_K)$ is free for several examples of such K. Finally in Sects. 8 and 9 we provide similar computations for several cyclic quartic and quintic number fields and for two noncyclic number fields of degree six. All examples in Sects. 7 through 9 were computed using the computer algebra systems Sage and/or Magma. For each of our examples of number fields K in Sects. 6 through 9 we show that the value of $\delta(\mathrm{Int}(\mathcal{O}_K))$ is determined by our conjectures stated in Sect. 2.

I would like to acknowledge the work of two of my undergraduate research assistants, Benjamin Sergent and Anthony Brice, who helped with Sect. 2 of this paper for a semester-long research project. Benjamin wrote a computer program to estimate densities, and both used the program to collect data, the analysis of which ultimately led to the conjectures in Sect. 2.

2 Arithmetic Conjectures

For a Dedekind domain D such that $PO(D)$ is finite, the set of all nonnegative integers n such that the D-module $\mathrm{Int}_n(D)$ is free can be expressed in terms of certain arithmetic functions u_S for multisets $S : \mathbb{Z}_{>1} \longrightarrow \mathbb{Z}_{\geq 0}$. In this section we discuss various density properties of these functions u_S that are relevant for computing $\delta(\mathrm{Int}(D))$.

Following [1, Chap. II], for any nonnegative integer n and any integer $k > 1$, we let

$$w_k(n) = \sum_{i=1}^{\infty} \left\lfloor \frac{n}{k^i} \right\rfloor.$$

Alternatively, by [1, Exercise II.8 and Lemma II.2.4], one has

$$w_k(n) = \sum_{i=1}^{n} v_k(i) = \frac{n - s_k(n)}{k - 1},$$

where $v_k(i)$ for any positive integer i is the largest nonnegative integer t such that k^t divides i and where $s_k(n)$ is the sum of the digits in the base k expansion of n. Define

$$u_k(n) = \sum_{i=1}^{n} w_k(i) = \sum_{i=1}^{n} (n - i + 1) v_k(i),$$

and define

$$u_S(n) = \sum_{k \in S} u_k(n)$$

for any subset S of $\mathbb{Z}_{>1}$. More generally, to allow for repetition, for any multiset $S : \mathbb{Z}_{>1} \longrightarrow \mathbb{Z}_{\geq 0}$ (in which each $k \in \mathbb{Z}_{>1}$ appears a finite number $S(k)$ times), we may define

$$u_S(n) = \sum_{k=2}^{\infty} S(k) u_k(n).$$

Note that $w_k(n) = u_k(n) = 0$ if $k > n$, so $u_S(n)$ is well defined even if S is infinite.
    We begin with the following problem.

Problem 2.1. Let d be a positive integer and $S : \mathbb{Z}_{>1} \longrightarrow \mathbb{Z}_{\geq 0}$ a multiset. Compute the density $\delta(d, S)$ of the subset $\{n \in \mathbb{Z}_{\geq 0} : d \mid u_S(n)\}$ of $\mathbb{Z}_{\geq 0}$.

One of our initial approaches to the problem was to write a program (in C#) that can compute $\delta_N(d, S) = |\{n < N : d \mid u_S(n)\}|/N$ for any sufficiently small d, S, and N. The program is efficient enough that on our computers the program can easily handle $N \leq 10^{10}$ for small d and S in a reasonable amount of time. After collecting data on many triples d, S, and N, we first observed that $\delta_N(d, S)$ for sufficiently large N is close to a rational number. Thus we conjectured that

$$\delta(d, S) = \lim_{n \to \infty} \delta_N(d, S) = \delta(\{n \in \mathbb{Z}_{\geq 0} : d \mid u_S(n)\})$$

exists and is rational. Then, after collecting and analyzing further data and proving some special cases, we made following conjectures.

Conjecture 2.2. Let d be a positive integer and $S : \mathbb{Z}_{>1} \longrightarrow \mathbb{Z}_{\geq 0}$ a multiset, and let $g = \gcd(d, S(2), S(3), S(4), \ldots)$ and $l = \gcd(d, 2S(2), 3S(3), 4S(4), \ldots)$.

(1) $\delta(d, S)$ exists, is rational, and is at least $1/d$.
(2) $\delta(d, S) = \delta(a, S)\delta(b, S)$ if $d = ab$ and $\gcd(a, b) = 1$.
(3) $\delta(d, S) = 1/d$ if and only if $l = 1$.
(4) If $g = 1$, then

$$\frac{1}{d} \leq \delta(d, S) \leq \frac{1}{d^2} \sum_{i=1}^{d} \gcd(i, d).$$

(5) Let $\gcd(S) = \gcd\{k : S(k) \neq 0\}$. If $g = 1 = \gcd(\gcd(S), d/\gcd(d, \gcd(S)))$, then

$$\delta(d, S) = \frac{1}{dl} \sum_{a=1}^{l} \gcd(a, l).$$

The following conjecture generalizes statements (1)–(3) of Conjecture 2.2.

Conjecture 2.3. For $i = 1, 2, \ldots, r$, let d_i be a positive integer and $S_i : \mathbb{Z}_{>1} \longrightarrow \mathbb{Z}_{\geq 0}$ a multiset. Let Δ denote the density of the set of nonnegative integers n such that $u_{S_i}(n)$ is divisible by d_i for $i = 1, 2, \ldots, r$.

(1) Δ exists, is rational, and is at least $1/(d_1 d_2 \cdots d_r)$.
(2) $\Delta \geq \delta(d_1, S_1)\delta(d_2, S_2) \cdots \delta(d_r, S_r)$, and equality holds if the d_i are pairwise relatively prime.
(3) $\Delta = 1/(d_1 d_2 \cdots d_r)$ if and only if, for each prime divisor p of $d_1 d_2 \cdots d_r$, the characteristic functions $S_i|_{\mathbb{Z}_{>1} \setminus p\mathbb{Z}} : \mathbb{Z}_{>1} \setminus p\mathbb{Z} \longrightarrow \mathbb{Z}_{\geq 0}$ of the multisets $S_i \setminus p\mathbb{Z}$ for those i such that $p \mid d_i$ are linearly independent over $\mathbb{F}_p$.

The following proposition provides a restatement of Conjecture 2.3(1).

Proposition 2.4. *Conjecture 2.3(1) is equivalent to the following. Let G be a finite abelian group of order d, and for each $g \in G$ let S_g be a subset of $\mathbb{Z}_{>1}$. Suppose that the sets S_g are pairwise disjoint. Then the density of the set of all $n \in \mathbb{Z}_{\geq 0}$ such that $\sum_{g \in G} u_{S_g}(n)g = 0$ in G exists, is rational, and is at least $1/d$.*

The following is a weaker version of Conjecture 2.3(1) that is sufficient for studying Galois number fields.

Conjecture 2.5. Let G be a finite abelian group of order d, and for each $g \in G$ let S_g be a finite set of prime powers. Suppose that the sets S_g are pairwise disjoint and that no two elements of $\bigcup_{g \in G} S_g$ are powers of the same prime. Then the density of the set of all $n \in \mathbb{Z}_{\geq 0}$ such that $\sum_{g \in G} u_{S_g}(n)g = 0$ in G exists, is rational, and is at least $1/d$.

3 Some Special Cases of the Conjectures

In this section we prove some simple special cases of the conjectures provided in the previous section. First, we have the following.

Proposition 3.1. *Let $S : \mathbb{Z}_{>1} \longrightarrow \mathbb{Z}_{\geq 0}$ be a multiset and d a positive integer such that $d \mid k$ for all $k \in S$. Then $d \mid u_S(n)$ for all $n \in \mathbb{Z}_{\geq 0}$ such that $n \equiv -1 \pmod{d}$. In particular,*

$$\underline{\delta}(\{n \in \mathbb{Z}_{\geq 0} : d \mid u_S(n)\}) \geq 1/d.$$

Proof. We may assume without loss of generality that $S = \{k\}$. Now, since $v_k(i) = 0$ if $k \nmid i$, hence if $d \nmid i$, one has

$$u_k(n) = \sum_{i=1}^{n}(n+1-i)v_k(i) = \sum_{d \mid i \leq n}(n+1-i)v_k(i) \equiv \sum_{d \mid i \leq n} 0 \cdot v_k(i) \equiv 0 \pmod{d}$$

for all $n \equiv -1 \pmod{d}$. The proposition follows. $\square$

Next, as our main result on the density conjectures, we compute the density of n such that d divides $u_d(n)$. Let $s_d(n)$ denote the sum of the base d digits of n.

Lemma 3.2. *One has $n + w_d(n) \equiv s_d(n) \pmod{d}$ for all d and n.*

Proof. The lemma follows easily from the equation $w_d(n) = (n - s_d(n))/(d - 1)$. $\square$

Proposition 3.3. *Let $d \in \mathbb{Z}_{>1}$, and let $n \in \mathbb{Z}_{\geq 0}$. Let $n = qd + r$ with $0 \leq r < d$. Then $u_d(n) \equiv (r + 1)s_d(q) \pmod{d}$. In particular, $d \mid u_d(n)$ if and only if $d/\gcd(d, r + 1) \mid s_d(q)$, if and only if $s_d(n) \equiv n \pmod{d/\gcd(d, n + 1)}$.*

Proof. One has

$$
\begin{aligned}
u_d(n) &= n v_d(1) + (n-1) v_d(2) + \cdots + v_d(n) \\
&\equiv (r+1) v_d(d) + (r+1) v_d(2d) + \cdots + (r+1) v_d(qd) \\
&\equiv (r+1)(q + v_d(1) + v_d(2) + \cdots + v_d(q)) \\
&\equiv (r+1)(q + w_d(q)) \\
&\equiv (r+1) s_d(q) \ (\text{mod } d),
\end{aligned}
$$

where the last congruence holds by Lemma 3.2. The first statement of the lemma follows. Finally, modulo $d/\gcd(d, r+1)$, the integer $s_d(qd) = s_d(q)$ is congruent to 0 if and only if $s_d(n) = s_d(qd + r) = s_d(q) + r$ is congruent to r, or equivalently, n. This completes the proof. $\qquad\square$

Proposition 3.4. *Let $d \in \mathbb{Z}_{>1}$. Then $\delta(d, \{d\})$ is equal to*

$$
\frac{1}{d^2} \sum_{i=1}^{d} \gcd(i, d) = \frac{1}{d^2} \sum_{e|d} e\phi(d/e) = \frac{1}{d^2} \sum_{e|d} e\tau(e)\mu(d/e)
$$

$$
= \frac{1}{d} \prod_{p|d} \frac{(1 + v_p(d))p - v_p(d)}{p},
$$

where ϕ is Euler's phi function, τ is the number of divisors function, and μ is the Möbius function. More generally, $\delta(\{n \in \mathbb{Z}_{\geq 0} : u_d(n) \equiv a \ (\text{mod } d)\})$ is equal to

$$
\frac{1}{d^2} \sum_{\gcd(i,d)|a} \gcd(i, d) = \frac{1}{d^2} \sum_{e|\gcd(a,d)} e\phi(d/e)
$$

for any $1 \leq a \leq d$.

Proof. Choose i with $1 \leq i \leq d$ and consider the nonnegative integers $n = qd + i - 1$ congruent to $i - 1$ mod n. By Proposition 3.3 one has $u_d(n) \equiv a \ (\text{mod } d)$ if and only if $i s_d(q) \equiv a \ (\text{mod } d)$. Thus if the density is nonzero we must have $\gcd(i, d) \mid a$. For each such i there are $\gcd(i, d)$ possibilities modulo d for $s_d(q)$. Since the $s_d(q)$ are uniformly distributed among the d congruence classes modulo d, it follows that

$$
\delta(\{n \in \mathbb{Z}_{\geq 0} : u_d(n) \equiv a \ (\text{mod } d)\}) = \frac{1}{d} \sum_{\gcd(i,d)|a} \frac{\gcd(i, d)}{d}.
$$

The rest of the proposition follows from this equality and from well-known properties of Dirichlet convolution and of the multiplicative functions ϕ, τ, and μ. $\qquad\square$

Corollary 3.5. *Let p be a prime and n a nonnegative integer. Then $p \mid u_p(n)$ if and only if (1) $n \equiv -1 \pmod{p}$, or (2) $n \not\equiv -1 \pmod{p}$ and $s_p(n) \equiv n \pmod{p}$. Moreover, $\delta(p, \{p\})$ exists and is equal to $\frac{2p-1}{p^2}$.*

Finally we investigate $\delta(2, \{k\})$ for odd k.

Proposition 3.6. *Let $k \in \mathbb{Z}_{>1}$ be odd. If $\delta(2, \{k\})$ exists, then it is equal to $1/2$.*

Proof. One has

$$u_k(n) = nv_k(1) + (n-1)v_k(2) + \cdots + 1v_k(n)$$
$$\equiv v_k(2) + v_k(4) + \cdots + v_k(n)$$
$$= v_k(1) + v_k(2) + \cdots + v_k(n/2) = w_k(n/2) \pmod{2}$$

if n is even, and similarly

$$u_k(n) \equiv t_k(n) := v_k(1) + v_k(3) + \cdots + v_k(n) \pmod{2}$$

if n is odd. We show that the set $S = \{n \in \mathbb{Z}_{\geq 0} : u_k(n) \text{ is even}\}$ has density $1/2$ in both the set of even integers and the set of odd integers.

Suppose that $n = 2(2k^s - 1)$, where s is an odd integer, and let $m = k^s - 1$. Note that $v_k(k^s + r) = v_k(k^s - r)$ for all $r \leq m$. Therefore, by induction on r,

$$w_k(k^s + r) \equiv v_k(k^s) + w_k(k^s - r - 1) \equiv 1 + w_k(k^s - r - 1) \pmod{2},$$

that is, $w_k(k^s + r)$ and $w_k(k^s - r - 1)$ have opposite parity, for all $r \leq m$. Therefore the probability that $w_k(i)$ is even for $i \leq k^s + m = n/2$ is exactly $1/2$. It follows that the probability that $u_k(i) \equiv w_k(i/2) \pmod{2}$ is even for even $i \leq n$ is exactly $1/2$. Now take a limit as $s \to \infty$.

Now instead suppose that $n = 2m - 1$, where $m = k^s - 1$ for some odd integer s. Note that $v_k(k^s + 2r) = v_k(k^s - 2r)$ if $2r \leq m$. Therefore

$$t_k(k^s + 2r - 2) \equiv v_k(k^s) + t_k(k^s - 2r) \equiv 1 + t_k(k^s - 2r) \pmod{2}$$

if $2r \leq m$, so the probability that $u_k(i) \equiv t_k(i) \pmod{2}$ is even for odd $i \leq k^s + m - 2 = n$ is exactly $1/2$. Now take a limit as $s \to \infty$. $\qquad\square$

4 Dedekind Domains

Let D be an integral domain with quotient field K. For any $n \in \mathbb{Z}_{\geq 0}$, let

$$\mathfrak{I}_n(D) = \{a \in K : a \text{ is the leading coefficient of some } f \in \mathrm{Int}_n(D)\}.$$

Then $\mathfrak{I}_n(D)$ is a fractional ideal of D containing D. Therefore $n!_D = (D :_K \mathfrak{I}_n(D))$ is an ideal of D. If D is a Dedekind domain, then $\Pi_q(D) = \prod_{N(\mathfrak{p})=q} \mathfrak{p}$ denotes the product of prime ideals $\mathfrak{p}$ of D of norm q for any (prime power) q. We write $\mathfrak{I}_n = \mathfrak{I}_n(D)$ and $\Pi_q = \Pi_q(D)$ when the domain D is understood.

Proposition 4.1 ([1, Proposition II.3.1 and Corollary II.3.6]). *Let D be a Dedekind domain. For all $n \in \mathbb{Z}_{\geq 0}$ one has*

$$n!_D = \prod_{\mathfrak{p}} \mathfrak{p}^{w_{N(\mathfrak{p})}(n)} = \prod_q \Pi_q^{w_q(n)}.$$

Moreover, the following conditions are equivalent for any $n \in \mathbb{Z}_{\geq 0}$:

(1) *The D-module $\text{Int}_n(D)$ is free.*
(2) *The fractional ideal $\prod_{k=0}^{n} \mathfrak{I}_k$ of D is principal.*
(3) *The ideal $\prod_q \Pi_q^{u_q(n)} = \prod_{k=0}^{n} k!_D$ of D is principal.*

Proposition 4.2 ([1, 4]). *Let D be a Dedekind domain. The group $\text{PO}(D)$ is the subgroup of $\text{Cl}(D)$ generated by any of the following sets: $\{\Pi_q : q$ is a prime power$\}$; $\{n!_D : n$ is a positive integer$\}$; $\{q!_D : q$ is a prime power$\}$; $\{\mathfrak{I}_n : n$ is a positive integer$\}$; and $\{\mathfrak{I}_q : q$ is a prime power$\}$. Moreover, the following conditions are equivalent:*

(1) $\text{Int}(D)$ *has a regular basis.*
(2) *The D-module $\text{Int}_n(D)$ is free for all n.*
(3) $\text{PO}(D)$ *is trivial.*
(4) Π_q *is principal for all prime powers q.*
(5) $n!_D$ *is principal for all n.*
(6) $q!_D$ *is principal for all prime powers q.*
(7) $\mathfrak{I}_n$ *is principal for all n.*
(8) $\mathfrak{I}_q$ *is principal for all prime powers q.*

Proposition 4.3. *Let D be a Dedekind domain such that $\text{PO}(D)$ is cyclic of order d with generator g; for each i with $0 < i < d$ let S_i denote the set of all prime powers q such that $\Pi_q = g^i$ in $\text{PO}(D)$, and let S denote the multiset $\sum_{i=1}^{d-1} i S_i$. Then the D-module $\text{Int}_n(D)$ is free if and only if $u_S(n) = \sum_{i<d} i u_{S_i}(n)$ is divisible by d. In particular, one has $\delta(\text{Int}(D)) = \delta(d, S)$.*

Proof. One has

$$\prod_q \Pi_q^{u_q(n)} = \prod_{i<d} \prod_{q \in S_i} \Pi_q^{u_q(n)},$$

which reduces in $\text{PO}(D)$ to

$$\prod_{i<d} \prod_q g^{i u_q(n)} = g^{\sum_{i<d} i u_{S_i}(n)}.$$

Since g is a generator of $PO(D)$, this element is equal to the identity if and only if $u_S(n) = \sum_{i<d} i u_{S_i}(n)$ is divisible by d. $\square$

Corollary 4.4. *Conjecture 2.2(1) implies that Conjecture 1.1 holds for all Dedekind domains D such that $PO(D)$ is finite and cyclic.*

We generalize to the case where $PO(D)$ is not necessarily cyclic as follows.

Proposition 4.5. *Let D be a Dedekind domain such that $PO(D)$ is finite. Choose an isomorphism $\varphi : PO(D) \longrightarrow \mathbb{Z}/m_1\mathbb{Z} \times \cdots \times \mathbb{Z}/m_k\mathbb{Z} = G$. For each $I \in G$, let S_I denote the set of all prime powers q such that $\varphi(\Pi_q) = I$. Then $\mathrm{Int}_n(D)$ is free if and only if $\sum_{I \in G} u_{S_I}(n)I = 0$ in G.*

Proof. The proof is similar to that of Proposition 4.3. $\square$

Corollary 4.6. *Conjecture 2.3(1) implies Conjecture 1.1.*

Proof. This follows from Propositions 2.4 and 4.5. $\square$

5   Galois Number Fields

In this section we examine the consequences of the conjectures of Sect. 2 for computing $\delta(\mathrm{Int}(\mathcal{O}_K))$ for finite Galois extensions K of $\mathbb{Q}$. We denote the radical of an ideal I of a ring R by $\sqrt{I}$. If p is a prime, then f_p and e_p denote the inertial degree and ramification index, respectively, of p in K over $\mathbb{Q}$, and Δ_K the discriminant.

Proposition 5.1 ([1, Proposition II.4.2]). *Let K be a finite Galois extension of $\mathbb{Q}$ and let $D = \mathcal{O}_K$. Then for any power $q = p^r$ of a prime p, one has the following:*

(1) $\Pi_q = \sqrt{pD}$ *if p is ramified in K and $r = f_p$.*
(2) $\Pi_q = pD$ *is principal if p is unramified in K and $r = f_p$.*
(3) $\Pi_q = D$ *if $r \neq f_p$.*

In particular, $PO(D)$ is generated by $\sqrt{pD}$ for the set of primes p dividing Δ_K, and $\mathrm{Int}(D)$ has a regular basis if and only if $\sqrt{pD}$ is principal for all such p.

As a consequence of this result, if K is a finite Galois extension of $\mathbb{Q}$ and $D = \mathcal{O}_K$, then (1) the set of all prime powers q such that Π_q is not principal is finite, and (2) if Π_{q_1} and Π_{q_2} are two distinct nonprincipal generators of $PO(D)$, then q_1 and q_2 are powers of distinct primes. Thus, we have the following corollary.

Corollary 5.2. *Conjecture 2.5 implies Conjecture 1.2.*

Moreover, Propositions 4.1, 5.1, and 3.4 immediately yield the following.

Proposition 5.3. *Let K be a finite Galois extension of $\mathbb{Q}$ and $D = \mathcal{O}_K$. Then the D-module $\mathrm{Int}_n(D)$ is free if and only if $\prod_{p|\Delta_K} \sqrt{pD}^{u_{p^{f_p}}(n)}$ is principal.*

Corollary 5.4. *Let K be a finite Galois extension of $\mathbb{Q}$ and $D = \mathcal{O}_K$, and suppose that $K/\mathbb{Q}$ has a unique ramified prime p such that $\sqrt{pD}$ is not principal. Then $\mathrm{PO}(D)$ is cyclic, generated by $\sqrt{pD}$, and $\mathrm{Int}_n(D)$ is free if and only if $u_{p^{f_p}}(n)$ is divisible by $|\mathrm{PO}(D)|$. Suppose, moreover, that $q = p^{f_p} = |\mathrm{PO}(D)|$. Then*
$$\delta(\mathrm{Int}(D)) = \delta(q, \{q\}) = \frac{1}{q^2} \sum_{i=1}^{q} \gcd(i, q) = \frac{(f_p+1)p - f_p}{qp} > \frac{1}{q} = \frac{1}{|\mathrm{PO}(D)|}.$$

The following theorem provides a useful tool for studying the Pólya-Ostrowski group of cyclic and Galois number fields.

Theorem 5.5 ([9]). *Let K be a finite Galois extension of $\mathbb{Q}$ of degree n with Galois group G. Then $\mathrm{PO}(\mathcal{O}_K)$ is isomorphic to a quotient of $H = \prod_{p \mid \Delta_K} \mathbb{Z}/e_p\mathbb{Z}$ by a subgroup isomorphic to $H^1(G, \mathcal{O}_K^*)$, and the generators $\prod_{p^{f_p}}$ of $\mathrm{PO}(\mathcal{O}_K)$ for $p \mid \Delta_K$ correspond respectively to the elementary unit vectors in H. Moreover, if K is a cyclic extension of $\mathbb{Q}$, then $H^1(G, \mathcal{O}_K^*)$ has order $2n$ if K is real and $N(\mathcal{O}_K^*) = \{1\}$, and it has order n otherwise.*

Corollary 5.6. *Let K be a finite Galois extension of $\mathbb{Q}$ of degree n. Let S be the set of all integer primes p ramified in K such that $\prod_{p^{f_p}} = \sqrt{p\mathcal{O}_K}$ is not principal. Then $\mathrm{PO}(\mathcal{O}_K)$ is isomorphic to a quotient group of $H = \prod_{p \in S} \mathbb{Z}/e_p\mathbb{Z}$ and in particular has exponent dividing n. Moreover, the generators $\prod_{p^{f_p}}$ of $\mathrm{PO}(\mathcal{O}_K)$ for $p \in S$ correspond respectively to the elementary unit vectors in H.*

Proof. If $\prod_{p^{f_p}}$ is principal, then $\prod_{p^f}$ is trivial in $\mathrm{PO}(\mathcal{O}_K)$, and therefore the corresponding elementary unit vector in $\prod_{p \mid \Delta_K} \mathbb{Z}/e_p\mathbb{Z}$ lies in the image of the subgroup $H^1(G, \mathcal{O}_K^*)$, whence the factor $\mathbb{Z}/e_p\mathbb{Z}$ vanishes in the quotient ring. □

Corollary 5.7 ([2, Corollary 3.11]). *Let K be a cyclic extension of $\mathbb{Q}$ of degree n, let $D = \mathcal{O}_K$, and let $e = \prod_{p \mid \Delta_K} e_p$.*

(1) *If K is real and $N_{K/\mathbb{Q}}(D^*) = \{1\}$, then $|\mathrm{PO}(D)| = e/2n$.*
(2) *Otherwise, $|\mathrm{PO}(D)| = e/n$.*

If $n = q \neq 2$ is prime, then $\mathrm{PO}(D) \cong (\mathbb{Z}/q\mathbb{Z})^{s-1}$, where s is the number of distinct primes dividing Δ_K. If n is a power of a prime q, then $\mathrm{PO}(D)$ is a q-group.

Proposition 5.8. *Let K be a cyclic extension of $\mathbb{Q}$ of prime degree $q \neq 2$. Let $p_1, p_2, \ldots, p_s$ denote the prime divisors of Δ_K. For some rearrangement of the primes p_i, there exist $0 \leq c_1, c_2, \ldots, c_{s-1} < q$ such that $\prod_{p_s} = \prod_{i=1}^{s-1} \prod_{p_i}^{c_i}$ in $\mathrm{Cl}(\mathcal{O}_K)$. Then $\mathrm{Int}_n(\mathcal{O}_K)$ is free if and only if q divides $u_{p_i}(n) + c_i u_{p_s}(n)$ for all $i < s$.*

Proof. By Corollary 5.7 the group $\mathrm{PO}(D)$ is isomorphic to a quotient of $(\mathbb{Z}/q\mathbb{Z})^s$ by a subgroup of order q. By rearranging the factors we may assume that the subgroup is generated by an element of the form $(c_1, c_2, \ldots, c_{s-1}, -1)$, where $0 \leq c_1, c_2, \ldots, c_{s-1} < q$ for all i. The proposition then follows from Proposition 4.5. □

Proposition 5.9. *Let K be a finite Galois extension of $\mathbb{Q}$. Choose an isomorphism $\mathrm{PO}(\mathcal{O}_K) \cong G = \prod_{i=1}^{r} \mathbb{Z}/d_i\mathbb{Z}$. For each prime $p \mid \Delta_K$ let $I_p = (i_{p,1}, i_{p,2}, \ldots, i_{p,r})$,*

denote the image of $\Pi_{p^{f_p}}$ *in* G. *Then* $\mathrm{Int}_n(\mathcal{O}_K)$ *is free if and only if* $\sum_p u_{p^{f_p}} I_p = 0$
in G. *Moreover, if Conjecture 2.3(3) holds, then* $\delta(\mathrm{Int}(\mathcal{O}_K)) = 1/|G|$ *if and only*
if for each prime divisor p *of* $|G|$ *the vectors* $\sum_{q \neq p} i_{q,j} \mathbf{e}_q \in \prod_{q \neq p} \mathbb{F}_p$ *for* j *with*
$p \mid d_j$ *are linearly independent over* $\mathbb{F}_p$, *where the* $\mathbf{e}_q$ *are the elementary coordinate*
vectors.

Corollary 5.10. *Suppose that Conjecture 2.3(3) holds. Let* K *be a cyclic extension*
of $\mathbb{Q}$ *of prime degree* q. *Then* $\delta(\mathrm{Int}(\mathcal{O}_K)) = 1/|\mathrm{PO}(\mathcal{O}_K)|$ *if and only if either* q *is*
unramified in K *(i.e.,* $K/\mathbb{Q}$ *is tamely ramified) or* q *is ramified in* K *and* $\prod_{p \mid \Delta_K} \Pi_p^{c_p}$
is principal for some integers c_p *for* $p \mid \Delta_K$ *prime with* $q \nmid c_q$.

The following result, if Conjecture 2.2(5) holds, computes $\delta(\mathrm{Int}(\mathcal{O}_K))$ for "most"
Galois number fields such that $\mathrm{PO}(\mathcal{O}_K)$ is cyclic.

Proposition 5.11. *Let* K *be a finite Galois extension of* $\mathbb{Q}$ *such that* $\mathrm{PO}(\mathcal{O}_K)$ *is*
cyclic of order d. *Let* $p_1, p_2, \ldots, p_s$ *be the primes dividing* Δ_K *such that* Π_{q_i} *is not*
principal, where $q_i = p_i^{f_{p_i}}$. *Let*

$$l = \gcd\left(d, \frac{dq_1}{|\Pi_{q_1}|}, \frac{dq_2}{|\Pi_{q_2}|}, \ldots, \frac{dq_s}{|\Pi_{q_s}|}\right).$$

Then $l = l_1 l_2 \ldots l_s$, *where* $l_i = \gcd\left(q_i, \frac{d}{\mathrm{lcm}_{j \neq i} |\Pi_{q_j}|}\right)$ *for all* $i \leq s$, *and one has the*
following:

(1) *If Conjecture 2.2(5) holds, then*

$$\delta(\mathrm{Int}(\mathcal{O}_K)) = \frac{1}{dl} \sum_{i=1}^{l} \gcd(i, l) = \frac{1}{dl_1 \cdots l_s} \sum_{i=1}^{l_1} \gcd(i, l_1) \cdots \sum_{i=1}^{l_s} \gcd(i, l_s)$$

provided that either $s \neq 1$ *or* $s = 1$ *and* $v_{p_1}(d) \leq v_{p_1}(q_1)(= f_{p_1})$.
(2) *If Conjecture 2.2(4) holds, then*

$$\frac{1}{d} \leq \delta(\mathrm{Int}(\mathcal{O}_K)) \leq \frac{1}{d^2} \sum_{i=1}^{d} \gcd(i, d).$$

(3) *If Conjecture 2.2(3) holds, then* $\delta(\mathrm{Int}(\mathcal{O}_K)) = 1/d$ *if and only if* $l = 1$, *if and*
only if $l_1 = l_2 = \cdots = l_s = 1$, *if and only if for every prime* $p \mid d$ *there exists*
a prime $q \neq p$ *such that* $v_p(|\Pi_{q^{f_q}}|) = v_p(d)$ *(which holds, in particular, if*
$v_p(|\Pi_{p^{f_p}}|) < v_p(d)$ *for all* $p \mid d$).

Proof. Let d be a generator of $G = \mathrm{PO}(\mathcal{O}_K)$, and for each i, let t_i denote the unique
nonnegative integer less than $|G|$ such that $\Pi_{q_i} = g^{t_i}$ in G. By Proposition 4.3,
$\mathrm{Int}_n(\mathcal{O}_K)$ is free if and only if $u_S(n) = \sum_{i=1}^{s} t_i u_{q_i}(n)$ is divisible by d, where S is
the multiset with $S(n) = t_i$ if $n = q_i$ for some i and $S(n) = 0$ otherwise. Moreover,

$$\gcd(d, 2S(2), 3S(3), \ldots) = \gcd(d, t_1 q_1, t_2 q_2, \ldots, t_s q_s)$$
$$= \gcd(d, \gcd(d, t_1) q_1, \gcd(d, t_2) q_2, \ldots, \gcd(d, t_s) q_s)$$
$$= l = l_1 l_2 \ldots l_s.$$

Therefore, if Conjecture 2.2(5) holds, then $\delta(\text{Int}(\mathcal{O}_K)) = \frac{1}{dl} \sum_{i=1}^{l} \gcd(i, l)$ provided that $\gcd(\gcd(S), d / \gcd(d, \gcd(S))) = 1$, that is, provided that $s \neq 1$ or $s = 1$ and $v_{p_1}(d) \leq v_{p_1}(q_1) (= f_{p_1})$. Moreover, the arithmetic function $f(l) = \sum_{i=0}^{l} \gcd(i, l)$ is multiplicative and the l_i are pairwise relatively prime, so $f(l) = f(l_1) f(l_2) \cdots f(l_s)$. This proves statement (1) of the proposition, and (2) and (3) follow easily. $\qquad \square$

6 Quadratic Number Fields K with $|\text{PO}(\mathcal{O}_K)| = 2$

Cahen and Chabert [1, Corollary II.4.5] characterize all quadratic number fields K such that $\text{PO}(\mathcal{O}_K)$ is trivial. In a similar vein, the following result characterizes all quadratic number fields K such that $\text{PO}(\mathcal{O}_K)$ has order 2.

Proposition 6.1. *Let* $K = \mathbb{Q}(\sqrt{d})$ *be a quadratic extension of* $\mathbb{Q}$, *where* d *is a squarefree integer, let* $D = \mathcal{O}_K$, *and let* $N = N_{K/\mathbb{Q}}$. *Then* $\text{PO}(D)$ *has order two if and only if (1) exactly two prime numbers are ramified in* K *and* K *is imaginary or* $N(D^*) \neq \{1\}$ *or (2) exactly three prime numbers are ramified in* K *and* K *is real and* $N(D^*) = \{1\}$. *The following subcases are exhaustive:*

(1) $d = -p$, *where* $p \equiv 1 \pmod 4$ *is prime.*
(2) $d = -2p$, *where* p *is an odd prime.*
(3) $d = -pq$, *where* $p \equiv 1 \pmod 4$ *and* $q \equiv 3 \pmod 4$ *are prime.*
(4) $d = 2p$, *where* p *is an odd prime and* $N(D^*) \neq \{1\}$.
(5) $d = pq$, *where* $p, q \equiv 1 \pmod 4$ *are prime and* $N(D^*) \neq \{1\}$.
(6) $d = pq$, *where* $p \equiv 1 \pmod 4$ *and* $q \equiv 3 \pmod 4$ *are prime.*
(7) $d = 2pq$, *where* p, q *are distinct odd primes and* $N(D^*) = \{1\}$ *(which holds, for example, if* $p \equiv 3 \pmod 4$ *or* $q \equiv 3 \pmod 4$).
(8) $d = pqr$, *where* $pqr \equiv 1 \pmod 4$ *and* p, q, r *are distinct primes and* $N(D^*) = \{1\}$.

Proof. The first statement of the proposition follows from Corollary 5.7. Thus $\text{PO}(D)$ has order two if and only if Δ_K has two distinct prime factors or else K is real, $N(D^*) = \{1\}$, and Δ_K has three distinct prime factors. Moreover, Δ_K is equal to d if $d \equiv 1 \pmod 4$, and $4d$ otherwise. The eight cases of the proposition follow from these two facts. $\qquad \square$

For each of the cases in the proposition above, the following result determines necessary and sufficient conditions on n for $\text{Int}_n(\mathcal{O}_K)$ to be free.

Proposition 6.2. *Let* $K = \mathbb{Q}(\sqrt{d})$ *be a quadratic extension of* $\mathbb{Q}$, *where* d *is a squarefree integer, and let* $D = \mathcal{O}_K$. *Then* $\prod_{p\mid d} \Pi_p = \sqrt{d}\, D$ *is principal. Suppose that* $|\mathrm{PO}(D)| = 2$. *Let* S *be the set of prime numbers* p *that are ramified in* K *such that* $\Pi_p = \sqrt{pD}$ *is not principal. Then* $\mathrm{Int}_n(D)$ *is free if and only if the ideal* $\prod_{p\in S} \Pi_p^{u_p(n)}$ *is principal, if and only if* $u_S(n)$ *is even. Moreover,* Δ_K *and* S, *in cases (1) through (8), respectively, of Proposition 6.1 are given as follows:*

(1) $\Delta_K = -4p$ *and* $S = \{2\}$.
(2) $\Delta_K = -8p$ *and* $S = \{2, p\}$.
(3) $\Delta_K = -pq$ *and* $S = \{p, q\}$.
(4) $\Delta_K = 8p$ *and* $S = \{2, p\}$.
(5) $\Delta_K = pq$ *and* $S = \{p, q\}$.
(6) $\Delta_K = 4pq$, *and* $S = \{2\}$, $S = \{p, q\}$, *or* $S = \{2, p, q\}$. *Each of these three cases is possible and holds, respectively, if and only if:*

    (a) *There exist integers* a, b *such that* $pa^2 - qb^2 = \pm 1$. *For example, if* $d = 5 \cdot 11$, *then* $\Pi_5 = (15 + 2\sqrt{55})$ *is principal, and therefore* Π_3 *is also principal, and* $S = \{2\}$.

    (b) *There exist integers* a, b *such that* $a^2 - db^2 = \pm 2$. *For example, if* $d = 7 \cdot 17$, *then* $\Pi_2 = (11 + \sqrt{119})$ *is principal, and* $S = \{7, 17\}$.

    (c) *There exist integers* a, b *such that* $pa^2 - qb^2 = \pm 2$. *For example, if* $d = 3 \cdot 5$, *then* $\Pi_2 \Pi_3 = (3 + \sqrt{15})$ *is principal, and* $S = \{2, 3, 5\}$.

(7) $\Delta_K = 8pq$ *and* $S \subseteq \{2, p, q\}$, *where* $|S| = 2$. *The following examples show that one may or may not have* $2 \in S$, *and* S *may contain 0, 1, or 2 primes congruent to 1 mod 4, and likewise* S *may contain 0, 1, or 2 primes congruent to 3 mod 4:*

    (a) *If* $d = 2 \cdot 3 \cdot 5$, *then* $\Pi_5 = (5 + \sqrt{30})$ *is principal and* $S = \{2, 3\}$.
    (b) *If* $d = 2 \cdot 5 \cdot 11$, *then* $\Pi_{11} = (11 + \sqrt{110})$ *is principal and* $S = \{2, 5\}$.
    (c) *If* $d = 2 \cdot 3 \cdot 11$, *then* $\Pi_2 = (8 + \sqrt{66})$ *is principal and* $S = \{3, 11\}$.
    (d) *If* $d = 2 \cdot 11 \cdot 17$, *then* $\Pi_2 = (58 + 3\sqrt{264})$ *is principal and* $S = \{11, 17\}$.
    (e) *If* $d = 2 \cdot 17 \cdot 41$, *then* $\Pi_2 = (112 + 3\sqrt{1394})$ *is principal and* $S = \{17, 41\}$ *(and* $N(D^*) = \{1\}$ *since the continued fraction of* $\sqrt{d}$ *has (even) period length 6).*

(8) $\Delta_K = pqr$ *and* $S \subseteq \{p, q, r\}$, *where* $|S| = 2$. *The following examples show that* S *may contain 0, 1, or 2 primes congruent to 1 mod 4:*

    (a) *If* $d = 3 \cdot 5 \cdot 7$, *then* $\Pi_5 = (10 + \sqrt{105})$ *is principal and* $S = \{3, 7\}$.
    (b) *If* $d = 3 \cdot 5 \cdot 11$, *then* $\Pi_{11} = (77 + 6\sqrt{165})$ *is principal and* $S = \{3, 5\}$.
    (c) *If* $d = 5 \cdot 13 \cdot 29$, *then* $\Pi_{29} = (87 + 2\sqrt{1885})$ *is principal and* $S = \{5, 13\}$ *(and* $N(D^*) = \{1\}$ *since the continued fraction of* $\sqrt{d}$ *has (even) period length 4).*

Proof. The computation of S in each case follows readily from the fact that $\prod_{p\mid d} \Pi_p = \sqrt{d}\, D$ is principal. The examples in cases (6) through (8) are easily verified by showing that the ideal Π_p is principal for the relevant prime $p \mid \Delta_K$. $\quad\square$

Corollary 6.3. *Let* $K = \mathbb{Q}(\sqrt{d})$, *where either (1)* $d = -p$, *where* $p \equiv 1 \pmod 4$ *is prime, or (2)* $d = pq$, *where* $p \equiv 1 \pmod 4$ *and* $q \equiv 3 \pmod 4$ *are prime, and there exist integers* a, b *such that* $pa^2 - qb^2 = \pm 1$. *Then* $\mathrm{PO}(\mathcal{O}_K) = 2$ *and* $\delta(\mathrm{Int}(\mathcal{O}_K)) = 3/4$. *Moreover, if Conjecture 2.2(3) holds, then one has* $\delta(\mathrm{Int}(\mathcal{O}_K)) = 1/2$ *for any other quadratic extension* K *of* $\mathbb{Q}$ *with* $\mathrm{PO}(\mathcal{O}_K) = 2$.

7 Cyclic Cubic Number Fields

Recall by Corollary 5.7 that if K is a cyclic extension of $\mathbb{Q}$ of prime degree $q \neq 2$, then $\mathrm{PO}(\mathcal{O}_K)$ has order q^{s-1}, where s is the number of prime integers ramified in K. Using this fact, together with the following theorem, one can characterize all cyclic cubic extensions K for which $\mathrm{PO}(\mathcal{O}_K)$ has a given order.

Proposition 7.1 ([3, Lemma 6.4.5 and Theorem 6.4.11]). *For every cyclic cubic number field* K, *there exist unique integers* e, u, v *such that* e *is a product of distinct primes congruent to* 1 *mod* 3, u *is congruent to* 2 *mod* 3, $4e = u^2 + 3v^2$, $v > 0$, *and* $K = \mathbb{Q}(\theta)$, *where* θ *is a root of the polynomial* $f = X^3 - 3eX - eu$. *Conversely, every field of this type is a cyclic cubic number field. Moreover, the discriminant of* f *is* $(9ev)^2$; *if* $3 \nmid v$ *then* $\Delta_K = (9e)^2$; *and if* $3 \mid v$ *then* $\Delta_K = e^2$.

Using these results, [7] characterizes all cyclic cubic extensions K of $\mathbb{Q}$ such that $\mathrm{PO}(\mathcal{O}_K)$ is trivial (or equivalently $s = 1$): clearly $s = 1$ if and only if either $e = 1, u = -1, v = 1$, or e is prime and $3 \mid v$. In a similar vein, the following result characterizes all cyclic cubic extensions K of $\mathbb{Q}$ such that $\mathrm{PO}(\mathcal{O}_K)$ has order 3 (or equivalently $s = 2$). In this case, as in the quadratic case, we are also able to characterize all n such that $\mathrm{Int}_n(\mathcal{O}_K)$ is free.

Proposition 7.2. *The cyclic cubic number fields* K *such that* $|\mathrm{PO}(\mathcal{O}_K)| = 3$ *are precisely those of the form* $K = \mathbb{Q}(\theta)$, *where* θ *is a root of the polynomial* $X^3 - 3eX - eu$ *and* e, u, v *are integers such that* u *is congruent to* 2 *mod* 3 *and* $4e = u^2 + 3v^2$ *and either (1)* $3 \nmid v$ *and* e *is a prime congruent to* 1 *mod* 3 *or (2)* $3 \mid v$ *and* e *is a product of two distinct primes congruent to* 1 *mod* 3. *Choose a prime* $p \mid \Delta_K$ *such that* $\Pi_p = \sqrt{pD}$ *is not principal. For the other ramified prime* q, *set* ϵ *to be* 0 *if* Π_q *is principal,* 1 *if* Π_q *is in the same ideal class as* Π_p, *and* -1 *otherwise. Then* $\mathrm{Int}_n(D)$ *is free if and only if the ideal* $\Pi_p^{u_p(n)} \Pi_q^{u_q(n)}$ *is principal, if and only if* $u_p(n) + \epsilon u_q(n)$ *is divisible by* 3.

The following series of examples were computed with the computer algebra and number theory system Sage. We follow the notation of Proposition 7.2.

Example 7.3. In the following examples of cyclic cubic number fields K, one has $p = 3$ and $\epsilon = 0$ (i.e., Π_q is principal). Thus $\mathrm{Int}_n(\mathcal{O}_K)$ is free as an $\mathcal{O}_K$-module if and only if $u_3(n)$ is divisible by 3, if and only if either (1) $n \equiv 2 \pmod 3$ or (2) $n \not\equiv 2 \pmod 3$ and $s_3(n) \equiv n \pmod 3$. In particular, Corollary 5.4 implies that $\delta(\mathrm{Int}(\mathcal{O}_K)) = 5/9$.

(1) $e = 19$, $u = -1$, $v = 5$. Then $\Delta_K = (9 \cdot 19)^2$.
(2) $e = 109$, $u = -19$, $v = 5$. Then $\Delta_K = (9 \cdot 109)^2$.
(3) $e = 181$, $u = 26$, $v = 4$. Then $\Delta_K = (9 \cdot 181)^2$.

Example 7.4. In the following examples of cyclic cubic number fields K, one has $p \neq 3$ and $\epsilon = 0$ (i.e., Π_q is principal). Thus $\mathrm{Int}_n(\mathcal{O}_K)$ is free as an $\mathcal{O}_K$-module if and only if $u_p(n)$ is divisible by 3, and therefore Conjecture 2.2(3) implies that $\delta(\mathrm{Int}(\mathcal{O}_K)) = 1/3$.

(1) $e = 61$, $u = 14$, $v = 4$. Then $\Delta_K = (9 \cdot 61)^2$ and $p = 61$.
(2) $e = 67$, $u = -16$, $v = 2$. Then $\Delta_K = (9 \cdot 67)^2$ and $p = 67$.
(3) $e = 7 \cdot 13$, $u = -16$, $v = 6$. Then $\Delta_K = (7 \cdot 13)^2$ and $p = 7$.
(4) $e = 19 \cdot 37$, $u = -42$, $v = 6$. Then $\Delta_K = (19 \cdot 37)^2$ and $p = 19$.

Example 7.5. In the following examples of cyclic cubic number fields K, one has $p = 3$ and $\epsilon = 1$ (i.e., there is a prime $q \neq 3$ such that Π_3 and Π_q are not principal but $\Pi_3^2 \Pi_q$ is). Thus $\mathrm{Int}_n(\mathcal{O}_K)$ is free as an $\mathcal{O}_K$-module if and only if $u_3(n) + u_q(n)$ is divisible by 3, and Conjecture 2.2(3) implies that $\delta(\mathrm{Int}(\mathcal{O}_K)) = 1/3$.

(1) $e = 7$, $u = 5$, $v = 1$. Then $\Delta_K = (9 \cdot 7)^2$ and $q = 7$.
(2) $e = 31$, $u = 11$, $v = 1$. Then $\Delta_K = (9 \cdot 31)^2$ and $q = 31$.
(3) $e = 43$, $u = -13$, $v = 1$. Then $\Delta_K = (9 \cdot 43)^2$ and $q = 43$.
(4) $e = 73$, $u = 17$, $v = 1$. Then $\Delta_K = (9 \cdot 73)^2$ and $q = 73$.

Example 7.6. In the following examples of cyclic cubic number fields K, one has $p \neq 3$, $q \neq 3$, and $\epsilon = -1$ (i.e., there are primes $p, q \neq 3$ such that Π_p and Π_q are not principal but $\Pi_p \Pi_q$ is). Thus $\mathrm{Int}_n(\mathcal{O}_K)$ is free as an $\mathcal{O}_K$-module if and only if $u_p(n) - u_q(n)$ is divisible by 3, and Conjecture 2.2(3) implies that $\delta(\mathrm{Int}(\mathcal{O}_K)) = 1/3$.

(1) $e = 13 \cdot 19$, $u = -31$, $v = 3$. Then $\Delta_K = (13 \cdot 19)^2$ and $p = 13, q = 19$.
(2) $e = 7 \cdot 31$, $u = 29$, $v = 3$. Then $\Delta_K = (7 \cdot 31)^2$ and $p = 7, q = 31$.

Next we provide some examples of cyclic cubic number fields where the Pólya-Ostrowski group is not cyclic.

Example 7.7. In the following examples of cyclic cubic number fields K, there are exactly three prime integers that are ramified in K, and therefore, $\mathrm{PO}(\mathcal{O}_K) \cong G = (\mathbb{Z}/3\mathbb{Z})^2$. Moreover, Conjecture 2.3(3) implies that $\delta(\mathrm{Int}(\mathcal{O}_K)) = 1/9$.

(1) $e = 7 \cdot 19$, $u = 23$, $v = 1$. Then $\Delta_K = (3^2 \cdot 7 \cdot 19)^3$. Moreover, $\Pi_3^2 \Pi_7 \Pi_{19}$ is principal. Thus if Π_7 and Π_{19} are represented in G by $(1, 0)$ and $(0, 1)$, respectively, then Π_3 is represented by $(1, 1)$. Therefore $\mathrm{Int}_n(\mathcal{O}_K)$ is free if and only if 3 divides both $u_3(n) + u_7(n)$ and $u_3(n) + u_{19}(n)$.
(2) $e = 7 \cdot 13 \cdot 19$, $u = 83$, $v = 3$. Then $\Delta_K = (7 \cdot 13 \cdot 19)^3$. Moreover, $\Pi_7 \Pi_{13} \Pi_{19}$ is principal. Thus if Π_7 and Π_{13} are represented in G by $(1, 0)$ and $(0, 1)$, respectively, then Π_3 is represented by $(-1, -1)$. Therefore $\mathrm{Int}_n(\mathcal{O}_K)$ is free if and only if 3 divides both $u_7(n) - u_{19}(n)$ and $u_{13}(n) - u_{19}(n)$.
(3) $e = 7 \cdot 19 \cdot 31$, $u = 128$, $v = 6$. Then $\Delta_K = (7 \cdot 19 \cdot 31)^3$. Moreover, $\Pi_7^2 \Pi_{31}$ is principal. Thus if Π_7 and Π_{19} are represented in G by $(1, 0)$ and

(0, 1), respectively, then Π_{31} is represented by $(1, 0)$. Therefore $\text{Int}_n(\mathcal{O}_K)$ is free if and only if 3 divides both $u_7(n) + u_{31}(n)$ and $u_{19}(n)$.

Example 7.8. Let $e = 7 \cdot 13 \cdot 31$, $u = -106$, $v = 4$. Then $\Delta_K = (3^2 \cdot 7 \cdot 13 \cdot 31)^3$, so $\text{PO}(\mathcal{O}_K) \cong G = (\mathbb{Z}/3\mathbb{Z})^3$. Moreover, $\Pi_3 \Pi_7 \Pi_{13}^2$ is principal. Thus if Π_3, Π_7, and Π_{31} are represented in G by $(1, 0, 0)$, $(0, 1, 0)$, and $(0, 0, 1)$, respectively, then Π_{13} is represented by $(1, 1, 0)$. Therefore $\text{Int}_n(\mathcal{O}_K)$ is free if and only if 3 divides $u_3(n) + u_{13}(n)$, $u_7(n) + u_{13}(n)$, and $u_{31}(n)$. Moreover, Conjecture 2.3(3) implies that $\delta(\text{Int}(\mathcal{O}_K)) = 1/27$.

8 Cyclic Quartic and Quintic Examples

In this section we study several cyclic quartic and quintic number fields.

The cyclic quartic number fields in the following proposition are called the *simplest quartic fields*.

Proposition 8.1 ([6]). *Let $m \neq 3$ be a positive integer. Let $\theta_m \in \mathbb{C}$ be a root of $x^4 - mx^3 - 6x^2 + mx + 1$. Then $K_m = \mathbb{Q}(\theta_m)$ is a cyclic quartic extension of $\mathbb{Q}$.*

Example 8.2. Using Sage, we investigated $K = K_m$ for $m \leq 30$ and $m = 52$.

(1) If $m = 1, 2, 4, 5, 6, 8, 9, 10, 11, 14, 15, 16, 21, 22, 24, 25, 26, 29, 30$, then $\text{PO}(\mathcal{O}_K)$ is trivial: $\Pi_{p^{f_p}}$ is principal for all $p \mid \Delta_K$.
(2) If $m = 7, 12, 13, 17, 19, 20, 23, 27, 28$, then Δ_K has two prime factors p and q, and $\Pi_{p^{f_p}}^2$ and $\Pi_{p^{f_p}} \Pi_{q^{f_q}}$ are principal but $\Pi_{p^{f_p}}$ and $\Pi_{q^{f_q}}$ are not. In this case $\text{PO}(\mathcal{O}_K) \cong \mathbb{Z}/2\mathbb{Z}$ and $\text{Int}_n(\mathcal{O}_K)$ is free if and only if $u_{p^{f_p}}(n) + u_{q^{f_q}}(n)$ is even. In particular, if Conjecture 2.2(3) holds then $\delta(\text{Int}(\mathcal{O}_{K_m})) = 1/2$. Moreover, one has $f_p = f_q = 1$ and $e_p = e_q = 4$ for all such m except for $m = 28$, where one has $\Delta_K = 2^{11} \cdot 5^2$, $e_5 = f_5 = 2$, $e_2 = 4$, and $f_2 = 1$.
(3) If $m = 18$, then $\Delta_K = 2^4 \cdot 5^3 \cdot 17^3$ and $e_2 = f_2 = 2$. Moreover, Π_4 and $\Pi_5 \Pi_{17}$ are principal, but Π_5^2 is not. Therefore $\text{PO}(\mathcal{O}_K) \cong \mathbb{Z}/4\mathbb{Z}$, and $\text{Int}_n(D)$ is principal if and only if 4 divides $u_5(n) - u_{17}(n)$. In particular, if Conjecture 2.2(3) holds then $\delta(\text{Int}(\mathcal{O}_{K_m})) = 1/4$.
(4) If $m = 52$, then $\Delta_K = 2^{11} \cdot 5^3 \cdot 17^3$, and $\Pi_2 \Pi_5 \Pi_{17}$ and $\Pi_5^2 \Pi_{17}^2$ are principal, but $\Pi_5 \Pi_{17}$, Π_5^2, and Π_{17}^2 are not. Therefore $\text{PO}(\mathcal{O}_K) \cong \mathbb{Z}/2\mathbb{Z} \times \mathbb{Z}/4\mathbb{Z}$, with Π_2, Π_5, and Π_{17} represented by $(1, 0)$, $(0, 1)$, and $(1, -1)$, respectively. Therefore $\text{Int}_n(D)$ is principal if and only if 2 divides $u_2(n) + u_{17}(n)$ and 4 divides $u_5(n) - u_{17}(n)$. In particular, if Conjecture 2.3(3) holds then $\delta(\text{Int}(\mathcal{O}_{K_m})) = 1/8$.

Now we turn to a well-known family of cyclic quintic number fields.

Proposition 8.3 ([5,8]). *Let $m \in \mathbb{Z}$. Let $\theta_m \in \mathbb{C}$ be a root of*

$$x^5 + m^2 x^4 - (2m^3 + 6m^2 + 10m + 10)x^3 + (m^4 + 5m^3 + 11m^2 + 15m + 5)x^2$$
$$+ (m^3 + 4m^2 + 10m + 10)x + 1.$$

Then $K_m = \mathbb{Q}(\theta_m)$ is a cyclic quintic extension of $\mathbb{Q}$ with discriminant f^4, where $f = 5^b \prod_{p \in S} p$ is the conductor of $K/\mathbb{Q}$, where $b = 2$ if $5 \mid m$ and $b = 0$ otherwise, and where S is the set of all primes $p \equiv 1 \pmod 5$ such that $5 \nmid v_p(m^4 + 5m^3 + 15m^2 + 25m + 25)$.

Proposition 8.4. Assume the notation of Proposition 8.3. Suppose that $\Pi_5^b \prod_{p \in S} \Pi_p$ is principal.

(1) Suppose that $5 \nmid m$ (hence $b = 0$), and choose $p_0 \in S$. Then $\mathrm{PO}(\mathcal{O}_K) \cong (\mathbb{Z}/5\mathbb{Z})^{|S|-1}$, and $\mathrm{Int}_n(D)$ is free if and only if 5 divides $u_p(n) - u_{p_0}(n)$ for all $p \in S \setminus \{p_0\}$, if and only if 5 divides $u_p - u_q(n)$ for all $p, q \in S$.

(2) Suppose that $5 \mid m$ (hence $b = 2$). Then $\mathrm{PO}(\mathcal{O}_K) \cong (\mathbb{Z}/5\mathbb{Z})^{|S|}$, and $\mathrm{Int}_n(D)$ is free if and only if 5 divides $u_p(n) + 2u_5(n)$ for all $p \in S$, if and only if 5 divides $u_5(n) + 3u_p(n)$ for all $p \in S$.

In particular, Conjecture 2.3(3) implies that $\delta(\mathrm{Int}(\mathcal{O}_{K_m})) = 1/|\mathrm{PO}(\mathcal{O}_{K_m})|$.

Proof. This follows easily from Propositions 5.7 and 5.8.   □

Example 8.5. Using Sage, we investigated $K = K_m$ for $|n| \leq 10$. In all of these cases, $\Pi_5^b \prod_{p \in S} \Pi_p$ is principal (though we were unable to prove it generally), so Proposition 8.4 applies.

(1) If $m = 0, \pm 1, \pm 2, -3, \pm 4, -6, 7, 8, -9$, then Δ_K has a unique prime factor and therefore $\mathrm{PO}(\mathcal{O}_K)$ is trivial.
(2) If $m = 3, \pm 5, 6, -7, -8, 9, -10$, then Δ_K has exactly two distinct prime factors and therefore $\mathrm{PO}(\mathcal{O}_K) \cong \mathbb{Z}/5\mathbb{Z}$.
(3) If $m = 10$, then $\Delta_K = (5^2 \cdot 11 \cdot 61)^4$, and therefore $\mathrm{PO}(\mathcal{O}_K) \cong (\mathbb{Z}/5\mathbb{Z})^2$.

Problem 8.6. Assuming the notation of Proposition 8.3, is $\Pi_5^b \prod_{p \in S} \Pi_p$ necessarily principal in $\mathcal{O}_{K_m}$, for all $m \in \mathbb{Z}$?

9 Noncyclic Examples

Finally, we present two examples of noncyclic Galois number fields K. In the second example, $\mathrm{PO}(\mathcal{O}_K)$ is also noncyclic.

Example 9.1. Let K denote the Galois closure of $\mathbb{Q}(\theta)$, where θ is a root of $x^3 - 3x + 8$. Then $\Delta_K = -2^6 \cdot 3^8 \cdot 5^3$. Moreover, Π_2^2, Π_3^3, and Π_5 are principal (and $f_2 = f_3 = f_5 = 1$), while Π_2 and Π_3 are not. Therefore, $\mathrm{PO}(\mathcal{O}_K) \cong \mathbb{Z}_2 \times \mathbb{Z}_3 \cong \mathbb{Z}_6$, and $\mathrm{Int}_n(\mathcal{O}_K)$ is free if and only if $2 \mid u_2(n)$ and $3 \mid u_3(n)$, if and only if $6 \mid 3u_2(n) + 2u_3(n)$. In particular, if Conjecture 2.2(5) or 2.3(2) holds, then $\delta(\mathrm{Int}(\mathcal{O}_K)) = 5/12$.

Example 9.2. Let K denote the Galois closure of $\mathbb{Q}(\theta)$, where θ is a root of $x^3 - 210$. Then $\Delta_K = -2^4 \cdot 3^{11} \cdot 5^4 \cdot 7^4$. Moreover, Π_4^3, Π_{25}^3, $\Pi_4 \Pi_3^2$, and $\Pi_{25} \Pi_7$ are principal (and $f_2 = f_5 = 2$ and $f_3 = f_7 = 1$), while Π_4 and Π_{25} are not.

Therefore $\mathrm{PO}(\mathcal{O}_K) \cong \mathbb{Z}_3 \times \mathbb{Z}_3$, and $\mathrm{Int}_n(\mathcal{O}_K)$ is free if and only if 3 divides both $u_4(n) + u_3(n)$ and $u_{25}(n) - u_7(n)$. In particular, if Conjecture 2.3(3) holds, then $\delta(\mathrm{Int}(\mathcal{O}_K)) = 1/9$.

References

1. P.-J. Cahen, J.-L. Chabert, *Integer-Valued Polynomials*. Mathematical Surveys and Monographs, vol. 48 (American Mathematical Society, Providence, 1997)
2. J.-L. Chabert, Factorial groups and Pólya groups in Galoisian extension of $\mathbb{Q}$. In *Commutative Ring Theory and Applications: Proceedings of the Fourth International Conference*, ed. by M. Fontana, S-E. Kabbaj, S. Wiegand. Lecture Notes in Pure and Applied Mathematics, vol. 231 (Marcel Dekker, New York, 2003)
3. H. Cohen, *A Course in Computational Algebraic Number Theory*. Graduate Texts in Mathematics, vol. 138 (Springer, New York, 2000)
4. J. Elliott, Presentations and module bases of integer-valued polynomial rings. J. Algebra Appl. **12**(1), 1–25 (2013)
5. M.-N. Gras, Non monogénéité de l'anneau des entiers des extensions cycliques de $\mathbb{Q}$ de degré premier $l \geq 5$. J. Number Theory **23**, 347–353 (1986)
6. A. Hoshi, On the simplest quartic fields and related Thue equations. In *Joint Conference of ASCM 2009 and MACIS 2009*, ed. by S. Masakazu et al. COE Lecture Note Series, vol. 22 (Kyushu University, Faculty of Mathematics, Fukuoka, 2009), pp. 320–329
7. A. Leriche, Groupes, Corps et Extensions de Pólya: une Question de Capitulation, Ph.D. Thesis, Université de Picardie Jules Verne, 2010
8. R. Schoof, L.C. Washington, Quintic polynomials and real cyclotomic fields with large class numbers. Math. Comput. **50**, 543–556 (1988)
9. H. Zantema, Integer valued polynomials over a number field. Manuscr. Math. **40**, 155–203 (1982)

Some Closure Operations in Zariski-Riemann Spaces of Valuation Domains: A Survey

Carmelo Antonio Finocchiaro, Marco Fontana, and K. Alan Loper

Abstract In this survey we present several results concerning various topologies that were introduced in recent years on spaces of valuation domains.

Keywords Valuation domain • Semistar operation • *b*-operation • e.a.b. star operation • Spectral space • Constructible topology • Ultrafilter topology • Inverse topology • Kronecker function ring

Mathematics Subject Classification (MSC2010): 13A18; 13F05; 13G05.

1 Spaces of Valuation Domains

The motivations for studying from a topological point of view spaces of valuation domains come from various directions and, historically, mainly from Zariski's work on the reduction of singularities of an algebraic surface and a three-dimensional variety and, more generally, for establishing new foundations of algebraic geometry by algebraic means (see [44–46] and [47]). Other important applications with algebraic geometric flavor are due to Nagata [32, 33], Temkin [42], and Temkin and Tyomkin [43]. Further motivations come from rigid algebraic geometry started

C.A. Finocchiaro (✉) • M. Fontana
Dipartimento di Matematica e Fisica, Università degli Studi Roma Tre,
Largo San Leonardo Murialdo 1, 00146 Roma, Italy
e-mail: carmelo@mat.uniroma3.it; fontana@mat.uniroma3.it

K.A. Loper
Department of Mathematics, Ohio State University, Newark, OH 43055, USA
e-mail: lopera@math.ohio-state.edu

M. Fontana et al. (eds.), *Commutative Algebra: Recent Advances in Commutative Rings,*
Integer-Valued Polynomials, and Polynomial Functions, DOI 10.1007/978-1-4939-0925-4_9,
© Springer Science+Business Media New York 2014

by J. Tate [41] (see also the papers by Fresnel and van der Put [18], Huber and Knebusch [26], Fujiwara and Kato [19]) and from real algebraic geometry (see for instance Schwartz [38] and Huber [25]). For a deeper insight on these topics see [26].

In the following, we want to present some recent results in the literature concerning various topologies on collections of valuation domains.

Let K be a field, let A be an *arbitrary* subring of K, and let qf(A) denote the quotient field of A. Set

$$\mathrm{Zar}(K|A) := \{V \mid V \text{ is a valuation domain and } A \subseteq V \subseteq K = \mathrm{qf}(V)\}.$$

When A is the prime subring of K, we will simply denote by $\mathrm{Zar}(K)$ the space $\mathrm{Zar}(K|A)$. Recall that O. Zariski in [44] introduced a topological structure on the set $Z := \mathrm{Zar}(K|A)$ by taking, as a basis for the open sets, the subsets $B_F := B_F^Z := \{V \in Z \mid V \supseteq F\}$, for F varying in the family of all finite subsets of K (see also [47, Chapter VI, §17, page 110]). When no confusion can arise, we will simply denote by B_x the basic open set $B_{\{x\}}$ of Z. This topology is what that is now called *the Zariski topology on Z* and the set Z, equipped with this topology, denoted also by Z^{zar}, is usually called *the Zariski-Riemann space of K over A* (sometimes called abstract Riemann surface or generalized Riemann manifold).

In 1944, Zariski [44] proved the quasi-compactness of Z^{zar} and later it was proven and rediscovered by several authors, with a variety of different techniques, that if A is an integral domain and K is the quotient field of A, then Z^{zar} is a spectral space, in the sense of M. Hochster [24]. More precisely, in 1986–1987, Dobbs, Fedder, and Fontana in [4, Theorem 4.1] gave a purely topological proof of this fact and Dobbs and Fontana presented a more complete version of this result in [5, Theorem 2] by exhibiting a ring R (namely, the Kronecker function ring of the integral closure of A with respect to the b-operation) such that Z is canonically homeomorphic to $\mathrm{Spec}(R)$ (both endowed with the Zariski topology). Later, using a general construction of the Kronecker function ring developed by F. Halter-Koch [22], it was proved that the Zariski-Riemann space Z is a still a spectral space when K is not necessarily the quotient field of A (see [23, Proposition 2.7] or [9, Corollary 3.6]). In 2004, in the appendix of [28], Kuhlmann gave a model-theoretic proof of the fact that Z is a spectral space. Note also that a purely topological approach for proving that Z is spectral was presented by Finocchiaro in [8, Corollary 3.3]. Very recently, N. Schwartz [39], using the inverse spectrum of a lattice-ordered abelian group and its structure sheaf (see also Rump and Yang [37]), obtained, as an application of his main theorem (via the Jaffard-Ohm theorem), a new proof of the fact that Z is spectral.

Since Z is a spectral space, Z also possesses the constructible (or patch), the ultrafilter and the inverse topologies (definitions will be recalled later) and these other topologies turn out to be more useful than the Zariski topology in several contexts as we will see in the present survey paper.

2 The Constructible Topology

Let A be a ring and let $X := \mathrm{Spec}(A)$ denote the collection of all prime ideals of A. The set X can be endowed with the *Zariski topology* which has several attractive properties related to the "geometric aspects" of the set of prime ideals. As is well known, X^{zar} (i.e., the set X with the Zariski topology) is always quasi-compact, but almost never Hausdorff. More precisely, X^{zar} is Hausdorff if and only if $\dim(A) = 0$. Thus, many authors have considered a finer topology on the prime spectrum of a ring, known as *the constructible topology* (see [3, 21]) or as *the patch topology* [24].

In order to introduce this kind of topology in a more general setting, with a simple set-theoretic approach, we need some notation and terminology. Let $\mathscr{X}$ be a topological space. Following [40], we set

$$\mathring{\mathscr{K}}(\mathscr{X}) := \{U \mid U \subseteq \mathscr{X}, \, U \text{ open and quasi-compact in } \mathscr{X}\},$$
$$\overline{\mathscr{K}}(\mathscr{X}) := \{\mathscr{X} \setminus U \mid U \in \mathring{\mathscr{K}}(\mathscr{X})\},$$
$$\mathscr{K}(\mathscr{X}) := \text{the Boolean algebra of the subsets of } \mathscr{X} \text{ generated by } \mathring{\mathscr{K}}(\mathscr{X}).$$

As in [40], we call *the constructible topology on* $\mathscr{X}$ the topology on $\mathscr{X}$ whose basis of open sets is $\mathscr{K}(\mathscr{X})$. We denote by $\mathscr{X}^{\mathrm{cons}}$ the set $\mathscr{X}$, equipped with the constructible topology. In particular, when $\mathscr{X}$ is a spectral space, the closure of a subset Y of $\mathscr{X}$ under the constructible topology is given by

$$Cl^{\mathrm{cons}}(Y) = \bigcap\{U \cup (\mathscr{X} \setminus V) \mid U \text{ and } V \text{ open and quasi-compact in } \mathscr{X},$$
$$U \cup (\mathscr{X} \setminus V) \supseteq Y\}.$$

Note that, for Noetherian topological spaces, this definition of constructible topology coincides with the classical one given in [3].

When $X := \mathrm{Spec}(A)$, for some ring A, then the set $\mathring{\mathscr{K}}(X^{\mathrm{zar}})$ is a basis of open sets for X^{zar}, and thus the constructible topology on X is finer than the Zariski topology. Moreover, X^{cons} is a compact Hausdorff space and the constructible topology on X is the coarsest topology for which $\mathring{\mathscr{K}}(X^{\mathrm{zar}})$ is a collection of clopen sets (see [21, **I.7.2**]).

3 The Ultrafilter Topology

In 2008, the authors of [16] considered "another" natural topology on $X := \mathrm{Spec}(A)$, by using the notion of an ultrafilter and the following lemma.

Lemma 3.1 (Cahen-Loper-Tartarone [2, Lemma 2.4]). *Let Y be a subset of* $X := \mathrm{Spec}(A)$ *and let $\mathscr{U}$ be an ultrafilter on Y. Then* $\mathfrak{p}_{\mathscr{U}} := \{f \in A \mid V(f) \cap Y \in \mathscr{U}\}$ *is a prime ideal of A called* the ultrafilter limit point of Y, *with respect to $\mathscr{U}$.*

The notion of ultrafilter limit points of sets of prime ideals has been used to great effect in several recent papers [2, 30, 31]. If $\mathcal{U}$ is a trivial (or principal) ultrafilter on the subset Y of X, i.e., $\mathcal{U} = \{S \subseteq Y \mid \mathfrak{p} \in S\}$, for some $\mathfrak{p} \in Y$, then $\mathfrak{p}_{\mathcal{U}} = \mathfrak{p}$. On the other hand, when $\mathcal{U}$ is a nontrivial ultrafilter on Y, then it may happen that $\mathfrak{p}_{\mathcal{U}}$ does not belong to Y. This fact motivates the following definition.

Definition 3.2. Let A be a ring and Y be a subset of $X := \mathrm{Spec}(A)$. We say that Y is *ultrafilter closed* if $\mathfrak{p}_{\mathcal{U}} \in Y$, for each ultrafilter $\mathcal{U}$ on Y.

It is not hard to see that, for each $Y \subseteq X$,

$$Cl^{\mathrm{ultra}}(Y) := \{\mathfrak{p}_{\mathcal{U}} \mid \mathcal{U} \text{ ultrafilter on } Y\}$$

satisfies the Kuratowski closure axioms and the set of all ultrafilter closed sets of X is the family of closed sets for a topology on X, called *the ultrafilter topology on X*. We denote the set X endowed with the ultrafilter topology by X^{ultra}. The main result of [16] is the following.

Theorem 3.3 (Fontana-Loper [16, Theorem 8]). *Let A be a ring. The ultrafilter topology coincides with the constructible topology on the prime spectrum $\mathrm{Spec}(A)$.*

Taking as starting point the situation described above for the prime spectrum of a ring, the next goal is to define an ultrafilter topology on the set $Z := \mathrm{Zar}(K|A)$ (where K is a field and A is a subring of K) that is finer than the Zariski topology. We start by recalling the following useful fact.

Lemma 3.4 (Cahen-Loper-Tartarone [2, Lemma 2.9]). *Let K be a field and A be a subring of K. If Y is a nonempty subset of $Z := \mathrm{Zar}(K|A)$ and $\mathcal{U}$ is an ultrafilter on Y, then*

$$A_{\mathcal{U},Y} := A_{\mathcal{U}} := \{x \in K \mid B_x \cap Y \in \mathcal{U}\}$$

is a valuation domain of K containing A as a subring (i.e., $A_{\mathcal{U}} \in Z$), called the ultrafilter limit point of Y in Z, with respect to $\mathcal{U}$.

As before let Y be a nonempty subset of $Z := \mathrm{Zar}(K|A)$, when $V \in Y$ and $\mathcal{U} := \{S \subseteq Y \mid V \in S\}$ is the trivial ultrafilter of Y generated by V, then $A_{\mathcal{U}} = V$. But, in general, it is possible to construct nontrivial ultrafilters on Y whose ultrafilter limit point are not elements of Y. This leads to the following definition.

Definition 3.5. Let K be a field and A be a subring of K. A subset Y of $Z := \mathrm{Zar}(K|A)$ is *ultrafilter closed* if $A_{\mathcal{U}} \in Y$, for any ultrafilter $\mathcal{U}$ on Y.

For every $Y \subseteq Z$, we set

$$Cl^{\mathrm{ultra}}(Y) := \{A_{\mathcal{U}} \mid \mathcal{U} \text{ ultrafilter on } Y\}.$$

Theorem 3.6 (Finocchiaro-Fontana-Loper [10, Proposition 3.3, Theorems 3.4 and 3.9]). *Let K be a field, A be a subring of K, and $Z := \mathrm{Zar}(K|A)$. The following statements hold.*

(1) $\mathrm{Cl}^{\mathrm{ultra}}$ *satisfies the Kuratowski closure axioms and so the ultrafilter closed sets of Z are the closed sets for a topology, called the ultrafilter topology on Z.*

(2) *Denote by Z^{ultra} the set Z equipped with the ultrafilter topology. Then, Z^{ultra} is a compact Hausdorff topological space.*

(3) *The ultrafilter topology is the coarsest topology for which the basic open sets B_F of the Zariski topology of Z are clopen. In particular, the ultrafilter topology on Z is finer than the Zariski topology and coincides with the constructible topology.*

(4) *The surjective map $\gamma : \mathrm{Zar}(K|A)^{\mathrm{ultra}} \to \mathrm{Spec}(A)^{\mathrm{ultra}}$, mapping a valuation domain to its center on A, is continuous and closed.*

(5) *If A is a Prüfer domain, the map $\gamma : \mathrm{Zar}(K|A)^{\mathrm{ultra}} \to \mathrm{Spec}(A)^{\mathrm{ultra}}$ is a homeomorphism.*

Remark 3.7. (a) From Theorem 3.3 and the last statement in point (3) of the previous theorem it is obvious that points (4) and (5) of Theorem 3.6 hold when one replaces everywhere "ultra" with "cons."

(b) It is well known that points (4) and (5) of Theorem 3.6 hold when both spaces are endowed with the Zariski topology ([4, Theorem 2.5 and 4.1] and [5, Theorem 2 and Remark 3]).

We recall now another important notion introduced by Halter-Koch in [22] as a generalization of the classical construction of the Kronecker function ring.

Definition 3.8. Let T be an indeterminate over the field K. A subring S of $K(T)$ is called *a K-function ring* if (a) T and T^{-1} belong to S, and (b) $f(0) \in f(T)S$, for each nonzero polynomial $f(T) \in K[T]$.

We collect in the following proposition the basic algebraic properties of K−function rings [22, Remarks at page 47 and Theorem (2.2)].

Proposition 3.9 (Halter-Koch [22, Section 2, Remark (1, 2 and 3), Theorem 2.2, and Corollary 2.7]). *Let K be a field, let T be an indeterminate over K, and let S be a subring of $K(T)$. Assume that S is a K-function ring.*

(1) *If S' is a subring of $K(T)$ containing S, then S' is also a K-function ring.*

(2) *If $\mathscr{S}$ is a nonempty collection of K-function rings (in $K(T)$), then $\bigcap\{\Sigma \mid \Sigma \in \mathscr{S}\}$ is a K-function ring.*

(3) *S is a Bézout domain with quotient field $K(T)$.*

(4) *If $f := f_0 + f_1 T + \ldots + f_r T^r \in K[T]$, then $(f_0, f_1, \ldots, f_r)S = fS$.*

(5) *For every valuation domain V of K, the trivial extension or Gaussian extension $V(T)$ in $K(T)$ (i.e., $V(T) := V[T]_{M[T]}$, where M is the maximal ideal of V) is a K-function ring.*

Given a subring S of $K(T)$, we will denote by $\mathrm{Zar}_0(K(T)|S)$ the subset of $\mathrm{Zar}(K(T)|S)$ consisting of all the valuation domains of $K(T)$ that are trivial extensions of some valuation domain of K.

The following characterization of K-function rings provides a slight generalization of [23, Theorem 2.3], and its proof is similar to that given by O. Kwegna Heubo, which is based on the work by Halter-Koch [22].

Proposition 3.10 (Finocchiaro-Fontana-Loper [10, Propositions 3.2 and 3.3]). *Let K be a field, T an indeterminate over K, and S a subring of $K(T)$. Then, the following conditions are equivalent:*

(i) *S is a K–function ring.*
(ii) *S is integrally closed and $\mathrm{Zar}(K(T)|S) = \mathrm{Zar}_0(K(T)|S)$.*
(iii) *S is the intersection of a nonempty subcollection of $\mathrm{Zar}_0(K(T))$.*

We give next one of the main results in [10] which, for the case of the Zariski topology, was already proved in [23, Corollary 2.2, Proposition 2.7 and Corollary 2.9]. More precisely,

Theorem 3.11 (Finocchiaro-Fontana-Loper [10, Corollary 3.6, Proposition 3.9, Corollary 3.11]). *Let K be a field and T an indeterminate over K. The following statements hold.*

(1) *The natural map $\varphi : \mathrm{Zar}(K(T)) \to \mathrm{Zar}(K)$, $W \mapsto W \cap K$, is continuous and closed with respect to both the Zariski topology and the ultrafilter topology (on both spaces).*
(2) *If $S \subseteq K(T)$ is a K-function ring, then the restriction of φ to the subspace $\mathrm{Zar}(K(T)|S)$ of $\mathrm{Zar}(K(T))$ is a topological embedding, with respect to both the Zariski topology and the ultrafilter topology.*
(3) *Let A be any subring of K, and let*

$$\mathrm{Kr}(K|A) := \bigcap \{V(T) \mid V \in \mathrm{Zar}(K|A)\}.$$

Then $\mathrm{Kr}(K|A)$ is a K-function ring. Moreover, the restriction of the map φ to $\mathrm{Zar}(K(T)|\mathrm{Kr}(K|A))$ establishes a homeomorphism of $\mathrm{Zar}(K(T)|\mathrm{Kr}(K|A))$ with $\mathrm{Zar}(K|A)$, with respect to both the Zariski topology and the ultrafilter topology.
(4) *Let A be a subring of K, $S_A := \mathrm{Kr}(K|A)$, and let $\gamma : \mathrm{Zar}(K(T)|S_A) \to \mathrm{Spec}(S_A)$ be the map sending a valuation overring of S_A into its center on S_A. Then γ establishes a homeomorphism, with respect to both the Zariski topology and the ultrafilter topology; thus, the map*

$$\sigma := \gamma \circ \varphi^{-1} : \mathrm{Zar}(K|A) \to \mathrm{Zar}(K(T)|S_A) \to \mathrm{Spec}(S_A)$$

is also a homeomorphism. In other words, $\mathrm{Zar}(K|A)$ is a spectral space when endowed with either the Zariski topology or the ultrafilter topology.

Note that statement (4) of the previous theorem extends [5, Theorem 2] to the general case where A is an arbitrary subring of the field K.

4 The Inverse Topology

Let $\mathscr{X}$ be any topological space. Then, it is well known that the topology induces a natural *preorder* on $\mathscr{X}$ by setting

$$x \leq y :\Leftrightarrow y \in Cl(\{x\}).$$

Therefore,

$$x^{\uparrow} := \{y \in \mathscr{X} \mid x \leq y\} = Cl(\{x\});$$

in particular, if F is a closed subspace of $\mathscr{X}$ and $x \in F$, then $x^{\uparrow} \subseteq F$. The set $x^{\uparrow}$ is called the set of *specializations of x* in $\mathscr{X}$; on the other hand, the set

$$x^{\downarrow} := \{y \in \mathscr{X} \mid y \leq x\}$$

is called the set of *generizations of x*. Since the closed subspaces are closed under specializations, it follows easily that if U is an open subspace of $\mathscr{X}$ and $x \in U$, then $x^{\downarrow} \subseteq U$.

For a subset Y of $\mathscr{X}$ we denote by $Y^{\uparrow}$ (respectively, $Y^{\downarrow}$) the set of all specializations (respectively, generizations) of elements in Y.

If $\mathscr{X}$ is a T_0-space, then the preorder is a partial order on $\mathscr{X}$, and, for $x, y \in \mathscr{X}$, $x^{\uparrow} = y^{\uparrow}$ if and only if $x = y$.

Given a preordered set $(X, \leq)$, we say that *a topology $\mathscr{T}$ on X is compatible with the order $\leq$* if, for each pair of elements x and y in X, $y \in Cl^{\mathscr{T}}(x)$ implies that $x \leq y$. Obviously, in general, several different topologies on X may be compatible with the given order on X.

The following properties are easy consequences of the definitions (see, for instance, [6, Lemma 2.1, Proposition 2.3(b)]).

Lemma 4.1. *Let $(X, \leq)$ be a preordered set and let $Y \subseteq X$.*

(1) $Cl^{L}(Y) := Y^{\uparrow}$ *(respectively, $Cl^{R}(Y) := Y^{\downarrow}$) satisfies the Kuratowski closure axioms and so it defines a topological structure on X, called the* L(eft)-topology *(respectively, the* R(ight)-topology*) on X.*

(2) *The L-topology (respectively, R-topology) on X is the finest topology on X compatible with the given order (respectively, with the opposite order of the given order) on X.*

(3) *A subset U of X is open in the L-topology (respectively, R-topology) if and only if $U = U^{\downarrow}$ ($= Cl^{R}(U)$, i.e., it is closed in the R-topology) (respectively, $U = U^{\uparrow}$ ($= Cl^{L}(U)$, i.e., it is closed in the L-topology).*

(4) *Let $U \subseteq X$ be a nonempty open subspace of X endowed with the L-topology (respectively, R-topology). Then U is quasi-compact if and only if there exist $x_1, x_2, \ldots x_n$ in U, with $n \geq 1$, such that $U = x_1^{\downarrow} \cup x_2^{\downarrow} \cup \cdots \cup x_n^{\downarrow}$ (respectively, $U = x_1^{\uparrow} \cup x_2^{\uparrow} \cup \cdots \cup x_n^{\uparrow}$).*

Remark 4.2. In relation with Lemma 4.1(2), note that the COP (or closure of points) topology [29] is the coarsest topology on X compatible with a given order on X.

Recall that a topological space $\mathscr{X}$ is an *Alexandroff-discrete space* if it is T_0 and for each subset Y of $\mathscr{X}$ the closure of Y coincides with the union of the closures of its points [1, page 28]. Therefore, if $(X, \leq)$ is a partially ordered set, then the L-topology (or the R-topology) determines on X the structure of an Alexandroff-discrete space.

If $\mathscr{X}$ is a T_0 topological space, then the L-topology on $\mathscr{X}$, associated to the partial order defined by the given topology on $\mathscr{X}$, is finer than the original topology of $\mathscr{X}$, since for each $Y \subseteq \mathscr{X}$, $Cl^L(Y) \subseteq Cl(Y)$. Moreover, even if $\mathscr{X}$ is a spectral space, $\mathscr{X}^L$ (i.e., $\mathscr{X}$ equipped with the L-topology) is not spectral in general. For example, a spectral space having infinitely many closed points may not be quasi-compact with respect to the L-topology by Lemma 4.1(4) (e.g., $\mathscr{X} := \mathrm{Spec}(\mathbb{Z}) = \bigcup\{(p)^{\downarrow} \mid (p) \text{ is a nonzero prime ideal of } \mathbb{Z}\}$ is an open cover of $\mathscr{X}$ endowed with the L-topology without a finite open subcover).

Before stating a result providing a complete answer to the question of when the L-topology determines a spectral space (see Theorem 4.5), we recall a useful application of the L-topology showing that the constructible closure and the closure by specializations (or, L-closure) determine the structural closure in a spectral space (see, for instance, [24, Corollary to Theorem 1], [11, Lemma 1.1], or [6, Proposition 3.1(a)]).

Lemma 4.3. *Let $\mathscr{X}$ be a spectral space. For each subset Y of $\mathscr{X}$,*

$$Cl(Y) = Cl^L(Cl^{cons}(Y)).$$

Let $(X, \leq)$ be a preordered set and denote by $\mathrm{Max}(X)$ (respectively, $\mathrm{Min}(X)$) the set of all maximal (respectively, minimal) elements of X. In particular, if $\mathscr{X}$ is a topological space, we denote by $\mathrm{Max}(\mathscr{X})$ (respectively, $\mathrm{Min}(\mathscr{X})$) the set of all maximal (respectively, minimal) points of a topological space $\mathscr{X}$, with respect to the preorder $\leq$ induced by the topology of $\mathscr{X}$. It follows immediately by definition that

$$x \in \mathrm{Max}(\mathscr{X}) \quad \Leftrightarrow \quad \{x\} \text{ is closed in } \mathscr{X} \quad \Leftrightarrow \quad \{x\} \text{ is closed in } \mathscr{X}^L.$$

From the order-theoretic point of view, we have the following.

Lemma 4.4. *Let $(X, \leq)$ be a partially ordered set.*

(1) *The following conditions are equivalent:*

    (i) *x is a closed point in X^L.*

(ii) $x \in \mathrm{Max}(X)$.

(iii) x is an open point in X^R.

(2) The following conditions are equivalent:

(i) x is an open point in X^L.

(ii) $x \in \mathrm{Min}(X)$.

(iii) x is a closed point in X^R.

(3) X^L (respectively, X^R) is a T_1 topological space if and only if X^L (respectively, X^R) is a discrete space.

Theorem 4.5 (Dobbs-Fontana-Papick [6, Theorem 2.4]). Let X be a partially ordered set. Then X with the L-topology is a spectral space if and only if the following four properties hold:

(α) Each nonempty totally ordered subset Y of X has a sup.

(β) X satisfies the following condition:

(filtr^L) each nonempty lower-directed subset Y of X has a greatest lower bound $y := \inf(Y)$ such that $y^\uparrow = Y^\uparrow$.

(γ) $\mathrm{Card}(\mathrm{Max}(X))$ is finite.

(δ) For each pair of distinct elements x and y of X, there exist at most finitely many elements of X which are maximal in the set of common lower bounds of x and y.

The necessity of condition (α) follows from [27, Theorem 9], that of condition (β) uses [24, Proposition 5], and the necessity of condition (δ) is related to Lemma 4.1(4); condition (γ) holds in any Alexandroff-discrete space. The sufficiency of (α)–(δ) results by verifying the conditions of Hochster's characterization theorem [24].

Using the opposite order, from Theorem 4.5 we can easily deduce a characterization of when a partially ordered set with the R-topology is a spectral space.

Given a spectral space $\mathcal{X}$, the following proposition gives a complete answer to the question of when the continuous map $\mathcal{X}^L \to \mathcal{X}$ (where $\mathcal{X}^L$ denotes the topological space $\mathcal{X}$ equipped with the L-topology, associated to the partial order defined by the given topology on $\mathcal{X}$) is a homeomorphism. In particular, in this situation, $\mathcal{X}^L$ is a spectral space.

Proposition 4.6 (Picavet [35, V, Proposition 1], Dobbs-Fontana-Papick [6, Theorem 3.3]). Let $\mathcal{X}$ be a spectral space. The following are equivalent.

(i) $\mathcal{X}^L = \mathcal{X}$.

(ii) For each $x \in X$, $x^\downarrow$ is a quasi-compact open subset of $\mathcal{X}$.

(iii) For each family $\{U_\lambda \mid \lambda \in \Lambda\}$ of quasi-compact open subsets of $\mathcal{X}$, the set $\bigcap\{U_\lambda \mid \lambda \in \Lambda\}$ is still a quasi-compact open subset of $\mathcal{X}$.

(iv) Each increasing sequence of irreducible closed subsets of $\mathcal{X}$ stabilizes, and, for each family $\{U_\lambda \mid \lambda \in \Lambda\}$ of quasi-compact open subsets of the space $\mathcal{X}$, $\mathrm{Card}(\mathrm{Max}(\bigcap\{U_\lambda \mid \lambda \in \Lambda\}))$ is finite.

The previous proposition gives the motivation for studying the rings A such that, for each $P \in \text{Spec}(A)$, the canonical map $\text{Spec}(A_P) \hookrightarrow \text{Spec}(A)$ is open. This class of rings was introduced by G. Picavet in 1975 ([35] and [36]) under the name of *g-ring* and it can be shown that a spectral space $\mathscr{X}$ is such that $\mathscr{X}^L = \mathscr{X}$ if and only if $\mathscr{X}$ is homeomorphic to the prime spectrum of a g-ring [35, V, Proposition 1].

When $\mathscr{X}$ is a spectral space, Hochster [24] introduced a new topology on $\mathscr{X}$, called the inverse topology. If we denote by $\mathscr{X}^{\text{inv}}$ the set $\mathscr{X}$ equipped with the inverse topology, Hochster proved that $\mathscr{X}^{\text{inv}}$ is still a spectral space and the partial order on $\mathscr{X}$ induced by the inverse topology is the opposite order of that induced by the given topology on $\mathscr{X}$. More precisely:

Proposition 4.7 (Hochster [24, Proposition 8]). *Let $\mathscr{X}$ be a spectral space. For each subset Y of $\mathscr{X}$, set*

$$Cl^{\text{inv}}(Y) := \bigcap \{U \mid U \text{ open and quasi-compact in } \mathscr{X}, \ U \supseteq Y\}.$$

(1) *Cl^{inv} satisfies the Kuratowski closure axioms and so it defines a topological structure on $\mathscr{X}$, called the inverse topology; denote by $\mathscr{X}^{\text{inv}}$ the set $\mathscr{X}$ equipped with the inverse topology.*

(2) *The partial order on $\mathscr{X}$ induced by the inverse topology is the opposite order of that induced by the given topology on $\mathscr{X}$.*

(3) *$\mathscr{X}^{\text{inv}}$ is a spectral space.*

Let $\mathscr{X}$ be a spectral space. For each subset Y of $\mathscr{X}$, set

$$\text{Max}(Cl^{\text{inv}}(Y)) := \{x \in Cl^{\text{inv}}(Y) \mid x^{\uparrow} \cap Cl^{\text{inv}}(Y) = \{x\}\}.$$

B. Olberding in [34, Proposition 2.1(2)] has observed that

$$\text{Max}(Cl^{\text{inv}}(Y)) \subseteq Cl^{\text{cons}}(Y).$$

From the previous observation and from Lemmas 4.1, 4.3, and 4.4 and from Proposition 4.7, it is not very hard to prove the following (see also [6, Section 3], [40, Remark 2.2 and Proposition 2.3], [34, Proposition 2.3], and [9, Proposition 2.6]).

Corollary 4.8. *Let $\mathscr{X}$ be a spectral space.*

(1) *The constructible topology on $\mathscr{X}^{\text{inv}}$ coincides with the constructible topology on $\mathscr{X}$, i.e., $(\mathscr{X}^{\text{inv}})^{\text{cons}} = \mathscr{X}^{\text{cons}}$.*

(2) *For each subset Y of $\mathscr{X}$,*

$$Cl^{\text{inv}}(Y) = Cl^R(\text{Max}(Cl^{\text{inv}}(Y))) = Cl^R(Cl^{\text{cons}}(Y))\}.$$

(3) *$(\mathscr{X}^{\text{inv}})^L = \mathscr{X}^R$ and $(\mathscr{X}^{\text{inv}})^R = \mathscr{X}^L$.*

(4) *$(\mathscr{X}^{\text{inv}})^{\text{inv}} = \mathscr{X}$.*

(5) *For each $x \in \mathscr{X}$, $Cl^{\text{inv}}(x) = Cl^R(x)$ is an irreducible closed set of $\mathscr{X}^{\text{inv}}$ (and, obviously, $Cl(x) = Cl^L(x)$ is an irreducible closed set of $\mathscr{X}$).*

(6) *The topological space $\mathscr{X}$ is irreducible if and only if $\mathscr{X}^{inv}$ has a unique closed point.*

(7) *The topological space $\mathscr{X}$ has a unique closed point if and only if $\mathscr{X}^{inv}$ is irreducible.*

(8) *The following are equivalent.*

   (i) *Y is quasi-compact in $\mathscr{X}$.*

   (ii) *$Cl^R(Y)$ is quasi-compact in $\mathscr{X}$.*

   (iii) *$Cl^{inv}(Y) = Cl^R(Y)$.*

   (iv) *$Cl^{cons}(Cl^R(Y)) = Cl^R(Y)$.*

   (v) *$\mathrm{Max}(Cl^{inv}(Y)) \subseteq Y$.*

By using the inverse topology, we can state an easy corollary of Proposition 4.6 and Corollary 4.8 (see also [6, Theorem 3.3, Corollaries 3.4 and 3.5]), which provides in part (1) further characterizations of when $\mathscr{X}^L = \mathscr{X}$.

Corollary 4.9. *Let $\mathscr{X}$ be a spectral space.*

(1) *The following are equivalent.*

   (i) *$\mathscr{X}^L = \mathscr{X}$ (i.e., $\mathscr{X}$ is an Alexandroff-discrete topological space).*

   (ii) *Each open subset of $\mathscr{X}^{inv}$ is the complement of a quasi-compact open subset of $\mathscr{X}$.*

   (iii) *$\mathscr{X}^{inv}$ is a Noetherian space.*

(2) *The following are equivalent.*

   (i) *$\mathscr{X}^R = \mathscr{X}^{inv}$ (i.e., $\mathscr{X}^{inv}$ is an Alexandroff-discrete topological space).*

   (ii) *Each open subset of $\mathscr{X}$ is the complement of a quasi-compact open subset of $\mathscr{X}^{inv}$ (or, equivalently, each open subset of $\mathscr{X}$ is quasi-compact).*

   (iii) *$\mathscr{X}$ is a Noetherian topological space.*

(3) *The following are equivalent.*

   (i) *$\mathscr{X}$ is a Noetherian Alexandroff-discrete space.*

   (ii) *$\mathscr{X}^{inv}$ is a Noetherian Alexandroff-discrete space.*

   (iii) *$\mathrm{Card}(\mathscr{X})$ is finite.*

Recall that *a spectral map* of spectral spaces $f : \mathscr{X} \to \mathscr{Y}$ is a continuous map such that the preimage of every open and quasi-compact subset of $\mathscr{Y}$ under f is again quasi-compact. We say that a spectral map of spectral spaces $f : \mathscr{X} \to \mathscr{Y}$ is *a going-down map* (respectively, *a going-up map*) if for any pair of distinct elements $y', y \in \mathscr{Y}$ such that $y' \in \{y\}^{\downarrow}$ (respectively, $y' \in \{y\}^{\uparrow}$) and for any $x \in \mathscr{X}$ such that $f(x) = y$ there exists a point $x' \in \{x\}^{\downarrow}$ (respectively, $x' \in \{x\}^{\uparrow}$) such that $f(x') = y'$.

Lemma 4.10. *Let $f : \mathscr{X} \to \mathscr{Y}$ be a spectral map of spectral spaces.*

(1) *$f : \mathscr{X}^{cons} \to \mathscr{Y}^{cons}$ is a closed spectral map.*

(2) *The following are equivalent:*

   (i) f is a going-down (respectively, going-up) map.
  (ii) $f(x^{\downarrow}) = f(x)^{\downarrow}$ (respectively, $f(x^{\uparrow}) = f(x)^{\uparrow}$), for each $x \in \mathcal{X}$
 (iii) $f(\mathcal{X}'^{\downarrow}) = f(\mathcal{X}')^{\downarrow}$ (respectively, $f(\mathcal{X}'^{\uparrow}) = f(\mathcal{X}')^{\uparrow}$) for each $\mathcal{X}' \subseteq \mathcal{X}$.
  (iv) The continuous map $f : \mathcal{X}^R \to \mathcal{Y}^R$ (respectively, $f : \mathcal{X}^L \to \mathcal{Y}^L$) is closed.

(3) $f : \mathcal{X}^R \to \mathcal{Y}^R$ is closed (respectively, open) if and only if $f : \mathcal{X}^L \to \mathcal{Y}^L$ is open (respectively, closed).

(4) If $f : \mathcal{X} \to \mathcal{Y}$ is an open (respectively, closed) spectral map of spectral spaces, then f is a going-down (respectively, going-up) map.

(5) If $f : \mathcal{X} \to \mathcal{Y}$ is an open spectral map of spectral spaces, then $f : \mathcal{X}^{inv} \to \mathcal{Y}^{inv}$ is a closed spectral map.

Proof. (1) is an obvious consequence of the definitions. (2) Since a spectral map f is continuous then it is straightforward that $x' \leq x$ in $\mathcal{X}$ implies that $f(x') \leq f(x)$ and so $f(x^{\downarrow}) \subseteq f(x)^{\downarrow}$. Moreover, for each $y \in \mathcal{Y}$, $f^{-1}(y^{\downarrow}) = \bigcup\{x^{\downarrow} \mid x \in \mathcal{X}$ and $f(x) \leq y\}$ and so $f : \mathcal{X}^R \to \mathcal{Y}^R$ is also continuous. The various equivalences are now straightforward consequences of the definitions.

(3) is an easy consequence of (2).

(4) Let $x \in \mathcal{X}$. It is easy to see that $x^{\downarrow} = \bigcap\{U \mid U$ open and quasi-compact and $x \in U \subseteq \mathcal{X}\}$. Therefore, for any spectral map of spectral spaces $f : \mathcal{X} \to \mathcal{Y}$ and any $x \in \mathcal{X}$, the following holds

$$f(x^{\downarrow}) = f(\bigcap\{U \mid U \text{ open and quasi-compact and } x \in U \subseteq \mathcal{X}\})$$
$$\subseteq \bigcap\{f(U) \mid U \text{ open and quasi-compact and } x \in U \subseteq \mathcal{X}\}.$$

Conversely, assume that f is an open spectral map and take a point $y \in f(U)$, for any open and quasi-compact neighborhood U of $x \in \mathcal{X}$. Consider the following collection of subsets of $\mathcal{X}$:

$$\mathscr{F} := \mathscr{F}(y) := \{f^{-1}(\{y\}) \cap U \mid U \text{ open and quasi-compact and } x \in U \subseteq \mathcal{X}\}.$$

Note now that $\mathscr{F}$ is obviously closed under finite intersections, since the quasi-compact open sets of $\mathcal{X}$ are closed under finite intersections and, by assumption, each set belonging to $\mathscr{F}$ is nonempty. On the other hand, the set $f^{-1}(\{y\})$ is closed with respect to the constructible topology on $\mathcal{X}$ and thus is compact in $\mathcal{X}^{cons}$. Keeping in mind that each open and quasi-compact subspace of the given spectral topology on $\mathcal{X}$ is clopen in $\mathcal{X}^{cons}$, it follows immediately that $\mathscr{F}$ is a collection of closed subsets of the compact space $f^{-1}(\{y\})$ ($\subseteq \mathcal{X}^{cons}$), satisfying the finite intersection property. Therefore, by compactness, there exists a point $x' \in f^{-1}(\{y\}) \cap U$, for any open and quasi-compact neighborhood U of $x \in \mathcal{X}$. In particular, $x' \in \bigcap\{U \mid U$ open and quasi-compact and $x \in U \subseteq \mathcal{X}\} = \{x\}^{\downarrow}$ and so $x' \leq x$.

Therefore, $f(x') = y \leq f(x)$. We conclude that $f(x^\downarrow) = \bigcap\{f(U) \mid U$ open and quasi-compact and $x \in U \subseteq \mathscr{X}\}$. On the other hand, since f is open, we have

$$f(x)^\downarrow = \bigcap\{V \mid V \text{ open and quasi-compact and } f(x) \in V \subseteq \mathscr{Y}\}$$
$$\subseteq \bigcap\{f(U) \mid U \text{ open and quasi-compact and } x \in U \subseteq \mathscr{X}\}$$
$$= f(x^\downarrow).$$

Since the opposite inclusion holds in general, we have $f(x^\downarrow) = f(x)^\downarrow$ and so f is a going-down spectral map.

The parenthetical statement is easier to prove. Indeed, suppose that f is a closed spectral map, let $y', y \in \mathscr{Y}$ be such that $y' \in \{y\}^\uparrow$ and let $x \in \mathscr{X}$ be such that $f(x) = y$. By assumption, we have $f(Cl(\{x\})) = Cl(f(\{x\})) = Cl(\{y\})$ and thus, since $y' \in Cl(\{y\})$, there is a point $x' \in Cl(\{x\})$ such that $f(x') = y'$. This shows that f is a going-up map.

(5) If f is an open spectral map, then, by (3), f is going down and thus, by (2), $f : \mathscr{X}^R \to \mathscr{Y}^R$ is closed. Therefore, by using (1), for each $\mathscr{X}' \subseteq \mathscr{X}$, we have $Cl^{inv}(\mathscr{X}') = Cl^R(Cl^{cons}(\mathscr{X}'))$ and so $f(Cl^{inv}(\mathscr{X}')) = Cl^R(f(Cl^{cons}(\mathscr{X}'))) = Cl^R(Cl^{cons}(f(\mathscr{X}'))) = Cl^{inv}(f(\mathscr{X}'))$.

Example 4.11. We now show that it is not true that if $f : \mathscr{X} \to \mathscr{Y}$ is a closed spectral map of spectral spaces, then $f : \mathscr{X}^{inv} \to \mathscr{Y}^{inv}$ is an open spectral map. As a matter of fact, let K be a field and let $\mathscr{T} := \{T_i \mid i \in \mathbb{N}\}$ be an infinite and countable collection of indeterminates over K. Let $A := K[\mathscr{T}]$, let M be the maximal ideal of A generated by all the indeterminates, and let $B := A/M$. Set $\mathscr{X} := \mathrm{Spec}(B)$ and $\mathscr{Y} := \mathrm{Spec}(A)$. Of course, the inclusion $f : \mathscr{X} \to \mathscr{Y}$ (associated to the canonical projection $A \to B$) is a closed embedding, with respect to the Zariski topology. We claim that f is not open, if $\mathscr{X}$ and $\mathscr{Y}$ are endowed with the inverse topology. By contradiction, assume that $f : \mathscr{X}^{inv} \to \mathscr{Y}^{inv}$ is open. In this situation, $\mathscr{X}$ should be open in $\mathscr{Y}^{inv}$, (since $\mathscr{X}$ is trivially open in $\mathscr{X}^{inv}$). This implies that $\mathscr{Z} := \mathscr{Y} \setminus f(\mathscr{X}) = \mathrm{Spec}(A) \setminus \{M\}$ is closed in $\mathscr{Y}^{inv}$, i.e., $\mathscr{Z}$ is an intersection of a family of open and quasi-compact subspaces of $\mathscr{Y}$. Since $\mathscr{Z}$ differs from $\mathscr{Y}$ for exactly one point, it has to be quasi-compact, with respect to the Zariski topology of $\mathscr{Y}$. On the other hand, it is immediately verified that the open cover

$$\{\{P \in \mathscr{Y} \mid T_i \notin P\} \mid i \in \mathbb{N}\}$$

of $\mathscr{Z}$ has no finite subcovers, a contradiction.

5 Some Applications

The first application that we give is a topological interpretation of when two given collections of valuation domains are representations of the same integral domain.

Proposition 5.1 (Finocchiaro-Fontana-Loper [9, Proposition 4.1]). *Let K be a field. If Y_1, Y_2 are nonempty subsets of $\mathrm{Zar}(K)$ having the same closure in $\mathrm{Zar}(K)$, with respect to the ultrafilter topology, then*

$$\bigcap\{V \mid V \in Y_1\} = \bigcap\{V \mid V \in Y_2\}.$$

In particular,

$$\bigcap\{V \mid V \in Y\} = \bigcap\{V \mid V \in \mathrm{Cl}^{\mathrm{ultra}}(Y)\}.$$

The converse of the first statement in Proposition 5.1 is false (for an explicit example see Example 4.4 in [9]). More precisely, we will show that equality of the closures of the subsets Y_1, Y_2, with respect to the ultrafilter topology, implies a statement that, in general, is stronger than the equality of the (integrally closed) domains obtained by intersections. To see this, recall some background material about semistar operations.

Let A be an integral domain, and let K be the quotient field of A. As usual, denote by $\overline{F}(A)$ the set of all nonzero A–submodules of K and by $f(A)$ the set of all nonzero finitely generated A–submodules of K. As is well known, a nonempty subset Y of $\mathrm{Zar}(K|A)$ induces the *valuative semistar operation* $\wedge_Y$, defined by $F^{\wedge_Y} := \bigcap\{FV \mid V \in Y\}$, for each $F \in \overline{F}(A)$. A valuative semistar operation $\star$ is always e.a.b., that is, for all $F, G, H \in f(A)$, $(FG)^\star \subseteq (FH)^\star$ implies $G^\star \subseteq H^\star$ (for more details, see, e.g., [15]). Recall that we can associate to any semistar operation $\star$ on A a semistar operation $\star_f$ *of finite type* (on A), by setting $F^{\star_f} := \bigcup\{G^\star \mid G \in f(A),\ G \subseteq F\}$, for each $F \in \overline{F}(A)$; $\star_f$ is called *the semistar operation of finite type associated to* $\star$.

Since, for each $V \in \mathrm{Zar}(K|A)$ (equipped with the classical Zariski topology), $\mathrm{Cl}(\{V\}) = \{W \in \mathrm{Zar}(K|A) \mid W \subseteq V\}$ [47, Ch. VI, Theorem 38], the partial order associated to the Zariski topology of $\mathrm{Zar}(K|A)$ is defined as follows:

$$W \preceq V :\Leftrightarrow V \subseteq W.$$

For any subset $Y \subseteq \mathrm{Zar}(K|A)$, denote by $Y^\upharpoonright$ *the Zariski-generic closure of* Y, that is, $Y^\upharpoonright := \{W \in \mathrm{Zar}(K|A) \mid V \subseteq W,\ \text{for some}\ V \in Y\} = \mathrm{Cl}^R(Y)$. It is obvious that $\wedge_Y = \wedge_{Y^\upharpoonright}$. From Proposition 5.1 we also have $\wedge_Y = \wedge_{\mathrm{Cl}^{\mathrm{ultra}}(Y)}$.

Theorem 5.2 (Finocchiaro-Fontana-Loper [9, Theorem 4.9]). *Let A be an integral domain, K its quotient field, and Y_1, Y_2 two nonempty subsets of $\mathrm{Zar}(K|A)$. Then, the following conditions are equivalent:*

(i) *The semistar operations of finite type associated to $\wedge_{Y_1}$ and $\wedge_{Y_2}$ are the same, that is, $(\wedge_{Y_1})_f = (\wedge_{Y_2})_f$.*
(ii) *The subsets $\mathrm{Cl}^{\mathrm{ultra}}(Y_1), \mathrm{Cl}^{\mathrm{ultra}}(Y_2)$ of $\mathrm{Zar}(K|A)$ have the same Zariski–generic closure, that is, $\mathrm{Cl}^{\mathrm{ultra}}(Y_1)^\upharpoonright = \mathrm{Cl}^{\mathrm{ultra}}(Y_2)^\upharpoonright$.*

Let A be an integral domain, K its quotient field, and $Z := \text{Zar}(K|A)$. For any nonempty subset $Y \subseteq Z$, consider the K-function ring

$$\text{Kr}(Y) := \bigcap\{V(T) \mid V \in Y\}.$$

Note that, if we consider on the integral domain A the valuative (e.a.b.) semistar operation $\wedge_Y$ defined above, then the Kronecker function ring associated to $\wedge_Y$, $\text{Kr}(A, \wedge_Y)$, coincides with $\text{Kr}(Y)$ [13, Corollary 3.8].

We say that A is *a vacant domain* if it is integrally closed and, for any representation Y of A (i.e., $A = \bigcap\{V \mid V \in Y\}$), we have $\text{Kr}(Y) = \text{Kr}(Z)$; for instance, a Prüfer domain is vacant (see [7]).

Corollary 5.3. *Let A be an integrally closed domain and K its quotient field. The following conditions are equivalent:*

(i) *A is a vacant domain.*
(ii) *For any representation Y of A, $cl^{ultra}(Y)^{\flat} = \text{Zar}(K|A)$.*

Keeping in mind that it is known that the ultrafilter topology and the constructible topology on $\text{Zar}(K|A)$ coincide (Theorem 3.6(3)), the following result follows easily from Corollary 4.8(2).

Proposition 5.4. *Let K be a field and A be a subring of K. For any subset Y of $\text{Zar}(K|A)$, $cl^{inv}(Y) = cl^R(cl^{ultra}(Y))$.*

From the previous proposition, we can restate Corollary 5.3 as follows: A is a vacant domain if and only if for any representation Y of A, $cl^{inv}(Y) = \text{Zar}(K|A)$.

Recall that a *semistar operation* is *complete* if it is e.a.b. and of finite type. In order to state some characterizations of the complete semistar operations, we need some terminology.

For a domain A and a semistar operation $\star$ on A, we say that a valuation overring V of A is a *$\star$-valuation overring of A* provided $F^{\star} \subseteq FV$, for each finitely generated A-module F contained in the quotient field K of A. Set $\mathcal{V}(\star) := \{V \mid V \text{ is a } \star\text{-valuation overring of } A\}$ and let $b(\star) := \wedge_{\mathcal{V}(\star)}$. Finally, if $\star$ is an e.a.b. semistar operation on A, we can consider the Kronecker function ring $\text{Kr}(A, \star) := \bigcap\{V(T) \mid V \in \mathcal{V}(\star)\} = \text{Kr}(\mathcal{V}(\star))$ [14, Theorem 14(3)] and we can define a semistar operation $\text{Kr}(\star)$ on A by setting $E^{\text{Kr}(\star)} := E\text{Kr}(A, \star) \cap K$ for each nonzero A-module E contained in K (see, for instance, [17]). Then,

Proposition 5.5 (Fontana-Loper [12, Proposition 3.4], [13, Corollary 5.2], [15, Proposition 6.3], [17]). *Given a semistar operation $\star$, the following are equivalent:*

(i) *$\star$ is complete.*
(ii) *$\star = b(\star)$.*
(iii) *$\star = \text{Kr}(\star)$.*

The following result provides a topological characterization of when a semistar operation is complete.

Theorem 5.6 (Finocchiaro-Fontana-Loper [9, Theorem 4.13]). *Let A be an integral domain, K its quotient field, and $\star$ a semistar operation on A. Then, the following conditions are equivalent:*

(i) *$\star$ is complete.*

(ii) *There exists a closed subset Y of $\mathrm{Zar}(K|A)^{cons}$ such that $Y = Y^{\downarrow}$ and $\star = \wedge_Y$.*

(iii) *There exists a compact subspace Y' in $\mathrm{Zar}(K|A)^{cons}$ such that $\star = \wedge_{Y'}$.*

(iv) *There exists a quasi-compact subspace of Y'' of $\mathrm{Zar}(K|A)^{zar}$ such that $\star = \wedge_{Y''}$.*

From Theorems 5.2 and 5.6 we easily deduce the following:

Corollary 5.7. *Let A be an integral domain, K its quotient field, and Y a nonempty subset of $\mathrm{Zar}(K|A)$. Then, $(\wedge_Y)_f = \wedge_{Cl^{cons}(Y)}$.*

Finally, we can formulate some of the previous results in terms of Hochster's inverse topology [9, Theorem 4.9, Corollary 4.10].

Corollary 5.8. *Let A be an integral domain and K be its quotient field. The following statements hold.*

(a) *If Y_1, Y_2 are nonempty subsets of $\mathrm{Zar}(K|A)$, then the following are equivalent.*

   (i) *$(\wedge_{Y_1})_f = (\wedge_{Y_2})_f$.*

   (ii) *$\mathrm{Kr}(Y_1) = \mathrm{Kr}(Y_2)$.*

   (iii) *$Cl^{inv}(Y_1) = Cl^{inv}(Y_2)$.*

(b) *A is a vacant domain if and only if it is integrally closed and any representation Y of A is dense in $\mathrm{Zar}(K|A)$ with respect to the inverse topology.*

(c) *For any nonempty subset Y of $\mathrm{Zar}(K|A)$, $(\wedge_Y)_f = \wedge_{Cl^{inv}(Y)}$.*

B. Olberding in [34] calls a subset Y of $\mathrm{Zar}(K|A)$ an *affine subset of* $\mathrm{Zar}(K|A)$ if $A^{\wedge Y} := \bigcap\{V \mid V \in Y\}$ is a Prüfer domain with quotient field equal to K. Note that $Z := \mathrm{Zar}(K|A)$, equipped with the Zariski topology, can be viewed as a locally ringed space with the structure sheaf defined by

$$\mathscr{O}_Z(U) := A^{\wedge U} = \bigcap\{V \mid V \in U\}, \quad \text{for each nonempty open subset } U \text{ of } Z \,,$$

(for more details, see [34]). With this structure of locally ringed space, an affine subset Y of Z is not necessarily itself an affine scheme, that is, Y (endowed with the Zariski topology induced by Z) is not necessarily homeomorphic to $\mathrm{Spec}(\mathscr{O}_Z(Y))$, however, by Corollary 5.8(a), is an inverse-dense subspace of the affine scheme $(Cl^{inv}(Y), \mathscr{O}_Z|_{Cl^{inv}(Y)})$.

If T is an indeterminate over K and $A(T)$ is the Nagata ring associated to A [20, Section 33], for each $Y \subseteq Z$, we can consider

$$Y(T) := \{V(T) \mid V \in Y\} \subseteq \mathrm{Zar}_0(K(T)|A(T)) := \{V(T) \mid V \in Z\},$$

and, as above, $\mathrm{Kr}(Y) = \bigcap\{V(T) \mid V \in Y\}$.

The following statements were proved by Olberding [34, Propositions 5.6 and 5.10 and Corollaries 5.7 and 5.8]. We give next a proof based on some of the results contained in [9] and recalled above.

Proposition 5.9. *Let Y be a subset of $Z := \mathrm{Zar}(K|A)$. Then,*

(1) *Assume that A is a Prüfer domain with quotient field K. Then, $Y = Cl^{inv}(Y)$ if and only if $Y = \mathrm{Zar}(K|R)$ for some overring R of A.*

(2) *$Y = Cl^{inv}(Y)$ if and only if $Y(T) = \{W(T) \mid W \in Z \text{ and } W(T) \supseteq \mathrm{Kr}(Y)\}$.*

(3) *$Cl^{inv}(Y) = \{W \in Z \mid W(T) \supseteq \mathrm{Kr}(Y)\}$, hence $(Cl^{inv}(Y))(T) = \mathrm{Zar}(K(T)|\mathrm{Kr}(Y))$.*

(4) *$A^{\wedge Y} = \bigcap\{W \in Cl^{inv}(Y)\}$.*

(5) *If Y is an affine subset of Z, then*

$$Cl^{inv}(Y) = \{W \in Z \mid W \supseteq A^{\wedge Y}\} = \mathrm{Zar}(K|A^{\wedge Y}).$$

(6) *Assume that Y_1 and Y_2 are two affine subsets of Z, then*

$$\bigcap\{V \mid V \in Y_1\} = \bigcap\{V \mid V \in Y_2\} \iff Cl^{inv}(Y_1) = Cl^{inv}(Y_2).$$

(7) *$(Cl^{inv}(Y))(T) = Cl^{inv}(Y(T))$.*

(8) *Each ring of fractions of $A^{\wedge Y}$ can be represented as an intersection of valuation domains contained in a subset of $Cl^{inv}(Y)$; in other words, if S is a multiplicatively closed subset of $A^{\wedge Y}$, then $(A^{\wedge Y})_S = A^{\wedge \Sigma}$ for some $\Sigma \subseteq Cl^{inv}(Y)$.*

(9) *The canonical homeomorphism of topological spaces (all endowed with the Zariski topology)*

$$\tau : \mathrm{Spec}(\mathrm{Kr}(K|A)) \to \mathrm{Zar}(K|A), \quad Q \mapsto \mathrm{Kr}(K|A)_Q \cap K$$

(where $\tau = \sigma^{-1}$, see Theorem 3.11(4)) determines a continuous injective map $\mathrm{Spec}(\mathrm{Kr}(Y)) \to \mathrm{Zar}(K|A)$ which restricts to a homeomorphism of $\mathrm{Spec}(\mathrm{Kr}(Y))$ (respectively, $\mathrm{Max}(\mathrm{Kr}(Y))$) onto $Cl^{inv}(Y)$ (respectively, $\mathrm{Max}(Cl^{inv}(Y))$).

Proof. (1) Let Y be a nonempty, closed set with respect to the inverse topology, and let $R := A^{\wedge Y} := \bigcap\{V \mid V \in Y\}$. Since R is an overring of the Prüfer domain A, R is also a Prüfer domain; thus it is vacant. By Corollary 5.8(b), Y is a dense subspace of $\mathrm{Zar}(K|R)$, with the inverse topology, i.e., $Cl^{inv}(Y) \cap \mathrm{Zar}(K|R) = \mathrm{Zar}(K|R)$. Thus $\mathrm{Zar}(K|R) \subseteq Cl^{inv}(Y)$. On the other hand, the inclusion $Y \subseteq \mathrm{Zar}(K|R)$ implies $Cl^{inv}(Y) \subseteq \mathrm{Zar}(K|R)$, since $\mathrm{Zar}(K|R)$ is clearly inverse-closed. Therefore, $Y = Cl^{inv}(Y) = \mathrm{Zar}(K|R)$. The converse holds for any integral domain.

(2) Let $A(T)$ be Nagata ring associated to A. By Theorem 3.11(3), the natural map $\varphi : \text{Zar}(K(T)|\text{Kr}(K|A)) \subseteq \text{Zar}(K(T)|A(T)) \to \text{Zar}(K|A)$, $W \mapsto W \cap K$, is a homeomorphism with respect to the Zariski topology and to the constructible topology and thus also with respect to the inverse topology. By the previous homeomorphism Y is inverse-closed in $\text{Zar}(K|A)$ if and only if $Y(T)$ is inverse-closed in $\text{Zar}(K(T)|A(T))$. Therefore, if Y is inverse-closed, then $Y(T)$ is inverse-closed in $\text{Zar}(K(T)|A(T))$ and thus by (1), $Y(T) = \text{Zar}(K(T)|\text{Kr}(Y))$. Finally, by Proposition 3.10, $\text{Zar}_0(K(T)|\text{Kr}(Y)) = \text{Zar}(K(T)|\text{Kr}(Y))$.

Conversely, if $Y(T) = \text{Zar}_0(K(T)|\text{Kr}(Y))$, then $Y(T) = \text{Zar}(K(T)|\text{Kr}(Y))$ (Proposition 3.10); hence by (1) $Y(T)$ is inverse-closed.

(3) By (1), $\text{Zar}(K(T)|\text{Kr}(Y)) = Y(T)$ is an inverse-closed subspace of the space $\text{Zar}(K(T)|A(T))$; then $\{W \in Z \mid W(T) \supseteq \text{Kr}(Y)\}$ is an inverse-closed subspace of $\text{Zar}(K|A)$ (Theorem 3.11(3)) and it obviously contains Y. Let Y' be an inverse-closed subspace of $\text{Zar}(K|A)$ containing Y; then clearly $Y(T) \subseteq Y'(T) = \text{Zar}(K(T)|\text{Kr}(Y'))$ and so $\{W \in Z \mid W(T) \supseteq \text{Kr}(Y)\} \subseteq \{W' \in Z \mid W'(T) \supseteq \text{Kr}(Y')\} = \text{Cl}^{\text{inv}}(Y') = Y'$. The last part of the statement follows from Theorem 3.11(3).

(4) is a straightforward consequence of Corollary 5.8(c).

(5) Since Y is an affine set, $A^{\wedge Y}$ is a Prüfer domain with quotient field K. Therefore, by (1), $\text{Cl}^{\text{inv}}(Y) = \text{Zar}(K|R)$ for some overring R of A that, without loss of generality, we can assume integrally closed. Hence, $A^{\wedge Y} = A^{\wedge \text{Cl}^{\text{inv}}(Y)} = \bigcap\{V \in \text{Zar}(K|R)\} = R$ and so $\text{Cl}^{\text{inv}}(Y) = \text{Zar}(K|A^{\wedge Y})$.

(6) The implication ($\Leftarrow$) holds in general and is a straightforward consequence of Corollary 5.8(a). For ($\Rightarrow$), assume more generally that $R := \bigcap\{V \mid V \in Y_1\} = \bigcap\{V \mid V \in Y_2\}$ is a vacant domain with quotient field K then, by Corollary 5.3 (or Corollary 5.8(b)), $\text{Cl}^{\text{inv}}(Y_1) = \text{Zar}(K|R) = \text{Cl}^{\text{inv}}(Y_2)$.

(7) Since $\text{Cl}^{\text{inv}}(Y) = \text{Cl}^{\text{inv}}(\text{Cl}^{\text{inv}}(Y))$, then, by (2), we have $(\text{Cl}^{\text{inv}}(Y))(T) = \text{Zar}(K(T)|\text{Kr}(Y))$. The conclusion follows from (3).

(8) Note that $A^{\wedge Y} = \text{Kr}(Y) \cap K$ and so $(A^{\wedge Y})_S = \text{Kr}(Y)_S \cap K$. Since each overring of $\text{Kr}(Y)$ is a K-function ring, there exists $\Sigma \subseteq Y$ such that $\text{Kr}(Y)_S = \text{Kr}(\Sigma)$ (Proposition 3.9(1)). We conclude that $(A^{\wedge Y})_S = \text{Kr}(\Sigma) \cap K = A^{\wedge \Sigma}$.

(9) Observe that $\text{Kr}(Y)$ is an overring of the Prüfer domain $\text{Kr}(K|A)$. Thus $\text{Spec}(\text{Kr}(Y))$ is canonically embedded in $\text{Spec}(\text{Kr}(K|A))$. If $P \in \text{Spec}(\text{Kr}(Y))$, then, by (3), $\text{Kr}(Y)_P \cap K \in \text{Cl}^{\text{inv}}(Y)$. The conclusion follows from Theorem 3.11(4).

Remark 5.10. Another proof of Proposition 5.9(1) is based on the fact that when A is a Prüfer domain, it is easy to see that $\text{Zar}(K|S \cap T) = \text{Zar}(K|S) \cup \text{Zar}(K|T)$ for each pair of overrings S and T of A. Now, suppose $Y = \text{Cl}^{\text{inv}}(Y)$. When A is Prüfer, $\text{Cl}^{\text{inv}}(Y) = \bigcap\{\text{Zar}(K|A_\lambda) \mid \lambda \in \Lambda\}$, where A_λ is a finitely generated overring of A. Moreover, $\bigcap\{\text{Zar}(K|A_\lambda) \mid \lambda \in \Lambda\} = \text{Zar}(K|R)$, where R is the ring generated by $\bigcup\{A_\lambda \mid \lambda \in \Lambda\}$. Conversely, if $Y = \text{Zar}(K|R)$ for some overring R of A, then $Y = \bigcap\{\text{Zar}(K|A[r]) \mid r \in R\}$ and thus is inverse-closed since it is the intersection of a family of inverse-closed subsets, $\text{Zar}(K|A[r])$, of $\text{Zar}(K|A)$.

Theorem 5.11 (Olberding [34, Proposition 5.10]). *Let K be a field and A a subring of K. Let R be an integrally closed domain with quotient field K, containing as subring A. Given a subset $Y \subseteq \text{Zar}(K|A)$ such that $R = \bigcap\{V \mid V \in Y\}$ and $Y = \text{Cl}^{\text{inv}}(Y)$, then $\Phi(Y) := \text{Kr}(Y)$ is a K–function ring such that $\Phi(Y) \cap K = R$. Conversely, given a K–function ring Ψ, $A(T) \subseteq \Psi \subseteq K(T)$ such that $\Psi \cap K = R$, then $F(\Psi) := \{W \cap K \mid W \in \text{Zar}(K(T)|\Psi)\}$ is an inverse-closed subspace of $\text{Zar}(K|A)$ and $R = \bigcap\{V \mid V \in F(\Psi)\}$. Furthermore, $F(\Phi(Y)) = Y$ for each inverse-closed subspace Y of $\text{Zar}(K|A)$ and $\Phi(F(\Psi)) = \Psi$, for each K–function ring Ψ, $A(T) \subseteq \Psi \subseteq K(T)$.*

From the previous theorem, it follows that if Y_1 and Y_2 are two different subsets of $Z := \text{Zar}(K|A)$ such that $\text{Cl}^{\text{inv}}(Y_1) = \text{Cl}^{\text{inv}}(Y_2)$, then $\text{Kr}(Y_1)$ and $\text{Kr}(Y_2)$ are two different K–function rings such that $\text{Kr}(Y_1) \cap K = \bigcap\{V \mid V \in Y_1\} = \bigcap\{V \mid V \in Y_2\} = \text{Kr}(Y_2) \cap K$. Furthermore, if R is an integrally closed domain with quotient field K, then $\text{Zar}(K|R)$ is an inverse-closed subspace of $\text{Zar}(K|A)$, $R = \bigcap\{V \mid V \in \text{Zar}(K|R)\}$, and $\text{Kr}(\text{Zar}(K|R))$ is the smallest K–function ring such that $\text{Kr}(\text{Zar}(K|R)) \cap K = R$. If we assume that R is a Prüfer domain, then $\text{Kr}(\text{Zar}(K|R))$ is the unique K–function ring such that $\text{Kr}(\text{Zar}(K|R)) \cap K = R$ [20, Theorem 32.15 and Proposition 32.18].

We have already observed that $Z := \text{Zar}(K|A)$ (endowed with the Zariski topology) is always a spectral space, being canonically homeomorphic to $\text{Spec}(\text{Kr}(K|A))$. It is natural to investigate when the ringed space $(Z, \mathcal{O}_Z)$ is an affine scheme.

Theorem 5.12 (Olberding [34, Theorem 6.1 and Corollaries 6.2 and 6.3]). *Let K be a field, A a subring of K and Y a subspace of $Z := \text{Zar}(K|A)$ (endowed with the Zariski topology). Then,*

(1) *$(Y, \mathcal{O}_Y)$ is an affine scheme if and only if $\mathcal{O}_Y(Y)$ is a Prüfer domain and $Y = \text{Cl}^{\text{inv}}(Y)$ or, equivalently, if and only if Y is an inverse-closed affine subset of Z.*

(2) *$(Z, \mathcal{O}_Z)$ is an affine scheme if and only if the integral closure of A in K is a Prüfer domain with quotient field K.*

(3) *$Y = \text{Cl}^{\text{inv}}(Y)$ if and only if $(Y(T), \mathcal{O}_{Y(T)})$ is an affine scheme.*

Acknowledgements The authors thank the referee for providing helpful suggestions and pointing out to them the very recent paper by N. Schwartz [39].

References

1. Pavel S. Alexandroff, *Combinatorial Topology* (Graylock Press, Rochester, 1956)
2. P-J Cahen, K.A Loper, Francesca Tartarone, Integer-valued polynomials and Prüfer v—multiplication domains. J. Algebra **226**, 765–787 (2000)
3. C. Chevalley et Henri Cartan, Schémas normaux; morphismes; ensembles constructibles. Séminaire Henri Cartan **8**(7), 1–10 (1955–1956)
4. D. Dobbs, R. Fedder, M. Fontana, Abstract Riemann surfaces of integral domains and spectral spaces. Ann. Mat. Pura Appl. **148**, 101–115 (1987)
5. D. Dobbs, M. Fontana, Kronecker function rings and abstract Riemann surfaces. J. Algebra **99**, 263–274 (1986)
6. D. Dobbs, M. Fontana, I. Papick, On certain distinguished spectral sets. Ann. Mat. Pura Appl. **128**, 227–240 (1980)
7. A. Fabbri, Kronecker function rings of domains and projective models, Ph.D. Thesis, Università degli Studi "Roma Tre", 2010
8. C. Finocchiaro, Spectral spaces and ultrafilters. Comm. Algebra **42**(4), 1496–1508 (2014)
9. C. Finocchiaro, M. Fontana, K.A. Loper, The constructible topology on the spaces of valuation domains. Trans. Amer. Math. Soc. **365**(12), 6199–6216 (2013)
10. C. Finocchiaro, M. Fontana and K.A. Loper, Ultrafilter and constructible topologies on spaces of valuation domains. Comm. Algebra **41**(5), 1825–1835 (2013)
11. M. Fontana, Topologically defined classes of commutative rings. Ann. Mat. Pura Appl. **123**, 331–35 (1980)
12. M. Fontana, K.A. Loper, A Krull-type theorem for the semistar integral closure of an integral domain, Commutative algebra. AJSE Arab. J. Sci. Eng. Sect. C Theme Issues **26**, 89–95 (2001)
13. M. Fontana, K.A. Loper, Kronecker function rings: a general approach, in "Ideal theoretic methods in commutative algebra" ed. by D.D. Anderson, I.J. Papick, Columbia, MO, 1999, *Lecture Notes in Pure and Applied Mathematics*, vol. 220 (Dekker, New York, 2001) pp. 189–205
14. M. Fontana, K.A. Loper, in *Multiplicative Ideal Theory in Commutative Algebra: A Tribute to the Work of Robert Gilmer*, ed. by J. Brewer, S. Glaz, W. Heinzer, B. Olberding. An Historical Overview of Kronecker Function Rings, Nagata Rings and Related Star and Semistar Operations (Springer, Berlin, 2006)
15. M. Fontana, A. Loper, A generalization of Kronecker function rings and Nagata rings. Forum Math. **19**, 971–1004 (2007)
16. M. Fontana, K.A. Loper, The patch topology and the ultrafilter topology on the prime spectrum of a commutative ring. Comm. Algebra **36**, 2917–2922 (2008)
17. M. Fontana, K.A. Loper, Cancellation properties in ideal systems: a classification of e.a.b. semistar operations. J. Pure Appl. Algebra **213**(11), 2095–2103 (2009)
18. J. Fresnel, M. van der Put, Géométrie algébrique rigide et applications, Progress in Mathematics vol. 18 (Bikhäuser, Boston and Basel, 1981)
19. K. Fujiwara, F. Kato, Rigid geometry and applications, Advanced Studies in Pure Mathematics **45**, Moduli spaces and arithmetic geometry (Kyoto, 2004), (American Mathematical Society, 2006) pp. 327–386
20. R. Gilmer, *Multiplicative Ideal Theory* (Marcel Dekker, New York, 1972)
21. Alexander Grothendieck et Jean Dieudonné, Éléments de Géométrie (Algébrique I, IHES 1960, Springer, Berlin, 1970)
22. F. Halter-Koch, Kronecker function rings and generalized integral closures. Comm. Algebra **31**, 45–59 (2003)
23. O. Heubo-Kwegna, Kronecker function rings of transcendental field extensions. Comm. Algebra **38**, 2701–271 (2010)
24. Melvin Hochster, Prime ideal structure in commutative rings. Trans. Amer. Math. Soc. **142**, 43–60 (1969)

25. R. Huber, Bewertungsspektrum und rigide Geometrie, Regensburger Mathematische Schriften, vol. 23 (Universität Regensburg, Fachbereich Mathematik, Regensburg, 1993)
26. R. Huber, M. Knebusch, On valuation spectra, in "Recent advances in real algebraic geometry and quadratic forms: proceedings of the RAGSQUAD year", Berkeley, Contemp. Math. vol. 155, (American Mathematical Society, Providence, RI, 1994) pp. 1990–1991
27. I. Kaplansky, Commutative Rings, revised ed., (The University of Chicago Press, Illinois, 1974)
28. F.-V. Kuhlmann, Places of algebraic function fields in arbitrary characteristic. Adv. Math. **188**, 399–424 (2004)
29. W.J. Lewis, J. Ohm, The ordering of $\mathrm{Spec}(R)$. Canad. J. Math. **28**, 820–835 (1976)
30. K.A. Loper, Sequence domains and integer-valued polynomials. J. Pure Appl. Algebra **119**, 185–210 (1997)
31. K.A. Loper, A classification of all D such that $\mathrm{Int}(D)$ is a Prüfer domain. Proc. Amer. Math. Soc. **126**, 657–660 (1998)
32. M. Nagata, Imbedding of an abstract variety in a complete variety. J. Math. Kyoto Univ. **2**, 1–10 (1962)
33. M. Nagata, A generalization of the imbedding problem of an abstract variety in a complete variety. J. Math. Kyoto Univ. **3**, 89–102 (1963)
34. B. Olberding, Affine schemes and topological closures in the Zariski-Riemann space of valuation rings, preprint.
35. G. Picavet, Autour des idéaux premiers de Goldman dun anneau commutatif. Ann. Sci. Univ. Clermont Math. **57**, 73–90 (1975)
36. G. Picavet, Sur les anneaux commutatifs dont tout idéal premier est de Goldman. C.R. Acad. Sci. Paris Sér. I Math. **280**(25), A1719–A1721 (1975)
37. W. Rump, Y.C. Yang, Jaffard-Ohm correspondence and Hochster duality. Bull. Lond. Math. Soc. **40**, 263–273 (2008)
38. N. Schwartz, Compactification of varieties. Ark. Mat. **28**, 333–370 (1990)
39. N. Schwartz, , Sheaves of Abelian l-groups. Order **30**, 497–526 (2013)
40. N. Schwartz, M. Tressl, Elementary properties of minimal and maximal points in Zariski spectra. J. Algebra **323**, 698–728 (2010)
41. J. Tate, Rigid analytic spaces. Invent. Math. **12** , 257–269 (1971)
42. M. Temkin, Relative Riemann-Zariski spaces. Israel J. Math. 1–42 (2011)
43. M. Temkin, I. Tyomkin, Prüfer algebraic spaces (2011) arXiv:1101.3199
44. O. Zariski, The compactness of the Riemann manifold of an abstract field of algebraic functions. Bull. Amer. Math. Soc **50**, 683–691 (1944)
45. O. Zariski, The reduction of singularities of an algebraic surface. Ann. Math. **40**, 639–689 (2011)
46. O. Zariski, Reduction of singularities of algebraic three dimensional varieties. Ann. Math. **45**, 472–542 (1944)
47. O. Zariski, P. Samuel, Commutative Algebra, Volume 2, Springer, Graduate Texts in Mathematics New York, 1975, vol. 29, 1st edn. (Van Nostrand, Princeton, 1960)

Ten Problems on Stability of Domains

Stefania Gabelli

Abstract We survey the notions of (finite) stability, quasi-stability, and Clifford regularity of domains and illustrate some open problems.

Keywords Stable ideal • Divisorial ideal • Flat ideal • Clifford regular domain

Mathematics Subject Classification (2011): 13A15; 13F05; 13C10.

Introduction

Stability of ideals was explicitly introduced by J. Lipman in 1971 in order to study Arf rings [30]. However, this notion was already known and widely used in the context of one-dimensional Noetherian rings, in particular in relation with reflexive rings and decomposition of torsion-free modules [5, 31].

A stable ideal of a Noetherian ring is defined as an ideal that is projective over its ring of endomorphisms [51, 52]; extending this definition to arbitrary integral domains, one says that a nonzero ideal I of a domain R is stable if I is invertible in the overring $E(I) := (I : I)$ of R [1]. If each nonzero ideal (respectively, finitely generated ideal) of R is stable, one says that R itself is stable (respectively, finitely stable).

Since 1998, stability of domains has been thoroughly investigated by B. Olberding. In [39] he illustrated several ideal-theoretic and module-theoretic applications of this concept and announced some new results, then published in [40–42].

S. Gabelli (✉)
Dipartimento di Matematica e Fisica, Università degli Studi Roma Tre,
Largo S. L. Murialdo, 1, 00146 Roma, Italy
e-mail: gabelli@mat.uniroma3.it

M. Fontana et al. (eds.), *Commutative Algebra: Recent Advances in Commutative Rings,*   175
Integer-Valued Polynomials, and Polynomial Functions, DOI 10.1007/978-1-4939-0925-4_10,
© Springer Science+Business Media New York 2014

Invertible ideals are clearly stable, thus stability finds interesting applications in the setting of Prüfer domains (i.e., domains in which nonzero finitely generated ideals are invertible). Stable Noetherian domains are one-dimensional [52]. However, as showed by Olberding, a stable domain need not be coherent, nor one-dimensional, nor integrally closed [41, Sect. 3].

Weakening the notion of stability, in more recent years, other classes of domains were introduced, like Rutliff-Rush domains [33], quasi-stable domains [46], and Clifford regular domains [6]. All these notions coincide with stability in the Noetherian case, but not in general.

In this short survey, leaving aside the module-theoretic point of view, we focus on some ideal-theoretic aspects of stability and discuss some unresolved problems.

All the rings considered are commutative rings with unity that are not fields. A *local ring* is a ring with a unique maximal ideal and a *semilocal ring* is a ring with finitely many maximal ideals, not necessarily Noetherian.

If R is a ring with total quotient ring K, an *overring* of R is a ring between R and K. If I, J are R-submodules of K, we set $(I : J) := \{x \in K; \ xJ \subseteq I\}$ and $(I :_R J) := \{x \in R; \ xJ \subseteq I\}$.

1 Stable Notherian Rings

Stable ideals were introduced in 1971 by J. Lipman, in his paper [30] on *Arf rings*, which are local Noetherian rings satisfying certain conditions studied by Arf in [3].

Lipman worked in the setting of semilocal one-dimensional Macaulay rings, that is, semilocal one-dimensional Noetherian rings whose Jacobson radical contains a regular element. If R is such a ring, Lipman defined a regular ideal $I \subseteq R$ to be *stable* if $IR^I = I$ or, equivalently, $R^I = (I : I)$ [30, Definition 1.3], where $R^I := \bigcup_{n \geq 1} (I^n : I^n)$ is the ring obtained by *blowing up* I.

The main motivation for introducing this notion is that it furnishes a useful characterization of Arf rings.

Recall that if I is an ideal of the ring R, an element $x \in R$ is said to be *integral over I* if there exist a positive integer n and elements $a_k \in I^k$, $k = 1, \ldots, n$, such that

$$x^n + a_1 x^{n-1} + a_2 x^{n-2} + \cdots + a_{n-1} x + a_n = 0.$$

The ideal I is called *integrally closed* if all the elements of R which are integral over I belong to I.

Theorem 1.1 ([30, Theorem 2.2]). *A local one-dimensional Macaulay ring is an Arf ring if and only if each integrally closed regular ideal of R is stable.*

To the extent of proving this result, Lipman gave several characterizations of stable ideals. In particular, he proved the following:

Proposition 1.2 ([30, Lemma 1.11]). *Let R be a semilocal one-dimensional Macaulay ring and $I \subseteq R$ a regular ideal. The following conditions are equivalent:*

(i) I is stable (i.e., $IR^I = I$).
(ii) There exists an element $x \in I$ such that $I^2 = xI$.
(iii) There exists a regular element $x \in I$ such that $I = xR^I$.
(iv) There exists a regular element $x \in I$ such that $I = x(I : I)$.

In order to solve a problem posed by Bass, the notion of stability was then extended to Noetherian rings by J. Sally and W. Vasconcelos in a paper that was published in 1973. They called an ideal I of a Noetherian ring R *stable* if I is projective over its endomorphism ring $End_R(I)$ and called R stable if each ideal is stable [51, Sect. 1], [52, Sect. 2].

Note that when I is a regular ideal of the ring R, $End_R(I)$ is isomorphic to the overring $E(I) := (I : I)$ of R, and so I is stable if and only if it is an invertible ideal of $E(I)$. In particular, when R is a semilocal Noetherian ring, I is stable if and only if I is principal in $E(I)$. It follows from Proposition 1.2 that this more general notion of stability coincides with the one introduced by Lipman for regular ideals of semilocal one-dimensional Macaulay ring [52, Proposition 2.2].

Nevertheless, a local one-dimensional Noetherian ring which is stable according to Sally-Vasconcelos need not be a Macaulay ring. For example, one can take $R := k[[X, Y]]/\langle X^2, XY \rangle$, where k is a field and X, Y are indeterminates over k [52, page 324].

Proposition 1.3 ([52, Proposition 2.1]). *Let R be a Noetherian ring. If R is stable (i.e., each ideal is projective over its endomorphism ring), then R has dimension at most equal to one.*

Stability is related to the 2-generator property. An ideal of R is *2-generated* if it is generated by 2 elements, and R is 2-generated, or it has the *2-generator property*, if each *finitely generated* ideal is 2-generated. The 2-generator property plays an important part in the decomposition of torsion-free modules [32].

Bass proved that if R is a one-dimensional reduced Noetherian ring whose integral closure is a finitely generated R-module, the 2-generator property implies stability [5, Proposition 7.1 and Corollary 7.3]. Sally and Vasconcelos showed that, as conjectured by Bass, also the converse holds.

Theorem 1.4. *Let R be a Noetherian ring.*

(1) [52, Theorem 3.4] *Assume that R is a one-dimensional Macaulay ring whose maximal ideals are not minimal primes. If each regular ideal is 2-generated, then R is stable.*
(2) [51, Theorem 2.4] *Assume that R is one-dimensional reduced and that its integral closure is a finitely generated R-module. If R is stable, then R is 2-generated.*

However, even for Noetherian domains, the 2-generator property is strictly stronger than stability. The first example of a local Noetherian domain that is stable and not 2-generated was given in [52, Example 5.4]; several other examples are collected in [44, Sect. 3].

The relationships among the 2-generator property, stability of finitely generated regular ideals, and decomposition of finitely generated torsion-free modules were further investigated by D. Rush in two papers published in 1991 and 1995 [49, 50]. In particular he extended Bass' result to local rings.

Theorem 1.5 ([50, Proposition 2.5]). *Let R be a local ring. If R is 2-generated, each finitely generated regular ideal is stable.*

Rush gave also the following characterization of stability for Noetherian rings.

Theorem 1.6 ([49, Theorem 2.4]). *Let R be a one-dimensional Noetherian ring with integral closure R'. Then each regular ideal of R is stable if and only if the following conditions hold:*

(a) Each (or each finitely generated) R-submodule of R' containing R is a ring.
(b) Each maximal ideal of R has at most two maximal ideals of R' lying over it.

Together with other results of Rush, the last two theorems were later extended to general integral domains by B. Olberding in [40]. For an extension to rings, see Olberding's paper in this volume.

A notion weaker than stability, still useful to bound the number of generators of ideals, was introduced by P. Eakin and P. Sathaye in 1976. They observed that part of the Lipman's result given in Proposition 1.2 can be extended in the following way:

Proposition 1.7 ([15, Lemma, page 447]). *Let R be a local ring and I a finitely generated regular ideal of R. The following conditions are equivalent:*

(i) There exists an element $x \in I$ such that $I^2 = xI$.
(ii) There exists a regular element $x \in I$ such that $I = x(I : I)$.

Thus Eakin and Sathaye defined an ideal I of a *semilocal* ring to be *stable* if there is an element $x \in I$ such that $I^2 = xI$ and say that I is *prestable* if some power of I is stable, that is, for some $k \geq 1$ there is an $x \in I$ such that $I^{2k} = xI^k$ [15, Sect. 3].

Proposition 1.8 ([15, Corollary 1, page 446]). *Let R be a local ring and I a finitely generated ideal. The following conditions are equivalent:*

(i) I is prestable (i.e., $I^{2k} = xI^k$, for some $k \geq 1$).
(ii) There is a positive integer $b := b(I)$ such that I^n has b generators, for each $n \geq 1$.
(iii) There is a positive integer n such that I^n has n generators, for some $n \geq 1$.

Moreover, if I is regular and I^n has n generators, then $I^{2(n-1)} = xI^{n-1}$, for some $x \in I$.

2 Stable Domains

In a note of 1987, D.D. Anderson, J. Huckaba, and I. Papick considered the notion of stability for arbitrary integral domains [1]. Given a nonzero ideal I of a domain R, they say that I is *Lipman-stable* (for short, *L-stable*), if $R^I = (I : I)$ and say that I is stable according to Sally-Vasconcelos, or is *SV-stable*, if I is invertible in the overring $E(I) := (I : I)$. The domain R is called *L-stable* (respectively, *SV-stable*) if each nonzero ideal of R is L-stable (respectively, SV-stable).

As in [14, Sect. 7.4], one can also say that I is stable according to Eakin-Sathaye, or that I is *ES-stable*, if $I^2 = JI$ for some invertible ideal J contained in I. The ideal I is called *ES-prestable* (respectively, *SV-prestable*) if some power of I is ES-stable (respectively, SV-stable).

Proposition 2.1 ([1, Lemmas 2.1 and 2.2]). *Let R be a domain and I a nonzero ideal. Then*

$$I \ ES\text{-}stable \ \Rightarrow \ I \ SV\text{-}stable \Rightarrow I \ L\text{-}stable.$$

If I is finitely generated, often all these notions coincide.

Proposition 2.2 ([14, Corollary 7.4.2 and Proposition 7.4.3]). *Let R be a domain and I a nonzero finitely generated ideal. The following conditions are equivalent:*

(i) I is SV-stable.
(ii) $I R_M$ is SV-stable, for each maximal ideal $M \subseteq R$.
(iii) $I R_M$ is ES-stable, for each maximal ideal $M \subseteq R$.

In particular, if R is local, I is SV-stable if and only if I is ES-stable.

A domain R is integrally closed if and only if $R = (I : I)$ for each nonzero finitely generated ideal I. If $R = (I : I)$ for each nonzero ideal I, R is called *completely integrally closed*. Hence, if R is completely integrally closed, $R = (I : I) = R^I$, for each nonzero ideal; similarly, if R is integrally closed, $R = (I : I) = R^I$, for each nonzero finitely generated ideal. This shows that for completely integrally closed domains L-stability is a trivial concept.

Proposition 2.3. *(1) Let R be a completely integrally closed domain and I a nonzero ideal of R. Then I is L-stable; in addition,*

$$I \ is \ invertible \ \Leftrightarrow \ I \ is \ SV\text{-}stable.$$

(2) [14, Proposition 7.4.4] Let R be an integrally closed domain and I a nonzero finitely generated ideal of R. Then I is L-stable; in addition,

$$I \ is \ invertible \ \Leftrightarrow \ I \ is \ ES\text{-}(pre)stable \ \Leftrightarrow \ I \ is \ SV\text{-}(pre)stable.$$

SV-stability implies L-stability (Proposition 2.1). The converse is not true, even in the Noetherian case. In fact, by the proposition above, any Noetherian integrally closed domain is L-stable, but is SV-stable only if it is Dedekind. More generally, we have :

Proposition 2.4 ([1, Proposition 2.4]). *A Noetherian domain is SV-stable if and only if it is L-stable and one-dimensional.*

Proposition 2.3(2) furnishes also a characterization of Prüfer domains in terms of SV-stability. Recall that a domain R is called a *Prüfer domain* if R_P is a valuation domain, for each nonzero prime ideal P; this is equivalent to say that each nonzero finitely generated ideal is invertible. A Prüfer domain such that PR_P is a principal ideal, for each nonzero prime ideal P, is called *strongly discrete*.

Proposition 2.5. *(1) R is a Prüfer domain if and only if R is integrally closed and each finitely generated nonzero ideal of R is SV-stable (equivalently, ES-stable).*
(2) [1, Lemma 2.7] Each Prüfer domain is L-stable.
(3) [1, Proposition 2.10] A semilocal Prüfer domain (in particular, a valuation domain) is SV-stable if and only it is strongly discrete.

If the integral closure of R is a Prüfer domain, R is called *quasi-Prüfer* [14, Corollary 6.5.14]. Quasi-Prüfer domains can be characterized by the property that each nonzero finitely generated ideal is locally ES-prestable.

Theorem 2.6 ([15, Theorem 2]). *Let R be a local domain with integral closure R'. The following conditions are equivalent:*

(i) R' is a Prüfer domain.
(ii) Each nonzero finitely generated ideal of R is ES-prestable.

A global version of Theorem 2.6 is the following.

Theorem 2.7 ([14, Theorem 7.4.6]). *Let R be a domain with integral closure R'. The following conditions are equivalent:*

(i) R' is a Prüfer domain.
(ii) Each nonzero finitely generated ideal of R is SV-prestable.
(iii) Each nonzero 2-generated ideal of R is SV-prestable.

Since 1998, SV-stable domains have been thoroughly investigated by B. Olberding in a series of papers [38, 40–42]. Olberding calls an SV-stable ideal of a domain R simply a *stable ideal* and he says that R is *stable* (respectively, *finitely stable*) if each nonzero ideal (respectively, finitely generated ideal) of R is stable. We keep this notation; thus from now on "stable" means "SV-stable."

Stability and finite stability transfer to overrings. In addition, their study can be reduced to the local case.

Proposition 2.8 ([42, Lemma 2.4 and Theorem 5.1]). *Let S be an overring of the domain R. If R is (finitely) stable, then S is (finitely) stable.*

By the result above and Proposition 2.5, we get that finitely stable domains are quasi-Prüfer.

Corollary 2.9 ([50, Proposition 2.1]). *If the domain R is finitely stable (respectively, stable), its integral closure is a Prüfer domain (respectively, a strongly discrete Prüfer domain).*

Theorem 2.10. *Let R be a domain.*

(1) [42, Theorem 3.3] R is stable if and only if R_M is stable, for each maximal ideal M, and R has finite character (i.e., each nonzero element is contained at most in finitely many maximal ideals).

(2) [14, Proposition 7.3.4] R is finitely stable if and only if R_M is finitely stable.

Since any Prüfer domain is finitely stable, finitely stable domains need not have finite character. Also, finitely stable domains with finite character need not be stable. For example, any valuation domain with nonprincipal maximal ideal is finitely stable but not stable (Proposition 2.5).

Theorem 2.11 ([41, Theorem 2.3]). *A domain R is stable if and only if the following conditions hold: (a) R is finitely stable. (b) PR_P is a stable ideal of R_P, for each nonzero prime P. (c) R_P is a valuation domain for each nonzero nonmaximal prime P. (d) R has finite character.*

Thus the semilocal case given in [1] (Proposition 2.5(3)) can be generalized in the following:

Proposition 2.12 ([38, Theorem 4.6]). *Let R be an integrally closed domain. Then R is stable if and only if it is a strongly discrete Prüfer domain with finite character.*

A strongly discrete Prüfer domain such that each noninvertible element has finitely many minimal primes is called a *generalized Dedekind domain*. These domains were introduced by N. Popescu in [47] and have very good ring-theoretic and ideal-theoretic properties; an overview is given in [18].

By the previous proposition, integrally closed stable domains are generalized Dedekind. More precisely, we have:

Corollary 2.13. *The following conditions are equivalent for a domain R:*

(i) R is a generalized Dedekind domain with finite character.
(ii) R is integrally closed and stable.

It is also interesting to observe that stability of nonzero prime ideals forces a Prüfer domain to be generalized Dedekind.

Theorem 2.14 ([17, Theorem 5], [38, Theorem 4.7]). *The following conditions are equivalent for a domain R:*

(i) R is a generalized Dedekind domain.
(ii) R is a Prüfer domain and each nonzero prime ideal of R is stable.

However, a domain whose nonzero prime ideals are all stable need not be stable, as [41, Example 3.4] shows. An example of generalized Dedekind domain that is not stable is $R := \mathbb{Z} + X\mathbb{Q}[[X]]$; in fact R does not have finite character.

Olberding gave a complete characterization of stable domains [41]. In particular, he proved that each local stable domain is a suitable pullback of a local stable ring of dimension at most one.

Theorem 2.15 ([41, Corollary 2.7], [43]). *A local domain R is stable if and only if one of the following conditions is satisfied: (a) R is one-dimensional stable. (b) R is a strongly discrete valuation domain. (c) R arises from a pullback diagram of type:*

$$
\begin{array}{ccc}
R & \longrightarrow & D \\
\downarrow & & \downarrow \\
V & \xrightarrow{\ \varphi\ } & \dfrac{V}{I}
\end{array}
$$

where V is a strongly discrete valuation domain, I is an ideal of V, D is a local stable ring of dimension at most one having a prime ideal P such that P contains all the zero-divisors of D and $P^2 = (0)$, and V/I is isomorphic to the total quotient ring of D.

For example, let W be a one-dimensional discrete valuation domain with quotient field F and X be an indeterminate over F. With the notations of Theorem 2.15(c), setting $V := F[[X]]$, $I := X^2 V$, and $D := W[[X]]/X^2 W[[X]]$, we get that $R = W + XW + X^2 F[[X]]$ is a stable domain [43].

A stable Noetherian ring is one-dimensional (Proposition 1.3). An example of a local one-dimensional domain which is stable and not Noetherian was constructed by Olberding in [41, Proposition 5.2]. Generalizing this construction, Olberding then exhibited a whole class of examples, as a particular class of one-dimensional domains whose integral closure is not a finitely generated module [45, Theorems 4.1 and 4.4] (see also [44, Theorem 3.10]). In fact Theorem 1.4(2) can be extended in the following way:

Theorem 2.16. *Let R be a stable domain with integral closure R'.*

(1) [41, Proposition 4.5] If R is one-dimensional and $(R : R') \neq (0)$, R is Noetherian 2-generated and R' is a finitely generated R-module.

(2) [42, Corollary 4.17] If R is local and $(R : R') = (0)$, R' is a one-dimensional discrete valuation ring (in particular R is one-dimensional).

On the other hand, it is also possible for stable Noetherian domains, even Noetherian 2-generated domains, to have $(R : R') = (0)$ [44, Sect. 3].

By using Theorem 2.16, M. Roitman and the author of this paper recently proved that a stable one-dimensional domain is Mori and is precisely a Mori finitely stable domain.

Recall that if I is a nonzero ideal of the domain R, the *divisorial closure* of I is the ideal $I_v := (R : (R : I))$ and that I is called *divisorial* if $I = I_v$. A *Mori domain* is a domain satisfying the ascending chain condition on divisorial ideals. Clearly Noetherian domains are Mori. For the main properties of Mori domains, the reader is referred to [4].

Theorem 2.17 ([24], Theorem 2.34). *The following conditions are equivalent for a domain R:*

(i) R is stable and one-dimensional.
(ii) R is Mori and stable.
(iii) R is Mori and finitely stable.

In addition, under the previous conditions, for each nonzero ideal I of R, $I_v = \langle x, y \rangle_v$, for some $x, y \in I$.

This result shows that the local one-dimensional domains that are stable and not Noetherian constructed by Olberding in [45] are new examples of Mori domains.

It also shows that a one-dimensional stable domain R cannot arise a pullback like in Theorem 2.15, unless $R = V$ is a discrete valuation domain. In fact, in a pullback of that type, R is Mori if and only if V is a one-dimensional discrete valuation domain and D is a field [35, Theorem 9].

3 Divisorial Domains

The class of domains in which each ideal is divisorial has been investigated in the sixties of the last century by several authors and with different methods. Following S. Bazzoni and L. Salce, these domains are now called *divisorial domains*. If each overring of R is divisorial, R is called *totally divisorial* [12].

As for (finite) stability, the study of divisorial domains can be reduced to the local case. We recall that, with a terminology introduced by Matlis, a domain is called *h-local* if it has finite character and each nonzero prime ideal is contained in a unique maximal ideal.

Proposition 3.1 ([12, Proposition 5.4]). *A domain R is divisorial if and only if it is h-local and R_M is divisorial, for each maximal ideal M.*

The local Noetherian case was independently studied by H. Bass [5] and E. Matlis [31].

Theorem 3.2 ([5, Theorems 6.2, 6.3], [31, Theorem 3.8]). *Let R be a local Noetherian domain, with maximal ideal M. Then R is divisorial if and only if R is one-dimensional and $(R : M)$ is a 2-generated R-module.*

It was already known to W. Krull that an integrally closed domain such that each nonzero finitely generated ideal is divisorial is Prüfer. The following characterization of integrally closed divisorial domains was given by W. Heinzer in [25].

Theorem 3.3 ([25, Theorem 5.1]). *Let R be an integrally closed domain. Then R is divisorial if and only if R is an h-local Prüfer domain with invertible maximal ideals.*

In the general case, local divisorial domains were studied in [12, Sect. 5], [7, Sect. 2], [20, Sect. 1]. Recall that a nonzero ideal I of a domain R is called *m-canonical* if $(I : (I : J)) = J$, for each nonzero ideal J of R. With this terminology, the domain R is divisorial if and only if R itself is an m-canonical ideal.

Theorem 3.4. *Let R be a local domain, with maximal ideal M. Then*

(1) [12, Lemma 5.5] If M is principal, R is divisorial if and only if it is a valuation domain.

(2) [20, Theorem 1.2] If M is not principal and R is not a valuation domain, R is divisorial if and only if $(R : M) = (M : M)$ is a 2-generated R-module and M is an m-canonical ideal of $(M : M)$.

Divisoriality and stability are strictly related.

Theorem 3.5 ([40, Theorem 3.12 and Corollary 3.13]). *The following conditions are equivalent for a domain R:*

(i) R is stable and divisorial.

(ii) R is totally divisorial.

(iii) R is h-local and R_M is totally divisorial, for each maximal ideal M.

As always, the Noetherian case and the integrally closed case are of particular interest.

Theorem 3.6 ([12, Proposition 7.1 and Theorem 7.3]). *Let R be a Noetherian domain. The following conditions are equivalent:*

(i) R is stable and divisorial.

(ii) R is totally divisorial and one-dimensional.

(iii) R is 2-generated.

Thus a Noetherian stable domain is divisorial if and only if it is 2-generated. It follows from Theorem 2.16 that

Corollary 3.7. *Assume that R is a one-dimensional stable domain whose integral closure is a finitely generated R-module. Then R is (totally) divisorial.*

By Proposition 2.5(3) and Theorem 3.3, a stable valuation domain is (totally) divisorial. Globalizing we obtain:

Theorem 3.8 ([40, Theorem 3.1], [12, Proposition 7.6]). *Let R be an integrally closed domain. The following conditions are equivalent:*

(i) R is stable and divisorial.
(ii) R is an h-local strongly discrete Prüfer domain.
(iii) R is a divisorial generalized Dedekind domain.

In the local case, totally divisorial domains can be completely classified by using Theorem 2.15: they are either Noetherian 2-generated domains or strongly discrete valuation domains or arise from a suitable pullback diagram [40, Corollary 3.16].

4 Ratliff-Rush Domains

In a paper published in 1978, L. Ratliff and D. Rush associated to a regular ideal I of a Noetherian ring, the ideal $\tilde{I} := \bigcup_{n\geq 0}(I^{n+1} :_R I^n)$ [48]. W. Heinzer, D. Lantz, and K. Shah called $\tilde{I}$ the *Ratliff-Rush ideal associated to* I [27]. If $I = \tilde{I}$, I is called a *Ratliff-Rush ideal* and we can say that R is a *Ratliff-Rush ring* if each regular ideal is Ratliff-Rush.

Among other results, Ratliff and Rush proved that, for any regular ideal I of a Noetherian ring, there is a positive integer n such that, for $k \geq n$, $\widetilde{I^k} = I^k$, so that all sufficiently high powers of I are Rutliff-Rush. Indeed, if R is local with maximal ideal M and I is an M-primary ideal, $\tilde{I}$ is the unique largest ideal containing I and having the same Hilbert polynomial as I [48]. Stable ideals and integrally closed ideals of Noetherian rings are Ratliff-Rush [27].

An early survey on Ratliff-Rush ideals is [28]. In the setting of integral domains, Ratliff-Rush ideals were studied by A. Mimouni in 2009 [33, 34].

Proposition 4.1 ([33, Proposition 2.3 and Theorem 2.5]). *Let R be a domain. Then*

$$R \text{ stable } \Rightarrow R \text{ Ratliff-Rush } \Rightarrow R \text{ L-stable.}$$

In addition

Proposition 4.2 ([27, Proposition 3.1 and Theorem 2.9], [33, Corollary 2.8]). *A Noetherian domain is Ratliff-Rush if and only if it is stable.*

Ratliff-Rush domains are quasi-Prüfer.

Proposition 4.3 ([33, Lemma 2.4]). *Let R be a domain. If each nonzero finitely generated ideal of R is Rutliff-Rush, R' is a Prüfer domain.*

Theorem 4.4 ([33, Theorem 2.6]). *Let R be an integrally closed domain. The following conditions are equivalent:*

(i) $I = \tilde{I}$ *for each finitely generated nonzero ideal (respectively, each nonzero ideal) I of R.*
(ii) R *is a Prüfer domain (respectively, a strongly discrete Prüfer domain).*

Since any Prüfer domain is L-stable (Proposition 2.5(2)), an L-stable domain need not be Ratliff-Rush. Also, by Theorems 2.12 and 4.4, we get

Proposition 4.5. *An integrally closed Rutliff-Rush domain is stable if and only if it has finite character.*

5 Quasi-stable Domains

G. Picozza and F. Tartarone weakened the notion of stability in the following way. Observing that an invertible ideal of a domain is flat, they define a nonzero ideal I of a domain R to be *quasi-stable* if I is flat in its endomorphism ring $E(I) := (I : I)$ and they say that R is *quasi-stable* if each nonzero ideal is quasi-stable [46, Sect. 2].

A stable domain is clearly quasi-stable. In addition, quasi-stable domains are finitely stable. More precisely

Proposition 5.1 ([46, Proposition 2.4]). *A domain R is finitely stable if and only if each nonzero finitely generated ideal of R is quasi-stable.*

Thus (finite) stability and quasi-stability coincide for Mori domains (Theorem 2.17). The next result says that if R is integrally closed, quasi-stability is equivalent to R being Prüfer, so that a quasi-stable domain need not be stable (Proposition 2.5(3)).

Proposition 5.2 ([46, Proposition 2.7]). *The following conditions are equivalent for an integrally closed domain R.*

(i) R is quasi-stable.
(ii) R is finitely stable.
(iii) R is Prüfer.

A very tricky example of a finitely stable domain that is not quasi-stable is given in [46, Example 2.8]. Examples of local quasi-stable domains that are not integrally closed nor stable are constructed as pseudo-valuation domains in [46, Example 2.6(2)]. Precisely, let R be a pseudo-valuation domain with maximal ideal M and associated valuation domain $V := (M : M) = (R : M)$ and assume that $R \neq V$. If V is a 2-generated R-module, then R is quasi-stable and not integrally closed. If, in addition, M is not principal in V, then R is not stable.

By Theorem 2.16, a one-dimensional stable domain such that $(R : R') \neq (0)$ is Noetherian. This result cannot be extended to quasi-stable domains. Indeed, let R be a one-dimensional pseudo-valuation domain that is quasi-stable and not stable, as above. Then R is necessarily not Noetherian (nor Mori), but $V = R'$ and $(R : V) = M \neq (0)$.

It is not clear whether quasi-stability passes to overrings. However this happens in several cases, for example, quasi-stability transfers to localizations and fractional, flat, and Noetherian overrings. More generally

Proposition 5.3 ([46, Corollary 3.7]). *Let S be an overring of R and assume that each ideal of S is extended from a fractional ideal of R. If R is quasi-stable, then S is quasi-stable.*

Proposition 5.4 ([46, Corollary 3.8]). *Let R be a quasi-stable domain with integral closure R'. If $(R : R') \neq (0)$, each overring of R is quasi-stable.*

6 Clifford Regular Domains

Let S be a multiplicative commutative semigroup. An element $x \in S$ is called *von Neumann regular* (for short, *vN-regular*), if there exists an element $a \in S$ such that $x = x^2 a$. Idempotent and invertible elements are vN-regular. By a well-known theorem of Clifford, S is a disjoint union of groups if and only if all its elements are vN-regular: in this case, S is called a *Clifford semigroup*.

The set $\mathcal{F}(R)$ of nonzero fractional ideals of a domain R form a multiplicative semigroup, with unity R. The *class semigroup* of R is defined as the quotient semigroup of $\mathcal{F}(R)$ by the subgroup of nonzero principal ideals. A domain R is called a *Clifford regular domain* if its class semigroup is Clifford regular; this is equivalent to say that each nonzero fractional ideal is vN-regular in the semigroup $\mathcal{F}(R)$.

Dedekind domains are trivial examples of Clifford regular domains. S. Bazzoni and L. Salce showed that all valuation domains are Clifford regular and gave a complete description of the structure of the class semigroup in that case [11]. P. Zanardo and U. Zannier investigated the class semigroups of orders in number fields and showed that all orders in quadratic fields are Clifford regular domains [55]. The study of Clifford regular domains was then carried on by S. Bazzoni [6, 8–10].

Clifford regular domains are between stable and finitely stable domains.

Proposition 6.1 ([9, Proposition 2.3]). *A stable domain is Clifford regular and a Clifford regular domain is finitely stable.*

Hence, Clifford regularity and (finite) stability are equivalent for Mori domains (Theorem 2.17). Also, an integrally closed Clifford regular domain is Prüfer (Proposition 2.5(1)).

S. Bazzoni proved that a Clifford regular domain has finite character [10, Theorem 4.7]. This property characterizes Clifford regularity inside Prüfer domains and allows to show that the integral closure of a Clifford regular domain is still Clifford regular.

Theorem 6.2 ([9, Theorem 4.5]). *An integrally closed domain is Clifford regular if and only if it is a Prüfer domain with finite character.*

Proposition 6.3 ([10, Corollary 4.8]). *If R is a Clifford regular domain, its integral closure is Clifford regular.*

Even in the local case, Clifford regularity may not coincide with stability or finite stability. In fact, any valuation domain is Clifford regular [11, Theorem 3] but need not be stable (Proposition 2.5(3)). A local finitely stable domain that is not Clifford regular is exhibited in [9, Example 6.6].

The following result puts in relation stability and Clifford regularity.

Proposition 6.4 ([53, Theorem 2.6]). *Let R be a domain. The following conditions are equivalent:*

(i) R is stable.

(ii) R is Clifford regular and each nonzero idempotent fractional ideal of R is a ring.

Clifford regularity of overrings was investigated by L. Sega in [53]. Since ideals extended from vN-regular ideals are still vN-regular, the situation is similar to the one of quasi-stability.

Proposition 6.5 ([53, Proposition 4.1]). *Let S be an overring of R and assume that each ideal of S is extended from a fractional ideal of R. If R is Clifford regular, then S is Clifford regular.*

Proposition 6.6 ([53, Theorem 4.6]). *Let R be a Clifford regular domain with integral closure R'. If $(R : R') \neq (0)$, each overring of R is Clifford regular.*

7 Problems

As we summarize in the two tables below, all the stability conditions introduced in this paper are well understood when R is integrally closed or Noetherian. But in general there are several questions still unanswered; in this last section we illustrate some of them (Table 1).

Problem 7.1. A domain R is *Archimedean*, if $\bigcap_{n \geq 0} r^n R = (0)$, for each nonunit $r \in R$. Since Mori domains satisfy the ascending chain condition on principal ideals, they are Archimedean. The class of Archimedean domains includes also completely integrally closed domains and one-dimensional domains.

Question. Is a stable Archimedean domain one-dimensional?

This question has a negative answer. An example of a stable Archimedean domain of dimension 2 has been given in [24, Example 3.9].

The answer is positive in the semilocal case [24], so that a semilocal stable Archimedean domain is Mori (Theorem 2.17). However, in general the Archimedean property does not pass to localizations. For example, the ring of entire functions is an infinite-dimensional completely integrally closed (hence Archimedean) Bezout domain [14, Sect. 8.1] which is not locally Archimedean, because an Archimedean valuation domain is one-dimensional. Hence, a way of approaching this problem is trying to understand if the Archimedean property localizes under the hypothesis of stability (Table 2).

Table 1 The integrally closed case

Stable	⇔	Prüfer strongly discrete with finite character
⇓		⇓
Clifford regular	⇔	Prüfer with finite character
⇓		⇓
Quasi-stable	⇔	Prüfer
⇕		⇕
Finitely stable	⇔	Prüfer

Stable	⇔	Prüfer strongly discrete with finite character
⇓		⇓
Ratliff-Rush	⇔	Prüfer strongly discrete
⇓		⇓
Finitely stable	⇔	Prüfer
⇓		
L-stable		

Table 2 The Noetherian case

Stable	⇔	Clifford regular	⇔	Quasi-stable	⇔	Finitely stable
	⇔	Ratliff-Rush	⇔	L-stable one-dimensional		

Problem 7.2. A (one-dimensional) Mori domain whose nonzero finitely generated ideals are L-stable is not necessarily (finitely) stable. In fact if R is a local one-dimensional integrally closed Mori domain, each nonzero finitely generated ideal I of R is L-stable (Proposition 2.3(2)); but if R is (finitely) stable it must be a discrete valuation domain (Proposition 2.5(1)). Example of local one-dimensional integrally closed Mori domains that are not valuation domains can be constructed by means of pullbacks [4, Theorem 2.2]. Does Proposition 2.4 extend to Mori domains? That is

Question. Is a one-dimensional L-stable Mori domain stable?

Problem 7.3. For Noetherian domains, the Ratliff-Rush property is equivalent to (finite) stability and in the one-dimensional case also to L-stability (Propositions 2.4 and 4.2). Since stable domains are Ratliff-Rush, a Mori Ratliff-Rush domain need not be Noetherian (Sect. 2).

Question. Is a Mori Ratliff-Rush domain one-dimensional?

Apart from the Noetherian case, the answer is positive if either $(R : R') \neq (0)$ or R is seminormal [33, Corollary 2.10].

Problem 7.4. If R is a Mori stable domain (equivalently, a one-dimensional stable domain) and I is a nonzero ideal of R, we have $I_v = \langle x, y \rangle_v$, for some $x, y \in I$ (Theorem 2.17); thus we can say that in a stable Mori domain each divisorial ideal is *2-v-generated*. Since a divisorial Mori domain is Noetherian, this result generalizes (i) $\Rightarrow$ (iii) of Theorem 3.6.

Question. Assume that R is a one-dimensional Mori domain such that each divisorial ideal is 2-v-generated. Is it true that each divisorial ideal is stable?

Note that the answer to this question is negative if R has dimension greater than one. For example, let R be a Krull domain. Then each divisorial ideal of R is 2-v-generated [37, Proposition 1.2] and stability coincides with invertibility (Proposition 2.3(1)). Hence each divisorial ideal of R is stable (i.e., invertible) if and only if R is locally factorial [13, Lemma 1.1].

Recall that a nonzero ideal I of a domain R is *v-invertible* if $(I(R : I))_v = R$ and that I is called *v-stable* if I_v is v-invertible as an ideal of $E(I_v)$, that is, $(I_v(E(I_v) : I_v))_v = E(I_v)$. Clearly, v-invertibility implies v-stability. If each nonzero ideal of R is v-stable, we say that R is *v-stable* [21].

Each nonzero ideal of a Krull domain is v-invertible, thus a Krull domain is v-stable. However, a Krull domain is stable if and only if it is a Dedekind domain, that is, it has dimension one (Proposition 2.3(1)). An example of a one-dimensional Mori domain that is v-stable but not stable is given in [22, Example 2.6].

Question. Assume that R is a Mori domain such that each divisorial ideal is 2-v-generated. Is it true that R is v-stable?

Problem 7.5. The *t-closure* of a nonzero ideal I is defined by setting

$$I_t := \bigcup\{J_v;\ J \text{ finitely generated and } J \subseteq I\},$$

for each nonzero ideal I of R. If $I = I_t$, I is called a *t-ideal*. Invertible ideals are divisorial and divisorial ideals are t-ideals.

Olberding proved that R is a stable domain if and only if each nonzero ideal I is divisorial in $E(I)$ [42, Theorem 3.5]. When R is finitely stable, each finitely generated nonzero ideal I is a t-ideal of $E(I)$, being invertible. Does the converse hold?

Question ([46, Question 2.5]). Assume that each (finitely generated) nonzero ideal I is a t-ideal of $E(I)$. Is it true that R is finitely stable?

The answer is positive when, for each finitely generated nonzero ideal I, the ideal $(E(I) : I)$ is finitely generated in $E(I)$ [46, Proposition 2.4].

Problem 7.6. It is well known that if R has finite character, a locally invertible ideal is invertible. Conversely, if each locally invertible ideal is invertible R need not have finite character (e.g., a Noetherian domain need not have finite character). However, a Prüfer domain such that each locally invertible ideal is invertible does have finite character. This fact was conjectured by S. Bazzoni [6, p. 630] and proved by W. Holland, J. Martinez, W. McGovern and M. Tesemma in [29]. (A simplified proof is in [36]). F. Halter-Koch gave independently another proof, in the more general context of ideal systems [26]. Other contributions were given by M. Zafrullah in [54] and by C.A. Finocchiaro, G. Picozza and F. Tartarone in [16].

Following D.D. Anderson and M. Zafrullah, for short we call R an *LPI-domain* if each locally principal nonzero ideal of R is invertible [2].

Since a Prüfer domain is precisely a finitely stable integrally closed domain, one is lead to ask the following more general question.

Question ([10, Question 4.6]). Assume that R is a finitely stable LPI-domain. Is it true that R has finite character?

The answer is positive if and only if the LPI-property extends to fractional overrings [19, Corollary 15], in particular when R is Mori or integrally closed. For an exhaustive discussion of this problem see [19].

Problem 7.7. By Proposition 5.3, a quasi-stable domain is locally quasi-stable. What about the converse? Note that a locally quasi-stable domain, being locally finitely stable, is finitely stable.

Question ([46, Sect. 3]). Is it true that a domain which is locally quasi-stable is quasi-stable?

The answer is positive for integrally closed or Mori domains. Other than that, also when R is h-local [46, Corollary 3.13].

Problem 7.8. A similar question can be addressed for Clifford regular domains. Bazzoni proved that any localization of a Clifford regular domain is Clifford regular [9, Proposition 2.8] (Proposition 6.5) and that a Clifford regular domain has finite character [10, Theorem 4.7]. But it is not known if the converse is true in general.

Question ([9, Question 6.8]). Is it true that a domain which is locally Clifford regular and has finite character is Clifford regular?

We know that the answer is positive in the following cases: (a) When R is integrally closed. (This follows from Theorem 6.2.) (b) When R is Mori. In this case, each localization of R is Mori and Clifford regular, hence (finitely) stable. Thus R is locally stable and the finite character implies that R is stable. (c) When each nonzero prime ideal of R is contained in a unique maximal ideal, for example, if R is one-dimensional [23, Proposition 5.5].

Problem 7.9. A quasi-stable domain need not have finite character; thus a quasi-stable domain need not be Clifford regular.

Question. Is a Clifford regular domain quasi-stable?

The answer is positive when R is integrally closed or Mori.

Problem 7.10. It is easy to see that a vN-regular ideal is L-stable [9, Lemma 2.6], so that Clifford regular domains are L-stable. However, an L-stable domain need not be finitely stable; thus neither quasi-stable nor Clifford regular. For example, a Noetherian integrally closed domain is always L-stable, but it is stable if and only if it is a Dedekind domain (Proposition 2.3(2)).

Question. Is a finitely stable domain, or a quasi-stable domain, L-stable?

Again, the answer is positive when R is integrally closed or Mori.

References

1. D.D. Anderson, J.A. Huckaba, I.J. Papick, A note on stable domains. Houston J. Math. **13**, 13–17 (1987)
2. D.D. Anderson, M. Zafrullah, Integral domains in which nonzero locally principal ideals are invertible. Comm. Algebra **39**, 933–941 (2011)
3. C. Arf, Une interprétation algébrique de la suite des ordres de mulptiplicité d'une branche algébrique. Proc. London Math. Soc. **50**, 256–287 (1949)
4. V. Barucci, Mori domains, Non-Noetherian Commutative Ring Theory; Recent Advances, Chapter 3 (Kluwer Academic Publishers, Springer, 2000)
5. H. Bass, On the ubiquity of Gorenstein rings. Math Z. **82**, 8–28 (1963)
6. S. Bazzoni, Class semigroups of Prüfer domains. J. Algebra **184**, 613–631 (1996)
7. S. Bazzoni, Divisorial Domains. Forum Math. **12** , 397–419 (2000)
8. S. Bazzoni, Groups in the class semigroups of Prüfer domains of finite character. Comm. Algebra **28**, 5157–5167 (2000)
9. S. Bazzoni, Clifford regular domains. J. Algebra **238**, 703–722 (2001)
10. S. Bazzoni, Finite character of finitely stable domains. J. Pure Appl. Algebra **215**, 1127–1132 (2011)
11. S. Bazzoni, L. Salce, Groups in the class semigroups of valuation domains. Israel J. Math. **95**, 135–155 (1996)
12. S. Bazzoni, L. Salce, Warfield domains. J. Algebra **185**, 836–868 (1996)
13. A. Bouvier, The local class group of a Krull domain. Canad. Math. Bull. **26**, 13–19 (1983)
14. M. Fontana, J.A. Huckaba, I.J. Papick, Prüfer domains, Monographs and Textbooks in Pure and Applied Mathematics, vol. 203 (M. Dekker, New York, 1997)
15. P. Eakin, A. Sathaye, Prestable ideals. J. Algebra **41**, 439–454 (1976)
16. C.A. Finocchiaro, G. Picozza, F. Tartarone, Star-Invertibility and t-finite character in Integral Domains. J. Algebra Appl. **10**, 755–769 (2011)
17. S. Gabelli, *A Class of Prüfer Domains with Nice Divisorial Ideals*, Commutative Ring Theory (Fés, 1995), Lecture Notes in Pure and Applied Mathematics, **185** (Dekker, New York, 1997) pp. 313–318
18. S. Gabelli, *Generalized Dedekind Domains*, Multiplicative Ideal Theory in Commutative Algebra. A tribute to Robert Gilmer (Springer, New York, 2006)
19. S. Gabelli, Locally principal ideals and finite character. Bull. Math. Soc. Sci. Math. Roumanie (N.S.), Tome **56**(104), 99–108 (2013)
20. S. Gabelli, E. Houston, G. Picozza, w-Divisoriality in polynomial rings. Comm. Algebra **37**, 1117–1127 (2009)
21. S. Gabelli, G. Picozza, Star-stable domains. J. Pure Appl. Algebra **208**, 853–866 (2007)
22. S. Gabelli, G. Picozza, Star stability and star regularity for Mori domains. Rend. Semin. Mat. Padova **126**, 107–125 (2011)
23. S. Gabelli, G. Picozza, Stability and regularity with respect to star operations. Comm. Algebra **40**, 3558–3582 (2012)
24. S. Gabelli, M. Roitman, *On finitely Stable Domains* (submitted)
25. W. Heinzer, Integral domains in which each non-zero ideal is divisorial. Matematika **15**, 164–170 (1968)
26. F. Halter-Koch, Clifford semigroups of ideals in monoids and domains. Forum Math. **21**, 1001–1020 (2009)
27. W. Heinzer, D. Lantz, K. Shah, The Ratliff-Rush ideals in a Noetherian ring. Comm. Algebra **20**, 591–622 (1992)
28. W. Heinzer, B. Johnston, D. Lantz, K. Shah, *The Ratliff-Rush Ideals in a Noetherian Ring: A Survey*, Methods in module theory (Colorado Springs, CO, 1991), Lecture Notes in Pure and Applied Mathematics, vol. 140 (Dekker, New York, 1993) pp. 149–159
29. W.C. Holland, J. Martinez, W.Wm. McGovern M.Tesemma, Bazzoni's Conjecture. J. Algebra **320**, 1764–1768 (2008)

30. J. Lipman, Stable ideals and Arf rings. Amer. J. Math **93**, 649–685 (1971)
31. E. Matlis, Reflexive domains. J. Algebra **8**, 1–33 (1968)
32. E. Matlis, *Torsion-Free Modules* (University of Chicago press, Chicago, 1972)
33. A. Mimouni, Ratliff-Rush closure of ideals in integral domains. Glasg. Math. J. **51**, 681–689 (2009)
34. A. Mimouni, Ratliff-Rush closure of ideals in pullbacks and polynomial rings. Comm. Algebra **37**, 3044–3053 (2009)
35. A. Mimouni, *Pullbacks and Coherent-Like Properties*, Advances in commutative ring theory (Fez, 1997), Lecture Notes in Pure and Applied Mathematics, vol. 205 (Dekker, New York, 1999) pp. 437–459
36. W.Wm. McGovern, Prüfer domains with Clifford Class semigroup. J. Commut. Algebra **3**, 551–559 (2011)
37. J.L. Mott, M. Zafrullah, On Krull domains. Arch. Math. **56**, 559–568 (1991)
38. B. Olberding, Globalizing local properties of Prüfer domains. J. Algebra **205**, 480–504 (1998)
39. B. Olberding, *Stability of Ideals and its Applications*, Ideal theoretic methods in commutative algebra (Columbia, MO, 1999), Lecture Notes in Pure and Applied Mathematics, vol. 220 (Dekker, New York, 2001) pp. 319–341
40. B. Olberding, Stability, duality and 2-generated ideals, and a canonical decomposition of modules. Rend. Semin. Mat. Univ. Padova **106**, 261–290 (2001)
41. B. Olberding, On the classification of stable domains. J. Algebra **243**, 177–197 (2001)
42. B. Olberding, On the structure of stable domains. Comm. Algebra **30**, 877–895 (2002)
43. B. Olberding, *An Exceptional Class of Stable Domains*, Communication at AMS Meeting ♯991, (Special Session on Commutative Rings and Monoids, Chapel Hill, NC) October 2003.
44. B. Olberding, *Noetherian Rings Without Finite Normalizations*, Progress in commutative algebra 2 (Walter de Gruyter, Berlin, 2012) pp. 171–203
45. B. Olberding, One-dimensional bad Noetherian domains, Trans. AMS **366**, 4067–4095 (2014)
46. G. Picozza, F. Tartarone, Flat ideals and stability in integral domains. J. Algebra **324**, 1790–1802 (2010)
47. N. Popescu, On a class of Prüfer domains. Rev. Roumanie Math. Pures Appl. **29**, 777–786 (1984)
48. L.J. Ratliff, Jr., D.E. Rush, Two notes on reductions of ideals. Indiana Univ. Math. J. **27**, 929–934 (1978)
49. D.E. Rush, Rings with two-generated ideals. J. Pure Appl. Algebra **73**, 257–275 (1991)
50. D.E. Rush, Two-generated ideals and representations of abelian groups over valuation rings. J. Algebra **177**, 77–101 (1995)
51. J.D. Sally, W.V. Vasconcelos, Stable rings and a problem of Bass. Bull. AMS **79**, 574–576 (1973)
52. J.D. Sally, W.V. Vasconcelos, Stable rings. J. Pure Appl. Algebra **4**, 319–336 (1974)
53. L. Sega, Ideal class semigroups of overrings. J. Algebra **311**, 702–713 (2007)
54. M. Zafrullah, t-Invertibility and Bazzoni-like statements. J. Pure Appl. Algebra **214**, 654–657 (2010)
55. P. Zanardo, U. Zannier, The class semigroup of orders in number fields. Math. Proc. Cambridge Philos. Soc. **115**, 379–391 (1994)

The Development of Non-Noetherian Grade and Its Applications

Livia Hummel

Abstract Hochster and Barger were the first to introduce notions of grade for non-Noetherian rings. Their work laid the foundation for Alfonsi's work which unified and generalized these earlier definitions. The various notions of grade have played an important role in the development of the theory of coherent rings. This paper looks at the historical development of non-Noetherian grade, as well as its applications.

Keywords Non-Noetherian ring • Coherent ring • Grade • Depth • Polynomial grade

Mathematics Subject Classification (2010): 13-02, 13C15, 13D05

1 Introduction

The question of how to define grade in the context of non-Noetherian rings centers around the poor behavior of the classical notion of grade in the non-Noetherian context. Using Hochster's terminology [8], given a ring R and an R-module M, a sequence $x_1, \ldots, x_n$ forms a possibly improper M-sequence if for each i, $0 \leq i \leq n - 1$, x_{i+1} is a non-zero-divisor on $M/(x_1, \ldots x_i)M$. If $(x_1, \ldots x_n)M \neq M$, then $x_1, \ldots, x_n$ forms an M-sequence. Given an ideal I of R such that $IM \neq M$, the classical grade of I on M is defined

$$\text{grade}_R(I, M) = \sup\{n | \text{there exists a regular } M\text{-sequence } x_1, \ldots, x_n \in I\}.$$

If (R, m) is a local ring, we denote $\text{grade}_R(m, M)$ by $\text{depth}_R M$.

L. Hummel (✉)
University of Indianapolis, 1400 East Hanna Avenue, Indianapolis, IN 46227, USA
e-mail: hummell@uindy.edu

M. Fontana et al. (eds.), *Commutative Algebra: Recent Advances in Commutative Rings, Integer-Valued Polynomials, and Polynomial Functions*, DOI 10.1007/978-1-4939-0925-4_11,
© Springer Science+Business Media New York 2014

For a Noetherian ring R, an ideal I, and a finitely generated R-module M with $IM \neq M$, $\mathrm{grade}_R(I, M) > 0$ if and only if $(0 :_M I) = 0$. However, the following example presents a non-Noetherian ring with an ideal P that contain only zero-divisors such that $(0 :_R P) = 0$. As indicated in [16], Hochster appears to be the first to notice this pathological behavior of classical grade in the coherent case.

Example 1 ([24]). Let $R = k[[x, y]]$ be the power series ring in x and y over a field k. Let M be the R-module $M = \bigoplus\sum\limits_{\substack{\text{height } P=1, \\ P \text{ prime}}} k(P)$ where $k(P)$ is the quotient field of R/P. Let $A = R \oplus M$, where addition is component-wise and multiplication is given by

$$(r, m) \cdot (s, n) = (rs, rn + sm).$$

M is an ideal of A, $M^2 = (0)$, and A is a local ring with maximal ideal $P = A(x, 0) + A(0, y)$. As shown in [24] and [16] (page 117), $(0 :_R P) = 0$ and P contains only zero-divisors.

Hamilton and Marley [7] provide another example of a ring demonstrating this pathological behavior (see Example 3).

This observation led Hochster [8] to define the notion of polynomial grade, sometimes known as true grade, in the early 1970s. In his work, Barger [4] also investigated several different definitions of grade, seeking to understand the conditions and limitations under which these definitions were equivalent. McDowell [13] used homological definitions of grade, referred to in this paper as $M.\,\mathrm{depth}$ and $PN.\,\mathrm{depth}$, in his exploration of a class of coherent rings called pseudo-Noetherian rings. (In his paper, McDowell attributes this definition of depth to Auslander and Bridger [2].) In 1981, Alfonsi [1] generalized and unified these notions of non-Noetherian grade.

Since the development and refinements of the notion of grade in the non-Noetherian context, the work of Alfonsi, Barger, and Hochster has played an important role in the development of the theory of coherent rings. Recall that an R-module M admits a finite n-presentation if there is an exact sequence

$$F_n \to F_{n-1} \to \cdots \to F_0 \to M \to 0$$

of finitely generated free R-modules. Finitely presented modules are those modules that admit a finite 1-presentation. A ring R is called coherent if every finitely generated ideal is finitely presented. A ring is called stably coherent if every polynomial ring in a finite number of variables over R is a coherent ring. Alfonsi was able to prove the most general version of the Buchsbaum-Eisenbud Exactness Criteria for complexes over rings that aren't necessarily Noetherian. He also was able to simplify results in stable coherence [6]. Iroz and Rush [10] used grade in the context of associated prime ideals. More recently, Hamilton and Marley [7] have used polynomial grade in the theory of non-Noetherian Cohen-Macaulay rings. Hummel and Marley [9] generalized the Auslander-Bridger formula to the not necessarily Noetherian context, which provided the foundation for non-Noetherian Gorenstein rings.

In this paper we will look at the variety of definitions of grade within the non-Noetherian context, summarize their connections in the context of Alfonsi's unification, and present the applications of grade in non-Noetherian rings. In what follows all rings are commutative and not assumed to be Noetherian unless otherwise stated. In the naming of the various definitions of grade presented here, usually either the original notations or those used widely in the literature are chosen. In some cases (as in the notation for $A.$ depth) we use the naming convention of [15] and mimic this convention in notating other definitions (such as $PN.$ depth) that previously were notated no differently than from classical grade or depth.

2   Polynomial, Rees, and Koszul Grades

Hochster's original definition of polynomial grade relies upon the notion of admissible ideals.

Definition 1 ([8]). Let R be a ring, let be I an ideal, and let M be an R-module. The pair (I, M) is admissible if for each faithfully flat R-algebra S, and each possibly improper $(M \otimes_R S)$-sequence $x_1, \ldots, x_n \in I \otimes_R S$, it follows that $x_1, \ldots x_n$ is an $(M \otimes_R S)$-sequence.

Hochster shows that for M finitely generated, (I, M) is admissible if and only if $IM \neq M$ [8].

Definition 2 ([8]). Let R be a ring, let I be an ideal, and let M be an R-module such that (I, M) is admissible. Then

$$\text{p. grade}_R(I, M) = \sup\{\text{grade}_S(I \otimes_R S, M \otimes_R S) | S \text{ is a faithfully flat } R\text{-algebra}\}.$$

However, in recent literature ([7, 9, 15, 16]) the notion of polynomial grade is commonly defined using the following equivalent characterization.

Proposition 1 ([8]). *Let R be a ring, let I be an ideal, and let M be an R-module. Then*

$$\text{p. grade}_R(I, M) = \lim_{t \to \infty} \text{grade}(I R[x_1, \ldots, x_t], R[x_1, \ldots, x_t] \otimes_R M).$$

As a consequence of Definition 2 and Proposition 1, polynomial grade has the following properties.

Proposition 2 ([8, 16]). *Let R be a ring, let I be an ideal, and let M be an R-module. Polynomial grade has the following properties:*

1. $\text{grade}_R(I, M) \leq \text{p. grade}_R(I, M)$.
2. $\text{p. grade}_R(IR[x_1, \ldots, x_n], MR[x_1, \ldots, x_n]) = \text{p. grade}_R(I, M)$ *for all* $n > 0$.
3. *If* $J \subset I$, *then* $\text{p. grade}_R(J, M) \leq \text{p. grade}_R(I, M)$.
4. *If* R *is Noetherian,* M *is finitely generated, and* $IM \neq M$, *then* $\text{grade}_R(I, M) = \text{p. grade}_R(I, M)$.
5. *Given a faithfully flat* R-*algebra* S, $\text{p. grade}_S(I \otimes_R S, M \otimes_R S) = \text{p. grade}_R(I, M)$.
6. $\text{p. grade}_R(I, M) = 0$ *if and only if every finitely generated subideal* I_0 *of* I *has an element* $x \in M$ *such that* $Ix = 0$.
7. $\text{p. grade}_R(I, M) = \sup\{\text{p. grade}_R(J, M) \mid J \subset I \text{ is finitely generated}\}$.
8. *If* $\mathbf{x}$ *is a regular sequence on* M *contained in* I *of length* $\ell(\mathbf{x})$, *then*

$$\text{p. grade}_R(I, M) = \text{p. grade}_R(I, M/\mathbf{x}M) + \ell(\mathbf{x}).$$

9. $\text{p. grade}_R(I, M) = \text{p. grade}_R(P, M)$ *for some prime ideal* P *containing* I. *In particular,* $\text{p. grade}_R(I, M) = \text{p. grade}_R(\sqrt{I}, M)$.
10. *If* I *is generated by* n *elements, and* $IM \neq M$, *then* $\text{p. grade}_R(I, R) \leq n$. *Furthermore,*

$$\text{p. grade}_R(I, M) = \text{grade}(IR[t_1, \ldots, t_n], R[t_1, \ldots, t_n] \otimes_R M).$$

Note that in light of Proposition 2(6), in Example 1, we now see that $\text{p. grade}_R(P, M) > 0$ even though $\text{grade}_R(P, M) = 0$.

When (R, m) is a local ring with R-module M, $\text{p. grade}_R(m, M)$ is denoted by $\text{p. depth } M$. In Northcott [16] we see a generalization of the Auslander-Buchsbaum formula to the non-Noetherian setting. First we recall the original formula.

Theorem 1 ([3](Auslander-Buchsbaum Formula)). *Let* (R, m) *be a local Noetherian ring, and let* M *be an* R-*module with* $\text{pd}_R M < \infty$. *Then* $\text{pd}_R M + \text{depth}_R M = \text{depth}_R R$.

Northcott [16] retains projective dimension, but replaces classical grade with polynomial grade.

Theorem 2 ([16]). *Let* (R, m) *be a local ring and let* M *be an* R-*module with a finite length resolution of finitely generated free modules. Then* $\text{pd}_R M + \text{p. depth}_R M = \text{p. depth}_R R$.

Hochster's primary concern was grade sensitivity in the non-Noetherian case. Let R be a ring, let $I = (r_1, \ldots, r_n)$, and let M be a finitely generated R-module with $IM \neq M$. When R is Noetherian and $\mathscr{K}$ is the Koszul complex of M with respect to the generators of I, "grade sensitivity" corresponds to being able to recover the classical grade by counting the number of vanishing homology groups of the Koszul complex from the left [8]. Let $A = \mathbb{Z}[x_1, \ldots x_n]$, where $x_1, \ldots, x_n$ are indeterminants over $\mathbb{Z}$, and let $\mathscr{K}_0$ be the Koszul complex of A with respect to $x_1, \ldots, x_n$. Define $\mathscr{K} = \mathscr{K}_0 \otimes_A M$, treating M as an A-module via the homomorphism $A \to R$ that takes x_i to r_i. Since $\mathscr{K}_0$ is a free resolution of $N = A/(x_1, \ldots, x_n)$,

$$\mathrm{grade}_R(I, M) = n - \dim^A(N, M) = \mathrm{Tor\,dim}^A N - \dim^A(N, M),$$

where $\dim^A(N, M)$ is the greatest integer i such that $\mathrm{Tor}_i^A(N, M) \neq 0$ (or -1 if there is no such integer). Define $\mathrm{Tor\,dim}^A N = \sup_E \dim^A(N, E)$ for E finitely generated.

To investigate grade sensitivity in the non-Noetherian case, termed "G-sensitivity" in [8], Hochster proves the following results. In the following, given a ring R and an R-module M, define $a(M) = \{r \in R | M_r = 0\}$ where $M_r = R_r \otimes M$.

Theorem 3 ([8](Tor Inequality)). *Let R be a ring, let M be an R-module, and let $I = a(M)$. Let A be any R-algebra, and let N be any A-module. If $\dim^R(M, N) \geq 0$, then (IA, N) is admissible and*

$$\mathrm{p.\,grade}_A(IA, N) + \dim^R(M, N) \leq \mathrm{Tor\,dim}^R M.$$

Corollary 1 ([8](Extended Rees Inequality)). *Let R be a ring, let $M \neq 0$ be an R-module, and let $I = a(M)$. Then $\mathrm{p.\,grade}_R(I) \leq \mathrm{Tor\,dim}^R M$.*

Theorem 4 ([8](Characterization of G-sensitivity)). *Let R be a ring and let $M \neq 0$ be an R-module such that $1 \leq n = \mathrm{Tor\,dim}^R M < \infty$. Then M is G-sensitive if and only if the following conditions hold:*

1. *M is perfect, that is, $\mathrm{p.\,grade}(a(M)) = n$.*
2. *$a(M)$ is the radical of a finitely generated ideal.*
3. *For every prime ideal P of R containing $a(M)$, and each ideal J of A_P with $\sqrt{J} = PR_P$, $\mathrm{Tor}_n^R(M, R_P/J) \neq 0$.*

Proposition 2 (7) is a consequence of Hochster's G-sensitivity results.

According to [8], Hochster advised Barger to implicitly use the notion of polynomial grade to relate what are known as the classical, Koszul, and Rees grades. Barger's definitions follow below:

Definition 3 ([4]). Let R be a ring, let I be an ideal, and let M be an R-module with $IM \neq M$. For $x_1, \ldots, x_n \in I$, define $g(x_1, \ldots, x_n | M) = n - t$, where t is the largest integer such that the t-th homology module of the Koszul complex over M determined by $x_1, \ldots, x_n$ is not zero. The Koszul grade of I over M is defined

$$\mathrm{K.\,grade}_R(I, M) = \sup\{g(x_1, \ldots, x_n | M) | x_1, \ldots, x_n \in I\}.$$

Barger proves the following properties of Koszul grade:

Proposition 3 ([4]). *Let R be a ring, let I and J be ideals, and let M be an R-module with $JM \neq M$.*

1. *If $I \subset J$, then $\mathrm{K.\,grade}_R(I, M) \leq \mathrm{K.\,grade}_R(J, M)$.*
2. *If x is a non-zero-divisor, $\mathrm{K.\,grade}_R(J, M) = \mathrm{K.\,grade}_R(J, M/xM) + 1$.*
3. *If $(x_1, \ldots, x_n) = J$, then $\mathrm{K.\,grade}(J, M) \leq n$.*

4. If S is a faithfully flat extension of R, then

   a. $(J \otimes_R S)(M \otimes_R S) \neq (M \otimes_R S)$.
   b. $\operatorname{grade}_R(J, M) = \operatorname{K.grade}_R(J \otimes_R S, M \otimes_R S)$.
   c. If $I \subset J \subset \sqrt{I}$, *then* $\operatorname{K.grade}_R(I, M) = \operatorname{K.grade}_R(J, M)$.

The following properties of Koszul grade appear in [15]:

Proposition 4 ([15]). *Let R be a ring, let I be an ideal, and let M be an R-module.*

1. Let $f : R \to S$ *be a flat ring homomorphism; then*

$$\operatorname{K.grade}(I, M) \leq \operatorname{K.grade}(IS, M \otimes_R S).$$

2. Let $f : R \to S$ *be a ring homomorphism and let N be an S-module. Then*

$$\operatorname{K.grade}_R(I, N) = \operatorname{K.grade}_S(IS, N).$$

The classical and Koszul grades are related in the following ways:

Proposition 5 ([4]). *Let R be a ring, let I be an ideal, and let M be an R-module.*

1. $\operatorname{grade}_R(I, M) \leq \operatorname{K.grade}_R(I, M)$.
2. If $\operatorname{grade}_R(I, M) < \operatorname{K.grade}_R(I, M) = n$, *then there is a positive integer t such that for* $S = R[x_1, \ldots, x_t]$,

$$\operatorname{grade}_S(I \otimes_R S, M \otimes_R S) = \operatorname{K.grade}_S(I \otimes_R S, M \otimes_R S) = n.$$

Barger's definition of Rees grade is based upon Rees' ([18, 19]) characterization of Noetherian classical grade by the vanishing of Ext.

Definition 4 ([4]). Let R be a ring, let I be an ideal, and let M be an R-module. Then the Rees grade is given by

$$\operatorname{r.grade}_R(I, M) = \inf\{n \mid \operatorname{Ext}_R^n(R/I, M) \neq 0\}.$$

The following properties of Rees grade were first proved by Rees in [18]:

Proposition 6 ([4]). *Let R be a ring, let I be an ideal, and let M be an R-module.*

1. $\operatorname{grade}_R(I, M) \leq \operatorname{r.grade}_R(I, M)$.
2. If $x \in I$ *is a non-zero-divisor, then* $\operatorname{r.grade}_R(I, M) = 1 + \operatorname{r.grade}_R(I, M/xM)$.

A proof of the following characterization of r. grade can be found in [23]. (Azgharzadeh and Tousi [15] refer to this cohomological characterization as H. grade.) Let $\mathbf{x} = x_1, \ldots x_n$ be a finite sequence of elements in R. Given an R-module M and an ideal I, let $H_I^i(M)$ be the i-th local cohomology of M.

Proposition 7 ([23]). *Let R be a ring, let I be an ideal of R, and let M be an R-module. Then* $\operatorname{r.grade}_R(I, M) = \inf\{i \geq 0 \mid H_I^i(M) = \lim_{\substack{\longrightarrow \\ n}} \operatorname{Ext}_R^i(R/I^n, M) \neq 0\}$.

The following properties characterize the relationship between Rees and Koszul grades:

Proposition 8 ([4]). *Let R be a ring, let I be a finitely generated ideal, and let M be an R-module.*

1. The following statements are equivalent:

    a. r. $\mathrm{grade}_R(I, M) = 0$.
    b. K. $\mathrm{grade}_R(I, M) = 0$.
    c. there exists a nonzero element $m \in M$ such that $Im = 0$.

2. If M^∞ is the direct sum of countably many copies of M, then

$$\text{K. } \mathrm{grade}_R(I, M) - \inf\{n \mid \mathrm{Ext}_R^n(R/I, M^\infty)\}$$

and K. $\mathrm{grade}_R(I, M) \leq$ r. $\mathrm{grade}_R(I, M)$.
3. If K. $\mathrm{grade}_R(I, M) = 1$, then r. $\mathrm{grade}_R(I, M) = 1$ as well.
4. If R is coherent and K. $\mathrm{grade}_R(I, M) = n$, then r. $\mathrm{grade}_R(I, M) = n$.

Barger provides an example for when the first property fails when I is not finitely generated.

Example 2 ([4]). Let k be a field, let $R = k[x_1, \ldots, x_n, \ldots]/(x_1^1, x_2^2, \ldots, x_n^n, \ldots)$, and let $I = (x_1, \ldots, x_n, \ldots)$. Then $\mathrm{grade}_R(I, R) = $ K. $\mathrm{grade}_R(I, R) = 0$, but r. $\mathrm{grade}_R(I, R) > 0$.

3 Grade and Pseudo-Noetherian Rings

While attempting to extend Noetherian algebraic and homological results to the non-Noetherian context, McDowell [13] defined pseudo-Noetherian rings in 1976.

Definition 5 ([13]). A ring R is called pseudo-Noetherian if

1. R is coherent.
2. If $M \neq 0$ is a finitely presented R-module and I is a finitely generated ideal of R contained in the zero-divisors of M, then there exists $m \in M$, $m \neq 0$ such that $Im = 0$.

Note that by Proposition 8 (1), if R is pseudo-Noetherian, I is a finitely generated ideal, and M is a finitely presented R-module, r. $\mathrm{grade}_R(I, M) = $ K. $\mathrm{grade}_R(I, M) = 0$. Hence, McDowell uses the following definition of M. depth, as defined by [2].

Definition 6 ([2]). Let R be a ring and let M and N be R-modules. Define $M. \mathrm{depth}\, N = \inf\{n \mid \mathrm{Ext}_R^n(M, N) \neq 0\}$. If there is no such integer such that $\mathrm{Ext}_R^n(M, N) \neq 0$, then $M. \mathrm{depth}_R N = \infty$.

McDowell connects M. depth to the length of M-sequences.

Lemma 1 ([13]). *Let R be a pseudo-Noetherian ring, let I be a finitely generated ideal, and let M be a finitely presented R-module with $IM \neq M$. Then the length of any maximal M-sequence in I is equal to $(R/I).\operatorname{depth}_R M$.*

In fact, this result provides a characterization for local coherent rings that are also pseudo Noetherian: a local coherent ring is pseudo-Noetherian if and only if the above lemma holds [13].

In addition, if $I \subset J$ are finitely generated proper ideals of a local pseudo-Noetherian ring, $(R/I).\operatorname{depth}_R M \leq (R/J).\operatorname{depth}_R M$ for any finitely presented R-module M.

This leads to the definition of depth for pseudo-Noetherian rings.

Definition 7 ([13]). Let (R, m) be a local pseudo-Noetherian ring and let $M \neq 0$ be a finitely presented R-module. Define

$$PN.\operatorname{depth} M = \sup\{(R/I).\operatorname{depth}_R M \mid I \text{ a proper finitely generated ideal}\}.$$

If R is Noetherian, $PN.\operatorname{depth}_R M = (R/m).\operatorname{depth} M$.
Consequently, McDowell generalizes Lemma 1.

Theorem 5 ([13]). *Let (R, m) be a local pseudo-Noetherian ring and let $M \neq 0$ be a finitely presented R-module. Then the lengths of maximal M sequences are the same and equal to $PN.\operatorname{depth}$.*

If R is pseudo-Noetherian and I is a finitely generated ideal, then it follows that R/I is pseudo-Noetherian. Hence $PN.\operatorname{depth}$ also has the following properties:

Proposition 9 ([13]). *Let (R, m) be a local pseudo-Noetherian ring, let $M \neq 0$ be a finitely presented module, and let $x \in m$ be a non-zero-divisor of M. Then $PN.\operatorname{depth}_{R/(x)}(M/xM) = PN.\operatorname{depth}_R M - 1$.*

Theorem 6 ([13]). *Let (R, m) be a local pseudo-Noetherian ring. The following integers are equivalent:*

1. *$PN.\operatorname{depth}_R R$*
2. *The length of any maximal R-sequence*
3. *$\sup\{\operatorname{pd}_R M \mid M$ a finitely presented R-module with $\operatorname{pd}_R M < \infty\}$*

4 Early Applications of Polynomial Grade

Sakaguchi [20] connected polynomial grade to the valuative dimension introduced by Jaffard [11]. We begin with the definition of valuative dimension.

Definition 8 ([11]). If R is an integral domain, then the valuative dimension of R is defined

$$\dim_v R = \sup\{\dim V \mid V \text{ is a valuation overring of } R\}.$$

In general, if R is a commutative ring,

$$\dim_v R = \sup\{\dim_v(R/p) | p \in \operatorname{Spec}(R)\}.$$

If R is a ring and $M \neq 0$ is an R-module, the valuative dimension of M is defined by

$$\dim_v M = \dim_v R/(0 :_R M).$$

If R is Noetherian, then it can be shown that $\dim_v M = \dim M$ [20].

With these definitions, along with the notion of polynomial height (see [20] for details), Sakaguchi proves the following result:

Theorem 7 ([20]). *Let (R, m) be a local ring and let $M \neq 0$ be a finitely generated R-module. Then* p. $\operatorname{grade}_R(m, M) =$ p. $\operatorname{depth}_R R \leq \dim_v M$.

Iroz and Rush [10] made connections between associated primes and polynomial grade. First, we start with some definitions.

Definition 9 ([10]). Let R be a ring, let P be a prime ideal of R, and let M be an R-module.

1. If P is minimal over $(0 :_R m)$ for some $m \in M$, then P is called a weak Bourbaki prime of M. The set of weak Bourbaki primes of M is denoted $\operatorname{Ass}_f(M)$.
2. P is called a strong Krull prime of M if for every finitely generated ideal I contained in P, there exists an $m \in M$ such that $I \subseteq (0 :_R m) \subseteq P$. Denote the set of strong Krull primes of M by $\operatorname{sK}(M)$.

Iroz and Rush use the following result as a basis to connect polynomial grade to weak Bourbaki and strong Krull primes.

Proposition 10 ([8, 12]). *Let R be a ring, let I be an ideal, and let M be an R-module. The following statements are equivalent:*

1. p. $\operatorname{grade}_R(I, M) = 0$.
2. *Each finitely generated ideal $J \subseteq I$ is contained in a member of* $\operatorname{Ass}_f(M)$.
3. *Each finitely generated ideal $J \subseteq I$ is contained in a member of* $\operatorname{sK}(M)$.

The following result of Northcott [17] also connects polynomial grade and strong Krull primes.

Theorem 8 ([17]). *Let R be a ring, and let I be an ideal with* p. $\operatorname{grade}_R(I, R) > 0$. *Then I is projective if and only if*

1. *I has a finite length resolution by finitely generated projective modules.*
2. *Every $P \in \operatorname{sK}(R/A)$ has* p. $\operatorname{grade}_R(P, R) = 1$.

By means of the following lemma, Iroz and Rush prove that Northcott's theorem remains true if $\operatorname{Ass}_f(R/I)$ is replaced by $\operatorname{sK}(R/I)$.

Lemma 2 ([10]). *Let R be a ring, let $I \neq R$ be an ideal, and let M be an R-module. Then*

$$\text{p. grade}_R(I, M) = \inf\{\text{p. grade}_R(P, M) | P \in \text{Ass}_f(R/I)\}$$
$$= \inf\{\text{p. grade}_R(P, M) | P \text{ is minimal over } A\}.$$

Finally, Iroz and Rush connect polynomial grade to projective ideals.

Theorem 9 ([10]). *Let R be a ring and let I be an ideal of R with p. grade$_R(I, R) > 0$. Then I is projective if and only if the following conditions hold:*

1. p. grade$_R(P, R) = 1$ for all $P \in \text{Ass}_f(R/I)$.
2. I has a finite length resolution by finitely generated projective R-modules.

5   Alfonsi's Generalization of Grade

This section will look at Alfonsi's [1] definition of grade and its connection to the previously presented definitions of grade and depth.

Definition 10 ([1]). Let R be a ring, let M be a finitely presented R-module, and let N be an R-module. Define A. grade$_R(M, N) \geq n$ if for every finite complex

$$P_\bullet = P_n \to P_{n-1} \to \cdots \to P_0 \to M \to 0$$

of finitely generated projective modules P_i $(0 \leq i \leq n)$, there exists a finite complex

$$Q_\bullet = Q_n \to Q_{n-1} \to \cdots \to Q_0 \to M \to 0$$

of finitely generated projective modules Q_j $(0 \leq j \leq n)$ and a chain map $P_\bullet \to Q_\bullet$ of complexes over M such that the induced maps $H_i(\text{Hom}_R(Q_\bullet, N)) \to H_i(\text{Hom}_R(P_\bullet, N))$ are zero for $0 \leq i < n$.

Define A. grade$_R(M, N)$ as the largest integer satisfying the above conditions; let A. grade$_R(M, N) = \infty$ if no such integer exists.

If M admits a finite n-presentation (for instance, if R is coherent), then [1]

$$\text{A. grade}_R(M, N) = \sup\{n | \text{Ext}_R^i(M, N) = 0, 0 \leq i < n\}.$$

Hence we immediately have the following connection between McDowell's notion of grade, Rees grade, and the grade of Alfonsi.

Proposition 11. *Let R be a ring, let I be a finitely generated ideal, let M be an R-module admitting a finite n-presentation, and let N be an R-module. Then*

$M. \operatorname{depth}_R N = \operatorname{A.grade}_R(M, N)$. If $M = R/I$, then $\operatorname{r.grade}_R(M, N) = \operatorname{A.grade}_R(M, N)$.

Alfonsi's definition of grade has the following properties:

Proposition 12 ([1]). *Let R be a ring, let M be a finitely presented R-module, and let N be an R-module.*

1. If $x_1, \ldots, x_n$ are finitely many variables over R, then

$$\operatorname{A.grade}_{R[x_1, \ldots, x_n]}(M \otimes_R R[x_1, \ldots, x_n], N \otimes_R R[x_1, \ldots, x_n]) = \operatorname{A.grade}_R(M, N).$$

2. If $\mathbf{x} = x_1, \ldots, x_n$ is a regular N-sequence contained in $(0 :_R M)$, then $\operatorname{A.grade}_R(M, N) = n + \operatorname{A.grade}_R(M, N/(\mathbf{x})N)$.

Proposition 13 ([1]). *Let R be a ring, let M be a finitely presented R-module, let N be an R-module, and let $n \geq 0$ be an integer. The following are equivalent:*

1. $\operatorname{A.grade}_R(M, N) \geq n$.
2. There exists a faithfully flat R-algebra S, and a sequence $v_1, \ldots, v_n \in (0 :_S M \otimes_R S)$ which forms an $N \otimes_R S$-regular sequence.
3. There exists a faithfully flat R-algebra S, and a sequence of elements $v_1, \ldots, v_t \in (0 :_S M \otimes_R S)$ satisfying $H_i(K_\bullet(v_1, \ldots, v_t, N \otimes_R S)) = 0$ for $i > t - n$.
4. For every sequence of elements $v_1, \ldots, v_t$ of R satisfying $(0 :_R (0 :_R M)) = (0 :_R (v_1, \ldots, v_t))$, we have $H_i(K_\bullet(v_1, \ldots, v_t, N)) = 0$ for $i > t - n$.
5. $\operatorname{Ext}^i_{R[x_1, \ldots, x_r]}(M \otimes_R R[x_1, \ldots, x_r], N \otimes_R R[x_1, \ldots, x_r]) = 0$ for $i < n$ and an integer $r \geq n$.

Corollary 2 ([1]). *Let R be a ring, let M be a finitely presented R-module, let N be an R-module, and let S be a flat extension of R. Then $\operatorname{A.grade}_S(M \otimes_R S, N \otimes_R S) \geq \operatorname{A.grade}_R(M, N)$. If S is faithfully flat over R, then $\operatorname{A.grade}_S(M \otimes_R S, N \otimes_R S) = \operatorname{A.grade}_R(M, N)$.*

Proposition 14 ([1]). *Let R be a ring, let N be an R-module, and let I and J be two finitely generated ideals of R.*

1. If $I \subset J$ and $\operatorname{A.grade}_R(R/I, N) \geq n$, then $\operatorname{A.grade}_R(R/J, N) \geq n$.
2. If $\operatorname{A.grade}_R(R/I, N) \geq n$ and $\operatorname{A.grade}_R(R/J, N) \geq n$, then $\operatorname{A.grade}_R(R/IJ, N) \geq n$.

Proposition 15 ([1]). *Let R be a ring, let I be an ideal, and let M and N be R-modules. If M is finitely presented, then $\operatorname{A.grade}_R(M, N) = \inf\{\operatorname{A.grade}_{R_P}(M_P, N_P) | P \in SuppM\}$.*

Thus Proposition 14 leads to an extension of Alfonsi's definition of grade that does not require M to be a finitely presented module.

Definition 11 ([1]). *Let R be a ring, and let M and N be R-modules. Then $\operatorname{A.grade}_R(M, N) \geq n$ if for every $x \in M$, $(0 :_R x)$ contains a finitely generated ideal I_x satisfying $\operatorname{A.grade}_R(R/I_x, N) \geq n$.*

Alfonsi also shows that the two different definitions of grade he has presented are equivalent over finitely presented modules.

Theorem 10 ([1]). *Let R be a ring, let M be a finitely presented R-module, and let N be an R-module. Then Definitions 10 and 11 agree.*

Finally, we have the following connections between most of the notions of grade previously presented.

Proposition 16 ([15]). *Let R be a ring, let I be an ideal, and let M be an R-module. The following relations hold:*

$$\mathrm{grade}(I, M) \leq \mathrm{p.\,grade}(I, M) = \mathrm{K.\,grade}(I, M) = \mathrm{A.\,grade}(I, M).$$

In addition to the Buchsbaum-Eisenbud Exactness Criteria (stated in Theorem 13 below), Alfonsi [1] also made connections between small finitistic projective dimension and A. grade. We begin with the definition of small finitistic projective dimension and its relevant properties.

Definition 12. Let R be a ring, let M be an R-module, and let n be an integer. The small finitistic projective dimension of M is at most n, denoted $\mathrm{fPdim}_R M \leq n$, if for every complex

$$P_\bullet = 0 \to P_{n+1} \xrightarrow{\rho} P_n \to \cdots \to P_0$$

of finitely generated projective R-modules P_i such that the sequence

$$0 \to P_{n+1} \otimes_R M \xrightarrow{\rho \otimes 1_M} P_n \otimes_R M \to \cdots \to P_0 \otimes_R M$$

is exact, then the map $P_{n+1} \to P_n$ is left invertible (i.e., there exists a homomorphism $\phi : P_n \to P_{n+1}$ such that $\phi\rho = 1_{P_{n+1}}$).

$\mathrm{fPdim}_R M$ is the smallest integer satisfying the above conditions; if no such integer exists, $\mathrm{fPdim}_R M = \infty$.

Theorem 11 ([1]). *Let R be a ring, let M and N be R-modules, and let n be an integer. The following are equivalent:*

1. *$\mathrm{A.\,grade}_R(M, N) > n$.*
2. *For every flat R-algebra S with $\mathrm{fPdim}_S(N \otimes_R S) \leq n$, $M \otimes_R S = 0$.*

It follows that

$$\mathrm{A.\,grade}_R(M, N) = \inf\{\mathrm{fPdim}_S(N \otimes_R S) \,|\, S \text{ is a ring that is a flat } R\text{-module and } M \otimes_R S \neq 0\}.$$

This last characterization of A. grade can be further generalized.

Theorem 12 ([1]). *Let R be a ring, let M be a finitely presented R-module, and let N be an R-module. Then* $\mathrm{A.\,grade}_R(M, N) = \inf\limits_{p \in Supp(M)} \{\mathrm{fPdim}_{R_p} N_p\}.$

Alfonsi defines the depth of a module in the following manner:

Definition 13 ([1]). Let (R, m) be a local ring, and let M be an R-module. We define the A.depth of M as A. $\text{depth}_R M = \text{A.}\,\text{grade}_R(R/m, M)$.

Alfonsi [1] proves several results analogous to their Noetherian depth counterpart. The first such result is similar to the result of Auslander.

Proposition 17 ([1]). *Let* (R, m) *be a local ring, let* N *be an* R-*module, let* $L_\bullet = 0 \to L_n \xrightarrow{d_n} \cdots \xrightarrow{d_1} L_0$ *be a complex of finite free* R-*modules, and let* $Q = \text{Coker}(d_1)$. *Suppose that the coefficient ideal of* d_n *is contained in* m *and that the complex* $L_\bullet \otimes_R N$ *is exact. Then we have the equality* A. $\text{depth}_R(N) = \text{A.}\,\text{depth}_R(N \otimes_R Q) + n$.

The following results, several which are well known in the Noetherian case, follow:

Corollary 3 ([1]). *Let* (R, m) *be a local ring, and let* M *be an* R-*module. Then the following statements hold:*

1. A. $\text{depth}_R M = \text{fPdim}_R M$.
2. *Let* A *be a finitely presented* R-*algebra, let* N *be an* A-*module that is also a finite* R-*flat module, let* Q *be a prime ideal of* A, *and let* P *be its image in* Spec R. *Then we have the equality*

$$\text{A.}\,\text{depth}_{A_Q}((N \otimes_R M)_Q) = \text{A.}\,\text{depth}_{R_P}(M_P) + \text{A.}\,\text{depth}_{A_Q \otimes_R k(P)}(N_Q \otimes_R k(P)).$$

Alfonsi [1] uses his notion of grade to further generalize the Buchsbaum-Eisenbud Exactness Criterion first proved in the Noetherian case in [5] and generalized to non-Noetherian rings by Northcott [16].

Theorem 13 ([1](Buchsbaum-Eisenbud Exactness Criterion)). *Let* R *be a ring and let* $F_\bullet = 0 \to F_n \xrightarrow{u_n} F_{n-1} \to \cdots \xrightarrow{u_1} F_0$ *be a complex of finitely generated free* R-*modules. Let* $r_i = \text{rank}\, F_i - \text{rank}(F_{i+1} + \cdots + (-1)^{n-1}\, \text{rank}\, F_n)$, *and denote the ideal generated by all the* $r_i \times r_i$ *minors of the matrix* u_i *by* $c(\wedge^{r_i} u_i)$. *Then the complex* $F_\bullet$ *is exact if and only if for every* $1 \leq i \leq n$, A. $\text{grade}_R(R/c(\wedge^{r_i} u_i), R) \geq i$.

The Buchsbaum-Eisenbud Exactness Criterion allowed Alfonsi to prove the following result, which simplified the proofs of many results in stable coherence (see [6]). Recall that an R-module has weak dimension n if n is the smallest nonnegative integer such that there is an exact sequence $0 \to F_n \to \cdots \to F_1 \to F_0 \to M \to 0$ of flat R-modules. The weak dimension of a ring is the supremum of the weak dimensions of the modules of R.

Theorem 14 ([1]). *Let* R *be a coherent ring of finite weak dimension. Then the polynomial ring* $R[x_1, \ldots, x_n]$ *is coherent if and only if* $R_P[x_1, \ldots, x_n]$ *is a coherent ring for every prime ideal* P *of* R.

6 Grade and Coherent Cohen-Macaulay and Gorenstein Rings

The results of this section arose out of the work of Hamilton and Marley [7] to define Cohen-Macaulay rings in the non-Noetherian context as well as the work of Hummel and Marley [9] to define the non-Noetherian Gorenstein counterpart. While Proposition 16 showed the equality of Alfonsi's grade and polynomial grade, in a nod to historical precedence, the following section will use the notation of polynomial grade. However, Alfonsi's characterization of grade was used to prove several of the following results.

We begin with additional characterizations of polynomial grade. Let $\mathbf{x} = x_1, \ldots x_n$ be a finite sequence of elements in R. Given an R-module (M), let $\check{H}_{\mathbf{x}}^i(M)$ be the i-th Čech cohomology of M with respect to $\mathbf{x}$.

Proposition 18 ([7]). *Let R be a ring, let $\mathbf{x}$ be a finite sequence of elements of R with length $\ell = \ell(\mathbf{x})$, let $I = (\mathbf{x})R$, and let M be an R-module. The following integers are equivalent:*

1. p. $\mathrm{grade}_R(I, M)$
2. K. $\mathrm{grade}_R(I, M)$
3. $\sup\{k \geq 0 | \check{H}_{\mathbf{x}}^i(M) = 0$ for all $i < k\}$

Moreover, $IM \neq M$ if and only if any one of the above integers is finite.

Asgharzadeh and Tousi [15] refer to part (3) of Proposition 18 as Čech grade, denoted Č. $\mathrm{grade}_R(I, M)$. Hamilton and Marley prove an additional connection of polynomial depth and weak Bourbaki primes.

Lemma 3 ([7]). *Let M be an R-module and $p \in \mathrm{Ass}_f(M)$. Then p. $\mathrm{depth}_{R_p} M_p = 0$.*

Hamilton and Marley [7] show that a ring may contain ideals of polynomial grade $j > 1$ but no ideals of polynomial grade i, where $0 < i < j$. This is one of the distinctions between classical and polynomial grades.

Proposition 19 ([7]). *Let (R, m) be a local ring of dimension d. For a fixed integer $i \geq 0$, let $M_i = \bigoplus\limits_{\substack{p \in \mathrm{Spec}\, R \\ \text{height } p \leq i}} k(p)$, where $k(p)$ is the residue field of R_p. Let $S = R \times M_i$ be the trivial extension of R by M_i, let $j : S \to R$ be the natural projection, and let I be a finitely generated ideal of S. Then $\mathrm{height}\, I = \mathrm{height}\, j(I)$ and*

$$\text{p. } \mathrm{grade}_R(I, R) = \begin{cases} 0 & \text{if } \mathrm{height}\, I \leq i \\ \text{p. } \mathrm{grade}_R(j(I), R) & \text{if } \mathrm{height}\, I > i. \end{cases}$$

In fact, applying this proposition provides an additional example of a ring displaying the pathological grade behavior first noticed by Hochster, as mentioned in the Introduction.

Example 3 ([7], Example 2.10). Let (R, m) be a Cohen-Macaulay local ring of dimension $d > 0$. Using the notation of Proposition 19, let $S = R \otimes M_{d-1}$. Then S is a local ring of dimension d with maximal ideal $n = m \times M_{d-1}$ with the following properties:

1. p. depth $S = \dim S$.
2. p. grade $I = 0$ for all ideals I of S such that $\sqrt{I} \neq n$; in particular, n consists entirely of zero-divisors.
3. p. depth $S_p = 0$ for all $P \in \operatorname{Spec} S \setminus \{n\}$.

Polynomial depth also plays an important role in the characterization of not necessarily Noetherian Cohen-Macaulay rings. To define Cohen-Macaulay rings, Hamilton and Marley use the definition of strong parameter sequences introduced by Schenzel (see [21] or [7] for details), which play the role of systems of parameters in the non-Noetherian context. Hamilton and Marley thus define Cohen-Macaulay rings in the following manner:

Definition 14 ([7]). A local ring is Cohen-Macaulay if every strong parameter sequence of R is a regular sequence.

Both classical and polynomial grade correspond to the length of strong parameter sequences in Cohen-Macaulay rings.

Proposition 20 ([7]). *Let R be a ring; the following statements are equivalent:*

1. *R is Cohen-Macaulay.*
2. *grade$((\mathbf{x})R, R) = \ell(\mathbf{x})$ for every strong parameter sequence $\mathbf{x}$.*
3. *p. grade$_R((\mathbf{x})R, R) = \ell(\mathbf{x})$ for every strong parameter sequence $\mathbf{x}$.*

Hummel and Marley [9] use the notion of Gorenstein dimension as the foundation for the theory of not necessarily Noetherian Gorenstein rings. Let $M^* = \operatorname{Hom}_R(M, R)$.

Definition 15 ([2]). Let R be a ring and let M be a finitely generated R-module.

1. M is in the class $G(R)$ if

   a. $\operatorname{Ext}_R^i(M, R) = \operatorname{Ext}_R^i(M^*, R) = 0$ for all $i \geq 0$.
   b. The canonical map $M \to M^{**}$ is an isomorphism.

2. M has Gorenstein dimension n, denoted $\operatorname{Gdim} M = n$, if there exists a minimal length exact resolution $0 \to G_n \to \cdots \to G_0 \to M \to 0$ such that $G_i \in G(R)$ for each i. If no finite resolution exists, then $\operatorname{Gdim} M = \infty$.

Gorenstein rings are defined in the following manner:

Definition 16 ([9]). A local ring R is Gorenstein if $\operatorname{Gdim} R/I < \infty$ for every finitely generated ideal I.

The proof connecting Gorenstein and Cohen-Macaulay rings requires a generalization of the Auslander-Bridger formula to non-Noetherian rings. In the Noetherian context, the Auslander-Bridger formula provides a link between depth and Gorenstein dimension.

Theorem 15 ([2]). *Let R be a local Noetherian ring and let M be an R-module with* Gdim $M < \infty$. *Then* Gdim M + depth M = depth R.

The Auslander-Bridger formula was first extended to pseudo-coherent rings by McDowell [14].

Theorem 16 ([14]). *Let R be a local pseudo-Noetherian ring, and let M be a nonzero finitely presented R-module with* Gdim $M < \infty$. *Then* Gdim M + PN. depth M = PN. depth R.

Hummel and Marley [9] provided the final generalization of the Auslander-Bridger formula in the coherent case. There is a version of this formula that doesn't require the ring to be coherent, but rather assumes additional conditions on the R-module (see [9] for details).

Theorem 17 ([9]). *Let R be a local coherent ring and let M be a finitely presented R-module with* Gdim $M < \infty$. *Then* Gdim M + p. depth M = p. depth R.

Using the Auslander-Bridger formula, Hummel and Marley [9] were able to show that coherent local Gorenstein rings are Cohen-Macaulay.

With this definition of Gorenstein rings, Hummel and Marley [9] provide a one-directional analogy of a well-known Noetherian characterization of Gorenstein rings in the coherent context.

Proposition 21 ([9]). *If R is a local coherent Gorenstein ring with* p. depth R = $n < \infty$, *then every n-generated ideal generated by a regular sequence is irreducible.*

Polynomial grade also provides another characterization of coherent Gorenstein rings involving the FP-injective dimension introduced by Stenström.

Definition 17 ([22]). Let R be a ring and let M be an R-module.

1. M is called FP-injective if $\text{Ext}_R^1(F, M) = 0$ for all finitely presented modules F.
2. The FP-injective dimension of M is defined as

$$\text{FP-id}_R M = \inf\{n \geq 0 | \text{Ext}_R^{n+1}(F, M) = 0, \text{ for every finitely presented } R\text{-module } F\}.$$

The following relation between polynomial depth and Gorenstein dimension seen in [9],

$$\text{p. depth } R \leq \sup\{\text{Gdim}(R/I)|I \text{ a finitely generated ideal}\}$$
$$= \sup\{n | \text{Ext}_R^i(R/I, R) = 0, \forall i \geq n, \forall \text{ finitely generated ideal } I\},$$

leads to the following characterization of Gorenstein rings.

Theorem 18 ([9]). *Let R be a local coherent ring. The following conditions are equivalent for $n \geq 0$:*

1. FP-id$_R$ $R \leq n$.
2. R is Gorenstein with depth $R = n$.

Acknowledgements The author would like to thank the referee for their thorough reading and helpful comments to improve this paper. In addition, the author thanks Sarah Glaz for her suggestion regarding the need for the current work, as well as for her assistance and continued encouragement.

References

1. B. Alfonsi, Grade non-noethérien, Comm. Algebra **9**(8), 811–840 (1981)
2. M. Auslander, M. Bridger, *Stable Module Theory*, Memoirs of the American Mathematical Society, vol. 94 (American Mathematical Society, Providence, R.I., 1969)
3. M. Auslander, D.A. Buchsbaum, Homological dimension in local rings. Trans. Amer. Math. Soc. **85**, 390–405 (1957)
4. S.F. Barger, A theory of grade for commutative rings. Proc. Amer. Math. Soc. **36**, 365–368 (1972)
5. D.A. Buchsbaum, D. Eisenbud, What makes a complex exact?. J. Algebra **25**, 259–268 (1973)
6. Sarah Glaz, Commutative coherent rings: historical perspective and current developments. Nieuw Arch. Wisk. (4) **10**(1–2), 37–56 (1992)
7. T.D. Hamilton, T. Marley, Non-Noetherian Cohen-Macaulay rings. J. Algebra **307**(1), 343–360 (2007)
8. M. Hochster,Grade-sensitive modules and perfect modules. Proc. London Math. Soc. (3) **29**, 55–76 (1974)
9. L. Hummel, T. Marley, The Auslander-Bridger formula and the Gorenstein property for coherent rings. J. Comm. Alg. **1** (2009)
10. J. Iroz, D.E. Rush, Associated prime ideals in non-Noetherian rings. Canad. J. Math. **36**(2), 344–360 (1984)
11. P. Jaffard, Théorie de la Dimension Dans les Anneaux de Polynomes, Mémorial Science Mathematics Fasc. vol. 146 (Gauthier-Villars, Paris, 1960)
12. D. Lazard, Autour de la platitude. Bull. Soc. Math. France **97**, 81–128 (1969)
13. K.P. McDowell, Pseudo-Noetherian rings. Canad. Math. Bull. **19**(1), 77–84 (1976)
14. K.P. Mcdowell, Commutative Coherent Rings, ProQuest LLC, Ann Arbor, MI, 1974, Thesis (Ph.D.)–McMaster University (Canada)
15. A. Mohsen, M. Tousi, On the notion of Cohen-Macaulayness for non-Noetherian rings. J. Algebra **322**, (2009)
16. D.G. Northcott, *Finite Free Resolutions*, Cambridge Tracts in Mathematics, vol. 71 (Cambridge University Press, Cambridge, 1976)
17. D.G. Northcott, Projective ideals and MacRae's invariant. J. London Math. Soc. (2) **24**(2), 211–226 (1981)
18. D. Rees, A theorem of homological algebra. Proc. Cambridge Philos. Soc. **52**, 605–610 (1956)
19. D. Rees, The grade of an ideal or module. Proc. Cambridge Philos. Soc. **53**, 28–42 (1957)
20. M. Sakaguchi, A note on the polynomial grade and the valuative dimension. Hiroshima Math. J. **8**(2), 327–333 (1978)
21. P. Schenzel, Proregular sequences, local cohomology, and completion. Math. Scand. **92**(2), 161–180 (2003)
22. B. Stenström, Coherent rings and FP-injective modules. J. London Math. Soc. 2(2), 323–329 (1970)
23. J.R. Strooker, *Homological Questions in Local Algebra*, London Mathematical Society Lecture Note Series, vol. 145 (Cambridge University Press, Cambridge, 1990)
24. W.V. Vasconcelos, Annihilators of modules with a finite free resolution. Proc. Amer. Math. Soc. **29**, 440–442 (1971)

Stable Homotopy Theory, Formal Group Laws, and Integer-Valued Polynomials

Keith Johnson

Abstract In this survey we describe some ways in which algebras of integer-valued polynomials arise in stable homotopy theory and in the study of formal group laws. For several generalized homology theories certain values of the theories have a natural description as such algebras and since these values are the ones arising in the construction of the Adams-Novikov spectral sequence for computing stable homotopy groups these algebras and their homological properties are of considerable interest.

Keywords Integer-valued polynomial • Stable homotopy theory • Formal group law • Hopf algebroid • Adams-Novikov spectral sequence

MSC(2010) classification: [2010]16S36 (13F20,11C08)

1 Introduction

The study of algebras of integer-valued polynomials is usually thought of in ring theory primarily as a source of examples and counterexamples of algebras with unusual properties ([11]). Such algebras do arise naturally in other branches of mathematics, however, and in this survey we will describe two related examples of this. The first is algebraic topology where values of certain generalized homology theories sometimes carry this structure. In the examples we will describe these algebras are the algebras of natural transformations of the homology theory to itself. These play an important role in computations in stable homotopy theory. The other area in which we will describe examples is the study of formal group laws which

K. Johnson (✉)
Department of Mathematics, Dalhousie University, Halifax, Nova Scotia, B3H 4R2, Canada
e-mail: johnson@mathstat.dal.ca

M. Fontana et al. (eds.), *Commutative Algebra: Recent Advances in Commutative Rings,*   213
Integer-Valued Polynomials, and Polynomial Functions, DOI 10.1007/978-1-4939-0925-4_12,
© Springer Science+Business Media New York 2014

are 2-variable power series with certain group-like properties. Here the algebra of self-isomorphisms of such series sometimes carries the structure of an algebra of integer-valued polynomials.

The survey is organized as follows. We begin with a sketch of some background and terminology from algebraic topology, followed by a description of how the algebra $Int(\mathbb{Z})$ of rational integer-valued polynomials on $\mathbb{Z}$ arises naturally in complex K-theory. This is followed by two sections giving some of the facts about formal group laws and stable homotopy theory which we will need. We then describe some of the algebras of integer-valued polynomials arising in these two areas.

2 Some Algebraic Topology

In broadest generality algebraic topology can be described as the study of topological spaces by algebraic means, and it proceeds by finding correspondences (functors in modern categorical language) from topological spaces and continuous maps to algebraic structures such as groups and homomorphisms. A familiar example of this is the fundamental group $\pi_1 : Top \to Grp$ which associates to a space the equivalence class of loops in the space beginning and ending at a fixed base point, the equivalence relation being continuous deformation (homotopy) and the group product being concatenation. In complex analysis, for example, this is useful when you need the winding number in Cauchy's integral formula (which is really an isomorphism between $\pi_1(\mathbb{C} \backslash pt)$ and $\mathbb{Z}$), and there is an efficient algorithm for its computation for most spaces of interest using Van Kampen's theorem. One can generalize this to define higher homotopy groups $\pi_n(X, x_0)$ which are equivalence classes of maps of the n-dimensional sphere, S^n, into the space X. These are important in that many problems of fundamental interest, in differential geometry in the large, for example, reduce to their calculation; however, these groups are not computable in the same way as π_1. Heinz Hopf showed, in the 1930s ([21]), that these groups are more complicated than you expect even for a space as simple as the 2-dimensional sphere, S^2. In fact computing $\pi_n(S^m)$ for all n and a given $m > 1$ is one of the central unsolved problems in algebraic topology.

To try and get around this problem of non-computability topologists formulated the definition of homology groups $H_n(X)$ and $H_n(X, A)$ for $A \subseteq X$. There is a geometric and combinatorial construction of these objects and its essential feature is that for tractable spaces such as simplicial or cell complexes these are efficiently computable. Under the influence of Emmy Noether during the 1930s this was recast in an algebraic form and the computability was described in terms of a long exact sequence relating $H_n(X)$, $H_n(X, A)$, and $H_n(A)$. This algebraic approach also prompted the construction of a dual version, cohomology, denoted $H^n(\ )$, which had the virtue of having the structure of a ring rather than just an abelian group, making it capable of detecting more topological phenomena. This was studied during the 1930s and 1940s and was axiomatized by Eilenberg and Steenrod ([18, 19]) with a set of 7 axioms, one of which was a "normalization" condition which asserted that for the one point space, pt, the group $H_n(pt)$ was $\mathbb{Z}$ if $n = 0$ and 0 for $n \neq 0$. These became quite powerful tools, leading, for example, to J.P. Serre's 1951 thesis in which he showed that $\pi_n(S^m)$ is finite unless $n = m$ or $n = 2n + 1$.

A significant advance in the 1960s was the discovery that there were other homology and cohomology theories besides $H_n(\ )$ and $H^n(\ )$ satisfying all of the axioms of Eilenberg and Steenrod except the "normalization" axiom. This means in particular that they have long exact sequences connecting $E_n(X)$, $E_n(X, A)$, and $E_n(A)$ if $A \subseteq X$. In fact there are lots of these theories. A theorem of E.H. Brown ([9]) says that any one of these arises as $E^n(X) = [X, E_n]$, i.e., homotopy classes of maps from X to a representing space E_n and if you pick a collection of spaces E_n and maps $E_n \to \Omega E_{n+1}$ from E_n to the space of loops in E_{n+1} which are all homotopy equivalences, then you always get one. In spite of this there are only a few families of such theories which are used in practice, all of which originate with specific geometric constructions.

One family is the various sorts of K-theory defined using equivalence classes of vector bundles on the space X. This was originally constructed for algebraic varieties by A. Grothendieck and extended to general topological spaces by Atiyah and Hirzebruch ([5]). Since the only variant of this we will be interested in uses complex vector bundles, we will denote it and the associated homology theory by $K^*(\ )$ and $K_*(\ )$. Some specific values of this theory are, first, for the one point space

$$K^*(pt) = K_*(pt) = \begin{cases} \mathbb{Z} & \text{if } * = 2n \\ 0 & \text{if } * = 2n + 1 \end{cases}$$

(note here that n may be negative) and, next, for the infinite projective space, $\mathbb{CP}^\infty$, which is the direct limit $\lim_{n \to \infty} \mathbb{CP}^n$ of the finite-dimensional complex projective spaces with respect to the usual inclusion maps:

$$K^*(\mathbb{CP}^\infty) = K^*(pt)[[x]] = \begin{cases} \mathbb{Z}[[x]] & \text{if } * = 2n \\ 0 & \text{if } * = 2n + 1 \end{cases}$$

(the computation of the ordinary cohomology of complex projective space as a truncated polynomial algebra is a standard exercise in algebraic topology. Passing to the direct limit gives $H^*(\mathbb{CP}^\infty) = \mathbb{Z}[[x]]$ and there is a spectral sequence, due to Atiyah and Hirzebruch ([6]), which computes the generalized cohomology of a space in terms of its ordinary cohomology and the value of the generalized cohomology theory at the one point space. In this case it has no nontrivial differentials and so reduces to a tensor product).

The second family of homology and cohomology theories is constructed by taking equivalence classes of manifolds with the equivalence relation of cobordism, i.e., two manifolds are equivalent if their disjoint union is the boundary of a manifold of one dimension higher. These equivalence classes form the elements of $E_*(pt)$

and for $E_n(X)$ one takes equivalence classes of maps from n-manifolds into X modulo a pair of these being equivalent if the map the pair defines over the disjoint union of the manifolds can be extended to $n + 1$-manifold with this disjoint union as boundary. That this defines a homology theory is due to Rene Thom ([33]), as are many of the early calculations in this area. A general reference for this area is [31]. One gets a variety of different homology theories in this way by imposing restrictions on the sort of manifolds considered (oriented, unoriented, complex, symplectic, framed, etc.). The one we will be interested in future is "almost complex," i.e., manifolds with a stably complex structure on their tangent spaces. The resulting homology theory is denoted $MU_*(X)$ and called complex cobordism. Two values of interest to us are:

$$MU_*(pt) = \mathbb{Z}[m_i : i = 1, 2, \ldots]$$

(The degrees of the generators are $deg(m_i) = 2i$.)

$$MU^*(\mathbb{CP}^\infty) = MU_*(pt)[[x]]$$

3 Formal Group Laws

In fact the construction of the previous section works for any generalized cohomology theory, $E^*(\ )$, for which $E^*(\mathbb{CP}^\infty) = E_*(pt)[[x]]$ and this is fairly common (such cohomology theories are called complex oriented). For such theories the map μ induces a map

$$E_*(pt)[[x]] = E^*(\mathbb{CP}^\infty) \to E^*(\mathbb{CP}^\infty \times \mathbb{CP}^\infty)$$
$$= E_*(pt)[[x]] \otimes E_*(pt)[[x]] = E_*(pt)[[x, y]]$$

and so $\mu_*(x)$ is a power series in two variables with coefficients in $E_*(pt)$:

$$\mu_*(x) = \sum_{i,j \geq 0} a_{i,j} x^i y^j = F(x, y).$$

Because μ is a classifying map for the tensor product of line bundles this power series has certain group-like properties reflecting those of the tensor product:

$$F(x, y) = \sum a_{i,j} x^i y^j = x + y + \sum_{i,j \geq 1} a_{i,j} x^i y^j$$

$$F(x, 1) = x$$

$$F(1, y) = y$$

$$F(x, y) = F(y, x)$$

$$F(x, F(y, z)) = F(F(x, y), z)$$

A power series with coefficients in a commutative ring R with these properties is called a (one-dimensional, commutative) formal group law over R and the study of these predates their occurrence in algebraic topology. The original definition is due to Bochner ([8]). A general reference for this topic is [20]. For the cohomology theories we have encountered so far these can be calculated:

For $H^*(\ )$ the associated formal group law is $F_H(x, y) = x+y$, called the additive formal group law, defined over $H_*(pt) = \mathbb{Z}$.

For $K^*(\ )$ the associated formal group law is $F_k(x, y) = x + y + txy$, called the multiplicative formal group law, defined over $K_*(pt) = \mathbb{Z}[t, t^{-1}]$ ([4]).

For $MU^*(\ )$ the associated formal group law is denoted $F_L(x, y)$ and is the universal formal group law, defined over the ring $L = MU_*(pt) = \mathbb{Z}[m_i : i = 1, 2, \ldots]$. This formal group law is universal in the sense that if $F(x, y)$ is a formal group law over a ring R, then there is a unique ring homomorphism $L \to R$ such that $F = \phi_*(F_L) = \sum \phi(a_{i,j}) x^i y^j$. That such a universal formal group exists, and the computation of the structure of the ring L, is due to Lazard ([29]). That the formal group law associated to MU is this universal formal group law is a theorem of Quillen ([30]).

Bochner's original definition of a formal group arose from the study of Lie groups and they arise naturally in many other areas besides cohomology theory, in particular in the study of elliptic curves. An elliptic curve over $\mathbb{C}$, being in particular a 1-dimensional Lie group, has both a group product and a differentiable structure, meaning that the group product can be developed as a power series in 2 variables. The fact that this power series comes from a group product implies that it satisfies the axioms of a formal group. Thus an elliptic curve has associated to a formal group for which there is an explicit formula. If the elliptic curve is given in Jacobi form as $y^2 = 1 - 2\delta x^2 + \epsilon x^4 = S_{\delta, \epsilon}(x)$, then the associated formal group law is given by

$$F_{\delta, \epsilon}(x, y) = (x \sqrt{S_{\delta, \epsilon}(y)} + y \sqrt{S_{\delta, \epsilon}(x)})/(1 + \epsilon x^2 y^2).$$

For any formal group law, F, defined over a torsion-free ring R with quotient field K, there is a power series $f(x) \in K[x]$, called the logarithm of F, with the property that

$$F(x, y) = f^{-1}(f(x) + f(y))$$

The name comes from the special case of the multiplicative formal group law, $F_K(x, y)$, for which

$$f_K(x) = \frac{1}{t} \log(1 + tx) \in \mathbb{Q}[t, t^{-1}][[x]]$$

and so

$$f_K^{-1}(x) = \frac{1}{t}(e^{tx} - 1).$$

More generally, an isomorphism between formal groups $F(x, y)$ and $G(x, y)$ is a power series $f(x)$ with the property that $f(F(x, y)) = G(f(x), f(y))$ and in this language a logarithm is an isomorphism between $F(x, y)$ and the additive formal group. The multiplicative formal group $F_K(x, y)$ depends on a unit, t, and different choices for this unit will give different, isomorphic, formal groups. The isomorphism is the power series $f_{K,v}^{-1} \circ f_{K,u}(x)$ which we may compute explicitly:

$$
\begin{aligned}
f_{K,v}^{-1} \circ f_{K,u}(x) &= \frac{1}{v} \exp(v f_{K,u}(x) - 1) \\
&= \frac{1}{v} \exp\left(\frac{v}{u} \log(1 + ux) - 1\right) \\
&= \frac{1}{v} \exp(\log((1 + ux)^{v/u}) - 1) \\
&= \frac{1}{v}((1 + ux)^{v/u} - 1) \\
&= \frac{1}{v} \sum_{n=1}^{\infty} \binom{v/u}{n} u^n x^n
\end{aligned}
$$

The occurrence of binomial polynomials in the coefficients of this general isomorphism between different multiplicative formal groups suggests a connection with integer-valued polynomials. In fact these coefficients generate an algebra of Laurent polynomials determined by an integrality condition that is of considerable interest to topologists. We describe this and its generalizations in the next section.

4 Some Stable Homotopy Theory

The coefficients of the power series $f_{K,v}^{-1} \circ f_{K,u}(x)$ generate a subalgebra of $\mathbb{Q}[u, v, u^{-1}, v^{-1}]$ over the ring $\mathbb{Z}[u, v, u^{-1}, v^{-1}]$ which plays a critical role in the study of the homology theory $K_*(\ )$. This algebra is usually denoted K_*K and is called the Hopf algebra of stable cooperations for K-theory. (Actually this is more properly called a bilateral Hopf algebra or a Hopf algebroid since the left and right counits don't coincide.) Before discussing the topological significance of this algebra let us note two different descriptions of it. First

$$K_*K = \{f(u, v) \in \mathbb{Q}[u, u^{-1}, v, v^{-1}] : f(kt, \ell t) \in \mathbb{Z}\left[\frac{1}{k}, \frac{1}{\ell}\right][t, t^{-1}]$$

$$\text{for all } k, \ell \in \mathbb{Z}\}$$

This description is due to J.F. Adams, R.M. Switzer, and A.S. Harris [3]. Also, letting $w = v/u$, there is the description

$$K_*K = \mathbb{Z}[t, t^{-1}] \otimes \{f(w) \in \mathbb{Q}[w] : w^n f(w) \in Int(\mathbb{Z}) \text{ for some } n\}$$

also due to Adams et al. although published later in [1]. Further details about its structure are given in [2,7,12,13,17] and [23]. It is because of this second description that elements of K_*K of degree 0 are sometimes referred to as stably integer-valued polynomials.

The topological significance of the Hopf algebra K_*K stems from a general result of J.F. Adams giving a method for using generalized homology or cohomology theories to compute homotopy groups in a certain range. It has been known since the 1930's, by a theorem of H. Freudenthal, that the homotopy groups of spheres, $\pi_m(S^n)$, had the stability property that if $m < 2n - 1$, then $\pi_m(S^n)$ is isomorphic to $\pi_{m+1}(S^{n+1})$. Therefore the limit group $\lim_{k\to\infty} \pi_{n+k}(S^k) = \pi_n^S(S)$ always exists and is called the nth stable homotopy group of spheres (or, more properly, of the sphere spectrum). Adams constructed a spectral sequence that begins with E_2 term depending on the Hopf algebra E_*E for $E_*(\ )$ a generalized homology theory, and converges to a certain localization (also dependent on E) of $\pi_*^S(S)$. The E_2 term is the (bigraded) Ext group

$$Ext_{E_*E}^{i,j}(E_*(pt), E_*(pt)).$$

Here these extension groups are computed with respect to the coalgebra structure of E_*E over which $E_*(pt)$ is a comodule. Almost all of our current knowledge of the stable homotopy groups of spheres derives from this spectral sequence for various choices of $E_*(\ )$. Usually things are arranged so that most of the work is in computing the Ext groups and the higher differentials in the spectral sequence vanish for dimensional reasons.

How is the description of K_*K in terms of integer-valued polynomials useful here? The aim is to compute $Ext_{K_*K}(K_*(pt), K_*(pt))$ and this is an Ext group with respect to the comodule structure. The action of the coproduct in K_*K on the elements u and v is very simple: $\Delta(u) = u \otimes 1$ and $\Delta(v) = 1 \otimes v$ and this determines Δ completely when the elements of K_*K are expressed as Laurent polynomials in u and v. (If K_*K is described in terms of generator and relations the expression is much more unwieldy.) This allows a complete resolution of $K_*(pt)$ as a K_*K comodule to be constructed. For example, if one uses the cobar resolution, then the nth stage is $\otimes_{i=1}^n K_*K \subseteq \mathbb{Q}[u_1, u_1^{-1}, v_1, v_1^{-1}, \ldots, v_n, v_n^{-1}]$, which can be described as Laurent polynomials in $2n$ variables satisfying an integrality condition, and the differentials can be given explicitly so that one obtains results such as the following:

$$Ext^{1,n} = \{c \in \mathbb{Q} : c(w^n - 1) \in Int(\mathbb{Z}\setminus p\mathbb{Z}) \text{ for all primes } p\}/\mathbb{Z}$$

and

$$Ext^{2,n} = \{f \in \mathbb{Q}[w] : f(w_1 w_2) - f(w_1) - w_1^n f(w_2) \in Int((\mathbb{Z}\backslash p\mathbb{Z}) \times (\mathbb{Z}\backslash p\mathbb{Z}))$$

$$\text{for all primes } p\}/Int(\mathbb{Z})$$

The first of these is of the form $\mathbb{Z}/m(n)\mathbb{Z}$ and the condition concerning membership in $Int(\mathbb{Z}\backslash p\mathbb{Z})$ determines the integer $m(n)$. If p is an odd prime then the largest integer, ℓ, for which $(w^{p^k(p-1)} - 1)/p^\ell \in Int(\mathbb{Z}\backslash p\mathbb{Z})$, is $\ell = k + 1$ and this gives the p-component of $m(n)$ if $p^k(p-1)$ divides n exactly. The Clausen–von Staudt theorem implies that $m(n)$ is also related to the denominator of the nth Bernoulli number. The second group is 0 if $n \neq 0$ and is $\mathbb{Q}/\mathbb{Z}$ if $n = 0$. It is generated by the polynomials $(w^{p^k(p-1)} - 1)/p^{2(k+1)}$, which are not integer valued but lie in the numerator of the quotient. Details and further information about the structure of $K_0 K$ are contained in [2, 14, 26] and [32].

To proceed further with this we need a bit more stable homotopy theory. We have so far the cohomology theories $H^*(\ )$, $K^*(\ )$, and $MU^*(\ )$, each represented by homotopy classes of maps into a sequence of spaces (a spectrum) $\underline{H} = \{H_n : n = 0, 1, 2 \ldots\}$, $\underline{K} = \{K_n : n = 0, 1, 2, \ldots\}$, and $\underline{MU} = \{MU_n : n = 0, 1, 2, \ldots\}$. Using this representation topologists have constructed other cohomology theories by making geometric constructions on these representing spaces. For example there is a geometric construction whose effect on ordinary homology groups is localization at a prime number p. If this construction is applied to each of the spaces in the spectrum $\underline{MU}$, then these spaces each decompose as one-point unions of simpler spaces, all of which are similar. Taking one of these in each dimension yields a spectrum which represents a new cohomology theory called Brown-Peterson cohomology and denoted $BP^*(\ )$ with $BP_*(pt) = \mathbb{Z}_{(p)}[v_1, v_2, \ldots]$ and $degree(v_i) = 2(p^i - 1)$. (This theory was originally built by Brown and Peterson ([10]) using a direct construction; however, part of Quillen's work connecting $MU_*(\ )$ with formal group laws gives an algebraic construction of this theory and identifies $BP_*(pt)$ with the representing $\mathbb{Z}_{(p)}$-algebra for p-typical formal group laws.) An advantage of this theory is that since the degrees of the generators grow exponentially the nonzero groups in the E_2 term of the Adams spectral sequence based on this theory are much more sparsely distributed making computations easier in many cases.

One may also perform constructions on the spaces of the spectrum $\underline{BP}$ to make some of the classes of the generators v_i homotopically trivial or to make others homotopically invertible. There are conditions restricting when such operations will result in a useful homology theory, called the Landweber exact functor theorem ([27]). One case of interest for us to which this theorem applies is that of making one of the generators, v_n, invertible and making all of the v_i's for $i > n$ trivial. The result, originally constructed by D. Johnson and W.S. Wilson ([22]), is denoted $E(n)_*(\ )$. By construction $E(n)_*(pt) = \mathbb{Z}_{(p)}[v_1, v_2, \ldots, v_n, v_n^{-1}]$. For small values of n this is related to theories we have already seen: $E(0)_*(\ )$ is ordinary homology localized at the prime p, and $E(1)_*(\ )$ is related to K theory. When localized at the prime p the theory $K_*(\ )$ decomposes as a direct sum of $p - 1$ copies of $E(1)_*(\ )$. These theories all are complex oriented and so have formal group laws associated to them. The conditions of the exact functor theorem may also be used to construct a complex-oriented cohomology theory, denoted $Ell^*(\ )$ and called elliptic cohomology, whose associated formal group is that of the Jacobi elliptic curve described in the previous section. Its coefficient ring is given by

$$Ell_*(pt) = \mathbb{Z}\left[\frac{1}{2}\right][\delta, \epsilon, (\delta^2 - \epsilon)^{-1}]$$

5 More Algebras of Integer-Valued Polynomials

The occurrence of integer-valued polynomials in the description of K_*K carries over to some of these new homology theories [15, 16, 24, 28]. Let A denote the ring of integers in a degree n unramified extension of $\mathbb{Q}$. The Hopf algebra $E(n)_*E(n)$ is torsion-free and the polynomials in its image under the natural map

$$E(n)_*E(n) \to E(n)_*E(n) \otimes \mathbb{Q} = E(n)_*(pt) \otimes E(n)_*(pt) \otimes \mathbb{Q}$$

$$= \mathbb{Q}[u_1, \ldots, u_n, u_n^{-1}, v_1, \ldots, v_n, v_n^{-1}]$$

satisfy the integrality condition ([24]) $f(a_1, \ldots, a_n, b_1, \ldots, b_n) \in A$ if $a_i, b_i \in A$, $a_n, b_n \equiv 1 \pmod{p}$, and $a_i, b_i \equiv 0 \pmod{p}$ for $i < n$.

In the case of elliptic cohomology there are two related occurrences of integer-valued polynomials in the description of Ell_*Ell. To see these first note that the logarithm of the Jacobi elliptic formal group law, which is given by

$$log_{\delta,\epsilon}(x) = \int_0^x \frac{dt}{\sqrt{S_{\delta,\epsilon}(t)}} = \sum_{n\geq0} P_n\left(\frac{\delta}{\sqrt{\epsilon}}\right) \epsilon^{n/2} \frac{x^{2n+1}}{2n + 1}$$

where $P_n(\ )$ is the nth Legendre polynomial, contains the parameters δ and ϵ and that, as in the case of the multiplicative formal group law, different choices for these parameters give different, isomorphic formal groups with the isomorphism given by

$$log_{\delta_L,\epsilon_L}^{-1}(log_{\delta_R,\epsilon_R}(x)) = \sum_{i=1}^{\infty} m_i(\delta_L, \delta_R, \epsilon_L, \epsilon_R)x^i$$

The coefficients $m_i(\delta_L, \delta_R, \epsilon_L, \epsilon_R)$ are polynomials in 4 variables and give a set of generators for Ell_*Ell as a module over $\mathbb{Z}[1/2][\delta_L, \delta_R, \epsilon_L, \epsilon_R]$. They can be computed recursively using the formula for $log_{\delta,\epsilon}(x)$. The first few are:

$$\frac{-\delta_L + \delta_R}{3}$$

$$\frac{\delta_L^2 + 10\delta_L\delta_R + 10\delta_R^2 + 3\epsilon_L - 3\epsilon_R}{30}$$

$$\frac{-\delta_L^3 + 35\delta_L^2\delta_R^-2598\delta_L\delta_R^2 + 45\delta_R^3 - 33\delta_L\epsilon_L + 105\epsilon_L\delta_R + 63\delta_L\epsilon_R - 135\delta_R\epsilon_R}{630}$$

As in the case of K-theory the fact that these polynomials are coefficients of an isomorphism of formal group laws implies that they satisfy an integrality condition. In this case it is that the polynomials $m_i(ku^2, kv^2, \ell u^4, \ell v^4) \in \mathbb{Q}[u, v]$ are stably $\mathbb{Z}[1/2]$-integer valued for any integers k and ℓ. Since these polynomials arise as coefficients of isomorphisms associated to elliptic curves they satisfy in addition a stronger integrality condition for certain values of k, ℓ, namely those for which the Jacobi elliptic curve $y^2 = 1 - 2kx^2 + \ell x^4$ has complex multiplication. For those values the polynomials are stably integer valued for a ring which is an order in a quadratic number field, namely the endomorphism ring of the elliptic curve. Values for which this occurs are given in [25].

In addition to this type of integrality condition for these polynomials there is another which stems from their relation to elliptic curves and to the fact that the coefficient ring $Ell_*(pt) = \mathbb{Z}[1/2][\delta, \epsilon, (\delta^2 - \epsilon)^{-1}]$ can be identified with a ring of modular forms, namely the ring of modular forms with respect to the congruence subgroup $\Gamma_0(2)$ of $SL_2(\mathbb{Z})$ whose q-series lie in $\mathbb{Z}[1/2][[q]])$. This identification is made by identifying δ and ϵ with two specific q-series. When the polynomials $m_i(\delta_L, \delta_R, \epsilon_L, \epsilon_R)$ are evaluated at these q-series the resulting series in $\mathbb{Q}[[q_L, q_R]]$ in fact lie in the subring $\mathbb{Z}[1/2][[q_L, q_R]]$ and this condition characterizes elements of Ell_*Ell ([15, 28]).

References

1. J.F. Adams, *Stable Homotopy and Generalised Homology* Chicago Lectures in Mathematics, (University of Chicago Press, Chicago and London, 1974)
2. J.F. Adams, F. Clarke, Stable operations on complex K-theory, Illinois J. Math **21**, 826–829 (1977)
3. J.F. Adams, A.S. Harris, R.M. Switzer, Hopf algebras of cooperations for real and complex K-theory. Proc. London Math. Soc. **23**(3), 385–408 (1971)
4. S. Araki, *Typical Formal Groups in Complex Cobordism and K-Theory*, Lectures in Mathematics (Kyoto University, Kinokuniya Bookstore, Tokyo, 1973)
5. M. Atiyah, F. Hirzebruch, Riemann-Roch theorems for differentiable manifolds. Bul. Amer. Math. Soc. **65**, 276–281 (1959)
6. M. Atiyah, F. Hirzebruch, Vector bundles and homogeneous spaces. Proc. Sympos. Pure Math. Amer. Math. Soc. **3**, 7–38 (1961)
7. A. Baker, F. Clarke, N. Ray, L. Schwartz, On the Kummer congruences and the stable homotopy of BU. Trans. Amer. Math. Soc. **316**, 385–432 (1989)
8. S. Bochner, Formal Lie groups. Ann. Math. **47**, 192–201 (1946)
9. E.H. Brown, Cohomology theories. Ann. Math. **75**, 467–484 (1962)

10. E.H. Brown, F. Peterson, A spectrum whose Zp cohomology is the algebra of reduced pth powers. Topology **5**, 149–154 (1966)
11. P.-J. Cahen, J.-L. Chabert, *Integer Valued Polynomials*. (American Mathematic Society, Providence, R.I., 1997)
12. F. Clarke, *On the Determination of $K_*(K)$ Using the Conner-Floyd Isomorphism*, preprint, Swansea (1972)
13. F. Clarke, Self Maps of BU. Math. Proc. Cam. Phil. Soc. **89**, 491–506 (1981)
14. F. Clarke, M. Crossley, S. Whitehouse, Bases for Co-operations in K-Theory. K-Theory **23**, 237–250 (2001)
15. F. Clarke, K. Johnson, Cooperations in Elliptic Homology. LMS Lect. Notes **176**, 131–143 (1992)
16. F. Clarke, K. Johnson, On the Classifying Ring for Abel Formal Group Laws, J. Pure Appl. Alg. **213**, 1290–1298 (2009)
17. P.E. Conner, E.E. Floyd, *The Relation of Cobordism to K-Theories*, Lecture Notes in Mathematics, vol. 28 (Springer, Berlin/New York, 1966)
18. S. Eilenberg, N. Steenrod, Axiomatic approach to homology theory. Proc. Nat. Acad. Sci. U.S.A. **31**, 117–120 (1945)
19. S. Eilenberg, N. Steenrod, *Foundations of Algebraic Topology* (Princeton University Press, Princeton, N.J., 1952)
20. M. Hazewinkel, *Formal Groups and Applications* (Academic Press, New York, 1973)
21. H. Hopf, Über die Abbildungen der dreidimensionalen Sphäre auf die Kugelfläche. Math. Ann. **104**, 637–665 (1931)
22. D.C. Johnson, W.S. Wilson, BP operations and Morava's extraordinary K-theories, Math. Z. **144**, 55–75 (1975)
23. K. Johnson, The action of the stable operations of complex K-theory on coefficient groups. Illinois J. Math. **28**, 57–63 (1984)
24. K. Johnson, On the Structure of the Hopf Algebroid $E(n)_*E(n)$. Amer. J. Math **122**, 223–234 (2000)
25. K. Johnson,, Numerical polynomials and endomorphisms of formal group laws. London Math. Soc. Lect. Notes **342**, 204–213 (2007)
26. K. Johnson, *The invariant subalgebra and anti-invariant submodule of K_0K*. Jour. K-theory **1**, 123–145 (2008)
27. P. Landweber, Homological properties of comodules over $MU_*(MU)$ and $BP_*(BP)$. Amer. J. Math. **98**, 591–610 (1976)
28. G. Laures, The Topological q-expansion Principle. Topology **38**, 387–425 (1999)
29. M. Lazard, Sur les groupes de Lie formels á un paramétre. Bull. Soc. Math. France **83**, 251–274 (1955)
30. D. Quillen, On the formal group laws of unoriented and complex cobordism. Bull. Amer. Math. Soc. **75**, 1293–1298 (1969)
31. D. Ravenel, Complex Cobordism and Stable Homotopy Groups of Spheres, (Academic Press, New York, NY, 1986)
32. L. Schwartz, *Opérations d'Adams en K-homologie et applications*. Bull. Soc. Math. France **109**, 237–257 (1981)
33. R. Thom, Quelques propriétés globales des variétés différentiables. Comment. Math. Helv. **28**, 17–86 (1954)

How to Construct Huge Chains of Prime Ideals in Power Series Rings

Byung Gyun Kang and Phan Thanh Toan

Abstract Let R be a commutative ring with identity. It is well known that if each chain of prime ideals in R has length at most n, then each chain of prime ideals in the polynomial ring $R[X]$ has length at most $2n+1$. For the power series ring $R[\![X]\!]$, there is however no similar upper bound on lengths of chains of its prime ideals. In fact, in some special cases, there may exist chains of prime ideals in $R[\![X]\!]$ with huge lengths (e.g., $2^{\aleph_1}$) even if each chain of prime ideals in R has length at most one. The purpose of this work is to give a brief review on known constructions of chains of prime ideals in $R[\![X]\!]$ in those cases. By taking into account the techniques which are used in the constructions and possibly by applying some new tools, we hope to construct huge chains of prime ideals in $R[\![X]\!]$ in more general cases.

Keywords Almost Dedekind domain • Krull dimension • Non-SFT ring • Power series ring • Valuation domain

Mathematics Subject Classification (2010): 13A15, 13C15, 13F05, 13F25, 13F30.

B.G. Kang (✉) • P.T. Toan
Department of Mathematics, Pohang University of Science and Technology,
Pohang 790-784, Korea
e-mail: bgkang@postech.ac.kr; pttoan@postech.ac.kr

M. Fontana et al. (eds.), *Commutative Algebra: Recent Advances in Commutative Rings,*   225
Integer-Valued Polynomials, and Polynomial Functions, DOI 10.1007/978-1-4939-0925-4_13,
© Springer Science+Business Media New York 2014

1   Countably Infinite Ascending Chains of Prime Ideals in Power Series Rings over Non-SFT Rings

1.1   Krull Dimension

In this paper, a ring R always means a commutative ring with identity and $\mathrm{Spec}(R)$ denotes the *spectrum* of R (the set of all prime ideals of R). A *chain* of prime ideals in R is a nonempty subset $\mathscr{C}$ of $\mathrm{Spec}(R)$ such that for any two prime ideals P, Q in $\mathscr{C}$ either $P \subseteq Q$ or $Q \subseteq P$. The *length* of a chain $\mathscr{C}$ of prime ideals is defined by $|\mathscr{C}| - 1$, where $|\mathscr{C}|$ denotes the cardinality of $\mathscr{C}$. The *Krull dimension* of a ring R, denoted by $\dim R$, is a "measure" of the lengths of chains of its prime ideals. It is the largest cardinal number α (if any) such that there exists a chain of prime ideals in R whose length is equal to α. We write $\dim R \geq \alpha$ if there is a chain of prime ideals in R with length $\geq \alpha$. We also write $\dim R = \infty$ if there is no finite upper bound on lengths of chains of prime ideals in R. Note that $\dim R = \infty$ does not imply $\dim R = \aleph_0$ or $\dim R \geq \aleph_0$. This happens when every chain of prime ideals in R has finite length and there is no finite upper bound for the lengths of those chains.

1.2   Krull Dimension of Polynomial Rings

In 1953, Seidenberg showed in [14] that if R is a ring with $\dim R = n$, then

$$n + 1 \leq \dim R[X] \leq 2n + 1. \tag{1}$$

He also showed that all intermediate values can be obtained by appropriately choosing R. More precisely, if m and n are nonnegative integers with $n + 1 \leq m \leq 2n + 1$, then there exists a ring R such that $\dim R = n$ and $\dim R[X] = m$ [15]. Therefore, the possibilities for $\dim R[X]$ were completely determined.

1.3   Krull Dimension of Power Series Rings over Non-SFT Rings

In the late 1960s, Gilmer and some of his students started to work on Krull dimension of the power series ring $R[\![X]\!]$. While it is easy to show that $n + 1 \leq \dim R[\![X]\!]$ if $\dim R = n$, $\dim R[\![X]\!]$ can be infinite even if $\dim R$ is finite. This result was shown by Arnold (one of Gilmer's students) in 1973 [1]. In his paper, Arnold proved that $\dim R[\![X]\!] = \infty$ when the ring R is not an SFT ring (the term SFT stands for "strong finite type"). He defined a ring R to be an *SFT ring* if for each ideal I of R, there exist a finitely generated ideal J of R with $J \subseteq I$ and a positive integer k such that $a^k \in J$ for each $a \in I$. In fact, he constructed a countably infinite ascending chain of prime ideals in $R[\![X]\!]$, i.e., $\dim R[\![X]\!] \geq \aleph_0$, when R is a ring which is not an SFT ring. Note that R being an SFT ring does not imply the finiteness of $\dim R[\![X]\!]$. In 2002, Coykendall gave an example of an SFT domain V_1 with $\dim V_1 = 1$ such that $\dim V_1[\![X]\!] = \infty$ [3].

1.3.1 Arnold's Construction

A ring R is called a *non-SFT ring* if R is not an SFT ring. We outline Arnold's construction of an infinite ascending chain of prime ideals in $R[\![X]\!]$ when R is a non-SFT ring. For more details, see [1].

If R is a non-SFT ring, then we can choose a sequence $\{a_i\}_{i=0}^{\infty}$ such that

$$a_{k+1}^{k+1} \notin (a_0, a_1, \ldots, a_k), \tag{2}$$

for each integer $k \geq 0$. For each positive integer m, choose a sequence $\{a_{m,i}\}_{i=0}^{\infty}$ as follows. For $m = 1$, take $a_{1,i} = a_i$. Inductively, take $a_{n,i} = a_{n-1,i^2+1}$. Let m, n, μ, r be integers such that $m \geq n \geq 1$ and $r \geq 0$. For $f = \sum_{i=0}^{\infty} b_i X^i \in R[\![X]\!]$, the tuple (f, m, μ, r) has property (n) if for $i \geq r$ there exists an integer t_i such that the following hold (where $a_{m,i} = a_{n,k_i} = a_{1,s_i}$):

1. $b_{t_i} \equiv a_{m,i}^{\mu} \pmod{A_{s_i-1}}$.
2. $t_i \leq \mu k_i$.
3. $b_j \in A_{s_i-1}$ for $0 \leq j < t_i$.

Set $S_n = \{f \in R[\![X]\!] \mid (f, m, \mu, r)$ has property (n) for some $m, \mu,$ and $r\}$. Then S_n is a multiplicatively closed subset of $R[\![X]\!]$. Furthermore, if $n > n_1$, then $S_n \subset S_{n_1}$. Hence, there exists a chain of multiplicatively closed subsets of $R[\![X]\!]$,

$$S_1 \supset S_2 \supset \cdots \supset S_n \supset \cdots . \tag{3}$$

Let $A = (a_0, a_1, \ldots, a_n, \ldots)$. Then $AR[\![X]\!] \cap S_1 = \emptyset$. Hence, there exists a prime ideal P_1 in $R[\![X]\!]$ such that $AR[\![X]\!] \subseteq P_1$ and $P_1 \cap S_1 = \emptyset$. Suppose that exists a chain $P_1 \subset P_2 \subset \cdots \subset P_n$ of prime ideals in $R[\![X]\!]$ such that $P_i \cap S_i = \emptyset$ for $i = 1, 2, \ldots, n$. Let $C_n = P_n + (f_{(n)})$, where $f_{(n)} = \sum_{i=0}^{\infty} a_{n,i} X^i$. Then $C_n \cap S_{n+1} = \emptyset$. Therefore, there exists a prime ideal P_{n+1} such that $P_n \subset C_n \subseteq P_{n+1}$ and $P_{n+1} \cap S_{n+1} = \emptyset$. By induction, there is an infinite ascending chain of prime ideals in $R[\![X]\!]$,

$$P_1 \subset P_2 \subset \cdots \subset P_n \subset \cdots . \tag{4}$$

Therefore, we have the following theorem.

Theorem 1. *If R is a non-SFT ring, then $\dim R[\![X]\!] = \infty$. In fact, in this case, there exists an infinite ascending chain of prime ideals in $R[\![X]\!]$ and hence $\dim R[\![X]\!] \geq \aleph_0$.*

1.3.2 Coykendall's Example

Note that the converse of Theorem 1 does not hold. Coykendall gave an example of an SFT domain V_1 such that dim $V_1[\![X]\!] \geq \aleph_0$. The construction of V_1 is as follows. Let V be a one-dimensional nondiscrete valuation domain with value group $\mathbb{Q}$ and residue field $\mathbb{F}_2$ (the choice of the value group and the residue field is in fact not important). Then V can be written as

$$V = \mathbb{F}_2[x^\alpha]_{\mathcal{M}}, \tag{5}$$

where $\mathbb{F}_2[x^\alpha] = \{\sum_{i=0}^{n} \epsilon_i x^{\alpha_i} \mid \epsilon_i \in \mathbb{F}_2, \alpha_i \in (\mathbb{Q}^+ \cup \{0\})\}$ and $\mathcal{M} \subseteq \mathbb{F}_2[x^\alpha]$ is the maximal ideal of $\mathbb{F}_2[x^\alpha]$ generated by the monomials. Let

$$V_1 = \mathbb{F}_2 + xV. \tag{6}$$

Then V_1 is a one-dimensional SFT domain. An infinite chain of prime ideals in $V_1[\![X]\!]$ is obtained from an infinite chain of prime ideals in $V[\![X]\!]$. Consider an infinite descending chain of prime ideals lying over (0) in $V[\![X]\!]$,

$$P_1 \supset P_2 \supset \cdots \supset P_n \supset \cdots. \tag{7}$$

For the existence of the chain of prime ideals, see Theorem 4 below. The choice of V_1 ensures that $P_n \cap V_1[\![X]\!] \neq P_{n+1} \cap V_1[\![X]\!]$ for all n. Hence, we get an infinite descending chain of prime ideals in $V_1[\![X]\!]$,

$$P_1 \cap V_1[\![X]\!] \supset P_2 \cap V_1[\![X]\!] \supset \cdots \supset P_n \cap V_1[\![X]\!] \supset \cdots. \tag{8}$$

Thus, we have the following theorem.

Theorem 2. *There exists an SFT domain V_1 such that* dim $V_1[\![X]\!] \geq \aleph_0$, *i.e., the converse of Theorem 1 does not hold.*

1.4 The Finite Case

Theorems 1 and 2 say that dim $R[\![X]\!]$ is generally large. This still holds when it is finite (comparing to the polynomial ring case). Indeed, even when dim $R[\![X]\!] < \infty$, it is not always true that dim $R[\![X]\!] \leq 2n + 1$. In 2009, Kang and Park showed that there exists a ring R with dim $R = n$ such that $2n + 1 < $ dim $R[\![X]\!] < \infty$ [9]. Note that R must be an SFT ring. The construction of R is from mixed extensions. A *mixed extension* is an extension of the form

$$R[X_1][\![X_2]\!] \cdots [X_n], \tag{9}$$

where $[X_i]$ can be either $[X_i]$ or $[\![X_i]\!]$. Kang and Park showed the following formula for Krull dimension of mixed extensions.

Theorem 3. *If R is an m-dimensional SFT Prüfer domain, then*

$$\dim(R[X_1][X_2]\cdots[X_n]) = \begin{cases} mn + 1 & \textit{if } [X_i] = [\![X_i]\!] \textit{ for some } 1 \le i \le n \\ m + n & \textit{otherwise.} \end{cases} \quad (10)$$

Corollary 1. *Let R be a commutative ring with identity. Then $\dim R[\![X]\!] < \infty$ does not imply that $\dim R[\![X]\!] \le 2 \dim R + 1$.*

Proof. For an integer $n > 2$, let V be a discrete valuation domain with $\dim V = d > 2(n-1)/(n-2)$. Put $R = V[X_1, \ldots, X_{n-1}]$. Then $\dim R = d + (n-1)$ since V is an SFT Prüfer domain. We have $\dim R[\![X_n]\!] = \dim V[X_1]\cdots[X_{n-1}][\![X_n]\!] = dn + 1 = 2d + (n-2)d + 1 > 2d + 2(n-1) + 1 = 2\dim R + 1$. $\qquad\square$

2 Countably Infinite Descending Chains of Prime Ideals in Power Series Rings over Valuation Domains

A valuation domain V is called a *discrete valuation domain* if every primary ideal of V is a power of its radical. Suppose that V is a one-dimensional valuation domain. Let $V^* = V \setminus \{0\}$. If V is discrete, then it is a Noetherian ring (in fact a PID) and hence $\dim V[\![X]\!] = 2$. By the result of Arnold [1], $\dim V[\![X]\!] = \infty$ if V is nondiscrete (note that a finite-dimensional valuation domain is nondiscrete if and only if it is non-SFT [2]). Note that Krull dimension of the ring of entire functions is uncountable (in fact $2^{\aleph_1}$ under the continuum hypothesis [7]) and there are similarities between $V[\![X]\!]_{V^*}$ and the ring of entire functions. For example, one can define zeros and multiplicities for elements in $V[\![X]\!]_{V^*}$ (or $V[\![X]\!]$) if V is a one-dimensional nondiscrete valuation domain. Indeed, in 1999, using these concepts and introducing the concept of infinite product of power series, Kang and Park showed that $\dim V[\![X]\!]_{V^*} = \infty$ for a one-dimensional nondiscrete valuation domain V [8]. In fact, they constructed a countably infinite descending chain of prime ideals in $V[\![X]\!]$ which do not contain any nonzero elements of V. In their paper, the chain of prime ideals $P_1 \supset P_2 \supset \cdots \supset P_n \supset \cdots$ comes from a chain $I_1 \supset I_2 \supset \cdots \supset I_n \supset \cdots$ of ideals in $V[\![X]\!]$, where each I_n is a collection of all power series that have the prescribed zeros and multiplicities. By induction, the existence of the chain of prime ideals follows from the fact that a prime ideal P_n minimal over I_n is never minimal over I_{n+1}, which allows them to get a smaller prime ideal P_{n+1}. We now show the construction the ideals I_n and the existence of the prime ideals P_n.

In the remaining of this section, V always denotes a one-dimensional nondiscrete valuation domain with maximal ideal M.

Definition 1 (Zeros). Let f be a power series in $V[\![X]\!]$, an element $b \in M$ is called a *zero* of f provided that $X - b$ divides f in $V[\![X]\!]$.

For each $f \in V[\![X]\!]$, let $Z(f)$ be the set of zeros of f. By definition, $Z(f)$ is a subset of the maximal ideal M of V.

Definition 2 (Multiplicities). Let f be a nonzero power series in $V[\![X]\!]$ and b be an element in M. The *multiplicity* of f at b, denoted by $m(f, b)$, is the largest nonnegative integer n such that $(X - b)^n$ divides f. If $f = 0$, then we set $m(f, b) = \infty$ for all $b \in M$.

For a power series $f = \sum_{i=0}^{\infty} a_i X^i$ in $V[\![X]\!]$, we denote by $\delta(f)$ the smallest index $n \geq 1$ such that $a_n \neq 0$ (if f is just a constant in V, then we set $\delta(f) = \infty$), and we say $\delta(f)$ is the *initial degree* of f. We say that a sequence of power series $\{f_i = \sum_{j=0}^{\infty} a_{ij} X^j\}_{i=1}^{\infty}$ is *upper triangular (resp. echelon)* if $\delta(f_n) \geq n$ for all n (resp. $\lim_{n \to \infty} \delta(f_n) = \infty$).

Definition 3 (Infinite product of power series). Let $\{f_i = \sum_{j=0}^{\infty} a_{ij} X^j\}_{i=1}^{\infty}$ be an upper triangular sequence of power series in $V[\![X]\!]$ such that $\sum_{i=1}^{\infty} v(a_{i0}) \leq v(a) < \infty$ for some $a \in V$. Define an infinite product $f = \sum_{i=0}^{\infty} a_i X^i$ of all f_i by

$$a_0 = a,$$

$$a_1 = a_{11} \frac{a}{a_{10}},$$

$$a_2 = a_{12} \frac{a}{a_{10}} + a_{22} \frac{a}{a_{20}},$$

$$a_3 = a_{13} \frac{a}{a_{10}} + a_{23} \frac{a}{a_{20}} + a_{33} \frac{a}{a_{30}} + a_{22}a_{11} \frac{a}{a_{10}a_{20}},$$

$$\vdots$$

$$a_n = (f_1 f_2 \cdots f_n)_n \frac{a}{a_{10}a_{20} \cdots a_{n0}},$$

$$\vdots$$

where $(f_1 f_2 \cdots f_n)_n$ means the coefficient of X^n in the product $f_1 f_2 \cdots f_n$. The resulting product f is denoted by $(\prod_{i=1}^{\infty} f_i; a)$.

For an echelon sequence, we can define f similarly. For more information about infinite product of power series, we refer the readers to [8].

Let v be the valuation associated with V. Using infinite product of power series, Kang and Park showed that there exist $g_1, g_2, \ldots, g_k, \ldots$ in $V[\![X]\!]$ and $b_1, b_2, \ldots, b_n, \ldots$ in M such that

1. $v(b_1) > v(b_2) > \cdots > v(b_n) > \cdots$,
2. $\{b_1, b_2, \ldots, b_n, \ldots\} \subseteq Z(g_i)$ for each i,
3. $m(g_i, b_n) = n^i$ for each n and i.

Let

$$A_k = \{(b_j, j^k) \mid j \in \mathbb{N}\}. \tag{11}$$

Write $f((b_j, j^k)) = 0$ or $(b_j, j^k) \in Z(f)$ if b_j is a zero of f with multiplicity $\geq j^k$. Let

$$I_k = \{f \in V[\![X]\!] \mid f \equiv 0 \text{ almost everywhere on } A_k\}. \tag{12}$$

Then we get a chain of ideals

$$I_1 \supset I_2 \supset \cdots \supset I_n \supset \cdots. \tag{13}$$

Take a prime ideal P_1 minimal over I_1. Then $P_1 \cap V = (0)$. By the construction of the ideals I_k, P_1 is never minimal over I_2. Hence, there exists a prime ideal P_2 properly contained in P_1 and minimal over I_2. By induction, suppose we can chose prime ideals $P_1 \supset P_2 \supset \cdots \supset P_n$ such that P_i is minimal over I_i for $i = 1, 2, \ldots, n$. Then P_n is not minimal over I_{n+1}. Thus, we can find a prime ideal P_{n+1} minimal over I_{n+1} such that $P_n \supset P_{n+1}$. Therefore, we get an infinite descending chain of prime ideals

$$P_1 \supset P_2 \supset \cdots \supset P_n \supset \cdots. \tag{14}$$

Note that $I_1 \subseteq M[\![X]\!]$. By shrinking $M[\![X]\!]$ to a prime ideal minimal over I_1, we can choose from the beginning that $P_1 \subseteq M[\![X]\!]$. This completes the proof of the following theorem.

Theorem 4. *Let V be a one-dimensional nondiscrete valuation domain with maximal ideal M. Then $\dim V[\![X]\!]_{V^*} = \infty$. In fact, there exists an infinite descending chain of prime ideals inside $M[\![X]\!]$ which do not contain any nonzero elements of V.*

3   Uncountable Chains of Prime Ideals in Power Series Rings over Valuation Domains

In 1982, Eakin and Sathaye posed the question whether $\dim V[\![X]\!]_{V^*}$ is uncountable for a one-dimensional nondiscrete complete valuation domain V [5]. In 2013, Kang and Park gave an affirmative answer to this question without using the assumption that V is complete [10]. They showed that if V is a one-dimensional nondiscrete valuation domain, then there exists a chain $\{Q_\alpha\}_{\alpha \in \mathbb{R}^+}$ of prime ideals in $V[\![X]\!]$ such that $Q_\alpha \cap V = (0)$ for each $\alpha \in \mathbb{R}^+$, and hence $\dim V[\![X]\!]_{V^*} \geq \aleph_1$ [10]. This result also gives an answer to the conjecture posed by Coykendall that there is a chain of prime ideals in $V[\![X]\!]$ that is order-isomorphic to the positive half-line [4].

While prime ideals in [8] are obtained by picking ones among prime ideals minimal over some given ideals, prime ideals in [10] are in a different way. The fact that was used in [10] is [6, Theorem 19.6], which says that if P is a prime ideal of a domain D, then there exists a valuation domain W containing D such that $M \cap D = P$, where M is the maximal ideal of W. Note that in a valuation domain W, the radical of any proper ideal is a prime ideal. Hence, instead of constructing a chain of prime ideals in $V[X]$, one can construct a chain of prime ideals (by taking the radical of any chain of ideals) in a valuation domain W containing $V[X]$ and then take contractions of them to $V[X]$. The chain $\{Q_\alpha\}_{\alpha \in \mathbb{R}^+}$ of prime ideals exactly comes in this way. Of course, some technical things needed to be done to guarantee that the prime ideals obtained from taking contractions are all distinct. Even though their main interest was in the one-dimensional case, Kang and Park were able to show that $\dim W[X]$ is uncountable for any nondiscrete valuation domain W [10].

Let V be a one-dimensional nondiscrete valuation domain with maximal ideal M and let v be the valuation associated with V. Since the value group of V is a dense subgroup of $\mathbb{R}$, we can choose $b_n \in M$ so that $v(b_n) \le n^{-n^2-1}e^{-n}$. For each $\alpha \in \mathbb{R}^+$, define

$$I_\alpha = \{f \in V[X] \mid \text{there exists a constant } c \in \mathbb{R}^+ \text{ such that } m(f, b_n) \ge cn^{\lfloor n\alpha \rfloor}\}. \tag{15}$$

Let $0 \ne d_\alpha \in V$ such that $\sum_{n=1}^\infty n^{\lfloor n\alpha \rfloor + 1}v(b_n) \le v(d_\alpha)$. Let

$$f_\alpha = \left(\prod_{i=1}^\infty (X^n - b_n^n)^{n^{\lfloor n\alpha \rfloor}}; d_\alpha \right). \tag{16}$$

Then $m(f_\alpha, b_n) = n^{\lfloor n\alpha \rfloor}$ for each $n \ge 1$. Thus $f_\alpha \in I_\alpha$ and hence I_α is nonempty. In fact, I_α is a proper ideal of $V[X]$. We have a chain $\{I_\alpha\}_{\alpha \in \mathbb{R}^+}$ of proper ideals of $V[X]$. We now want a valuation domain W containing $V[X]$ such that $\sqrt{I_\alpha W} \cap V[X] \ne \sqrt{I_\beta W} \cap V[X]$ whenever $\alpha \ne \beta$. For this purpose let $D = V[X]$ and let

$$S = \{g/h \mid g, h \in D, \text{ there exists } \alpha \in \mathbb{R}^+ \text{ such that } g \in I_\alpha \text{ and}$$

$$h \notin P_\beta \text{ for any } \beta \ge \alpha \text{ and any prime ideal } P_\beta \text{ minimal over } I_\beta\}. \tag{17}$$

An important fact is that $SD[S]$ is a proper ideal of $D[S]$. Hence, we can choose a prime ideal P of $D[S]$ containing $SD[S]$. By [6, Theorem 19.6], there exists a valuation domain W containing $D[S]$ with maximal ideal Q such that $Q \cap D[S] = P$. For each $\alpha \in \mathbb{R}^+$, put

$$Q_\alpha = \sqrt{I_\alpha W} \cap D. \tag{18}$$

Then $\{Q_\alpha\}_{\alpha \in \mathbb{R}^+}$ is chain of prime ideals in $D = V[\![X]\!]$. Suppose that $\alpha < \beta$. We have $f_\alpha \in Q_\alpha$. We show that $f_\alpha \notin Q_\beta$. Suppose on the contrary that $f_\alpha \in Q_\beta = \sqrt{I_\beta W} \cap D$. Then $f_\alpha^l \in I_\beta W$ for some $l \geq 1$. Then $f_\alpha^l \in g_\beta W$ for some $g_\beta \in I_\beta$. So $f_\alpha^l / g_\beta \in W$. However, $g_\beta / f_\alpha^l \in S \subseteq P \subseteq Q$, a contradiction. Therefore, $\{Q_\alpha\}_{\alpha \in \mathbb{R}^+}$ is chain of prime ideals in $V[\![X]\!]$ with length $|\mathbb{R}^+|$. Thus we have the following.

Theorem 5. *If V is a one-dimensional nondiscrete valuation domain, then* $\dim V[\![X]\!] \geq$ *the cardinality of the continuum. Hence,* $\dim V[\![X]\!] \geq \aleph_1$.

By changing I_α to $\{f \in V[\![X]\!] \mid$ there exists a constant $c \in \mathbb{R}^+$ such that $m(f, b_n) \geq cn^{\lfloor n\alpha \rfloor}$ for $n \gg 1\}$. Kang and Park could manage to show that the obtaining Q_α satisfies $Q_\alpha \subseteq M[\![X]\!]$ and $Q_\alpha \cap V = (0)$.

Theorem 6. *Let V be a one-dimensional nondiscrete valuation domain with maximal ideal M. Then there exists an uncountable chain $\{Q_\alpha\}_{\alpha \in \mathbb{R}^+}$ of prime ideals inside $M[\![X]\!]$ such that $Q_\alpha \cap V = (0)$ for each $\alpha \in \mathbb{R}^+$. Hence,* $\dim V[\![X]\!]_{V^*} \geq \aleph_1$.

For arbitrary nondiscrete valuation domain W, Kang and Park showed that $\dim W[\![X]\!]$ is still uncountable.

Theorem 7. *If W is a nondiscrete valuation domain, then* $\dim W[\![X]\!]$ *is uncountable.*

4 Uncountable Chains of Prime Ideals in Power Series Rings over Non-SFT Rings

In 2013, Kang, Loper, Lucas, Park, and Toan showed that $\dim R[\![X]\!]$ is uncountable if R is a non-SFT ring [11], i.e., $\dim R[\![X]\!] \geq \aleph_1$, a one-step improvement of Arnold's result in the sense that $\aleph_1$ is the very next cardinal number to $\aleph_0$. To construct an infinite ascending chain of prime ideals in $R[\![X]\!]$, Arnold first started with a chain of multiplicatively closed subsets $S_1 \supset S_2 \supset \cdots \supset S_n \supset \cdots$ in $R[\![X]\!]$. Then, using induction, he constructed a chain of prime ideals $P_1 \subset P_2 \subset \cdots \subset P_n \subset \cdots$ in $R[\![X]\!]$ such that $P_n \cap S_n = \emptyset$ for each n. Here prime ideals P_n are obtained by extending ideals, missing the mutiplicatively closed subsets S_n using Zorn's Lemma. In this construction, the chain of multiplicatively closed subsets is indexed over $\mathbb{N}$ (and hence is countable). The authors of [11] generalized this result by showing that there exists an uncountably infinite chain of prime ideals in $R[\![X]\!]$ when R is a non-SFT ring. They constructed a chain of multiplicatively closed subsets $\{T_s\}_{s \in \mathscr{A}}$ in $R[\![X]\!]$ (where $\mathscr{A}$ is an uncountable set). Then they showed that there exist an uncountable subset $\{T_s\}_{s \in \mathscr{B}}$ of $\{T_s\}_{s \in \mathscr{A}}$ and an uncountable chain of prime ideals $\{P_s\}_{s \in \mathscr{B}}$ in $R[\![X]\!]$ such that $P_s \cap T_s = \emptyset$ for each $s \in \mathscr{B}$. Since the chain of prime ideals is uncountable, the mathematical induction does not work. The existence of the chain of prime ideals was proved by using a nice property of the set $\mathscr{A}$ (which was called the *fathomless property*).

We first give the construction (without proof) of the fathomless set $\mathscr{A}$. Let $\mathbb{N} = \{1, 2, \ldots\}$ be the set of positive integers, and let $\mathscr{U}$ be the set of all subsets U of $\mathbb{N}$ such that $U = \{n, n+1, \ldots\}$ for some $n \in \mathbb{N}$. For two strictly increasing sequences $s = \{s_n\}$ and $t = \{t_n\}$ of positive integers, we set $s \gg t$ (we also write $t \ll s$) if for each positive integer k, there is a set $U \in \mathscr{U}$ (depending on k) such that $s_n > k t_n$ for each $n \in U$, i.e., $s_n > k t_n$ for all large n. Let $\mathscr{S}$ be the collection of all $\mathscr{A}$ such that $\mathscr{A}$ has the following properties:

1. $\mathscr{A}$ is a nonempty collection of strictly increasing sequences $s = \{s_n\}$ of positive integers.
2. If $s \in \mathscr{A}$, then $s \gg b$, where b is the sequence defined by $b_n := n$ for all n.
3. If $s, t \in \mathscr{A}$ and $s \neq t$, then $s \gg t$ or $t \gg s$.

If u is the sequence defined by $u_n := b_n^2$ for each n, then it is easy to see that u is a strictly increasing sequence of positive integers and that $u \gg b$. It follows that the set $\mathscr{S}$ is nonempty. We order $\mathscr{S}$ by set-theoretic inclusion. By Zorn's Lemma, there exists a maximal element in $\mathscr{S}$.

Definition 4. A totally ordered set $(\mathscr{Y}, \ll)$ is called a *fathomless set* if for every countable nonempty subset $\mathscr{C}$ of $\mathscr{Y}$, there exists an element $y \in \mathscr{Y}$ such that $y \ll c$ for all $c \in \mathscr{C}$.

Theorem 8. *The set $(\mathscr{A}, \ll)$ is a fathomless set.*

Let R be a non-SFT ring. There exists a sequence $\{a_i\}_{i=0}^{\infty}$ of elements of R such that $a_m^m \notin (a_0, a_1, \ldots, a_{m-1})$ for each $m \geq 1$. For each m, let

$$I_m := (a_0, a_1, \ldots, a_m). \tag{19}$$

For each $s = \{s_n\} \in \mathscr{A}$, define a power series $a_{(s)}$ in $R[\![X]\!]$ by

$$a_{(s)} = a_0 + a_1 x^{s_1} + \cdots + a_n x^{s_n} + \cdots. \tag{20}$$

Definition 5. For $s \in \mathscr{A}$, we say that a power series $g = \sum_{j=0}^{\infty} g_j X^j$ has the property (s) if there exist a sequence $\{q_m\}$ of positive integers, a positive integer μ, and a set $U \in \mathscr{U}$ such that the following hold for each $m \in U$.

1. $q_m \leq \mu s_m$.
2. $g_{q_m} \equiv a_m^k \pmod{I_{m-1}}$ for some $1 \leq k \leq \mu$.
3. $g_j \in I_{m-1}$ for all $j < q_m$.

For $s \in \mathscr{A}$, let T_s denote the set of power series in $R[\![X]\!]$ having the property (s). It is easy to see that $a_{(s)} \in T_s$ (with $\{q_m\} = s$, $\mu = 1$, and $U = \mathbb{N}$). Hence, T_s is nonempty. In fact, T_s is a multiplicatively closed subset and $\{T_s\}_{s \in \mathscr{A}}$ is a chain of multiplicatively closed subsets since if $s \ll v$, then $T_s \subset T_v$ (see [11]). For a multiplicatively closed set T in $R[\![X]\!]$, let $Q(T)$ be the set of prime ideals P in $R[\![X]\!]$ that are maximal with respect to missing T (i.e., $P \cap T = \emptyset$). In particular, for $s \in \mathscr{A}$, $Q(T_s)$ is the set of prime ideals in $R[\![X]\!]$ that are maximal with respect to missing T_s. It was shown that $Q(T_s)$ is nonempty and if $v \ll s$, then each prime ideal in $Q(T_s)$ is properly contained in some prime ideal in $Q(T_v)$. Let $\mathscr{P}$ be the set of elements $E = \{(P_i, s_i)\}_{i \in I}$ (for some nonempty set I) satisfying the following:

1. $s_i \in \mathscr{A}$ and $P_i \in Q(T_{s_i})$ for each $i \in I$.
2. $\{P_i\}$ is a strict chain of prime ideals, i.e., $\{P_i\}$ is a chain of prime ideals and $P_i = P_j$ if and only if $i = j$.

Order $\mathscr{P}$ by set-theoretic inclusion. By Zorn's Lemma, there exists a maximal element in $\mathscr{P}$. Let $E = \{(P_i, s_i)\}_{i \in I}$ be a maximal element in $\mathscr{P}$. We show that the chain of prime ideals $\{P_i\}_{i \in I}$ is uncountable. Suppose on the contrary that $\{P_i\}_{i \in I}$ is countable. Hence, the set $\{s_i\}_{i \in I}$ is countable. Since the set $(\mathscr{A}, \ll)$ is a fathomless set, there is a sequence $v \in \mathscr{A}$ such that $s_i \gg v$ for each i. By Zorn's Lemma, one can find a prime ideal $N' \in Q(T_v)$ such that $N' \supset P_i$ for all $i \in I$. The set $\{(P_i, s_i)\}_{i \in I} \cup \{(N', v)\}$ properly contains the set $\{(P_i, s_i)\}_{i \in I}$ in $\mathscr{P}$, a contradiction. This proves the following theorem.

Theorem 9. *If R is a non-SFT ring, then there exists an uncountably infinite chain of prime ideals in $R[\![X]\!]$, i.e., $\dim R[\![X]\!] \geq \aleph_1$.*

5 Huge Chains of Prime Ideals in Power Series Rings over Almost Dedekind Domains

Another technique of constructing chains of prime ideals in power series rings appeared in Loper and Lucas's papers [12, 13]. Their constructions are great applications of (nonprincipal) ultrafilters. With the aid of some set theory, it was shown that $\dim R[\![X]\!] \geq 2^{\aleph_1}$ if R is a one-dimensional nondiscrete valuation domain or a non-Noetherian almost Dedekind domain. In each paper, a totally order $\gg$ on $R[\![X]\!]$ was given and a prime ideal in $R[\![X]\!]$ is obtained by collecting all power series f such that $f \gg g$ for all $g \in S$ (where S is given). For an ultrafilter $\mathscr{U}$ over a set $\mathbf{X}$, if $A \cup B = \mathbf{X}$, then either $A \in \mathscr{U}$ or $B \in \mathscr{U}$. This property of ultrafilters is extremely important in showing that the constructed ideals are prime (if ab belongs to the constructed ideal, then either a or b belongs to that ideal).

Because of the same approach of the two constructions, we are not going to elaborate one-dimensional nondiscrete valuation domain case [13]. We however note that in [13], for a one-dimensional nondiscrete valuation domain V (with maximal ideal M), each prime ideal in the constructed chain is between $MV[\![X]\!]$ and $M[\![X]\!]$, which contrasts to the result of Kang and Park, where the prime ideals in the chain all contract to zero.

Let D be an almost Dedekind domain. If D is Dedekind, then D is Noetherian and hence $\dim D[\![X]\!] = 2$. So in this section we always assume that D is an almost Dedekind domain that is not Dedekind. The construction of chains of prime ideals in $D[\![X]\!]$ is based on two countably infinite set $\{M_n\}_{n=1}^{\infty}$ and $\{\rho_n\}_{n=1}^{\infty}$, satisfying the following conditions:

1. Each M_n is a maximal ideal of D and each ρ_n is an element of D.
2. For each n, ρ_n belongs to every M_m except M_n.

For a power series $f = \sum_{n=0}^{\infty} f_n X^n \in D[X]$, let ϕ_f be the function defined by

$$\phi_f(n) = \min\{i \mid f_i \notin M_n\}. \tag{21}$$

If $f \in M_n[X]$, then define $\phi_f(n) = \infty$. Let $\mathcal{U}$ be a nonprincipal ultrafilter on $\mathbb{N}$. For two power series f and g in $D[X]$, define

1. $f \sim_{\mathcal{U}} g$ if $\phi_f \sim_{\mathcal{U}} \phi_g$, i.e., there exist a positive integer m and a set $W \in \mathcal{U}$ such that $m\phi_f(n) \geq \phi_g(n)$ and $m\phi_g(n) \geq \phi_f(n)$ for each $n \in W$.
2. $f \gg_{\mathcal{U}} g$ if $\phi_f \gg_{\mathcal{U}} \phi_g$, i.e., for each positive integer m, there exists a set $U \in \mathcal{U}$ such that $\phi_f(n) > m\phi_g(n)$ for each $n \in U$.

Let $\mathcal{J}^* = \{f \in D[X] \mid 0 \gg_{\mathcal{U}} f \gg_{\mathcal{U}} X\}$. For each nonempty subset S of $\mathcal{J}^*$, let

$$P_S := \{f \in D[X] \mid f \gg_{\mathcal{U}} s \text{ for each } s \in S\}. \tag{22}$$

Using the fact that $\phi_{fg}(n) = \phi_f(n) + \phi_g(n)$ and $\phi_{f+g}(n) \geq \min\{\phi_f(n), \phi_g(n)\}$, we can show that P_S is a prime ideal of $D[X]$. Let $\mathcal{P}(\mathcal{J}^*)$ be the set of all prime ideals of the form P_S. Then $\mathcal{P}(\mathcal{J}^*)$ is a chain of prime ideals in $D[X]$. The following Sierpiński restrictions hold:

1. For each $f \in \mathcal{J}^*$, there are power series $k, h \in \mathcal{J}^*$ such that $k \gg_{\mathcal{U}} f \gg_{\mathcal{U}} h$.
2. For $f, h \in \mathcal{J}^*$ with $f \gg_{\mathcal{U}} h$ and $g \in D[X]$, if either $f \sim_{\mathcal{U}} g$ or $f \gg_{\mathcal{U}} g \gg_{\mathcal{U}} h$, then $g \in \mathcal{J}^*$.
3. If $\{f_n\}$ and $\{g_m\}$ are countable subsets of $\mathcal{J}^*$ with $f_n \gg_{\mathcal{U}} g_m$ for all f_n and g_m, then there is a power series $b \in \mathcal{J}^*$ such that $f_n \gg_{\mathcal{U}} b \gg_{\mathcal{U}} g_m$ for all f_n and g_m. In particular, for $f, g \in \mathcal{J}^*$ with $f \gg_{\mathcal{U}} g$, there is a power series $b \in D[X]$ such that $f \gg_{\mathcal{U}} b \gg_{\mathcal{U}} g$.
4. If $\{f_n\} \subseteq \mathcal{J}^*$ is a countably infinite sequence such that $f_{n+1} \gg_{\mathcal{U}} f_n$ for all n, then there is a power series $k \in \mathcal{J}^*$ such that $k \gg_{\mathcal{U}} f_n$ for all n.
5. If $\{f_n\} \subseteq \mathcal{J}^*$ is a countably infinite sequence such that $f_n \gg_{\mathcal{U}} f_{n+1}$ for all n, then there is a power series $h \in \mathcal{J}^*$ such that $f_n \gg_{\mathcal{U}} h$ for all n.

These Sierpiński restrictions play an important role in showing that the cardinality of $\mathcal{P}(\mathcal{J}^*)$ is at least $2^{\aleph_1}$.

Theorem 10. *If D is almost Dedekind domain that is not Dedekind, then* $\dim D[X] \geq 2^{\aleph_1}$, *i.e., there exists a chain of prime ideals in $D[X]$ with length $\geq 2^{\aleph_1}$.*

Note that $2^{\aleph_1}$ is the best lower bound for $\dim D[X]$ that one can get (under the continuum hypothesis $2^{\aleph_0} = \aleph_1$) since if D is countable, then $D[X]$ has the size $2^{\aleph_0} = \aleph_1$ and hence the power set of $D[X]$ has cardinality $2^{\aleph_1}$, which implies that every chain of prime ideals in $D[X]$ has length at most $2^{\aleph_1}$.

6 Problems

Two typical examples of non-SFT rings R are finite-dimensional nondiscrete valuation domains and non-Noetherian almost Dedekind domains. For these classes of rings R, dim $R[\![X]\!] \geq 2^{\aleph_1}$. We conjecture that this still holds for the larger class of rings, the class of non-SFT rings.

Conjecture 1. If R is a non-SFT ring, then dim $R[\![X]\!] \geq 2^{\aleph_1}$.

If R is a non-SFT ring, then dim $R[\![X]\!] = \infty$. However, the converse does not hold. At this moment, it seems to be very difficult to answer the following question.

Question 1. If dim $R[\![X]\!] = \infty$, then is it true that dim $R[\![X]\!] \geq 2^{\aleph_1}$?

7 Conclusion

While the possibilities for the Krull dimension of polynomial rings are completely determined, those for power series rings have not yet been answered in general. The Krull dimension of a power series ring $R[\![X]\!]$ is generally large (when it is either finite or infinite). For the finite case, this was shown by Kang and Park in [9]. For the infinite case, after the work of Arnold [1], a lot of constructions have been made ranging from using the famous Zorn's Lemma (to extend an ideal disjoint from a multiplicatively closed subset to a prime ideal) to picking up a prime ideal minimal over a given ideal or inventing nice tools that make use of zeros, multiplicities, infinite products of power series, ultrafilters, etc. We hope that these skillful techniques could be used in construction of (large chains of) prime ideals in $R[\![X]\!]$ in more general cases (e.g., when R is a non-SFT ring).

References

1. J.T. Arnold, Krull dimension in power series rings, Trans. Amer. Math. Soc. **177**, 299–304 (1973)
2. J.T. Arnold, Power series rings over Prüfer domains, Pacific J. Math. **44**, 1–11 (1973)
3. J. Coykendall, The SFT property does not imply finite dimension for power series rings. J. Algebra **256**(1), 85–96 (2002)
4. J. Coykendall, in *Progress on the Dimension Question for Power Series Rings*, ed. by J.W. Brewer, S. Glaz, W.J. Heinzer, B.M. Olberding, Multiplicative Ideal Theory in Commutative Algebra, (Springer, New York, 2006) pp. 123–135
5. P. Eakin, A. Sathaye, *Some Questions about the Ring of Formal Power Series*, in: Commutative Algebra (Fairfax Va., 1979), Lecture Notes in Pure and Appl. Math. vol. 68, (Dekker, New York, 1982) pp. 275–286
6. R. Gilmer, *Multiplicative Ideal Theory* (Marcel Dekker, New York, 1972)
7. M. Henriksen, On the prime ideals of the ring of entire functions. Pacific J. Math. **3**, 711–720 (1953)

8. B.G. Kang, M.H. Park, A localization of a power series ring over a valuation domain. J. Pure Appl. Algebra **140**(2), 107–124 (1999)
9. B.G. Kang, M.H. Park, *Krull Dimension of Mixed Extensions*. J. Pure Appl. Algebra **213**(10), 1911–1915 (2009)
10. B.G. Kang, M.H. Park, Krull-dimension of the power series ring over a nondiscrete valuation domain is uncountable. J. Algebra **378**, 12–21 (2013)
11. B.G. Kang, K.A. Loper, T.G. Lucas, M.H. Park, P.T. Toan, *The Krull dimension of power series rings over non-SFT rings*. J. Pure Appl. Algebra **217**(2), 254–258 (2013)
12. K.A. Loper, T.G. Lucas, Constructing chains of primes in power series rings. J. Algebra **334**, 175–194 (2011)
13. K.A. Loper, T.G. Lucas, Constructing chains of primes in power series rings II. J. Algebra Appl. **12**(1), 1250123, 30 (2013)
14. A. Seidenberg, A note on the dimension theory of rings. Pacific J. Math. **3**, 505–512 (1953)
15. A. Seidenberg, On the dimension theory of rings II. Pacific J. Math. **4**, 603–614 (1954)

Localizing Global Properties to Individual Maximal Ideals

Thomas G. Lucas

Abstract We consider three related questions. Q_1: Given a global property **G** of a domain R, what does a particular maximal ideal M of R "know" about the property with regard to the ideals $I \subseteq M$ and elements $t \in M$? Suppose **P** is such a property corresponding to **G**. Q_2: If each maximal ideal knows it has property **P**, does R have the corresponding global property **G**? Q_3: If at least one maximal ideal knows it has property **P**, does R have the global property **G**? We assume that if $I \subseteq M$, then M can tell when a particular element $t \in M$ is contained in I and when it isn't. Thus for a pair of ideals I and J contained in M, M knows when $I \subseteq J$. In addition, this allows M to understand the intersection of ideals it contains. In some cases, if a single maximal ideal knows **P**, then R will satisfy **G**. For example, there are such **P**s for **G** $\in$ {PIDs, Noetherian domains, Domains with ACCP, Domains with finite character}.

Keywords Integral domain • Maximal ideal

MSC(2010) classification: [2010]Primary: 13A15, 13G05

1 Introduction

Throughout this paper, R represents an integral domain that is properly contained in its quotient field K. We let $\text{Max}(R)$ denote the set of maximal ideals of R. For each nonzero ideal I we let $\text{Max}(R, I)$ denote the set of maximal ideals that contain I and let $\Theta(I) := \bigcap\{R_N \mid N \in \text{Max}(R)\backslash\text{Max}(R, I)\}$ with $\Theta(M) = K$ when I is contained in the Jacobson radical of R.

T.G. Lucas (✉)
Department of Mathematics and Statistics, University of North Carolina Charlotte,
Charlotte, NC 28223, USA
e-mail: tglucas@uncc.edu

M. Fontana et al. (eds.), *Commutative Algebra: Recent Advances in Commutative Rings,*   239
Integer-Valued Polynomials, and Polynomial Functions, DOI 10.1007/978-1-4939-0925-4_14,
© Springer Science+Business Media New York 2014

The original inspiration for this work was from the papers "Overrings of Prüfer domains. II" by Robert Gilmer and Bill Heinzer [6] and "Globalizing local properties of Prüfer domains" by Bruce Olberding [8]. In [4], Gilmer introduced the notion of an integral domain R being a #-domain meaning that for each pair of distinct sets of maximal ideals $\mathcal{M}$ and $\mathcal{N}$, $\bigcap\{R_M \mid M \in \mathcal{M}\} \neq \bigcap\{R_N \mid N \in \mathcal{N}\}$, the sets $\mathcal{M}$ and $\mathcal{N}$ are not required to be disjoint. Gilmer and Heinzer declared R to be a ##-domain if each overring of R is #. If R is a Prüfer domain, they showed that R is a ##-domain if and only if each nonzero prime P contains an invertible ideal I such that each maximal ideal that contains I also contains P. They also showed that for an individual prime Q of R, again with R Prüfer, R_Q does not contain $\Theta(Q)$ if and only if Q contains an invertible ideal I such that each maximal ideal that contains I also contains Q. A nonzero prime ideal P is (now) said to be *sharp* if R_P does not contain $\Theta(P)$ (see, for example, [2]). If P is sharp, then necessarily, $\Theta(P)$ properly contains R and for each $t \in \Theta(P) \backslash R_P$, P contains $(R :_R t)$ but no maximal ideal that is comaximal with P contains this ideal. Conversely, if there is an element $s \in K \backslash R$ such that P contains $(R :_R s)$ but no maximal ideal that is comaximal with P contains this ideal, then $s \in \Theta(P) \backslash R_P$. In the Prüfer case, each ideal of the form $(R :_R b)$ is an invertible (two-generated) ideal of R. In Theorems 3.13 and 3.14, we show that a maximal ideal M can "know" when it is sharp.

One of the main topics considered in [8] is characterizing when a Prüfer domain is h-local (each nonzero prime is contained in a unique maximal ideal and each nonzero nonunit is contained in only finitely many maximal ideals). Matlis proved that a domain R is h-local if and only if $\Theta(M) \cdot R_M = K$ for each maximal ideal M of R (see [7, Theorem 8.5]). We define a maximal ideal M to be h-local if $\Theta(M) \cdot R_M = K$ (see [3, Sect. 6.1]) and proceed to characterize when M is h-local based solely on what M can know about the nonzero ideals it contains [Theorem 3.12].

A more recent inspiration comes from the craft of writing fiction. Authors of fiction employ various types of narrators. Most common are some type of third person narrator (almost always singular) with some degree of omniscience. A somewhat distant second is a first person singular narrator. On rare occasions, an author makes use of a first person plural narrator. Rarer still is a second person narrator. A common exercise in a creative writing class is to rewrite a short story using a different narrator and then analyze the changes this makes to how the characters and story are perceived.

So what happens if we change the "narrator" of the story about a particular integral domain? Specifically, imagine we have found a collection of books, each written about a single domain where the narrator of each chapter is a single maximal ideal M of the domain, each maximal ideal writing one chapter. All that M can write about is what it knows about the ideals (of that particular domain) and the elements it contains, with absolutely no knowledge of the ideals and elements it does not contain (not even of the existence of such things). Our job is to read one of the books and deduce as many global properties for that particular domain from what its maximal ideals have written. An underlying idea is that we, the readers, can deduce more about the domain than the authors/narrators $= \mathrm{Max}(R)$ (at least individually) seem to know.

A basic assumption is that each maximal ideal M of R not only knows (in some way) what it means for a nonempty subset $X \subseteq M$ to be an ideal of R (and can write about it). In addition, we make a pair of related assumptions with regard to what a particular maximal ideal can know about containment properties among the ideals and elements it contains. Also, we assume M has at least some ability to factor products.

(A1) If $t \in M$ and $I \subseteq M$ is an ideal of R, then M knows when t is in I and when t is not in I.

(A2) If I and J are ideals contained in M, then M knows when I is contained in J and when I is not contained in J.

(A3) M can "factor" nonzero products of principal ideals provided each factor is contained in M: if $(0) \neq tR \subseteq aR \subseteq M$, then M knows if there is a $b \in M$ such that $tR = aRbR$ or that no such b exists.

Later we add the assumption that each $M \in \text{Max}(R)$ can distinguish between finitely many elements (ideals) and infinitely many elements (ideals).

(A4) M can tell the difference between "finitely many things" and "infinitely many things."

Using only (A1) and (A2), it is possible to discover when R is a PID [Theorem 2.2]. Also, each maximal ideal M can understand what is meant by intersecting a family of ideals contained in M – such an intersection $\bigcap J_\alpha$ is simply the ideal J that is both contained in each J_α and contains each ideal I that is contained in each J_α [Theorem 2.1]. When we add (A3), it is possible to discover that the domain in question is a Prüfer domain [Theorem 2.8]. We can also discover when R is local [Theorem 2.9]. Perhaps somewhat surprisingly, it is possible for us to determine when a given nonzero ideal Q is primary (or not) [prime, or not] based solely on what a single maximal ideal that contains Q knows about Q (and some related ideals) [see Theorems 3.8 and 3.9]. With all four, we can discover when R is Noetherian, when it satisfies ACCP (ascending chain condition on principal ideals), when it has finite character, and even when it is h-local [Theorems 3.1, 3.2, 3.6 and 3.10, respectively].

2   Using (A1), (A2) and (A3)

For this section we assume only (A1), (A2) and (A3) for each maximal ideal $M \in \text{Max}(R)$.

(A1) If $t \in M$ and $I \subseteq M$ is an ideal of R, then M knows when t is in I and when t is not in I.

(A2) If I and J are ideals contained in M, then M knows when I is contained in J and when I is not contained in J.

(A3) If $(0) \neq tR \subseteq aR \subseteq M$, then M knows if there is a $b \in M$ such that $tR = aRbR$ or that no such b exists.

For a nonzero ideal $I \subseteq M$, M recognizes a generating set for I as a nonempty set $X \subseteq I$ such that each ideal $J \subseteq M$ that contains X also contains I.

Theorem 2.1. *Let M be a maximal ideal of R.*

1. *If I is a nonzero ideal contained in M, then I is principal if and only if there is an element $t \in I$ such that each ideal $J \subseteq M$ that contains t also contains I.*
2. *If $\{J_\alpha\}$ is a family of ideals with each $J_\alpha \subseteq M$, then M knows $\bigcap J_\alpha$ as the only ideal that is both contained in each J_α and contains every ideal I that is contained in each J_α.*

Proof. Let I be a nonzero ideal that is contained in M. If $t \in I$ is such that each ideal $J \subseteq M$ that contains t also contains I, then we know that $I = tR$.

For the second statement, let $J = \bigcap J_\alpha$. Since J is an ideal and each J_α is contained in M, M would know that each J_α contains J by (A2). Also by (A2), if $I \subseteq J_\alpha$ for each α, then certainly $I \subseteq M$ and so M would know $I \subseteq J$ by (A2). $\square$

Theorem 2.2. *The following are equivalent for the domain R.*

1. *R is a PID.*
2. *For each maximal ideal $M \in \mathrm{Max}(R)$, M knows that each ideal $I \subseteq M$ contains an element t such that each ideal $J \subseteq M$ that contains t also contains I.*
3. *There is a maximal ideal $M \in \mathrm{Max}(R)$ that knows that for each ideal $I \subseteq M$, there is an element $t \in I$ such that each ideal $J \subseteq M$ that contains t also contains I.*

Proof. It is clear that (1) implies (2) and that (2) implies (3).

Suppose there is a maximal ideal $M \in \mathrm{Max}(R)$ such that for each ideal $I \subseteq M$, there is an element $t \in I$ such that each ideal $J \subseteq M$ that contains t also contains I. We know that tR is one such ideal. Hence $I = tR$ and at least each ideal that is contained in M is principal. For a generic ideal A of R, simply multiply A by a nonzero $s \in M$ to obtain an ideal $sA \subseteq M$. We have $sA = xR$ for some $x \in M$ and from this we see that $A = (x/s)R$ is principal. Hence R is a PID when at least one maximal ideal satisfies (3). $\square$

Using the factoring property (A3), it is also possible for M to know enough that we can interpret this knowledge as R is integrally closed. The key is that for nonzero nonunits $a, b \in R$, a/b is integral over R if and only if there is a positive integer n such that $(a/b)a^n \in I$ where $I = a^n R + a^{n-1}bR + \cdots + b^n R$.

Theorem 2.3. *Let M be a fixed maximal ideal of R and let $a, b \in M \setminus \{0\}$.*

1. *$a/b \in R$ if and only if M knows $a \in bR$.*
2. *$a/b \in K \setminus R$ is integral over R if and only if M knows $a \notin bR$ and there is a positive integer n such that $a^{n+1} \in I$ where $I = a^n bR + a^{n-1}b^2 R + \cdots + b^{n+1}R$.*

Proof. It is clear that a is in bR if and only if $a/b \in R$.

Suppose $a/b \in K \setminus R$. Then a is not in bR. For each positive integer n, let $I_n = a^n bR + a^{n-1}b^2 R + \cdots + b^{n+1}R$ and $A_n = a^n R + a^{n-1}bR + \cdots + b^n R$. For M, I_n is the ideal that contains the set $X_n := \{a^n b, a^{n-1}b^2, \ldots, b^{n+1}\}$ and each ideal $J \subseteq M$ that contains X_n contains I_n.

It is trivial that $(a/b)a^i b^j = a^{i+1}b^{j-1} \in A_n$ for $1 \leq j \leq n$ and $i + j = n$. Also $a^{n+1} \in I_n$ if and only if $(a/b)a^n \in A_n$. Thus for integrality, all we need to consider is whether there is an n such that $a^{n+1} \in I_n$. Hence the following are equivalent.

(i) a/b integral over R.
(ii) There is a positive integer n such that $(a/b)A_n \subseteq A_n$.
(iii) There is a positive integer n such that $a^{n+1} \in I_n$. $\square$

We have two corollaries. In the first we give necessary and sufficient conditions for R to fail to be integrally closed and in the second, necessary and sufficient conditions for R to be integrally closed.

Corollary 2.4. *The following are equivalent for an integral domain R that is not a field.*

1. *R is not integrally closed.*
2. *For each maximal ideal M, there is a pair of nonzero elements $a, b \in M$ such that aR is not contained in bR, and there is a positive integer n such that $a^{n+1} \in I_n$ where $I_n = a^n bR + a^{n-1}b^2 R + \cdots + b^{n+1}R$.*
3. *There is a maximal ideal M and a pair of nonzero elements $a, b \in M$ such that aR is not contained in bR, and there is a positive integer n such that $a^{n+1} \in I_n$ where $I_n = a^n bR + a^{n-1}b^2 R + \cdots + b^{n+1}R$.*

Corollary 2.5. *The following are equivalent for an integral domain R that is not a field.*

1. *R is integrally closed.*
2. *For each maximal ideal M, if $a, b \in M \setminus \{0\}$ are such that a^{n+1} is in the ideal $I_n = a^n bR + a^{n-1}b^2 R + \cdots + b^{n+1}R$ for some positive integer n, then $a \in bR$.*
3. *The statement in (2) holds for at least one maximal ideal M of R.*

For each nonzero nonunit x of R, let $\mathscr{P}(x) = \{yR \mid (0) \subsetneq yR \subseteq xR\}$ be the set of nonzero principal ideals that are contained in xR. For each maximal ideal M that contains x, the set $\mathscr{P}(x)$ can be split into two disjoint sets, $\mathscr{F}_M(x) = \{zR \mid zR = xRwR$ for some $w \in M\}$ and $\mathscr{N}_M(x) = \{tR \mid tR \subseteq xR$ and there is no $b \in M$ such that $tR = xRbR\}$. By (A3), M knows which ideals are of the type in $\mathscr{F}_M(x)$ and which are of the type in $\mathscr{N}_M(x)$. In particular, it knows that $x^2 R \in \mathscr{F}_M(x)$ and $xR \in \mathscr{N}_M(x)$.

Theorem 2.6. *Let M be a maximal ideal of R and let $I = xR + yR$ where x and y are nonzero elements of M. Then M does not contain II^{-1} if and only if M knows at least one of $\mathscr{N}_M(x) \bigcap \mathscr{P}(y)$ and $\mathscr{N}_M(y) \bigcap \mathscr{P}(x)$ is nonempty.*

Proof. Suppose M does not contain II^{-1}. Then without loss of generality we may assume there is an element $f \in I^{-1}$ such that $fx \notin M$. Let $a = fx$ and $b = fy$;

both of these elements are in R. Hence the ideal $bxR = ayR$ is contained in both $\mathcal{P}(x)$ and $\mathcal{P}(y)$. Since $a \in R \backslash M$, $ayR \in \mathcal{N}_M(y) \bigcap \mathcal{P}(x)$.

Suppose $\mathcal{N}_M(x) \bigcap \mathcal{P}(y)$ is nonempty, equivalently, M knows there is an ideal that is in both $\mathcal{N}_M(x)$ and $\mathcal{P}(y)$. Each ideal in $\mathcal{N}_M(x)$ has the form $wR = dxR$ for some $d \in R \backslash M$. If wR is also in $\mathcal{P}(y)$, then $wR = fyR$ for some $f \in R$. It follows that $dx = gy$ for some $g \in R$. Since $dx/y = g \in R$, we see that $d/y \in I^{-1}$ is such that $(d/y)y = d \in II^{-1} \backslash M$. $\qquad \square$

Corollary 2.7. *Let M be a maximal ideal of the domain R. If $I = xR + yR$ is a nonzero ideal such that M is the only maximal that contains I, then I is invertible if and only if M knows at least one of $\mathcal{P}(x) \bigcap \mathcal{N}_M(y)$ and $\mathcal{P}(y) \bigcap \mathcal{N}_M(x)$ is nonempty.*

Recall that if each nonzero two-generated ideal of a domain is invertible, then the domain is Prüfer. From the previous theorem we have a way for the maximal ideals of R to provide us enough information to conclude that R is a Prüfer domain. We say that M knows **Pru1** if for $x, y \in M \backslash \{0\}$, M knows at least one of $\mathcal{P}(x) \bigcap \mathcal{N}_M(y)$ and $\mathcal{P}(y) \bigcap \mathcal{N}_M(x)$ is nonempty. An alternate characterization of a Prüfer domain is an integrally closed domain such that for each pair of elements $a, b \in R \backslash \{0\}$, $ab \in a^2 R + b^2 R$ [5, Theorem 24.3]. We say that M knows **Pru2** if for each pair of elements $a, b \in M \backslash \{0\}$, $ab \in a^2 R + b^2 R$ and whenever there is a positive integer n such that $a^{n+1} \in a^n bR + a^{n-1}b^2 R + \cdots + b^{n+1} R$, then it is also the case that $a \in bR$.

Theorem 2.8. *The following are equivalent for an integral domain R.*

1. *R is Prüfer domain.*
2. *Each maximal ideal of R knows **Pru1** holds.*
3. *Each maximal ideal of R knows **Pru2** holds.*
4. *There is a maximal ideal of R that knows **Pru2** holds.*

Proof. If R is a Prüfer domain, then each two-generated ideal is invertible. Thus by Theorem 2.6 each maximal ideal of R knows **Pru1**. Conversely if each maximal ideal knows **Pru1**, then no maximal ideal contains II^{-1} when I is a nonzero two-generated ideal. It follows that each such ideal I is invertible and thus R is a Prüfer domain.

By Corollary 2.5, if at least one maximal ideal knows **Pru2**, then R is integrally closed. If R is integrally closed, then a two-generated ideal $rR + sR$ is invertible if and only if $rs \in r^2 R + s^2 R$ [5, Proposition 24.2]. Thus if M knows **Pru2**, then R is integrally closed and each two-generated ideal contained in M is invertible. For a two-generated ideal $rR + sR$, simply multiply by a nonzero $c \in M$ and note that we will then have $cr, cs \in M$ with $(cr)(cs) \in (cr)^2 R + (cs)^2 R$. Simply cancel c^2 to obtain $rs \in r^2 R + s^2 R$. $\qquad \square$

The factoring property can also be used to establish that a particular domain has a unique maximal ideal.

Theorem 2.9. *For an integral domain R, R is local if and only if at least one $M \in$ Max(R) knows that for all (some) $x \in M \setminus \{0\}$, xR is the only ideal of the type in $\mathcal{N}_M(x)$.*

Proof. Suppose R is local. Then it has a unique maximal ideal M. In this case if $x, y \in M \setminus \{0\}$ with $yR \subsetneq xR$, then $y = xz$ for some nonunit z which must be in M. Hence xR is the only ideal of the type in $\mathcal{N}_M(x)$.

For the converse, suppose M is a maximal ideal and there is an element $z \in M \setminus \{0\}$ such that zR is the only ideal of the type in $\mathcal{N}_M(z)$. If $w \in R \setminus M$, the principal ideal wzR is in $\mathcal{N}_M(z)$ and thus $wzR = zR$. Since $z \neq 0$, w is a unit of R and we have R is local. □

3 Finiteness

In this final section we add the "finite" assumption to the other three. We do not assume M can count, only that it can distinguish between finitely many things and infinitely many things where the "things" are either elements or ideals.

(A4) If X is a set of elements that are contained in M, then M knows when the set is finite and when it is not finite. It can make a similar distinction for any collection of ideals it contains.

Property (A4) gives us a way to see when R is Noetherian based on the knowledge of a single maximal ideal.

Theorem 3.1. *The following are equivalent for a domain R that is not a field.*

1. *R is Noetherian.*
2. *For each maximal ideal M, M knows that for each ideal I contained in M, there is a finite set of elements $X \subseteq I$ such that each ideal $J \subseteq M$ that contains X contains I.*
3. *For each maximal ideal M, if $I_1 \subseteq I_2 \subseteq I_3 \subseteq \cdots$ is a chain of ideals contained in M, then M knows the chain is finite.*
4. *There is a maximal ideal M such that M knows that for each ideal I contained in M, there is a finite set of elements $X \subseteq I$ such that each ideal $J \subseteq M$ that contains X contains I.*
5. *There is a maximal ideal M such that if $I_1 \subseteq I_2 \subseteq I_3 \subseteq \cdots$ is a chain of ideals contained in M, then M knows the chain is finite.*

Proof. The equivalence of (1), (2) and (3) is clear. Also, (2) implies (4), and (3) implies (5). The implications (4) $\Rightarrow$ (2) and (5) $\Rightarrow$ (3) follow from cancellation. For an ideal J that is comaximal with M, choose any nonzero $t \in M$. Then $tJ \subseteq M$ is a nonzero ideal. If tJ is finitely generated, say by $ta_1, ta_2, \ldots, ta_n$, then by cancellation, J is generated by $a_1, a_2, \ldots, a_n$. Similarly for a chain of ideals not necessarily contained in M, multiplying each ideal by t gives a chain of ideals contained in M. The original chain stabilizes in R if and only if the new chain in M stabilizes.   □

A similar characterization can be given for ACCP (ascending chain condition on principal ideals). For nonzero principal ideals $x_1 R \subseteq yR$, $xR = yR$ if and only if $txR = tyR$ for each nonzero $t \in R$. Thus, if each ascending chain of principal ideals contained in a particular maximal ideal M stabilizes, each ascending chain of principal ideals stabilizes in R.

Theorem 3.2. *The following are equivalent for an integral domain R that is not a field.*

1. *R satisfies ACCP.*
2. *Each maximal ideal M knows that if $x_1 R \subseteq x_2 R \subseteq x_3 R \subseteq \cdots$ is an ascending chain of principal ideals contained in M, then the chain is finite.*
3. *There is a maximal ideal M that knows each ascending chain of principal ideals contained in M is finite.*

We next consider finite character. The goal is to describe how a given maximal ideal can recognize what we know as finite character without mentioning any other maximal ideals.

First we introduce the idea of a "M-closed set" and then an "(I, M)-complete set." The latter will allow us to provide a way for M to know a property for a particular nonzero ideal $I \subseteq M$ that we can interpret as I being in only finitely many maximal ideals.

For a given maximal ideal M, let $\mathscr{X}$ be a finite list (set) of ideals each contained in M. We say that $\mathscr{X}$ is M-*closed* if M knows the following about $\mathscr{X}$.

1. M is included in the list $\mathscr{X}$.
2. The ideal J that is contained in each ideal in $\mathscr{X}$ and contains each ideal that is contained in each ideal in $\mathscr{X}$, is an ideal in $\mathscr{X}$.
3. Each ideal that contains J and is contained in M is in the list $\mathscr{X}$.
4. For comparable ideals $A \subsetneq B$ where both A and B are in $\mathscr{X}$, if $d \in B \backslash A$, then no power of d is in A.

Theorem 3.3. *Let M be a maximal ideal of a domain R. If $\mathscr{X} = \{J_1, \ldots J_n\}$ is M-closed with $M = J_n$ and $J_1 = \bigcap \{J_i \mid 1 \leq i \leq n\}$, then there is an integer $k \geq 0$ such that $n = 2^k$ and there is a finite set of k maximal ideals $\mathscr{Y} = \{N_1, N_2, \ldots, N_k\}$ such that (i) M is not in $\mathscr{Y}$, (ii) $J_1 = \bigcap \{N_i \mid 0 \leq i \leq k\}$ where $N_0 = M$, (iii) each ideal in $\mathscr{X}$ is the intersection of M and finitely many maximal ideals in the set $\mathscr{Y}$, and (iv) each finite intersection of maximal ideals in the set $\mathscr{Y}$ with M is an ideal in $\mathscr{X}$.*

Proof. Suppose $\mathscr{X} = \{J_1, \ldots J_n\}$ is an M-closed set with $M = J_n$ and $J_1 = \bigcap \{J_i \mid 1 \leq i \leq n\}$. We first show that each minimal prime of J_1 is a maximal ideal of R. For this suppose P is a prime ideal that is properly contained in M, then for each $t \in M \backslash P$, the family $\{t^j R + P \mid j \geq 1\}$ is an infinite set of distinct ideals such that each properly contains P. As each ideal that contains J_1 is listed in $\mathscr{X}$ (and $\mathscr{X}$ is finite), M is minimal over J_1. If $N \neq M$ is another maximal ideal that contains J_1 and P' is a prime ideal that is properly contained in N, then for each $s \in N \setminus P'$, the family $\{s^j R + P' \mid j \geq 1\}$ is an infinite set of distinct ideals such that each properly contains P'. If $P' \subseteq M$, then P' does not contain J_1. In the case $P' \not\subseteq M$, the family $\{M \cap (s^j R + P') \mid j \geq 1\}$ is an infinite set of distinct ideals each properly containing $M \cap P'$. Hence P' does not contain J_1. Therefore each minimal prime of J_1 is a maximal ideal of R.

If $Q_1, Q_2, \ldots, Q_m$ are maximals ideal that contain J_1, then $J_1 \subseteq Q := \bigcap \{Q_i \mid 1 \leq i \leq m\}$. Hence $J_1 \subseteq M \cap Q$ and we have $M \cap Q \in \mathscr{X}$. As $\mathscr{X}$ is finite and each such Q is a finite intersection of maximal ideals, only finitely many maximal ideals contain J_1. Thus we have a finite set $\mathscr{Y}' = \{N_0, N_1, \ldots, N_k\}$ which consists of all maximal ideals that contain J_1. Without loss of generality we may assume $N_0 = M$, in which case $\mathscr{Y} = \{N_1, N_2, \ldots, N_k\}$. The set $\{0, 1, 2, \ldots, k\}$ has 2^k subsets that contain 0. Each such subset A gives rise to a distinct radical ideal $J_A = \bigcap \{N_i \mid i \in A\}$ that is in $\mathscr{X}$ (since each of these ideals contains J_1 and is contained in M). In particular the intersection of all the N_is is in $\mathscr{X}$. This ideal is simply the radical of J_1. Note that if $A \subseteq B$, then the only ideals between J_A and J_B are the ideals of the form J_C where $A \subseteq C \subseteq B$. Moreover $J_B \subsetneq J_A$ if and only if $A \subsetneq B$.

To complete the proof we show that $J_1 = \bigcap \{N_i \mid 0 \leq i \leq k\}$. From above, we have that $\mathscr{Y}'$ is the complete set of minimal primes of J_1. So for each $b \in \bigcap \{N_i \mid 0 \leq i \leq k\}$, there is a positive integer j such that $b^j \in J_1$. But this implies $b \in J_1$ since $\mathscr{X}$ is M-closed. Therefore $J_1 = \bigcap \{N_i \mid 0 \leq i \leq k\}$ and $\mathscr{X}$ contains 2^k ideals, each an ideal of the type J_A for some (unique) subset A of $\{0, 1, \ldots, k\}$ that contains 0. $\qquad\square$

For a nonzero ideal $I \subseteq M$ we say that a finite list $\mathscr{X}$ is (I, M)-*complete* if M knows the following about $\mathscr{X}$:

1. $\mathscr{X}$ is M-closed,
2. $I \subseteq A$ for each A in the list $\mathscr{X}$, and
3. if $I \subseteq H \subseteq M$, there is a (unique) A in the list $\mathscr{X}$ such that $H \subseteq A$ and $H \subseteq B$ for some B in the list $\mathscr{X}$ implies $A \subseteq B$.

The notion of an (I, M)-complete list provides a way for M to know enough for us to conclude that I is contained in only finitely many maximal ideals.

Theorem 3.4. *Let $I \subseteq M$ be nonzero ideals with M maximal. Then I is contained in only finitely many maximal ideals if and only if M knows there is an (I, M)-complete list $\mathscr{X}$ such that for each (I, M)-complete list $\mathscr{Y}$, each ideal in the list $\mathscr{Y}$ is also in the list $\mathscr{X}$.*

Proof. If $N_0 = M, N_1, \ldots, N_k$ is the complete list of maximal ideals that contain I, then the list $\mathscr{X}$ of ideals $\{J_A = \bigcap \{N_i \mid i \in A\} \mid A \subseteq \{0, 1, \ldots, k\}$ with $0 \in A\}$ is (I, M)-complete and no (I, M)-complete list contains an ideal not in the list $\mathscr{X}$.

For the converse, if X is an (I, M)-complete list, then there is a finite set of maximal ideals $\mathcal{N} = \{N_0 = M, N_1, \ldots, N_k\}$ such that $\mathcal{X} = \{J_A = \bigcap\{N_i \mid i \in A\} \mid A \subseteq \{0, 1, \ldots, k\}$ with $0 \in A\}$. If $\mathcal{Y}$ is an (I, M)-complete list for which each ideal in $\mathcal{Y}$ is in $\mathcal{X}$, then the underlying set of maximal ideals corresponding to $\mathcal{Y}$ is a subset of $\mathcal{N}$. It follows that if each (I, M)-complete list is a sublist of $\mathcal{X}$, then $\mathcal{N}$ consists of the complete set of maximal ideals that contains I.   □

While we have not assumed M can count, at least it can recognize when there is only one of something.

Corollary 3.5. *For a nonzero ideal I contained in a maximal ideal M, M is the only maximal ideal that contains I if and only if M knows that $\{M\}$ is the only (I, M)-complete set.*

Recall that a domain R is h-local if it has finite character and each nonzero prime ideal is contained in a unique maximal ideal. By the previous theorem, it is possible for use to deduce that R has finite character based on the knowledge of single maximal ideal.

Theorem 3.6. *The following are equivalent for a domain R.*

1. *R has finite character.*
2. *For each maximal ideal M, M knows that for each nonzero ideal I it contains, there is an (I, M)-complete list $\mathcal{X}$ such that each (I, M)-complete list is a sublist of $\mathcal{X}$.*
3. *There is a maximal ideal M that knows that for each nonzero ideal $I \subseteq M$, there is an (I, M)-complete list $\mathcal{X}$ such that each (I, M)-complete list is a sublist of $\mathcal{X}$.*

Proof. The implications (1) $\Rightarrow$ (2), and (2) $\Rightarrow$ (3) follow easily from Theorem 3.4. To see that (3) implies (1), simply start with a nonzero ideal B of R. Then for each nonzero $t \in M$, tB is a nonzero ideal contained in M. By Theorem 3.4, tB and thus B are each contained in only finitely many maximal ideals. Hence R has finite character.   □

By Corollary 3.5, if we know that P is a prime ideal, then we can determine when it is contained in a unique maximal ideal. The challenge in getting a characterization for h-local domains based solely on what the maximal ideals know about the ideals of R is to find a way for a given maximal ideal to know enough about a particular nonzero ideal that we can tell exactly when the ideal is prime.

We say that a nonzero subideal Q of a maximal ideal M is an M-*prime* if for elements $x, y \in M$, $xy \in Q$ implies at least one of x and y is in Q. Certainly a prime ideal that is contained in M is M-prime, but so is $M \bigcap N$ for any maximal ideal $N \neq M$. We define MP-*primary ideals* in a similar manner. A nonzero ideal $J \subseteq M$ is an MP-primary ideal if there is an M-prime P such that (i) $J \subseteq P \subseteq M$, (ii) for each $t \in P$, there is a positive integer n such that $t^n \in J$, and (iii) for elements $a, b \in M$, if $ab \in J$ and $a \notin P$, then $b \in J$. As with M-primes, it is possible to have an MP-primary ideal that is neither a primary ideal of R nor an M-prime. For example, if I is a proper N-primary ideal where $N \neq M$ is a maximal ideal, then $J = M \bigcap I$ is a proper MP-primary ideal for the M-prime $P = M \bigcap N$. Clearly J is neither a primary ideal of R nor an M-prime. Note that both definitions are such that M can know when an ideal is an M-prime and when an ideal is an MP-primary ideal.

For a nonzero ideal $I \subseteq M$, we let $I_{(M)} = \{x \in M \mid yR \in \mathcal{N}_M(x)$ for some $y \in I\}$ and $\sqrt[M]{I} = \{x \in M \mid yR \in \mathcal{N}_M(x^n)$ for some $y \in I$ and positive integer $n\}$. The first set is the same as $I R_M \bigcap R$ and the second is the same as $\sqrt{I R_M} \bigcap R = \sqrt{I} R_M \bigcap R$. The advantage in using these definitions is that M can understand both ideals (without needing to know what "localization" is).

Theorem 3.7. *Let I be a nonzero ideal that is contained in the maximal ideal M. Then $I_{(M)} = I R_M \bigcap R$ and $\sqrt[M]{I} = \sqrt{I R_M} \bigcap R$.*

Proof. Suppose $x \in I_{(M)}$ and let $y \in I$ be such that $yR \in \mathcal{N}_M(x)$. Then $yR = xRaR$ for some $a \in R \backslash M$. Without loss of generality we may assume $y = xa$. It follows that $x = y/a \in I R_M \bigcap R$.

The argument is reversible. If $z \in I R_M \bigcap R$, then we have $z = w/b$ for some $w \in I$ and $b \in R \backslash M$. It follows that $wR = zRbR \in \mathcal{N}_M(z)$. Hence $z \in I_{(M)}$.

For each element s of $\sqrt[M]{I}$, there is a positive integer n such that $s^n \in I_{(M)}$. Thus $s^n \in I R_M$ and we have $s \in \sqrt{I R_M} \bigcap R$. Conversely, if $t \in \sqrt{I R_M} \bigcap R$, then $t^n \in I R_M$ for some n and we have $t^n \in I_{(M)}$. It follows that $t \in \sqrt[M]{I}$. □

Theorem 3.8. *Let M be a maximal ideal of a domain R. Then the following are equivalent for a nonzero ideal $Q \subseteq M$.*

1. *Q is a prime ideal of R.*
2. *(M knows) $Q = Q_{(M)}$ is an M-prime.*
3. *(M knows) $Q = \sqrt[M]{Q}$ is an M-prime.*

Proof. In general we have $Q \subseteq Q_{(M)} \subseteq \sqrt[M]{Q}$. Thus $Q = \sqrt[M]{Q}$ implies $Q = Q_{(M)}$. Also, if Q is a prime ideal, then $\sqrt{Q R_M} = Q R_M$ and $Q R_M \bigcap R = Q$. Thus it suffices to show that (1) and (2) are equivalent.

If Q is a prime ideal of R, then $Q R_M \bigcap R = Q$ and certainly Q is M-prime.

For the converse assume Q is an M-prime and $Q = Q_{(M)}$. To see that Q is a prime ideal of R, suppose $x, y \in R$ are such that $xy \in Q$. Since M is a maximal ideal, at least one of x and y is contained in M. If both are, we simply use that Q is an M-prime to get that at least one of x and y is contained in Q. Thus we may assume $x \in M$ and $y \in R \backslash M$. Since $w = xy \in Q$, $wR \in \mathcal{N}_M(x)$ and so we have $x \in Q_{(M)} = Q$. Therefore Q is a prime ideal of R. □

Theorem 3.9. *Let M be a maximal ideal of a domain R. Then the following are equivalent for a nonzero ideal $Q \subseteq M$.*

1. *Q is a primary ideal of R.*
2. *M knows there is an M-prime P such that Q is MP-primary and $Q = Q_{(M)}$.*
3. *M knows there is an M-prime $P \subseteq M$ such that Q is MP-primary and $P = P_{(M)}$.*

Proof. It is clear that (1) implies both (2) and (3).

To see that (2) implies (1), assume there is an M-prime P such that $Q = Q_{(M)}$ is MP-primary. To see that Q is a primary ideal of R, let $x, y \in R$ be such that $xy \in Q$. If $x \notin M$, then we have $y \in Q_{(M)} = Q$. Similarly $x \in Q$ if $y \notin M$. Thus we may assume both x and y are in M. Since P is an M-prime, at least one of x and y is in P. If both are, then a power of each is in Q since Q is MP-primary. On the other hand, if $x \in P$ and $y \in M \backslash P$, then $x \in Q$. Thus Q is a primary ideal of R.

To finish the proof we show (3) implies (1). For this, we again assume Q is MP-primary, but replace the assumption that $Q = Q_{(M)}$ by $P = P_{(M)}$. By Theorem 3.8, P is a prime ideal of R. We also have $\sqrt{Q} = P$. If $P = M$, then Q is M-primary as its radical is a maximal ideal.

On the other hand, if P is properly contained in M, then there is an element $m \in M \backslash P$. Let $x, y \in R$ be such that $xy \in Q$. Then $xy \in P$. If $x \notin M$, then $y \in P$. We also have $mx \in M \backslash P$ with $mx \notin P$ and $(mx)y \in Q$. Thus $y \in Q$. Similarly, $x \in Q$ if $y \notin M$.

Finally if both $x, y \in M$, then we use the definition of MP-primary to conclude that either some power of both x and y is in Q or at least one of the two is in Q. Therefore Q is a primary ideal of R (with $\sqrt{Q} = P$). $\square$

We now have enough to characterize when R is h-local based entirely on what its maximal ideals know.

Theorem 3.10. *Let R be a domain. Then R is h-local if and only if each maximal ideal M knows the following:*

1. *for each nonzero ideal $I \subseteq M$, there is an (I, M)-complete list $\mathscr{X}$ such that each (I, M)-complete list is a sublist of $\mathscr{X}$, and*
2. *if $P \subseteq M$ is a nonzero ideal such that $P = P_{(M)}$ and P is an M-prime, then $\{M\}$ is the only (P, M)-complete list.*

Proof. If R is h-local and M is a maximal ideal, then M knows (1) for each nonzero ideal it contains. If $P \subseteq M$ is a nonzero prime ideal of R, then $P = P_{(M)}$ is an M-prime (Theorem 3.8) and $\{M\}$ is the only (P, M)-complete set (since M is the only maximal ideal that contains P).

Next assume each maximal ideal knows that both (1) and (2) hold for the ideals it contains.

Let Q be a nonzero prime ideal of R and let M be a maximal ideal that contains Q. Then $Q = Q_{(M)}$ and Q is an M-prime. Hence M knows that $\{M\}$ is the only (Q, M)-complete set. We deduce that M is the only maximal ideal that contains Q.

For finite character, we simply apply Theorem 3.6. Thus R is h-local. $\square$

We will make use of the following lemma in the proof of the next theorem.

Lemma 3.11. *Let I be a nonzero ideal of a domain R. Then for each maximal ideal M containing I, each minimal prime of $I_{(M)}$ is contained in M.*

Proof. Let M be a maximal ideal that contains I and let P be a minimal prime of $I_{(M)}$. Then for each $q \in P$, there is a positive integer n and an element $a \in R \backslash P$ such that $aq^n \in I_{(M)}$. If q is not in M, then we have $a \in IR_M \cap R = I_{(M)}$, a contradiction. $\square$

Recall from above that R is h-local if and only if $\Theta(M) \cdot R_M = K$ for each maximal ideal M of R [7]. Based on this characterization, the notion of an h-local maximal ideal was introduced in [3] as a maximal ideal M of R such that $\Theta(M) \cdot R_M = K$. Next we show that M can know enough that we can deduce exactly when it is h-local. The characterization below is related to the fact that R is h-local if and only if each nonzero prime ideal contains an invertible ideal that is contained in a unique maximal ideal (see [1, Corollary 3.4]).

Theorem 3.12. *Let M be a maximal ideal of a domain R. Then M is h-local if and only if M knows the following:*

(a) *for each nonzero ideal $I \subseteq M$, $\{M\}$ is the only $(I_{(M)}, M)$-complete set, and*

(b) *if P is a nonzero M-prime such that $P = P_{(M)}$, then there is a pair of nonzero elements $x, y \in P$ such that $\{M\}$ is the only (J, M)-complete set of the ideal $J = xR + yR$ and at least one of $\mathscr{P}(x) \cap \mathscr{N}_M(y)$ and $\mathscr{P}(y) \cap \mathscr{N}_M(x)$ is nonempty.*

Proof. Assume M knows both (a) and (b). If $P \subseteq M$ is a nonzero prime, then $P = P_{(M)}$ is an M-prime (Theorem 3.8), so M is the only maximal ideal that contains P. For the ideal $J = xR + yR$, the fact that $\{M\}$ is the only (J, M) complete set tells us that M is the only maximal ideal that contains J. By Corollary 2.7, we see that J is invertible.

By way of contradiction, suppose $S = \Theta(M) \cdot R_M$ is properly contained in K and let Q be a nonzero prime ideal of S. Then $Q \cap R_M$ is a nonzero prime ideal of R_M, necessarily contained in M. Hence $P = Q \cap R$ is a nonzero prime ideal of R that is contained in M. By the above, there is a two-generated invertible ideal $B = aR + bR$ that is contained in P such that M is the only maximal ideal that contains B. Hence B^{-1} is contained in R_N for each maximal ideal $N \in \text{Max}(R) \backslash \{M\}$. It follows that $B^{-1} \subseteq \Theta(M)$, but this implies $1 \in B\Theta(M) \subseteq BS \subseteq Q$, a contradiction. Hence M is h-local.

For the converse assume M is h-local. There is nothing to prove if M is the only maximal ideal of R so we may assume R is not local. We first show that if P is a nonzero prime ideal that is contained in M, then M is the only maximal ideal that contains P. By way of contradiction assume $N \neq M$ is another maximal ideal that contains P. Then $R_P \supseteq R_N$ and $R_P \supseteq R_M$ implies $R_P \supseteq R_N \cdot R_M \supseteq \Theta(M) \cdot R_M = K$, which is impossible since P is not zero. Thus M is the only maximal ideal that contains P.

Next we show that $P\Theta(M) = \Theta(M)$. By way of contradiction, assume $P\Theta(M)$ is a proper ideal of $\Theta(M)$ and let Q' be a prime of $\Theta(M)$ that contains $P\Theta(M)$. Then $Q = Q' \cap R$ is a prime ideal of R that contains P. It follows that M is the only maximal ideal of R that contains Q and thus $R_M \subseteq R_Q \subseteq \Theta(M)_{Q'} \subsetneq K$ which implies $R_M \cdot \Theta(M)_{Q'} \subsetneq K = R_M \cdot \Theta(M)$, a contradiction.

By Lemma 3.11, if I is a nonzero ideal that is contained in M, then each minimal prime of $I_{(M)}$ is contained in M. It follows that M is the only maximal ideal that contains $I_{(M)}$. In addition we have that $I_{(M)}\Theta(M) = \Theta(M)$ for otherwise a minimal prime of $I_{(M)}\Theta(M)$ will contract to prime ideal of R that is not contained in M.

Continuing with P a nonzero prime contained in M, let x be a nonzero element contained in P. Then from above, the ideal $J = xR_{(M)}$ blows up $\Theta(M)$ and M is the only maximal ideal that contains J. Thus there is a finite subset $\{x, a_1, a_2, \ldots, a_n\}$ of J and elements $s, s_1, s_2, \ldots, s_n \in \Theta(M)$ such that $sx + s_1a_1 + s_2a_2 + \cdots + s_na_n = 1$. We have $JR_M = xR_M$ and $JR_N = R_N$ is generated by $\{x, a_1, a_2, \ldots, a_n\}$ for each maximal ideal $N \neq M$. Hence $J = xR + a_1R + \cdots + a_nR$ is invertible. To complete the proof we show that $J = xR + yR$ for some y.

For each $1 \leq i \leq n$, there are elements $b_i \in R$ and $t_i \in R \setminus M$ such that $a_i = b_ix/t_i \in R_M$. Since there are only finitely many a_is, we may assume $t_i = t_j = t$ for all i and j. As in the proof of Theorem 2.6, $t/x \in J^{-1}$. Since $t \notin M$ and M is the only maximal ideal that contains J, there is an element $y \in J$ and an element $w \in R$ such that $wt + y = 1$. For each a_i we have $a_i = a_iwt + a_iy = b_ixw + a_iy \in xR + yR \subseteq J$. Hence $J = xR + yR$. $\qquad\square$

The next two results provide ways for a given maximal ideal to know enough that we can characterize it as being sharp. We start with the general case and then provide a simpler characterization for the case R is a Prüfer domain.

Theorem 3.13. *Let M be a maximal ideal of a domain R. Then M is sharp if and only if M knows there is a pair of nonzero elements $a, b \in M$ and a corresponding ideal $J \subseteq M$ where a is not in bR, $aRcR \subseteq bR$ for each $c \in J$ and $\{M\}$ is the only (J, M)-complete set.*

Proof. Assume M is sharp and let $t \in \Theta(M) \setminus R_M$. Then $t \in R_N$ for each maximal ideal $N \neq M$ and thus M is the only ideal that contains $J = (R :_R t) = (R : R + tR)$. It follows that $t = a/b$ for some $b \in M$ and $a \in R$. We may further assume $a \in M$ since $ba/b^2 = t$. We have $J = (R : R + tR) = (bR : bR + aR)$ with $\{M\}$ the only (J, M)-complete set. Also, (by definition) J is the set of elements $\{f \in R \mid aRfR \subseteq bR\} = \{f \in M \mid arfR \subseteq bR\}$.

For the converse, suppose M knows there is a pair of nonzero elements $a, b \in M$ and a corresponding ideal $J \subseteq M$ where a is not in bR, $aRcR \subseteq bR$ for each $c \in J$ and $\{M\}$ is the only (J, M)-complete set. For the element $t = a/b$ we have that $J \subseteq (R :_R t) \subseteq M$. It follows that $t \in R_N$ for each maximal ideal $N \neq M$, but t is not in R. Thus $t \in \Theta(M) \setminus R_M$ and therefore M is sharp. $\qquad\square$

As noted in the Introduction, if R is a Prüfer domain and $s \in K \setminus R$, then the ideal $(R :_R s)$ is a two-generated invertible ideal of R.

Theorem 3.14. *Let M be a maximal ideal of a Prüfer domain R. Then M is sharp if and only if M knows it contains a pair of nonzero elements x and y such that*

$\{M\}$ is the only (I, M)-complete set where $I \subseteq M$ is the ideal that contains both x and y and is contained in each ideal $J \subseteq M$ that contains both x and y.

For a nonzero ideal I, if $I \subsetneq I_v \neq R$, there may be a maximal ideal M that contains I but not I_v. To investigate divisorial ideals and divisorial closure from the perspective of the maximal ideals, we first need a way to provide an internal description of $(R : I)$. For this, start with a maximal ideal M and let I be a nonzero ideal that is contained in M. Next let X_M be the set of ordered pairs (a, b) where $a, b \in M \setminus \{0\}$ are such that $a \notin bR$ and $aI \subseteq bR$. Routine calculations show that $(a, b) \in X_M(I)$ if and only if there is a $t \in (R : I) \setminus R$ such that $t = a/b$. Note that for $t \in (R : I)$, we can always write $t = c/d = mc/md$ for some $d \in I$, $c \in R$ and arbitrary $m \in M \setminus \{0\}$. Clearly, the set $X_M(I)$ is empty if and only if $(R : I) = R$. If $J \subseteq M$ and $X_M(J) = X_M(I)$, then $X_M(I + J) = X_M(I)$. It follows that there is a largest ideal $B \subseteq M$ such that $X_M(B) = X_M(I)$. We refer to this ideal as the M-divisorial closure of I. While M relies on its ability to determine containment relations between the ideals and elements it contains to obtain B, we see that $B = I_v \cap M$. In the case $X_M(I)$ is empty, $B = M$. Note that we also have $B = M$ in the case $I_v = M$.

Lemma 3.15. *Let I be a nonzero ideal of a domain R.*

1. *If I is not maximal, then it is divisorial if and only if it is M-divisorial for each maximal ideal $M \in \operatorname{Max}(R, I)$.*
2. *If I is maximal, then it is divisorial if and only $X_I(I)$ is nonempty.*

Proof. If I is maximal, then either $X_I(I)$ is empty in which case $I_v = R$ or $X_I(I)$ is nonempty in which case $I = I_v$. For the nonmaximal case, it is clear that I is M-divisorial for each $M \in \operatorname{Max}(R, I)$ when I is divisorial. For the converse in the nonmaximal case, suppose I is M-divisorial for each maximal ideal $M \in \operatorname{Max}(R, I)$. Then from the discussion above, $I = I_v \cap M$ for each maximal ideal $M \in \operatorname{Max}(R, I)$. Since I is not a maximal ideal, having $I = I_v \cap M$ tells us that I_v is properly contained in at least one maximal ideal N. It follows that $I = I_v \cap N = I_v$. □

Recall that a domain is a *Mori domain* if ACC holds for divisorial ideals. We say that $M \in \operatorname{Max}(R)$ is *Mori maximal* if ACC holds for M-divisorial ideals.

Theorem 3.16. *The following are equivalent for a domain R.*

1. *R is a Mori domain.*
2. *Each maximal ideal of R is Mori maximal.*
3. *At least one maximal ideal of R is Mori maximal.*

Proof. Obviously, (2) implies (3). To see that (1) implies (2), assume R is a Mori domain. Let M be a maximal ideal of R and let $I_1 \subseteq I_2 \subseteq I_3 \subseteq \cdots$ be an ascending chain of M-divisorial ideals. Then we have $(I_1)_v \subseteq (I_2)_v \subseteq (I_3)_v \subseteq \cdots$. Since R is a Mori domain, there is an n such that $(I_n)_v = (I_k)_v$ for all $k \geq n$. We also have $I_j = (I_j)_v \cap M$ for each j. It follows that $I_n = I_k$ for all $k \geq n$.

Finally, suppose $M \in \mathrm{Max}(R)$ is Mori maximal and let $A_1 \subseteq A_2 \subseteq A_3 \subseteq \cdots$ be an ascending chain of divisorial ideals of R. Let $t \in M \setminus \{0\}$ and consider the chain $tA_1 \subseteq tA_2 \subseteq tA_3 \subseteq \cdots$. Each tA_j is a divisorial ideal that is contained in M and so each is M-divisorial. Hence there is an integer n such that $tA_n = tA_k$ for all $k \geq n$. By cancellation, we have $A_n = A_k$ for all $k \geq n$. Hence R is a Mori domain. $\square$

References

1. D.D. Anderson, M. Zafrullah, Independent locally-finite intersections of localizations. Houston J. Math. **25**, 433–452 (1999)
2. M. Fontana, E. Houston, and T. Lucas, Toward a classification of prime ideals in Prüfer domains. Forum Math. **22**, 741–766 (2010)
3. M. Fontana, E. Houston, T. Lucas, Factoring Ideals in Integral Domains, Lecture Notes of the Unione Matematica Italiana, vol. 14 (Springer, Berlin, 2013)
4. R. Gilmer, Overrings of Prüfer domains, J. Algebra **4**, 331–340 (1966)
5. R. Gilmer *Multiplicative Ideal Theory*, Queen's Papers in Pure and Applied Mathematics, vol. 90 (Queen's University Press, Kingston, 1992)
6. R. Gilmer, W. Heinzer, Overrings of Prüfer domains. II. J. Algebra **7**, 281–302 (1967)
7. E. Matlis, *Cotorsion Modules*, Mem. American Mathematical Society No, vol. 49, (Providence RI, 1964)
8. B. Olberding, Globalizing local properties of Prüfer domains. J. Algebra **205**, 480–504 (1998)

Prime Ideals That Satisfy Hensel's Lemma

Stephen McAdam

Abstract Nagata proved that (R, P) is a Henselian domain if and only if every integral extension domain of R is quasi-local. We explore, with partial success, how to generalize that result.

Keywords Henselian • Prime ideals • Integral extensions • Integral domains

Subject Classifications: 13A15, 13B22, 13G05, 13J15

1 Introduction

Notation. Throughout, R will be a commutative domain with integral closure R' and Jacobson radical $J(R)$. P will be a nonzero prime ideal of R.

Definition 1. We call P an H-prime if the following holds. For any non-constant monic polynomial $f(X) \in R[X]$, if there exist non-constant monic polynomials $g(X)$ and $h(X)$ in $R[X]$ such that $f(X) = g(X)h(X) \bmod P$ and such that $g(X)$ and $h(X)$ are comaximal (i.e., $g(X)R[X] + h(X)R[X] = R[X]$), then $f(X)$ is reducible in $R[X]$.

The following crucial result is proven in [1, (2.2)].

S. McAdam (✉)
Mathematics Department, RLM 8.100, The University of Texas at Austin,
2515 Speedway - Stop C1200, Austin, TX 78712-1202, USA
e-mail: mcadam@math.utexas.edu

M. Fontana et al. (eds.), *Commutative Algebra: Recent Advances in Commutative Rings,*   255
Integer-Valued Polynomials, and Polynomial Functions, DOI 10.1007/978-1-4939-0925-4__15,
© Springer Science+Business Media New York 2014

Theorem 1. *Let* $P \subseteq J(R)$. *The following are equivalent.*

(i) *P is an H-prime.*

(ii) *For all non-constant monic polynomials* $f(X) \in R[X]$, *if there exist non-constant monic polynomials* $g(X)$ *and* $h(X)$ *in* $R[X]$ *such that* $f(X) \equiv g(X)h(X) \bmod P$ *and such that* $g(X)$ *and* $h(X)$ *are comaximal, then there are monic polynomials* $g'(X)$ *and* $h'(X)$ *in* $R[X]$ *such that* $f(X) = g'(X)h'(X)$ *and* $g(X) \equiv g'(X) \bmod P$ *and* $h(X) \equiv h'(X) \bmod P$.

Remark 1. 1. We do not know whether some version of Theorem 1 (i) $\Rightarrow$ (ii) holds when P is not contained in $J(R)$, although (ii) $\Rightarrow$ (i) is trivially true.

2. Hensel's lemma says that if R is complete in the P-adic topology, then P satisfies condition (ii) of Theorem 1 and so is an H-prime. (Hence, H-primes do exist.)

3. We will see that when P is not contained in $J(R)$, it is in some sense unlikely for P to be an H-prime. In particular, we will see that if R is Noetherian, and P is an H-prime, then $P \subseteq J(R)$.

4. In the above, when we wrote $g(X)$ and $h(X)$, we assumed they were comaximal. In some references, that is modified to say, $PR[X] + g(X)R[X] + h(X)R[X] = R[X]$. However, the bulk of our interest here will be in the case that $P \subseteq J(R)$, and when that is true, the two conditions are equivalent. This is easily seen, using the fact that if M is a maximal ideal of $R[X]$ and M contains a monic polynomial $k(X)$, then $M \cap R$ is maximal in R. That fact, [5, Lemma 1.1(v)], is an easy consequence of the fact that the integral extension $R \subseteq R[X]/k(X)R[X]$ satisfies going up.

Lemma 1. *Let* $P \subseteq Q$ *be prime ideals of* R. *If* Q *is an* H-*prime, then so is* P.

Proof. Suppose P is not an H-prime. Then there is an irreducible non-constant monic $f(X) \in R[X]$ and non-constant monic polynomials $g(X)$ and $h(X)$ in $R[X]$ such that $f(X) \equiv g(X)h(X) \bmod P$ and such that $g(X)$ and $h(X)$ are comaximal. However, we also have $f(X) \equiv g(X)h(X) \bmod Q$, and that implies Q is not an H-prime. $\square$

The inspiration for this paper is the following well-known result of Nagata [6, (43.12)].

Theorem 2. *Let* (R, P) *be a quasi-local domain. Then, P satisfies condition (ii) of Theorem 1 (i.e., an H-prime) if and only if every integral extension domain of R is quasi-local. (When those equivalent conditions hold, (R, P) is called a Henselian domain.)*

The goal of this paper is to try to globalize that and to see if some similar result holds for H-primes that are not the sole maximal ideal their ring R. The first guess might be that P is an H-prime if and only if for every integral extension domain T of R, there is a unique prime of T lying over P. However, when R is Noetherian, that guess is hopelessly wrong, as we now show.

By [2, Theorem 1.1(ii)], if R is Noetherian and if in every integral extension domain of R only one prime ideal lies over P, then R is local and P is its maximal ideal. Hence, if our above guess were correct, it would imply that if P is an H-prime (with R Noetherian), then P would be maximal. However, Lemma 1 shows that is not always the case for H-primes. As our first guess is wrong, we need a more appropriate (possible) extension of Nagata's result. That leads us to our next definition.

Definition 2. We call P a K-prime if there does not exist an integral extension domain T of R such that exactly two primes of T lie over P and those two primes are comaximal in T.

Question 1. How closely related are H-primes and K-primes? Specifically, if $P \subseteq J(R)$, are the concepts of H-prime and K-prime equivalent?

We will prove the following two propositions.

Proposition 1. *If R is integrally closed, then P is an H-prime if and only if it is a K-prime.*

Proposition 2. *Suppose that for all nonzero non-units $y \in R'$, there is a prime ideal Q' of R' containing y such that either $Q' \neq Q'^2$, or R'/Q' is not integrally closed, or the quotient field of R'/Q' is not algebraically closed. If P is a K-prime of R, then $P \subseteq J(R)$ and P is an H-prime.*

Proposition 2 shows that in a very large class of domains, K-primes are H-primes, considerably strengthening the work in [5], in which R' was the integral closure of Noetherian domain. Much less is known about the converse of Proposition 2, Proposition 1 being the most significant case in which it is known to hold.

Example 1. If P is a prime in a Henselian domain (R, Q), then P is both an H-prime and a K-prime. Since every integral extension of R is quasi-local, P must be a K-prime. Since Q is an H-prime, Lemma 1 shows P is an H-prime.

Example 2 (Heitmann). Let T is Noetherian integrally closed non-Henselian domain, and (with Y an indeterminate) let $R = T[[Y]]$ and $P = YR$. Hensel's lemma shows that P is an H-prime. Also, Proposition 1 shows that since R is integrally closed, P is also a K-prime. Finally, [6, (43.4)] shows R is not Henselian.

The present paper constitutes a streamlining and extension of Sects. 2 and 3 of [5]. The improvement of this work over the earlier work is due to the availability of Theorems 1 (above) and 4 (below), both proved in [1] (as well as a new construction given in Sect. 5 below). Section 1 of [5] contains some related facts of interest. Specifically, [5, (1.5(i) $\Leftrightarrow$ (ii)] shows that if R is Noetherian and if P is not a K-prime, then for any $m \geq 1$, there is an integral extension domain T of R in which there are exactly m primes lying over P and those m primes are pairwise comaximal.

2   Proposition 1

Definition 3. Recall that if Q is a prime ideal in a ring R, then a prime q in the polynomial ring $R[X]$ is called an upper to Q if $q \cap R = Q$, but $q \neq QR[X]$. Furthermore, if q is an upper to Q and q contains a monic polynomial, then q is called an integral upper to Q. (All of the facts we use about uppers and integral uppers are easily proven and can be found in [5, Lemma 1.1].)

Lemma 2. *Let $R \subseteq T$ be rings, and let P be a prime ideal of R. Let Q be a prime ideal of T with $Q \cap R = P$, and let $t \in Q$. Then $Q \cap R[t] = (P, t)R[t]$.*

Proof. One inclusion is obvious. For the other, assume that $f(t) \in Q \cap R[t]$ (with f a polynomial with coefficients in R). Since $t \in Q$, we must have the constant coefficient of f in $Q \cap R = P$. Hence $f(t) \in (P, t)R[t]$. $\square$

Lemma 3. *Let P be a prime ideal in a domain R. The following are equivalent.*

(a) P is not a K-prime.

(b) There is an integral extension domain T of R in which the set V of prime ideals lying over P can be partitioned into two nonempty subsets, say $V = V_1 \cup V_2$, such that $\cap\{p \mid p \in V_1\}$ and $\cap\{q \mid q \in V_2\}$ are comaximal in T.

(c) There is an integral upper K to 0 in $R[X]$ such that K is contained in the uppers $(P, X)R[X]$ and $(P, X+1)R[X]$, but in no other uppers to P except those two.

(d) There is an integral extension domain $R[t]$ of R such that the only prime ideals of $R[t]$ that lie over P are $(P, t)R[t]$ and $(P, t+1)R[t]$.

Proof. (d) $\Rightarrow$ (a) $\Rightarrow$ (b): These are obvious from the definition of a K-prime.

(b) $\Rightarrow$ (d): Assuming (b) and using comaximality, pick $t \in T$ with $t \equiv 0 \bmod \cap \{p \mid p \in V_1\}$ and $t \equiv -1 \bmod \cap \{q \mid q \in V_2\}$. As t is contained in each prime in V_1, Lemma 2 shows that every prime ideal in V_1 intersects $R[t]$ at $(P, t)R[t]$. Similarly, since $t + 1 \in \cap\{q \mid q \in V_2\}$, we see that every prime in V_2 intersects $R[t+1] = R[t]$ at $(P, t+1)R[t+1] = (P, t+1)R[t]$. Finally, since all primes in V contract to one of these two primes, lying over in $R[t] \subseteq T$ shows they are the only primes of $R[t]$ lying over P.

(c) $\Leftrightarrow$ (d): For an integral extension of domains $R \subseteq R[t]$, let K be the kernel of the map $R[X] \rightarrow R[t]$. Thus K is an integral upper to 0 and $R[X]/K$ is isomorphic to $R[t]$. The prime ideals of $R[X]/K$ that lie over P all have the form L/K where L is an upper to P in $R[X]$ with L containing K. The equivalence of (c) and (d) follows easily. $\square$

Lemma 4. *(a) Let R' be an integrally closed domain, and let L be an ideal of $R'[X]$. Then L is an integral upper to 0 if and only if $L = f(X)R'[X]$ for some non-constant monic irreducible polynomial $f(X) \in R'[X]$.*

(b) Let R be an arbitrary domain. If $f(X)$ is a non-constant monic polynomial in $R[X]$ which is irreducible in $R'[X]$, then $f(X)R[X]$ is an integral upper to 0 in $R[X]$.

(c) Let R be an arbitrary domain. If $g(X)$ is a non-constant polynomial, then some upper to 0 in $R[X]$ contains $g(X)$.

Proof. (a) This is well known. (A proof is recorded in [5, Lemma 2.4].)

(b) Suppose R and $f(X)$ are as in (b). By part (a), $f(X)R'[X]$ is an integral upper to 0 in $R'[X]$, and so $f(X)R'[X] \cap R[X]$ is an integral upper to 0 in $R[X]$. However, since $f(X)$ is monic in $R[X]$, an easy exercise shows $f(X)R'[X] \cap R[X] = f(X)R[X]$.

(c) Let F be the quotient field of R. Since $g(X)$ is not a unit of $F[X]$, it is contained in some prime ideal H of $F[X]$. Let $L = H \cap R[X]$. We have $L \cap R = (H \cap R[X]) \cap (F \cap R) = (H \cap F) \cap R = 0 \cap R = 0$. Thus, L is an upper to 0 in $R[X]$, and $g(X) \in L$. $\qquad\square$

Proposition 1. *Let R be integrally closed. Then P is an H-prime if and only if it is a K-prime.*

Proof. Suppose P is not an H-prime. Then there exists a non-constant monic irreducible $f(X) \in R[X]$ and comaximal non-constant monic polynomials $g(X)$ and $h(X)$ in $R[X]$ such that $f(X) \equiv g(X)h(X) \bmod P$. By part (a) of the previous lemma, $R \subseteq R[X]/f(X)R[X]$ is an integral extension of domains. The primes of the larger domain that lie over P in R all have the form $L/f(X)R[X]$, with L an upper to P in $R[X]$ that contains $f(X)$. In other words, they are the images in $R[X]/f(X)R[X]$ of those uppers L to P that contain $f(X)$. As $f(X) \equiv g(X)h(X) \bmod P$ with $g(X)$ and $h(X)$ comaximal, that set of L can be partitioned into those L that contain $g(X)$ and those L that contain $h(X)$. Thus, the set of primes lying over P is $V = V_1 \cup V_2$, with $V_1 = \{L/f(X)R[X] \mid L$ is an upper to P containing $f(X)$ and $g(X)\}$ and $V_2 = \{L/f(X)R[X] \mid L$ is an upper to P containing $f(X)$ and $h(X)\}$. The comaximality of $g(X)$ and $h(X)$ shows that union is disjoint and also shows that the comaximality of $\cap\{q \mid q \in V_1\}$ and $\cap\{q \mid q \in V_2\}$. We claim that neither set in that union is empty. For that, it will suffice (by symmetry) to show that there does exist an upper L to P in $R[X]$ with $f(X) \in L$, such that $g(X) \in L$. Letting g' represent $g(X) \bmod P$, part (c) of the previous lemma shows there is an upper L' to 0 in $(R/P)[X]$ with $g'(X) \in L'$. Now it is easily seen that L' has the form $L/PR[X]$ for some upper L to P in $R[X]$, with $g(X) \in L$. Since $f(X) - g(X)h(X) \in PR[X] \subseteq L$, we also have $f(X) \in L$. That proves the claim. Finally, using Lemma 3((b) $\Rightarrow$ (a)), P is not a K-prime.

Conversely, suppose P is not a K-prime. By Lemma 3((a) $\Rightarrow$ (c)), there is an integral upper K to 0 in $R[X]$ such that K is contained in the uppers $(P, X)R[X]$ and $(P, X + 1)R[X]$, but in no other uppers to P except those two. By Lemma 4(a), $K = f(X)R[X]$ for some non-constant monic irreducible polynomial $f(X) \in R[X]$. Thus, the only uppers to P in $R[X]$ that contain $f(X)$ are (P, X) and $(P, X + 1)$. It easily follows that the factorization of $f(X) \bmod P$ has the form $X^n(X + 1)^m$ (since if there was another factor, Lemma 4(c) applied to R/P would show that a third upper to P also contains $f(X)$). That shows P is not an H-prime. $\qquad\square$

3   Concerning K-Primes P Not Contained in $J(R)$

Lemma 5. *Let D be a domain between R and its quotient field, and let $C = \{r \in R \mid rd \in R$ for all $d \in D\}$ (the conductor of D to R). Suppose Q is a prime ideal in R comaximal to C, and let q be a prime ideal of D lying over Q. Then the following are true.*

(a) *For all $n \geq 1$, $q^n \cap R = Q^n$.*
(b) *For all $n \geq 1$, the following are equivalent:*

   (i) $Q^n \neq Q^{n+1}$;
   (ii) $q^n \neq q^{n+1}$;
   (iii) $Q^n \not\subseteq q^{n+1}$.

(c) $R/Q = D/q$.

Proof. (a) Suppose $q^n \cap R$ properly contains Q^n. Then there exist $s_{ij} \in q$ with $r = \sum_{j=1}^m \prod_{i=1}^n s_{ij} \in (q^n \cap R) - Q^n$. Now $(Q^n : r) = \{x \in R \mid xr \in Q^n\}$ is a proper ideal of R and consists of zero divisors modulo Q^n. By Zorn's lemma, it can be enlarged to an ideal N maximal with respect to consisting of zero divisors modulo Q^n, and by a standard argument [3, Theorem 1], N is a prime ideal of R. As $Q^n \subseteq (Q^n : r) \subseteq N$, we have $Q \subseteq N$, so that C is not contained in N. Pick $c \in C - N$. Now $c^n r = \sum_{j=1}^m \prod_{i=1}^n (c s_{ij}) \in Q^n$, since each $c s_{ij} \in q \cap R = Q$. Thus $c^n \in (Q^n : r) \subseteq N$. That contradicts that c is not in N. Thus $q^n \cap R = Q^n$.

(b) Obviously (iii) implies (ii). Suppose (ii) holds, and let $y \in q^n - q^{n+1}$. As C and Q are comaximal, write $1 = c + z$ with $c \in C$ and $z \in Q$. Raising both sides to the nth power, we can write $1 = c^n + w$ with $w \in Q$. We have $y = c^n y + wy$. Now $wy \in Q q^n \subseteq q^{n+1}$, and since y is not in q^{n+1} we must have $c^n y \notin q^{n+1}$. Thus, $c^n y \notin Q^{n+1}$. However, since $y \in q^n$ and $Cq \subseteq Q$, we have $c^n y \in Q^n$. Thus (i) holds. Finally, suppose (i) holds. Then (iii) follows, since part (a) shows $q^{n+1} \cap R = Q^{n+1}$.

(c) We have the natural embedding $R/Q \subseteq D/q$. In order to show equality, it will suffice to show that for all $y \in D$, there is a $t \in R$ with $t - y \in q$. By comaximality, there is a $c \in C$ with $c - 1 \in Q \subseteq q$. We have $yc - y \in q$, and so we let $t = yc$, which is in R.   □

Lemma 6. *The following are equivalent for a domain D.*

(i) *D is integrally closed, and its quotient field is algebraically closed.*
(ii) *Every non-constant monic polynomial in $D[X]$ can be factored into a product of monic linear polynomials in $D[X]$.*

Proof. Suppose (i) is true, and let $f(X)$ be a non-constant monic polynomial in $D[X]$. With Ω the algebraically closed quotient field of D, in $\Omega[X]$ we see that $f(X)$ factors into a product of linear polynomials. Let $X - b$ be one of them. Since $f(b) = 0$, b is integral over D, and so $X - b$ is in $D[X]$. Thus (ii) holds.

Now suppose (ii) holds. Let Ω be an algebraic closure of the quotient field of D, and let T be the integral closure of D in Ω. Since Ω is algebraic over the quotient field of D, a standard argument shows Ω is the quotient field of T. Therefore, it will suffice to show $D = T$. Pick any $t \in T$. There is a monic polynomial in $D[X]$ having t as a root. By (ii), that monic polynomial factors into a product of monic linear factors in $D[X]$. Clearly one of those factors must be $X - t$, showing $t \in D$. Thus $D = T$. $\qquad\qquad\qquad\qquad\qquad\qquad\qquad\qquad\qquad\qquad\qquad\qquad\qquad\quad$ $\square$

We come to the main result of this section.

Theorem 3. *Suppose P is not contained in the Jacobson radical of R, and let Q be a prime of R comaximal to P. Consider the following three statements.*

(i) *$Q \neq Q^2$;*
(ii) *R/Q is not integrally closed;*
(iii) *the quotient field of R/Q is not algebraically closed.*

    (a) *If any of (i), (ii), or (iii) is true, then P is not an H-prime.*
    (b) *If the conductor C of R' to R is comaximal to Q, and if any of (i), (ii), or (iii) is true, then P is not a K-prime.*

Proof. (a) Suppose first that $Q \neq Q^2$. Pick $d \in Q - Q^2$. Since P is comaximal to Q and also to Q^2, by the Chinese remainder theorem, pick $b \in R$ with $b \equiv d \bmod Q$ and $b \equiv 1 \bmod P$, and pick $c \in R$ with $c \equiv d \bmod Q^2$ and $c \equiv 0 \bmod P$. Let $f(X) = X^2 + bX + c$. Clearly $f(X) \equiv X(X + 1) \bmod P$. Thus, to show P is not an H-prime, it will suffice to show $f(X)$ is irreducible in $R[X]$. That follows from Eisenstein's criterion, since $d \in Q - Q^2$, implies $b \in Q$ and $c \in Q - Q^2$.

Next, suppose either (ii) or (iii) is true. Then Lemma 6 shows there is some monic irreducible polynomial $\alpha(X) \in (R/Q)[X]$ of degree $n \geq 2$. Let $k(X)$ be a monic pre-image of $\alpha(X)$ in $R[X]$. As P and Q are comaximal, by the Chinese remainder theorem, there is a monic polynomial $f(X) \in R[X]$ with $f(X) \equiv k(X) \bmod Q$ and $f(X) \equiv X^{n-1}(X + 1) \bmod P$. The image of $f(X)$ in $(R/Q)[X]$ is $\alpha(X)$ which is irreducible in $(R/Q)[X]$, and so $f(X)$ is irreducible in $R[X]$. The factorization of $f(X) \bmod P$ therefore shows that P is not an H-prime.

(b) The proof is similar to that of (a), except we must move matters from R up to R', since the $f(X) \in R[X]$ mentioned in the proof of (a) will now need to be irreducible in $R'[X]$.

First suppose that $Q \neq Q^2$. Let Q' be a prime ideal of R' lying over Q. Using Lemma 5(b)((i) $\Rightarrow$ (iii)), we see that Q is not contained in Q'^2. Pick $d \in Q - Q'^2$, and pick b and c as in the proof of (a). Let $f(X) = X^2 + bX + c$. We have $b \in Q \subseteq Q'$ and (since $c - d \in Q^2 \subseteq Q'^2)c \in Q' - Q'^2$. Eisenstein's criterion shows $f(X)$ is irreducible in $R'[X]$. By Lemma 4(b), $K = f(X)R[X]$ is an integral upper to 0 in $R[X]$. However, we also have $f(X) \equiv X(X + 1) \bmod P$, showing that K is contained in $(P, X)R[X]$ and $(P, X + 1)R[X]$, but in no other uppers to P in $R[X]$. By Lemma 3((c) $\Rightarrow$ (a)), P is not a K-prime.

Now suppose that either R/Q is not integrally closed or its quotient field is not algebraically closed. Let $\alpha(X)$, $k(X)$, and $f(X)$ be as in the second half of the proof of part (a). Let Q' be a prime ideal of R' lying over Q. Using Lemma 5(c), the image of $f(X)$ in $(R/Q)[X] = (R'/Q')[X]$ is $\alpha(X)$, which is irreducible, and so $f(X)$ is irreducible in $R'[X]$. Thus $K = f(X)R[X]$ is an integral upper to 0 in $R[X]$. Since $f(X) \equiv X^{n-1}(X+1) \bmod P$, Lemma 3((c) $\Rightarrow$ (a)) shows P is not a K-prime. $\square$

Heuristic Remark: If P is an H-prime not contained in $J(R)$, then for every ideal Q comaximal to P, we must have (i), (ii), and (iii) of Theorem 3 all be false. We feel that justifies saying that H-primes not contained in the Jacobson radical are rather rare. In particular, since the Krull intersection theorem shows that for any prime $Q \neq 0$ in a Noetherian domain we have $Q \neq Q^2$, we see that in a Noetherian domain, P can only be an H-prime if $P \subseteq J(R)$. Similarly, K-primes not contained in the Jacobson radical are somewhat rare. However, Example 3 below shows both H-primes and K-primes not contained in $J(R)$ do exist.

The next corollary is the first of three key pieces in the proof of Proposition 2.

Corollary 1. *Suppose P is not contained in the Jacobson radical of R, and suppose P is also an ideal of R'. Let Q be a prime of R comaximal to P, and let Q' be a prime ideal of R' lying over Q. If any one of the following three conditions holds, then P is neither an H-prime nor a K-prime.*

 (i) $Q' \neq Q'^2$;
(ii) R'/Q' *is not integrally closed;*
(iii) *the quotient field of R'/Q' is not algebraically closed.*

Proof. Since P is an ideal in R', we have $PR' \subseteq P \subseteq R$, so that $P \subseteq C$, the conductor of R' to R. Therefore, Q is also comaximal to C. Using Lemma 5, we see that $Q' \neq Q'^2$ if and only if $Q \neq Q^2$, and also $R/Q = R'/Q'$. The corollary now follows from the theorem.

The hitch in the corollary is the need to have P be an ideal in R'. In Sect. 5, we deal with that problem by mimicking P with a prime we will call $P^{\#}$. $\square$

Example 3. Suppose R is the integral closure of the integers in the algebraic closure of the rationals. If $P \neq 0$ is a prime ideal of R, then P is an H-prime and a K-prime.

Proof. Suppose P is not an H-prime. Then there is a monic irreducible $f(X) \in R[X]$ such that $f(X)$ is reducible modulo P. That last implies the degree of $f(X)$ is at least 2. However, as $f(X)$ is irreducible, Lemma 6 shows the degree of $f(X)$ is 1, a contradiction. Thus P is an H-prime, and so by Proposition 1, it is also a K-prime. $\square$

4 Going Down from Maximals

We begin with another crucial result proven in [1, (2.3)]. (As in Theorem 1, we do not know if the assumption $P \subseteq J(R)$ is required.)

Theorem 4. *Let* $P \subseteq J(R)$. *The following are equivalent.*

(i) *P is an H-prime.*

(ii) *For all non-constant monic polynomials $f(X) \in R[X]$, if there exist non-constant monic polynomials $g(X)$ and $h(X)$ in $R[X]$ such that $f(X) \equiv g(X)h(X) \bmod P$ and such that $g(X)$ and $h(X)$ are comaximal, then for any upper K to 0 in $R[X]$ with $f(X) \in K$, either K and $g(X)$ are comaximal or K and $h(X)$ are comaximal.*

Definition 4. We say that P is a GDM prime if for all integral extension domains T of R and all maximal ideals N of T, there is a prime ideal Q of T such that $Q \subseteq N$ and $Q \cap R = P$. (By letting $T = R$, we see that a GDM prime must be contained in $J(R)$.)

Remark 2. GDM stands for "going down from maximals." In [5], GDM was defined in terms of finitely generated integral extensions. However, by Lemma 8, it is easily seen that it does not matter if we allow T to be arbitrary, or insist that it be finitely generated, or even insist that it be generated by a single element over R. All are equivalent.

In this section, we will show that if P is both a K-prime and a GDM prime, then P is an H-prime. (Later, we will see that in many domains, K-primes are GDM primes and so are H-primes.) The next result is the second key piece in the proof of Proposition 2.

Theorem 5. *If P is a K-prime and a GDM prime, then P is an H-prime.*

Proof. It will suffice for us to assume that P is a GDM prime but not an H-prime and to prove that P is not a K-prime. Since we know GDM primes are contained in the Jacobson radical, Theorem 4 shows there are non-constant monic polynomials $f(X)$, $g(X)$, and $h(X)$ in $R[X]$ and an upper, K, to 0 in $R[X]$ such that $f(X) \equiv g(X)h(X) \bmod P$, with $g(X)$ and $h(X)$ comaximal and with $f(X) \in K$, such that K is not comaximal to either $g(X)$ or $h(X)$.

Let $V = \{p \in \operatorname{Spec} R[X] \mid p$ is an upper to P and $K \subseteq p\}$. If $p \in V$, then $f(X) \in K \subseteq p$, and since $f(X) \equiv g(X)h(X) \bmod P$ (and $P \subseteq p$), we see that either $g(X) \in p$ or $h(X) \in p$. Thus if $V_g = \{p \in V \mid g(X) \in p\}$ and $V_h = \{p \in V \mid h(X) \in p\}$, then $V = V_g \cup V_h$. Since $g(X)$ and $h(X)$ are comaximal in $R[X]$, clearly V_g and V_h partition V.

We claim neither V_g nor V_h is empty. (The argument used in the analogous claim in the proof of Proposition 1 will not work here, since we only have $f(X) \in K$ instead of $K = f(X)R[X]$.) Since K is not comaximal to $g(X)$, there is a maximal ideal N of $R[X]$ that contains both K and $g(X)$. Now N/K is a maximal ideal in $R[X]/K$, and this last ring is an integral extension domain of R. Since P is assumed to be a GDM prime, there must be a prime ideal p'/K of $R[X]/K$ with $p'/K \subseteq N/K$ and with $(p'/K) \cap R = P$. We easily see that p' is an upper to P in $R[X]$ such that $K \subseteq p' \subseteq N$. Thus $p' \in V = V_g \cup V_h$. Suppose $p' \in V_h$. Then by definition, $h(X) \in p' \subseteq N$. However, N also contains $g(X)$, which contradicts that $g(X)$ and $h(X)$ are comaximal. Therefore, p' is not contained in V_h and so must be contained in V_g, which is therefore not empty. Similarly, V_h is not empty.

We easily see that in the integral extension domain $R[X]/K$ of R, the set of primes lying over P is $\{p/K \mid p \in V\} = \{p/K \mid p \in V_g\} \cup \{p/K \mid p \in V_h\}$. Neither subset in this partition is empty (by the preceding paragraph). Also, if g' and h' represent $g(X)$ and $h(X)$ taken modulo K, then since $g(X)$ and $h(X)$ are comaximal in $R[X]$, g' and h' are comaximal in $R[X]/K$. It follows that $\cap\{p/K \mid p \in V_g\}$ and $\cap\{p/K \mid p \in V_h\}$ are comaximal. It now follows from Lemma 3((b) $\Rightarrow$ (a)) that P is not a K-prime. □

Although we do not need it, the following is perhaps worth recording.

Lemma 7. *Let* $R \subseteq T$ *be an integral extension of domains. Let* Q *be prime in* T *with* $Q \subseteq J(T)$, *and let* $P = Q \cap R$. *If* Q *is an* H-prime, *then* P *is an* H-prime.

Proof. Since $Q \subseteq J(T)$, we have $P \subseteq J(R)$. Assuming Q is an H-prime, we will use Theorem 4 to show P is an H-prime. Let $f(X)$, $g(X)$, and $h(X)$ be non-constant monic polynomials in $R[X]$ with $f \equiv g(X)h(X) \bmod P$ and with $g(X)$ and $h(X)$ comaximal. Let K be an upper to 0 in $R[X]$ with $f(X) \in K$. (We must show K is comaximal to either $g(X)$ or $h(X)$.) There is an upper L to 0 in $T[X]$ with $L \cap R[X] = K$. In $T[X]$, we have $f(X) \equiv g(X)h(X) \bmod Q$, and $f(X) \in L$. Since Q is an H-prime contained in $J(T)$, Theorem 4 shows L is comaximal to one of $g(X)$ or $h(X)$. We may suppose L and $g(X)$ are comaximal in $T[X]$. An easy exercise (using going up) shows K and $g(X)$ are comaximal in $R[X]$. □

5 A Useful Construction

Lemma 8. *Let* $R \subseteq T$ *be rings, and let* P *be a prime ideal of* R *and* M *be a prime ideal of* T. *Let* W *be the set of all prime ideals of* T *that lie over* P. *If* W *is not empty, then there is a* $p \in W$ *such that* $p \subseteq M$ *if and only if* $\cap\{p' \mid p' \in W\} \subseteq M$.

Proof. One direction is trivial. For the other, assume $\cap\{p' \mid p' \in W\} \subseteq M$. We will show there is some $p \in W$ with $p \subseteq M$. (This task is simple if W happens to be finite.) Let p be a prime of T contained in M and minimal over $\cap\{p' \mid p' \in W\}$. It is well known that p consists of zero divisors modulo that intersection [3, Theorem 84]. That is, if $x \in p$, then there is a y not contained in $\cap\{p' \mid p' \in W\}$ such that $xy \in \cap\{p' \mid p' \in W\}$. Therefore, for some $p' \in W$, we have $y \notin p'$ but $xy \in p'$. It follows that $x \in p'$. This shows that $p \subseteq \cup\{p' \mid p' \in W\}$. Now consider any $x \in p \cap R$. For some $p' \in W$ we have $x \in p' \cap R = P$. Thus, $p \cap R \subseteq P$, and obviously $P \subseteq \cap\{p' \mid p' \in W\}\} \subseteq p$, so that $P \subseteq p \cap R$. We now have $p \cap R = P$, showing $p \in W$. Since $p \subseteq M$, that completes the argument. □

Remark 3. Let $R \subseteq T$ be an integral extension of rings. In [4, Proposition 2], it is shown that $R \subseteq T$ satisfies going down if and only if $R \subseteq R[t]$ satisfies going down for all $t \in T$. We leave to the reader the exercise of giving a second proof of that fact, using Lemma 8. Although the two approaches have much in common, we feel that Lemma 8 throws a bit more light on the subject.

Notation. Let $P^{\#} = \cap\{p' \in \operatorname{Spec} R' \mid p' \cap R = P\}$. Also let $R^{\#} = R + P^{\#} = \{r + x \mid r \in R \text{ and } x \in P^{\#}\}$.

Lemma 9. $R^{\#}$ *is a domain between* R *and* R'. $P^{\#}$ *is a prime ideal in* $R^{\#}$ *(and an ideal in* R'*) and is the only prime ideal of* $R^{\#}$ *lying over* P *in* R.

Proof. Obviously $R \subseteq R^{\#} \subseteq R'$, and using that $P^{\#}$ is obviously an ideal in R', it is easily verified that $R^{\#}$ is a domain. The definition of $P^{\#}$ easily implies $P^{\#} \cap R = P$. Suppose $r + x$ and $s + y$ are two elements of $R^{\#}$, with $r, s \in R$ and $x, y \in P^{\#}$, such that $(r + x)(s + y) \in P^{\#}$. Since $sx + ry + xy \in P^{\#}$, we see $rs \in P^{\#} \cap R = P$, and so we may assume $r \in P \subseteq P^{\#}$, showing $r + x \in P^{\#}$. Thus $P^{\#}$ is a prime ideal of $R^{\#}$. Finally, suppose Q is any prime ideal of $R^{\#}$ lying over P in R. Then there is a prime ideal p' in R' with $p' \cap R^{\#} = Q$, so that $p' \cap R = P$. By definition, we have $P^{\#} \subseteq p'$, and so $P^{\#} \subseteq p' \cap R^{\#} = Q$. As $P^{\#}$ and Q are both in $R^{\#}$ and both lie over P, incomparability shows that $Q = P^{\#}$. Thus $P^{\#}$ is the unique prime of $R^{\#}$ lying over P. □

We come to the third and final key piece in our puzzle.

Lemma 10. *(a)* P *a* K-*prime if and only if* $P^{\#}$ *is a* K-*prime.*
(b) *The following are equivalent.*

   (i) P *is a GDM prime.*
   (ii) $P^{\#}$ *is a GDM prime.*
   (iii) $P^{\#} \subseteq J(R^{\#})$.
   (iv) $P^{\#} \subseteq J(R')$.

Proof. (a) Suppose $P^{\#}$ not a K-prime, so that there is an integral extension T of $R^{\#}$ in which exactly two primes, say p_1 and p_2, lie over $P^{\#}$, and p_1 and p_2 are comaximal. Obviously p_1 and p_2 lie over P, and since $P^{\#}$ is the unique prime of $R^{\#}$ lying over P, there are no other primes of T lying over P. Thus we see P is not a K-prime.

Conversely, if P is not a K-prime, then by Lemma 3((a) $\Rightarrow$ (c)) there is an integral upper K to 0 in $R[X]$ with K contained in $(P, X)R[X]$ and in $(P, X + 1)R[X]$, but in no other uppers to P. K can be lifted to an integral upper L to 0 in $R^{\#}[X]$. Since $K \subseteq (P, X)R[X]$ and L lies over K, by going up there is a prime ideal q of $R^{\#}[X]$ containing L and lying over $(P, X)R[X]$. It is easy to verify that q must be an upper to some prime of $R^{\#}$ lying over P. The only such prime is $P^{\#}$, and so q is an upper to $P^{\#}$. Since $X \in (P, X)R[X] \subseteq q$, we see that q must equal $(P^{\#}, X)R^{\#}[X]$. Thus $L \subseteq (P^{\#}, X)R^{\#}[X]$. Similarly, $L \subseteq (P^{\#}, X + 1)R^{\#}[X]$. Now any upper q' to $P^{\#}$ containing L contracts to an upper to P containing K. Thus $q' \cap R[X]$ is either $(P, X)R[X]$ or $(P, X + 1)R[X]$. Since q' contains either X or $X + 1$, it equals either $(P^\#, X)R^\#[X]$ or $(P^\#, X + 1)R^\#[X]$. Now Lemma 3$((c) \Rightarrow (a))$ shows $P^\#$ is not a K-prime.

(b) (i) $\Rightarrow$ (iii): Suppose (i) holds. Let M be a maximal ideal of $R^\#$. As $R \subseteq R^\#$ is an integral extension, the definition of GDM prime shows that M contains a prime of $R^\#$ lying over P. The only possibility is that M contains $P^\#$. Thus $P^\# \subseteq J(R^\#)$, and so (i) $\Rightarrow$ (iii).

(iii) $\Rightarrow$ (iv): Use that maximal ideals of R' contract to maximal ideals of $R^\#$.

(iv) $\Rightarrow$ (i): Suppose $P^\# \subseteq J(R')$. We will show P is a GDM prime. Let T be an integral extension domain of R, and let M be a maximal ideal of T. (We must show some prime of T contained in M lies over P.) Let S be the domain gotten by adjoining all the elements of R' to T. Thus $T \subseteq S$ is an integral extension, and so we can lift M to a maximal ideal N of S. As $R' \subseteq S$, $N \cap R'$ is a maximal ideal of R'.

Since (iv) shows $\cap\{p' \in \operatorname{Spec} R' \mid p' \cap R = P\} = P^\# \subseteq N \cap R'$, Lemma 8 shows there is a $p \in \operatorname{Spec} R'$ lying over P, with $p \subseteq N \cap R'$. Since R' (being integrally closed) satisfies the famous going down theorem, there is a prime q of S with $q \cap R' = p$ and $q \subseteq N$. Contracting to T, we see that $q \cap T$ is contained in $N \cap T = M$ and lies over P, showing P is a GDM prime.

(ii) $\Leftrightarrow$ (iii): We iterate, now finding $(R^\#)^\#$ and $(P^\#)^\#$. Since $P^\#$ is the unique prime ideal of $R^\#$ lying over P in R, we see that a prime ideal p' in R' lies over $P^\#$ in $R^\#$ if and only if it lies over P in R. Therefore, the definition shows $P^{\#\#} = P^\#$. Also, $R^{\#\#} = R^\# + P^{\#\#} = R^\# + P^\# = R^\#$. Using the equivalence of (i) and (iii) applied to $P^\#$, we now see $P^\#$ is a GDM prime if and only if $P^{\#\#} \subseteq J(R^{\#\#})$ if and only if $P^\# \subseteq J(R^\#)$.   □

Corollary 2. *Suppose P is a K-prime. If $P^\# \subseteq J(R^\#)$, then P is an H-prime and a GDM prime (so that $P \subseteq J(R)$).*

Proof. If $P^\# \subseteq J(R^\#)$, by Lemma 10(b), P is a GDM prime. By Theorem 5, P is an H-prime.   □

6   Proposition 2 (Slightly Augmented)

Proposition 2. *Suppose that for all nonzero non-units $y \in R'$, there is a prime ideal Q' of R' containing y such that at least one of the following is true: $Q' \neq Q'^2$, or R'/Q' is not integrally closed, or the quotient field of R'/Q' is not algebraically closed. If P is a K-prime of R, then $P \subseteq J(R)$, and P is an H-prime and a GDM prime.*

Proof. Assume P is a K-prime of R. By Corollary 2, it will suffice to show that $P^\# \subseteq J(R^\#)$. If not, let M be a maximal ideal of $R^\#$ not containing $P^\#$, and write $x + y = 1$ with $x \in P^\#$ and $y \in M$. Obviously y is a nonzero non-unit in $R^\#$ and so also in the integral extension R'. By hypothesis, there is a prime ideal Q' of R'

containing y such that either $Q' \neq Q'^2$, or R'/Q' is not integrally closed, or the quotient field of R'/Q' is not algebraically closed. Let $Q = Q' \cap R^{\#}$. Since $y \in Q$, we see that $P^{\#}$ and Q are comaximal. By Corollary 1 applied to $R^{\#}$, and its primes $P^{\#}$ and Q, we see that $P^{\#}$ is not a K-prime. That contradicts Lemma 10(a). $\quad \square$

The next corollary shows that Proposition 2 applies to a large class of domains.

Corollary 3. *Suppose R' satisfies any one of conditions (i) through (iv) below. If P is a K-prime of R, then $P \subseteq J(R)$, and P is an H-prime and a GDM prime.*

(i) *There is a subset S of $\operatorname{Spec} R'$ such that $R' = \cap\{R'_{Q'} \mid Q' \in S\}$ and such that for each $Q' \in S$, at least one of the following holds: (i) $Q'^2 \neq Q'$; (ii) R'/Q' is not integrally closed; or (iii) the quotient field of R'/Q' is not algebraically closed.*

(ii) *For every maximal ideal M of R', either $M \neq M^2$ or R/M is not algebraically closed.*

(iii) *R' is an intersection of some set W of quasi-local domains (D_α, N_α), each between R' and its quotient field, such that for each α, $\cap\{N_\alpha^n \mid n \geq 1\} = 0$.*

(iv) *R' is the intersection of a set of DVRs between R' and its quotient field. (This case includes Krull domains and so includes the case that R is Noetherian.)*

Proof. It will suffice to show that in each case, R' satisfies the hypothesis of Proposition 2.

(i) It will suffice to show that if y is a nonzero non-unit in R', then one of the Q' in S contains y. If not, then we would have $y^{-1} \in \cap\{R'_{Q'} \mid Q' \in S\} = R'$, a contradiction.

(ii) This follows from (i), since R equals the intersection of all of its maximal localizations.

(iii) If $N_\alpha = 0$, then (D_α, N_α) must be the quotient field of R' and can be ignored. Thus, we may assume $N_\alpha \neq 0$. Let $Q_\alpha = N_\alpha \cap R'$. For some $0 \neq z \in N_\alpha$, write $z = r/s$ with r and s nonzero in R'. Thus $r = sz \in N_\alpha \cap R = Q_\alpha$, showing $Q_\alpha \neq 0$. Therefore, $Q_\alpha \not\subseteq \cap\{N_\alpha^n \mid n \geq 1\}$. It follows that $Q_\alpha \neq Q_\alpha^2$. By (i), it will suffice to show that every nonzero non-unit y of R' is contained in some Q_α. Were that false, then y would be a unit in each D_α and so a unit in R', which is a contradiction.

(iv) This follows easily from (iii). $\quad \square$

Question 2. Modifying our earlier question, we ask if the concepts of H-prime and K-prime are equivalent for GDM primes?

References

1. W. Heinzer, D. Lantz, Factorization of Monic polynomials. Proc. Am. Math. Soc. **131**, 1049–1052 (2003)
2. W. Heinzer, S. Wiegand, Prime ideals in two-dimensional polynomial rings. Proc. Am. Math. Soc. **107**, 577–586 (1989)
3. I. Kaplansky, *Commutative Rings* (University of Chicago Press, Chicago, 1974)
4. S. McAdam, Going down and open extensions. Can. J. Math **27**, 111–114 (1975)
5. S. McAdam, Strongly comaximizable primes. J. Algebra **170**, 206–228 (1994)
6. N. Nagata, *Local Rings* (Interscience, New York, 1962)

Finitely Stable Rings

Bruce Olberding

Abstract A commutative ring R is finitely stable provided every finitely generated regular ideal of R is projective as a module over its ring of endomorphisms. This class of rings includes the Prüfer rings, as well as the one-dimensional local Cohen-Macaulay rings of multiplicity at most 2. Building on work of Rush, we show that R is finitely stable if and only if its integral closure $\overline{R}$ is a Prüfer ring, every R-submodule of $\overline{R}$ containing R is a ring and every regular maximal ideal of R has at most 2 maximal ideals in $\overline{R}$ lying over it. This characterization is deduced from a more general theorem regarding what, motivated by work of Knebusch and Zhang, we term a finitely stable subring R of a ring between R and its complete ring of quotients.

Keywords Stable ideal • Finitely stable ring • Prüfer ring • Prüfer extension

Mathematics Subject Classification (2011): 13F05; 13B22; 13C10

1 Introduction

Following Sally and Vasconcelos [24, 25], an ideal I of a commutative ring R is *stable* if I is projective as a module over its ring of endomorphisms End(I). This terminology originates with Lipman [13], who gave a different definition of stable ideals (one that reflects the stabilization of a certain chain of blow-up algebras) that Sally and Vasconcelos later observed was equivalent to the one given here when R

B. Olberding (✉)
Department of Mathematical Sciences, New Mexico State University,
Las Cruces, NM 88003-8001, USA
e-mail: olberdin@nmsu.edu

M. Fontana et al. (eds.), *Commutative Algebra: Recent Advances in Commutative Rings,*   269
Integer-Valued Polynomials, and Polynomial Functions, DOI 10.1007/978-1-4939-0925-4_16,
© Springer Science+Business Media New York 2014

is a one-dimensional local Cohen-Macaulay ring. When I is regular (meaning that I contains a nonzerodivisor), the ring $\text{End}(I)$ can be identified with a subring of the total quotient ring $\text{Quot}(R)$ of R, and hence, using a standard characterization of projective regular ideals, a regular ideal is stable if and only if it is an invertible ideal of a ring between R and $\text{Quot}(R)$. Thus stability generalizes the multiplicative notion of invertibility and has been studied from this point of view by many authors; see [15] and its references for older background, and for some recent examples of papers involving stable ideals in integral domains, see [2,4,6,7,9,14,21,26–28].

A ring R is *finitely stable* if every finitely generated regular ideal of R is stable. Thus a *Prüfer ring*, a ring in which every finitely generated regular ideal is invertible, is finitely stable. However, the class of finitely stable rings is broader than the class of Prüfer rings. For example, Bass [1, Corollary 7.3] showed that a reduced ring for which every ideal can be generated by two elements has the property that every ideal is stable. In particular, with k a field and X an indeterminate for k, the ring $k[X^2, X^3]$ is a finitely stable ring. More generally, Rush [23, Proposition 2.5] has shown that a ring for which every finitely generated ideal can be generated by two elements is finitely stable. The quasilocal reduced group rings with the two-generator property (and hence the property of being finitely stable) are characterized in [23]. We give several other sources of examples of finitely stable rings in Sect. 6.

The main goal of this article is to characterize a finitely stable ring R in terms of its integral closure $\overline{R}$. We show in Corollary 5.11 that a ring R is finitely stable if and only if $\overline{R}$ is a Prüfer ring; every R-module between R and $\overline{R}$ is a ring; and each regular maximal ideal of R has at most 2 maximal ideals in $\overline{R}$ lying over it. In fact, much of this characterization was proved already by Rush [23, Proposition 2.1], who showed that if R is a finitely stable ring, then $\overline{R}$ is a Prüfer ring and every R-module between R and $\overline{R}$ is a ring. (What we term a finitely stable ring, Rush calls "stable." We reserve *stable ring* for the case in which every regular ideal of R is stable.) Rush also proved that the converse holds whenever R is a quasilocal ring having at most 2 maximal ideals in $\overline{R}$ lying over it. Thus to combine these two results into a characterization of finitely stable rings two gaps remain: to remove the quasilocal hypothesis and prove that the integral closure of a finitely stable ring can have at most 2 maximal ideals lying over a maximal ideal of R. In Theorem 5.4, we fill in this gap, and in subsequent corollaries we draw a number of conclusions from this result.

Moving beyond the classical setting, we develop this theorem from a more general point of view based on that of the treatment of Prüfer subrings given by Knebusch and Zhang in [10]. This allows us to relativize the notion of a finitely stable ring: regular ideals are replaced by dense ideals, and rather than requiring all finitely generated regular (or dense) ideals to be invertible, we require only certain filters of them to have this property, those that blow up in a fixed ring between R and its complete ring of quotients. In Sect. 2, we recast the notion of stability in this more general setting, and in Sect. 5 we define finitely stable subrings. The idea is to consider a ring R and a ring S between R and its complete ring of quotients $Q(R)$ (a notion we review in Sect. 2); then R is finitely stable in S if every finitely generated ideal I of S such that $IS = S$ is stable. As in the theory of Prüfer subrings, a technical assumption—that R is "tight" in S—is also needed here; tight extensions, which are a special class of flat extensions, are reviewed in Sect. 4. In any case, it follows that a ring R is finitely stable (in the traditional sense) if and only if R is finitely stable in its total ring of quotients (in this new sense). Corollary 6.5 gives one application which the flexibility of this approach affords over the classical approach. Other motivations for this approach to multiplicative ideal theory can be found in [10].

In a future paper we will carry these ideas through for the class of stable rings, those rings for which every regular ideal is stable. We generalize this to the setting of stable subrings, and we classify such rings using pullback decompositions of their localizations at maximal ideals. In doing so, we correct also an error from [16], where it is asserted that every quasilocal stable domain is the pullback of a strongly discrete valuation domain V and a one-dimensional stable domain whose quotient field is the residue field of V. (The error is in Lemma 4.10 of [16], in the assertion that P is a primary ideal.) While it is the case that every such pullback is a stable domain, not every quasilocal stable domain can be decomposed as a pullback of a strongly discrete valuation domain and a one-dimensional stable *domain D*; instead, zerodivisors and other subtleties

Conventions. The total ring of quotients of a ring R is denoted Quot(R), while the complete ring of quotients of R is denoted $Q(R)$. Integral closure is denoted $\overline{R}$ but depends on the ambient ring, in the sense that when we work within a ring extension $R \subseteq S$, then $\overline{R}$ denotes the integral closure of R in S.

2 Preliminaries on Stable Ideals

In this section we recast some basic properties of stable ideals in the more general setting of dense ideals, and we show that as with regular ideals, dense stable ideals can be characterized by multiplicative properties. We recall first some terminology and notation.

(2.1) An ideal I of the ring R is *dense* if the only element r in R for which $rI = 0$ is $r = 0$. In particular, regular ideals are dense.

(2.2) When A and B are R-modules, we define $(B :_R A) = \{r \in R : rA \subseteq B\}$, and when R is clear from context, we write $(B : A)$ for $(B :_R A)$.

(2.3) The ring S is a *ring of quotients* of R if for all $x \in S$, $(R : x)S$ is a dense ideal of S. There exists a ring of quotients $Q(R)$, the *complete ring of quotients of R*, such that for every ring of quotients S of R, there is a unique ring homomorphism $S \rightarrow Q(R)$, and every such homomorphism is injective [11, Sect. 2.3]. Thus we may view a ring of quotients of R as a subring of $Q(R)$. In turn, every ring between R and $Q(R)$ is a ring of quotients of R.

(2.4) The *classical ring of quotients* is the localization of R at the multiplicatively closed set of nonzerodivisors of R. We denote this ring by Quot(R), and in line with the above convention we assume Quot(R) $\subseteq Q(R)$.

(2.5) When A and B are R-modules, we let $[B : A] = \{q \in Q(R) : qA \subseteq B\}$, and when S is a ring between R and $Q(R)$, we set $[B :_S A] = S \cap [B : A]$. In particular, $(B : A) = R \cap [B : A]$.

(2.6) Let $R \subseteq S \subseteq Q(R)$ be rings, and let A be an R-submodule of $Q(R)$. Then A is *invertible* if there exists an R-submodule B of $Q(R)$ such that $AB = R$; equivalently, $A[R : A] = R$. If also A and B are R-submodules of S, then A is an *S-invertible R-module*. An ideal I of R is invertible if and only if I is a dense ideal and a finitely generated projective R-module [10, Proposition 2.4, p. 99].

(2.7) We say that an R-submodule I of $Q(R)$ is a *fractional ideal* of R if there exists an invertible ideal J of R such that $IJ \subseteq R$. The fractional ideal I is *dense* if the only element r of R for which $rI = 0$ is $r = 0$. With J as above, then I is dense if and only if IJ is dense. Every invertible R-submodule of $Q(R)$ is a dense fractional ideal of R.

(2.8) For a fractional ideal I of R, let $E(I) = [I : I]$. When I is dense, then the canonical mapping $\phi : E(I) \to \mathrm{End}_R(I) : q \mapsto f_q$, where $f_q(x) = qx$ for all $x \in I$, is an isomorphism of R-algebras. When I is an ideal, this is a consequence of [11, Corollary, p. 99]. To see that it is also true when I is a dense fractional ideal, note first that since I is not annihilated by any nonzero element in $Q(R)$ [11, Corollary, p. 41], then the R-algebra homomorphism ϕ is injective. Thus we need only to verify that ϕ is surjective. Let $f \in \mathrm{End}_R(I)$, and let J be an invertible ideal such that $IJ \subseteq R$. Then since IJ is dense in R and $f(IJ) \subseteq IJ$, there exists by [11, Corollary, p. 99] an element q of $Q(R)$ such that $f(x) = qx$ for all $x \in IJ$. Since J is invertible, there exist $x_1, \ldots, x_n \in J$ and $y_1, \ldots, y_n \in [R : J]$ with $1 = \sum_i x_i y_i$. Let $x \in I$. Then, multiplying by $f(x)$, we have $f(x) = \sum_i x_i y_i f(x) = \sum_i y_i f(x_i x) = \sum_i y_i q x_i x = qx \sum_i x_i y_i = qx$, which proves that $f = \phi(q)$ and hence ϕ is an isomorphism.

(2.9) A dense fractional ideal I of R is *stable* if it is projective over its ring of endomorphisms. Thus by (2.8), I is stable if and only if I is projective as an $E(I)$-module. We show in Proposition 2.11 that a dense fractional ideal I is stable if and only if I is an invertible $E(I)$-module.

Lemma 2.10 (Knebusch–Zhang [10]). *Let $R \subseteq S \subseteq Q(R)$ be rings. The following statements are equivalent for an R-submodule I of S:*

(1) *I is an S-invertible R-module.*
(2) *$IS = S$ and $[J : I]I = J$ for all R-submodules J of S.*
(3) *$IS = S$ and I is an invertible R-module.*
(4) *$IS = S$ and I is a projective R-module.*

Proof. Since the proposition is not explicitly stated in [10], we indicate how it follows from standard arguments and results in [10].

(1) $\Rightarrow$ (2). That $[J : I]I = J$ for all R-submodules J of S is given by [10, Lemma 1.11, p. 90]. To see that $IS = S$, let J be an R-submodule of S such that $IJ = R$. Then $[R : I] = J \subseteq S$. Since $[S : I]I = S$, multiplying

both sides by $[R : I]$ gives $[S : I] = [R : I]S$. Thus since $S \subseteq [S : I]$ and $[R : I] \subseteq S$, we have $[R : I]S = S$. Multiplication by I then gives $S = IS$.

(2) $\Rightarrow$ (3). With $J = R$, (2) implies that $[R : I]I = R$, and hence I is invertible.

(3) $\Rightarrow$ (1). Since I is invertible, multiplying $IS = S$ by $[R : I]$ gives $S = [R : I]S$. Hence $[R : I] \subseteq S$, which proves that I is S-invertible.

(1) $\Leftrightarrow$ (4). This follows from [10, Proposition 2.3, p. 97]. □

From the lemma, we deduce a useful multiplicative characterization of stable fractional ideals.

Proposition 2.11. *A dense fractional ideal I of the ring R is stable if and only if I is an invertible $E(I)$-module; if and only if $1 \in [I : I^2]I$.*

Proof. By (2.8), I is stable if and only if I is projective as an $E(I)$-module. Thus by Lemma 2.10 (applied to the case where $S = Q(R)$), I is stable if and only if I is an invertible $E(I)$-module; if and only if $1 \in [E(I) : I]I = [I : I^2]$. □

One consequence of the criterion in Proposition 2.11 is that stability of ideals transfers to localizations and ring extensions. More generally, we have the following corollary.

Corollary 2.12. *Let $R_1 \subseteq S_1 \subseteq Q(R_1)$ and $R_2 \subseteq S_2 \subseteq Q(R_2)$ be ring extensions such that there exists a homomorphism of rings $\phi : Q(R_1) \to Q(R_2)$ with $\phi(R_1) \subseteq R_2$ and $\phi(S_1) \subseteq S_2$. If I is a stable fractional ideal of R_1 such that $IS_1 = S_1$, then $\phi(I)R_2$ is a stable fractional ideal of R_2 with $\phi(I)S_2 = S_2$.*

Proof. Since $IS_1 = S_1$, then I is dense and $\phi(I)S_2 = S_2$. Since I is a fractional ideal, there exists an invertible ideal J of R such that $IJ \subseteq R$. Then $\phi(J)R_2$ is an invertible ideal of R_2 (with inverse $\phi(J^{-1})R_2$) and $\phi(I)\phi(J)R_2 \subseteq R_2$. Hence $\phi(I)R_2$ is a fractional ideal of R_2 that is also dense since $\phi(I)S_2 = S_2$. Now since I is a stable dense fractional ideal of R, Proposition 2.11 implies that there exist $x_1, \ldots, x_n \in I$ and $y_1, \ldots, y_n \in [I : I^2]$ with $1 = \sum x_i y_i$. Thus

$$1 = \phi(1) = \sum \phi(x_i)\phi(y_i) \in [\phi(I)R_2 : \phi(I)^2 R_2]\phi(I)R_2,$$

and hence by Proposition 2.11, $\phi(I)R_2$ is a stable fractional ideal of R_2. □

The next proposition, which is well known for the case where I is a regular ideal, is a simple application of Proposition 2.11.

Proposition 2.13. *Suppose that I is a dense fractional ideal of the ring R:*

(1) *If $I^2 = AI$ for some invertible fractional ideal A of R with $A \subseteq I$, then $I = AE(I)$ and I is stable.*

(2) *If I is stable and J is also a stable dense fractional ideal, then IJ is stable.*

(3) *If J is a stable dense fractional ideal of R such that $E(J) \subseteq E(I)$ and IJ is stable, then I is stable.*

Proof. (1) Since $A \subseteq I$, then $AE(I) \subseteq I$. To verify the reverse inclusion, let $x \in I$. Then $xI \subseteq I^2 = AI$, so $x[R : A]I \subseteq I$, which shows that $x[R : A] \subseteq E(I)$ and hence $x \in AE(I)$. Therefore, $I = AE(I)$. Since A is an invertible ideal of R, we have by Corollary 2.12 that $I = AE(I)$ is stable.

(2) By Lemma 2.10(2),

$$[E(IJ) : IJ]IJ = [[E(IJ) : J] : I]IJ = [E(IJ) : J]J = E(IJ).$$

Thus IJ is an invertible $E(IJ)$-module, and since IJ is a dense fractional ideal, Proposition 2.11 implies that IJ is stable.

(3) We may assume without loss of generality that $E(J) = R$. Then by Proposition 2.11, J is an invertible R-module. Since IJ is stable and J is invertible, Lemma 2.10 and Proposition 2.11 imply that

$$1 \in [IJ : (IJ)^2]IJ = [[IJ : I^2J] : J]IJ = [IJ : I^2J]I = [I : I^2]I,$$

so that again by Proposition 2.11, I is stable.   □

3   Stable Ideals in Quadratic Extensions

Handelman [8, p. 147] shows that if R is a one-dimensional Noetherian domain having module-finite integral closure $\overline{R}$ in its quotient field, then R is a stable domain if and only if every ring between R and $\overline{R}$ is a Gorenstein ring. In proving this, Handelman showed that when R is a stable domain, every R-module B with $R \subseteq B \subseteq \overline{R}$ is a ring. We single out this last property and say that an extension of rings $R \subseteq S$ is *quadratic* if every R-submodule of S containing R is a ring. Thus $R \subseteq S$ is a quadratic extension if and only if for all $x, y \in S$, $xy \in xR + yR + R$; if and only if every finitely generated R-submodule A of S containing R satisfies $A^2 = A$. The terminology here is motivated by the fact that when $R \subseteq S$ is a quadratic extension, then $x^2 \in xR + R$ for all $x \in S$. In particular, quadratic extensions are also integral extensions. Moreover, if two is a unit in R, then $R \subseteq S$ is a quadratic extension if and only if every $t \in S$ is the root of a monic degree 2 polynomial with coefficients in R. Quadratic extensions are also studied in [18–20], where they play a key role in analyzing some classes of analytically ramified local Noetherian rings.

Lemma 3.1. *The following are equivalent for an extension of rings $R \subseteq S$:*

(1) $R \subseteq S$ *is a quadratic extension.*

(2) *For each multiplicatively closed subset X of R, $R_X \subseteq S_X$ is a quadratic extension.*

(3) *For each maximal ideal M of R, $R_M \subseteq S_M$ is a quadratic extension.*

Proof. To prove that (1) implies (2), let X be a multiplicatively closed subset of R, and let A be a finitely generated R_X-submodule of S_X containing R_X. Write $A = (a_1, \ldots, a_n)R_X$ for some $a_1, \ldots, a_x \in A$. Then for each i, $a_i = b_i/x_i$ for some $b_i \in S$ and $x_i \in X$. Define $B = (1, b_1, \ldots, b_n)R$. Then $A \subseteq B_X$. Also, if $x \in X$, then $1/x \in A$ since $R_X \subseteq A$, and $b_i/x = a_i(x_i/x) \in A$. Thus $B_X \subseteq A$, and hence $B_X = A$. Moreover, $R \subseteq B \subseteq S$, so since $R \subseteq S$ is quadratic, $B^2 = B$. Hence $A^2 = B_X^2 = B_X = A$, proving that A is a ring. This shows that (1) implies (2). That (2) implies (3) is clear. To see that (3) implies (1), let C be a finitely generated R-submodule of S containing R. Then for each maximal ideal M of R, $R_M \subseteq C_M \subseteq S_M$, so by (3), $C_M^2 = C_M$. Thus $C^2 = C$, since this equality holds locally, and this proves that C is a ring and $R \subseteq S$ is a quadratic extension. $\square$

We recall next Handelman's classification of quadratic extensions of finite-dimensional algebras in order to apply it in Proposition 3.3 and Theorem 5.4 to quadratic extensions of the residue field of a quasilocal ring. We denote by $\mathbb{F}_2$ the field $\mathbb{Z}/2\mathbb{Z}$.

Lemma 3.2 (Handelman [8, Lemma 5]). *Let F be a field and let S be a finite-dimensional F-algebra such that $F \subseteq S$ is a quadratic extension. Then S is isomorphic as an F-algebra to one of the following: F; a quadratic extension field of F; a quasilocal ring with square zero maximal ideal and residue field isomorphic to F; $F \times F$; or $F \times F \times F$. In the last case, $F = \mathbb{F}_2$.*

Proposition 3.3. *Let $R \subseteq S$ be a quadratic extension of rings. If P is a prime ideal of R, then S has at most 3 prime ideals lying over P.*

Proof. By Lemma 3.1, $R_P \subseteq S_P$ is a quadratic extension, so to simplify notation, we assume that R is a quasilocal ring with maximal ideal M, and we show that there are at most 3 prime ideals of S lying over M. In fact, since $R \subseteq S$ is an integral extension, it suffices to show that S has at most 3 maximal ideals. Suppose by way of contradiction that S has at least 4 maximal ideals, say M_1, M_2, M_3, M_4. Since $R \subseteq S$ is an integral, extension, these must lie over M. For each $i = 1, 2, 3, 4$, choose x_i in M_i but in no other maximal ideal M_j, $j \neq i$. Define $T = R + (x_1, x_2, x_3, x_4)R$. Since $R \subseteq S$ is a quadratic extension, T is a ring, so T/MT is a finite-dimensional R/M-algebra. Also, since $T \subseteq S$ is an integral extension, the contracted ideals $M_i \cap T$ are maximal in T, and the choice of the x_i implies that these ideals are distinct. Therefore, T/MT has at least 4 maximal ideals. But since $(R + MT)/MT \subseteq T/MT$ is also a quadratic extension and $(R + MT)/MT \cong R/M$ is a field, T/MT must satisfy one of the conditions of Lemma 3.2, an impossibility since T has 4 maximal ideals. Thus S has no more than 3 maximal ideals. $\square$

The next proposition, which is based on an argument from Rush [22, Lemma 2.1], provides a first example of the connection between quadratic extensions and stability.

Proposition 3.4. *Let $R \subseteq S \subseteq Q(R)$ be rings. Then $R \subseteq S$ is a quadratic extension if and only if $R \subseteq S$ is an integral extension for which every finitely generated R-submodule I of S containing R is an invertible $E(I)$-module.*

Proof. If $R \subseteq S$ is a quadratic extension, then $R \subseteq S$ is an integral extension and every R-submodule of S containing R is a ring, hence is invertible as a module over itself. Conversely, suppose that every finitely generated R-submodule I of S containing R is an invertible $E(I)$-module, and let A be a finitely generated R-submodule of S containing R. We claim that A is a ring. More precisely, we claim that $A = E(A)$. Since $1 \in A$, it follows that $E(A) \subseteq A$, so we need only to show that $A \subseteq E(A)$. Since A is an invertible $E(A)$-module, we have $1 \in [E(A) : A]A \subseteq [E(A) : A]S$. Also, since $AS = S$, it follows that $[E(A) : A] \subseteq S$. Thus $S = [E(A) : A]S$, and since $1 \in A$, $[E(A) : A]$ is an ideal of $E(A)$ that does not survive in S. Yet $E(A) \subseteq S$, and since S is integral over R, then S is integral over $E(A)$. Therefore, since $[E(A) : A]$ does not survive in S, it must be that $1 \in [E(A) : A]$, Hence $A \subseteq E(A)$. This proves $A = E(A)$, and hence $R \subseteq S$ is a quadratic extension.   □

Although not needed in later sections, we note in the next proposition that in the special case where $S = E(M)$ for a maximal ideal M, the condition in Proposition 3.4 that $R \subseteq S$ is integral is redundant.

Proposition 3.5. *Let R be a ring, and suppose that M is a maximal ideal of R. Then $R \subseteq E(M)$ is a quadratic extension if and only if every finitely generated R-submodule I of $E(M)$ containing R is an invertible $E(I)$-module.*

Proof. Suppose that every finitely generated R-submodule I of $E(M)$ containing R is $E(I)$-invertible. By Proposition 3.4, to prove that $R \subseteq E(M)$ is quadratic, it suffices to prove that $R \subseteq E(M)$ is an integral extension. If $R = E(M)$, there is nothing to prove, so we assume that $R \subsetneq E(M)$. Let $x \in E(M) \setminus R$, and set $A = xR + R$. We show that A is a ring (in particular, $E(A) = A$), so that $x^2 \in xR + R$, thus proving the claim. Since $1 \in A$, we have $E(A) \subseteq A \subseteq E(M)$. Also, $R/M \subseteq E(A)/M \subseteq A/M$. Since A/M has dimension at most 2 as an R/M-vector space, this forces $R = E(A)$ or $E(A) = A$. Suppose that $R = E(A)$. Since $AM \subseteq R$, we have $M \subseteq [R : A]$. On the other hand, since $1 \in A$ and $R \subsetneq A$, it must be that $[R : A] \subsetneq R$. Hence $M = [R : A]$. Now since A is an invertible $E(A)$-module and we have assumed that $E(A) = R$, it is the case that $[R : A]$ is an invertible ideal of R. Thus M is an invertible ideal of R. But this forces $E(M) = R$, contrary to assumption. Thus it must be that $A = E(A)$, and hence $R \subseteq E(M)$ is a quadratic extension. The converse is clear in light of Proposition 3.4.   □

The following proposition can be viewed as a kind of converse to Propositions 3.4 and 3.5. The argument is due to Rush and is adapted from the proof of Theorem 2.3 in [23].

Proposition 3.6 (cf. Rush [23, Theorem 2.3]). *Let $R \subseteq S \subseteq Q(R)$ be rings such that $R \subseteq S$ is a quadratic extension and S has at most 2 maximal ideals. If I is an*

*R-submodule of S such that IS is a principal regular ideal of S, then $I = yE(I)$
for some nonzerodivisor $y \in S$ and hence I is an invertible $E(I)$-module.*

Proof. By assumption there exists a nonzerodivisor $x \in S$ such that $IS = xS$.
We first show that x can be replaced by an element of I. To do this, it suffices
to prove that there exists $y \in I$ such that y is not in any ideal of the form IN,
N a maximal ideal of S. For then, since IS/IN is a one-dimensional S/N-vector
space, we have $xS = IS = yS + IN = yS + xN$, and since x is a nonzerodivisor
and this equality holds for each maximal ideal N of S, it follows that $xS = yS$.
Thus we show such a choice of y exists. Let N_1 be a maximal ideal of S, and observe
that $I \not\subseteq N_1 I$. For otherwise $IS \subseteq IN_1$, so that $x \in IN_1 = xN_1$, a contradiction
to the fact that x is a nonzerodivisor in S. (Since x is a nonzerodivisor in R, it is a
nonzerodivisor in $Q(R)$ also [11, Corollary, p. 41].) If N_1 is the only maximal ideal
of R_1, then we choose $y \in I \setminus IN_1$. Otherwise, by assumption, S has only one other
maximal ideal N_2. As above, $I \not\subseteq IN_2$. Furthermore, $I \neq (IN_1 \cap R) \cup (IN_2 \cap R)$,
since an abelian group is not the union of two proper subgroups, so there exists
$y \in I \setminus (IN_1 \cup IN_2)$. Therefore, in either case, we have $y \in I$ with $IS = xS = yS$.
Finally, y is a nonzerodivisor of S and $R \subseteq y^{-1}I \subseteq S$. Hence, since $R \subseteq S$ is a
quadratic extension, $y^{-1}I$ is a ring, which implies that $I = yE(I)$. □

4 Prüfer Subrings

In this section we review the notion of a Prüfer subring from [10], and we make a
few observations regarding these rings that are needed in the next section:

(4.1) An extension of rings $R \subseteq S$ is *weakly surjective* if for every prime ideal
P of R with $PS \neq S$, the canonical mapping $R_P \to S_P$ is surjective;
equivalently, the inclusion map $R \to S$ is flat and an epimorphism in the
category of rings [10, Theorem 4.4, p. 42]. We rely on the following ideal-
theoretic characterization: An extension $R \subseteq S$ is weakly surjective if and
only if $(R : x)S = S$ for all $x \in S$ [10, Theorem 3.13, p. 37]. Thus when
$R \subseteq S$ is weakly surjective, then S is a ring of quotients of R and we may
assume that $R \subseteq S \subseteq Q(R)$.

(4.2) A *tight extensions* is a special case of a weakly surjective extension. Following [10], the subring R of S is *tight* in S if for each $s \in S$, there exists
an S-invertible ideal I of R such that $I \subseteq (R : s)$; equivalently, for every
finitely generated R-submodule I of S, there exists an S-invertible ideal J
of R such that $IJ \subseteq R$. By contrast, $R \subseteq S$ is weakly surjective if and
only if the ideal I here is only asserted to be finitely generated and S-regular.
Colloquially, "fractions" in a weakly surjective extension can be cleared by
finitely generated S-regular ideals, while those in a tight extension can be
cleared by S-invertible ideals. In any case, since a tight extension $R \subseteq S$ is
weakly surjective, and hence is a ring of quotients, we can view S as a subring
of $Q(R)$.

(4.3) When R has only finitely many maximal ideals, then R is tight in S if and only if $S = R_X$ for some multiplicatively closed subset X of nonzerodivisors of R [10, Proposition 4.16, p. 116]. In particular, when R is tight in S and has only finitely many maximal ideals, then $R \subseteq S \subseteq \mathrm{Quot}(R)$.

(4.4) In [10], a Prüfer subring is defined in terms of Manis pairs. For our purposes we take a multiplicative characterization of Prüfer subrings [10, Theorem 2.1, p. 94] as our definition: The subring R of S is *Prüfer in S* if $R \subseteq S$ is a tight extension and every finitely generated S-regular ideal of R is S-invertible, where an R-submodule I of S is S-*regular* if $IS = S$. This notion encompasses the traditional concept of a Prüfer ring as one in which every finitely generated regular ideal of R is invertible: A ring R is a Prüfer ring if and only if R is a Prüfer subring of $\mathrm{Quot}(R)$.

We collect in the next proposition two characterizations of Prüfer subrings from [10], and we add two more that are needed later. Many other interesting characterizations, examples, and consequences can be found in [10].

Lemma 4.5. *The following statements are equivalent:*

(1) *R is Prüfer in S.*
(2) *$R \subseteq S$ is weakly surjective and every finitely generated S-regular ideal I of R is invertible.*
(3) *R is integrally closed in S and $R[s] = R[s^2]$ for all $s \in S$.*
(4) *R_X is Prüfer in S_X for each multiplicatively closed subset X of R.*
(5) *R_M is Prüfer in S_M for each maximal ideal M of R.*

Proof. The equivalence of (1) and (3) can be found in [10, Theorem 5.2, p. 47], and the equivalence of (1) and (2) is in [10, Theorem 2.1, p. 94].

(3) $\Rightarrow$ (4). Since (3) is equivalent to (1), to verify (4) we need only to show that the extension $R_X \subseteq S_X$ satisfies the criterion in (3). Let $s \in S$ and $b \in X$. Then by (3), $s = \sum_i r_i s^{2i}$ for some $r_i \in R$. Thus

$$\frac{s}{b} = \sum_i \frac{b^{2i} r_i}{b} \frac{s^{2i}}{b^{2i}} \in R_X[s^2/b^2],$$

which shows that $R_X[s/b] = R_X[s^2/b^2]$. Since R is integrally closed in S, then R_X is integrally closed in S_X, and hence R_X is Prüfer in S_X.

(4) $\Rightarrow$ (5). This is clear.

(5) $\Rightarrow$ (3). Since (1) is equivalent to (3), then R_M is integrally closed in S_M for each maximal ideal M of R. Thus R is integrally closed in S. Moreover, the equivalence of (1) and (3) implies that for each $s \in S$, $s/1 \in R_M[s^2/1]$ for all maximal ideals M of R, and hence $s \in R[s^2]$. Thus (3) is verified.   □

The next proposition, which is needed in the next section, generalizes to the setting of Prüfer subrings a few well-known facts about prime ideals in overrings of Prüfer domains.

Proposition 4.6. *Suppose R is Prüfer in S and A is a ring between R and S. Then the following statements hold for A:*

(1) *If I is an ideal of A, then $I = (I \cap R)A$; if I is S-regular, then so is $I \cap R$.*
(2) *The set of S-regular prime ideals of A is a tree with respect to set inclusion.*
(3) *If R has at most n S-regular maximal ideals, then A has at most n incomparable S-regular prime ideals.*

Proof. (1) Since R is Prüfer in S, the extension $R \subseteq A$ is weakly surjective [10, Theorem 5.2, p. 47], and so every ideal I of A satisfies $I = (I \cap R)A$ [10, Proposition 4.6, p. 43], from which it follows that if I is S-regular, then so is $I \cap R$.

(2) Since R is Prüfer in S, A is also Prüfer in S [10, Corollary 5.3, p. 50]. Suppose that P and Q are S-regular prime ideals of A contained in a maximal ideal M of A. Since $PQS = S$, there exists a finitely generated S-regular ideal $I \subseteq PQ$. Thus since A is Prüfer in S and I is S-regular, A/I is an arithmetical ring [10, Theorem 2.8, p. 101], and hence $P_M \subseteq Q_M$ or $Q_M \subseteq P_M$ [12, Example 18(b), p. 151]. Since P and Q are prime ideals, it follows that P and Q are comparable, and this verifies (2).

(3) Suppose that $P_1, \ldots, P_{n+1}$ are $n + 1$ incomparable S-regular prime ideals of A. Then by (1), $P_1 \cap R, \ldots, P_{n+1} \cap R$ are $n + 1$ incomparable S-regular prime ideals of R with $P_i = (P_i \cap R)A$. Yet by (2), the S-regular prime ideals of R form a tree with respect to inclusion, so since there are at most n maximal ideals of R, it is necessary that $P_i \cap R = P_j \cap R$ for some $i \neq j$, a contradiction to the fact that $P_1 \cap R, \ldots, P_{n+1} \cap R$ are incomparable. Therefore, A has at most n incomparable prime ideals. □

Sega [26, Proposition 3.10] uses Peskine's version of Zariski's Main Theorem to prove that if a domain R has a Prüfer integral closure $\overline{R}$ in its quotient field, then every overring A of R is flat over $A \cap \overline{R}$. In Proposition 4.7 we give a more general version of Sega's theorem, one that is adjusted to the generality of our setting and one whose relevance is that finitely stable rings have Prüfer integral closures (see Theorem 5.4). To do so, we use Evans' version of Zariski's Main Theorem [5, p. 45], since it permits zerodivisors: Let $R \subseteq S$ be rings with R integrally closed in S. Suppose that there exist $s_1, \ldots, s_n \in S$ with S integral over $R[s_1, \ldots, s_n]$. If there exists a prime ideal P of S that is maximal and minimal with respect to the prime ideals of S lying over $P \cap R$, then there exists $b \in R \setminus P$ such that $R_b = S_b$, where the subscript denotes localization at the set $\{1, b, b^2, \ldots\}$.

Proposition 4.7. *Let R be a subring of S. When the integral closure $\overline{R}$ of R in S is Prüfer in S, then every ring T between R and S is a weakly surjective extension of $T \cap \overline{R}$.*

Proof. Define $A = \overline{R} \cap T$, and let B be a ring between A and T. We first show that no two distinct comparable prime ideals of B lie over the same prime ideal in A. Suppose that there exist distinct prime ideals $P \subset Q$ of B such that $P \cap A = Q \cap A$.

Let $\overline{B}$ denote the integral closure of B in S. By Going Up, there exist distinct prime ideals $P_1 \subset Q_1$ of $\overline{B}$ such that $P_1 \cap B = P$ and $Q_1 \cap B = Q$. Now $\overline{A} = \overline{R}$ is a Prüfer subring of S, so since $\overline{A} \subseteq \overline{B} \subseteq S$, $\overline{B}$ is a weakly surjective extension of $\overline{A}$ [10, Theorem 5.2, p. 47]. Therefore, ideals of $\overline{B}$ are extended from their contractions to $\overline{A}$ [10, Proposition 4.6, p. 43], so that $P_1 = (P_1 \cap \overline{A})\overline{B}$ and $Q_1 = (Q_1 \cap \overline{A})\overline{B}$. Since $P_1 \neq Q_1$, it follows that $P_1 \cap \overline{A} \subset Q_1 \cap \overline{A}$ are distinct comparable prime ideals of $\overline{A}$ lying over the same prime ideal of A. Since this contradicts the fact that $\overline{A}$ is an integral extension of R, we conclude that no two distinct comparable prime ideals of B lie over the same prime ideal of A.

To prove next that $A \subseteq T$ is a weakly surjective extension, it suffices by [10, Theorem 3.13, p. 37] to show that for every prime ideal P of T, when $T_{[P]} = \{x \in S : tx \in T$ for some $t \in T \setminus P\}$ and $A_{[P \cap A]} = \{x \in S : ax \in A$ for some $a \in A \setminus P\}$, then $T_{[P]} \subseteq A_{[P \cap A]}$. Let P be a prime ideal of T, and let $x \in T_{[P]}$. Then there exists $t \in T \setminus P$ such that $tx \in T$. Now $A[t, tx]$ is a finitely generated A-subalgebra of T, and we have established above that no two distinct comparable prime ideals of $A[t, tx]$ lie over the same prime ideal of A. Hence we may apply Zariski's Main Theorem to obtain $b \in A \setminus P$ such that $A_b = A[t, tx]_b$. Thus there exists $k > 0$ such that $b^k t, b^k tx \in A$. Since $b^k t \in A \setminus P$, it follows that $x \in A_{[P \cap A]}$. Therefore, $T_{[P]} \subseteq A_{[P \cap A]}$, and we conclude that $A \subseteq T$ is a weakly surjective extension. ◻

Corollary 4.8. *Suppose $R \subseteq A \subseteq S$ are rings and the integral closure $\overline{R}$ of R in S is Prüfer in S. If $P \subseteq Q$ are prime ideals of A with $P \cap R = Q \cap R$, then $P = Q$.*

Proof. Let P and Q be prime ideals of A such that $P \subseteq Q$ and $P \cap R = Q \cap R$. Since $B := A \cap \overline{R}$ is the integral closure of R in A, this forces $P \cap B = Q \cap B$. By Proposition 4.7, $B \subseteq A$ is a weakly surjective extension, so $P = (P \cap B)A = (Q \cap B)A = Q$ [10, Proposition 4.6, p. 43]. ◻

5 Finitely Stable Rings

When R is a subring of the ring S, we say that R is *finitely stable in S* if R is tight in S and every finitely generated S-regular ideal of R is stable. It follows that R is a finitely stable ring if and only if R is finitely stable in $\mathrm{Quot}(R)$. The main goal of this section is to prove in Theorem 5.4 a characterization of finitely stable subrings and draw some consequences.

We observe first that the property of being a finitely stable subring is inherited by each intermediate ring.

Proposition 5.1. *If R is finitely stable in S, then every ring between R and S is finitely stable in S.*

Proof. Let A be a ring with $R \subseteq A \subseteq S$, and let I be an S-regular finitely generated ideal of A. Then there exist $y_1, \ldots, y_m \in I$ such that $I = (y_1, \ldots, y_m)A$. Since

$R \subseteq S$ is a tight extension, there exists an S-invertible ideal L of R such that $J := (y_1, \ldots, y_m)L \subseteq R$. Since J is an S-regular finitely generated ideal of R, J is stable. Moreover, since L is an invertible ideal of R, Proposition 2.13(3) implies that the dense fractional ideal $(y_1, \ldots, y_m)R$ is stable. Consequently, by Corollary 2.12, I is a stable ideal of A.

Finally, since R is tight in S, so is A. Indeed, let $x \in S$. Then there exists an S-invertible ideal H of R such that $xH \subseteq R$, and hence $xHA \subseteq A$. Since H is invertible in R, then HA is invertible in A. By Lemma 2.10, $HS = S$, and hence $(HA)S = S$. Thus by Lemma 2.10, HA is S-invertible, and hence A is tight in S.□

The next lemma is a technical consequence of Handelman's lemma (Lemma 3.2) that is needed in the proof of Theorem 5.4 to reduce from an arbitrary algebra over a field to a finite-dimensional one.

Lemma 5.2. *Let F be a field and let S be an F-algebra such that $F \subseteq S$ is a quadratic extension. If there is an F-subalgebra S' of S such that S' and S have the same number k of maximal ideals with $k > 1$, then $S' = S$ and S is isomorphic as an F-algebra to $\prod_{i=1}^{k} F$. If also $k = 3$, then $F \cong \mathbb{F}_2$.*

Proof. By Proposition 3.3, k is either 2 or 3. Let $N_1, \ldots, N_k$ denote the maximal ideals of S', and let $x_1, \ldots, x_k$ be elements of S' such that each x_i is contained in N_i but in no other maximal ideal of S'. Define a subspace T of S' by $T = x_1 F + \cdots + x_k F + F$, and observe that $F \subseteq T \subseteq S' \subseteq S$. We claim that $T = S$ and hence that $S' = S$. Since $F \subseteq S$ is a quadratic extension, T is a ring and S is an integral extension of F. Thus $T \cap N_1, \ldots, T \cap N_k$ are the maximal ideals of the ring T. Moreover, the choice of the x_i forces these maximal ideals to be distinct. Since $F \subseteq T$ is a quadratic extension with T finite-dimensional over F and $k > 1$, we apply Lemma 3.2 to obtain that if $k = 2$, then T is isomorphic to $F \times F$, or, if $k = 3$, then T is isomorphic to $F \times F \times F$, where in the latter case F is isomorphic to $\mathbb{F}_2$.

Now let $y \in S$, and define $T' = yF + T$. We show that $T = T'$, since this will show that every element of S is in T; i.e., $T = S$. By assumption, $F \subseteq S$ is quadratic, so T' is a ring, and thus $T \subseteq T' \subseteq S$ is an integral extension of rings. Hence, since T and S each have exactly k maximal ideals, then so does T'. Moreover, T' is a finite dimensional F-algebra and $F \subseteq T'$ is a quadratic extension. Thus if $k = 2$, then by Lemma 3.2, T' is isomorphic to $F \times F$ and hence isomorphic also to T; while if $k = 3$, then T' is isomorphic to $F \times F \times F$, and hence T' is isomorphic to T. In either case, the finite-dimensional F-vector space T' is isomorphic to its F-subspace T, and hence $T = T'$, which forces $y \in T$. The choice of $y \in S$ was arbitrary, so we conclude that $S = T$, and hence $S' = S$. The other claims in the lemma now follow from the fact that $T = S$.   □

For lack of a reference, we state a standard observation that is needed in the proofs of Theorem 5.4 and Corollary 5.10.

Lemma 5.3. *Let $R \subseteq S$ be rings, let A and B be R-submodules of S, where A is a finitely generated R-module, and let X be a multiplicatively closed subset of R. Then $[B :_S A]R_X = [B_X :_{S_X} A_X]$.*

Proof. Clearly, $[B :_S A]R_X \subseteq [B_X :_{S_X} A_X]$. To prove the reverse inclusion, let $q \in [B_X :_{S_X} A_X]$. Then $q = s/y$, for some $s \in S$ and $y \in X$, and $(s/y)A_X \subseteq B_X$. Since y is a unit in R_X, this implies that $(sA)_X \subseteq B_X$, and since sA is a finitely generated R-submodule of S, there exists $x \in X$ such that $xsA \subseteq B$. Hence $q = s/y = (xs/1)(1/xy) \in [B :_S A]R_X$, which proves that $[B :_S A]R_X = [B_X :_{S_X} R_X]$.   □

Theorem 5.4. *Let R be a subring of S, and let $\overline{R}$ denote the integral closure of R in S. The ring R is finitely stable in S if and only if*

(a) *R is tight in S,*
(b) *$\overline{R}$ is Prüfer in S,*
(c) *$R \subseteq \overline{R}$ is a quadratic extension, and*
(d) *each S-regular maximal ideal of R has at most 2 maximal ideals of $\overline{R}$ lying over it.*

Proof. Suppose that R is finitely stable in S. That R is tight in S is clear. To prove (c), that $R \subseteq \overline{R}$ is a quadratic extension, it suffices by Proposition 3.4 to show that every finitely generated R-submodule I of $\overline{R}$ containing R is an invertible $E(I)$-module. Let I be a finitely generated R-submodule of $\overline{R}$ containing R. Since R is tight in S, there exists an S-invertible ideal J of R such that $IJ \subseteq R$; in particular, I is a dense fractional ideal of R. Since IJ is a finitely generated S-regular ideal of R, IJ is stable, and hence by Proposition 2.13(3), I is stable. Thus by Proposition 2.11, I is an invertible $E(I)$-module, which proves that $R \subseteq \overline{R}$ is a quadratic extension.

Next we verify that $\overline{R}$ is a Prüfer subring of S. Since $R \subseteq S$ is weakly surjective, $\overline{R} \subseteq S$ is also weakly surjective [10, Proposition 3.7, p. 35]. Thus by Lemma 4.5, $\overline{R}$ is a Prüfer subring of S if and only if every finitely generated S-regular ideal of $\overline{R}$ is S-invertible. Let I be an S-regular finitely generated ideal of $\overline{R}$. By Proposition 5.1, $\overline{R}$ is finitely stable in S, so I is projective as an $E(I)$-module. The fact that I is a finitely generated ideal of the integrally closed subring $\overline{R}$ of S with $IS = S$ implies that $E(I) = \overline{R}$ [3, Example 4, page 305]. Therefore, I is a projective R-module, so that by Lemma 2.10, I an invertible ideal of $\overline{R}$. This shows that every finitely generated S-regular ideal of $\overline{R}$ is invertible, which proves that $\overline{R}$ is a Prüfer subring of S, hence verifying (b).

Now we prove that (d) holds. Let M be an S-regular maximal ideal of R. We claim that $\overline{R}$ has at most 2 maximal ideals lying over M. By way of contradiction, suppose that $\overline{R}$ has more than two maximal ideals lying over M. Then by Proposition 3.3, $\overline{R}$ has exactly three maximals ideals lying over M. We show this is impossible by exhibiting a finitely generated S-regular ideal of R that is not a stable ideal, contrary to assumption.

Denote the three distinct maximal ideals of $\overline{R}$ lying over M by $N_1, N_2,$ and N_3. Let $J = N_1 \cap N_2 \cap N_3$. Then since $J \cap R = M$, it follows that J is a maximal ideal of $R + J$. By Proposition 5.1, the ring $R + J$ is finitely stable in S. Also, the integral closure of $R + J$ in S is $\overline{R}$, and $\overline{R}$ has three maximal ideals lying over the maximal ideal J of $R + J$. Thus $R + J$ inherits all the assumptions on R and M, and we may assume without loss of generality that $J \subseteq R$; that is, we assume throughout the rest of the verification of (d) that R is finitely stable in S, M is a maximal ideal of R having three distinct maximal ideals N_1, N_2, N_3 of $\overline{R}$ lying over it, and $M = N_1 \cap N_2 \cap N_3$.

Claim 1. Each maximal ideal N_i of $\overline{R}$ lying over M has residue field $\mathbb{F}_2$.

Since $M = N_1 \cap N_2 \cap N_3$ is an ideal of $\overline{R}$ and $R \subseteq \overline{R}$ is a quadratic extension, we have that $R/M \subseteq \overline{R}/M$ is a quadratic extension, with R/M a field, such that $\overline{R}/M$ has three maximal ideals. Thus by Lemma 5.2, $\overline{R}/M$ is isomorphic as an R/M-algebra to $\mathbb{F}_2 \times \mathbb{F}_2 \times \mathbb{F}_2$. In particular, each of the maximal ideals N_i of $\overline{R}$ lying over M has residue field $\mathbb{F}_2$.

Claim 2. If I is a finitely generated S-regular R-submodule of S, then there exists $x \in I$ such that $I_M^2 = xI_M$.

Since $R \subseteq S$ is tight and I is a finitely generated R-submodule of S, there exists an invertible S-regular ideal B of R such that $B \subseteq R : I$, and hence I is a dense fractional ideal of R and BI is a finitely generated S-regular ideal of R, so that by assumption BI is stable. Since B is an S-invertible ideal of R, Proposition 2.13(3) implies that I is a stable fractional ideal of R. Therefore, by Corollary 2.12, I_M is a stable fractional ideal of R_M. Moreover, since BI_M is a finitely generated S_M-regular ideal of R_M, then $E(IB_M) \subseteq \overline{R}_M$, and since $\overline{R}_M$ has only finitely many maximal ideals (namely, those extended from $N_1, N_2,$ and N_3), it follows that $E(BI_M)$ has only finitely many maximal ideals. Since B is an invertible ideal of R, $E(I_M) = E(BI_M)$, and hence since $E(I_M)$ has only finitely many maximal ideals and I_M is an invertible $E(I_M)$-submodule of S_M, there exists x in I such that $I_M = xE(I_M)$. Thus $I_M^2 = xI_M$.

Claim 3. For each i, there exists n_i in N_i, but not in either of the other two maximal ideals of $\overline{R}$, such that $n_i S_M = S_M$.

Since M is S-regular, there exists a finitely generated S-regular ideal A of R contained in M. Let a_i be an element of N_i that is not in either of the other two maximal ideals of $\overline{R}$ lying over M. Then $A_i := A + a_i R$ is a finitely generated S-regular R-submodule of N_i. Therefore, by Claim 2, there exists $n_i \in A_i$ such that $(A_i^2)_M = n_i (A_i)_M$. Since $AS = S$, we have $A_i S = S$, and hence from $(A_i^2)_M = n_i (A_i)_M$, we deduce that $n_i S_M = S_M$. Moreover, $a_i^2 R_M \subseteq n_i R_M$, so n_i is in N_i but not in either of the other two maximal ideals of $\overline{R}$ lying over M.

Claim 4. With $x = n_1 n_2^2 n_3$, $y = n_1 n_2 n_3^2$, and A as in Claim 3, then $I := A^8 + (x, y)R$ is a finitely generated S-regular ideal of R that is not stable.

Since $AS = S$, it follows that $IS = S$, and hence I is a finitely generated S-regular ideal of R. Therefore, by Claim 2, there exists $z \in I_M$ such that $I_M^2 = zI_M$. Now since for each $i = 1, 2, 3$, $A^2 R_M \subseteq n_i R_M$, we have $A_M^8 \subseteq n_1 n_2^2 n_3 R_M = x R_M$, and hence $I_M = (x, y) R_M$. In fact, I_M is an S_M-regular ideal, since by Claim 3, $n_i S_M = S_M$ for each i. Thus by Lemma 4.5(4), $I \overline{R}_M = (x, y) \overline{R}_M$ is an invertible ideal of the Prüfer subring $\overline{R}_M$ of S_M. Therefore, since $I^2 \overline{R}_M = zI \overline{R}_M$ and $I \overline{R}_M$ is invertible, we have $(x, y) \overline{R}_M = I \overline{R}_M = z \overline{R}_M$.

Since $z R_M \subseteq (x, y) R_M$, we have $bz = xr + ys$ for some $b \in R \setminus M$ and $r, s \in R$. Now since the images of b, n_2, and n_3 in $\overline{R}_{N_1}$ are units,

$$
\begin{aligned}
n_1 (n_2 r + n_3 s) \overline{R}_{N_1} &= n_1 (n_2 r + n_3 s) n_2 n_3 \overline{R}_{N_1} \\
&= (n_1 n_2^2 n_3 r + n_1 n_2 n_3^2 s) \overline{R}_{N_1} \\
&= (xr + ys) \overline{R}_{N_1} = bz \overline{R}_{N_1} = z \overline{R}_{N_1} \\
&= (x, y) \overline{R}_{N_1} = n_1 (n_2^2 n_3, n_2 n_3^2) \overline{R}_{N_1} \\
&= n_1 \overline{R}_{N_1}.
\end{aligned}
$$

Thus since the image of n_1 in $\overline{R}_{N_1}$ is a nonzerodivisor (for by Claim 3, $n_i S_M = S_M$) we have $(n_2 r + n_3 s) \overline{R}_{N_1} = \overline{R}_{N_1}$. Suppose that neither r nor s is in M. Then since N_1 lies over M, neither r nor s is in N_1, and hence neither $n_2 r$ nor $n_3 s$ are in N_1. Thus the images of $n_2 r$ and $n_3 s$ in $\overline{R}/N_1$ are both nonzero. But by Claim 3, $\overline{R}/N_1$ has only two elements, so the sum of two nonzero elements is zero. Consequently, $n_2 r + n_3 s \in N_1$, so that $\overline{R}_{N_1} = (n_2 r + n_3 s) \overline{R}_{N_1} \subseteq N_1 \overline{R}_{N_1}$, a contradiction that implies that at least one of r and s is a member of M. Thus, if $r \in M$, then since N_3 lies over M, $n_2 r + n_3 s \in N_3$. On the other hand, if $s \in M$, then $n_2 r + n_3 s \in N_2$. Therefore, since at least one of r, s is in M, we conclude that $n_2 r + n_3 s \in N_2 \cup N_3$. We argue next that this conclusion prevents the ideal $(x, y) R$ from being stable, which in turn will contradict the assumption that R is finitely stable in S.

Using the fact that the images of n_1 and n_3 in $\overline{R}_{N_2}$ are units, we obtain with an argument similar to the one above:

$$
\begin{aligned}
(n_2^2 r + n_2 n_3 s) \overline{R}_{N_2} &= (n_2^2 r + n_2 n_3 s) n_1 n_3 \overline{R}_{N_2} \\
&= (n_1 n_2^2 n_3 r + n_1 n_2 n_3^2 s) \overline{R}_{N_2} = z \overline{R}_{N_2} \\
&= (x, y) \overline{R}_{N_2} = (n_1 n_2^2 n_3, n_1 n_2 n_3^2) \overline{R}_{N_2} \\
&= n_2 \overline{R}_{N_2}.
\end{aligned}
$$

Thus, since the image in $\overline{R}_{N_2}$ of n_2 is a nonzerodivisor, we have $(n_2 r + n_3 s) \overline{R}_{N_2} = \overline{R}_{N_2}$. Hence $n_2 r + n_3 s \notin N_2$. Similarly, we see that $n_2 r + n_3 s \notin N_3$. Indeed,

$$
\begin{aligned}
(n_2 n_3 r + n_3^2 s) \overline{R}_{N_3} &= (n_2 n_3 r + n_3^2 s) n_1 n_2 \overline{R}_{N_3} \\
&= (n_1 n_2^2 n_3 r + n_1 n_2 n_3^2 s) \overline{R}_{N_3} = z \overline{R}_{N_3}
\end{aligned}
$$

$$= (x, y)\overline{R}_{N_3} \;=\; (n_1 n_2^2 n_3, n_1 n_2 n_3^2)\overline{R}_{N_3}$$
$$= n_3 \overline{R}_{N_3}.$$

Thus $(n_2 n_3 r + n_3^2 s)\overline{R}_{N_3} \;=\; n_3 \overline{R}_{N_3}$, and since the image of n_3 in $\overline{R}_{N_3}$ is a nonzerodivisor, the image of $n_2 r + n_3 s$ in $\overline{R}_{N_3}$ is a unit. Therefore, $n_2 r + n_3 s \notin N_3$, which forces us to conclude that $n_2 r + n_3 s$ is not in N_2 or N_3, a contradiction to our prior conclusion that this element is in one of these two maximal ideals. Therefore, it cannot be that both $(x, y)R$ is a stable ideal of R and N_1, N_2, N_3 are distinct maximal ideals of $\overline{R}$ lying over M. Since R is finitely stable in S, this forces the conclusion that $\overline{R}$ has at most 2 maximal ideals lying over M, which verifies (d).

Conversely, suppose that (a)–(d) hold. Then by assumption R is tight in S, so it remains to prove that every finitely generated S-regular ideal of R is stable. Let I be a finitely generated S-regular ideal of R, and let M be a maximal ideal containing I. Then necessarily M is S-regular and I_M is a finitely generated S_M-regular ideal of R_M. By Lemma 4.5(4), $\overline{R}_M$ is Prüfer in S_M, and by (c), $\overline{R}_M$ has at most 2 maximal ideals. Thus, as an invertible ideal in a ring with only finitely many maximal ideals, $I\overline{R}_M$ is a principal regular ideal of $\overline{R}_M$. Moreover, by (b), $R \subseteq \overline{R}$ is a quadratic extension, so by (d) and Proposition 3.6, I_M is a stable ideal of R_M. Thus by Proposition 2.11 and Lemma 5.3, $1 \in [I_M :_{S_M} I_M^2]I_M = [I :_S I^2]I_M$. This shows that for all maximal ideals M of R containing I, $1 \in [I :_S I^2]I_M$. On the other hand, if N is a maximal ideal not containing I, then clearly $1 \in [I :_S I^2]I_N$. We conclude that $1 \in [I :_S I^2]I$ since this containment holds locally, and hence by Proposition 2.11, I is a stable ideal of R. Therefore, R is finitely stable in S. □

Corollary 5.5. *Suppose R is a quasilocal ring that is finitely stable in S. If the integral closure $\overline{R}$ of R in S is quasilocal, then every ring between R and S is quasilocal. Otherwise, every ring between R and S has at most 2 maximal ideals.*

Proof. Let A be a ring with $R \subseteq A \subseteq S$. By Theorem 5.4, there are at most 2 maximal ideals of $\overline{R}$ and $\overline{R}$ is a Prüfer subring of S. Thus by Proposition 4.6(3), the integral closure of A in S has at most 2 maximal ideals; hence A has at most 2 maximal ideals. Moreover, if $\overline{R}$ is quasilocal, then by Proposition 4.6(3), so is $\overline{A}$, and hence so is A. □

To motivate the next corollary, we give an example which shows that although an invertible ideal of a quasilocal ring is principal, a dense stable ideal I of a quasilocal ring need not be principal over $E(I)$.

Example 5.6. It is possible for a quasilocal domain R to have a stable maximal ideal M that is not principal as an ideal of $E(M)$. Let D be a Dedekind domain containing a field k and having an ideal I that is not principal, let F be the quotient field of D and let X be indeterminate for F. Define $R = k + XI + X^2 F[[X]]$. Then R is a quasilocal domain with maximal ideal $M := XI + X^2 F[[X]]$. Moreover, $M^2 = X^2 I^2 + X^3 F[[X]] = XIM$, so that since XIR is an invertible ideal of R, then M is by Proposition 2.13(1) a stable ideal of R with endomorphism ring

$E(M) = D + XF[[X]]$. If M is a principal ideal of $E(M)$, say $M = (aX + zX^2)E(M)$, where $a \in I$ and $z \in F[[X]]$, then $M = (aX + zX^2)(D + XF[[X]]) = (aD)X + X^2F[[X]]$, which forces $aD = I$, contrary to the choice of I. Therefore, M is not principal as an ideal of $E(M)$.

In the example, $E(M)$ has infinitely many maximal ideals. When R is finitely stable in S and quasilocal, then Corollary 5.5 shows such an example cannot occur, and hence we have the following corollary.

Corollary 5.7. *Suppose R is a quasilocal ring that is finitely stable in S. If I is a stable S-regular ideal of R, then I is a principal ideal of $E(I)$.* □

Along the same lines as Corollary 5.5, there are at most 2 prime ideals lying over an S-regular prime ideal.

Corollary 5.8. *If R is finitely stable in S and A is a ring with $R \subseteq A \subseteq S$, then each prime ideal of R has at most 2 prime ideals in A lying over it.*

Proof. Let P be a prime ideal of R. Suppose first that P is not S-regular. We claim that $R_P = A_P = S_P$. Let $x \in S$. Then since $R \subseteq S$ is tight, there exists an S-regular ideal I of R such that $xI \subseteq R$. Since I is S-regular and P is not, there exists $b \in I \setminus P$. Thus $bx \in R$, and hence $x/1 \in R_P$. It follows that $R_P = A_P = S_P$, and hence P has only prime ideal in A lying over it.

Next, suppose that P is an S-regular prime ideal of R. First we claim that there are at most 2 prime ideals of $\overline{R}$ lying over P. Suppose that P_1, P_2, P_3 are distinct prime ideals of $\overline{R}$ lying over P. Let M be a maximal ideal of R containing P. Then by Going Up, there exist maximal ideals M_1, M_2, M_3 of $\overline{R}$ lying over M such that for each i, $P_i \subseteq M_i$. Since P is S-regular, so is M, and hence by Theorem 5.4 there are at most 2 maximal ideals of $\overline{R}$ lying over M. Therefore, we may assume after relabeling that $M_1 = M_2$ and hence that P_1 and P_2 are incomparable S-regular prime ideals of $\overline{R}$ that are contained in the same maximal ideal. But by Theorem 5.4, $\overline{R}$ is a Prüfer subring of S, and hence by Proposition 4.6(2) the S-regular prime ideals of R form a tree with respect to inclusion. This contradiction implies that there are at most 2 prime ideals of $\overline{R}$ lying over P.

Now, with A as in the corollary, suppose by way of contradiction that there exist three distinct prime ideals P_1, P_2, P_3 of A lying over P. By Theorem 5.4, $\overline{R}$ is Prüfer in S, so by Corollary 4.8, the prime ideals P_1, P_2, P_3 are incomparable. Let $\overline{A}$ denote the integral closure of A in S, and for each i, let Q_i be a prime ideal of $\overline{A}$ lying over P_i. Then Q_1, Q_2, Q_3 are incomparable prime ideals of $\overline{A}$, and hence by Proposition 4.6(1), $Q_1 \cap \overline{R}, Q_2 \cap \overline{R}, Q_3 \cap \overline{R}$ are incomparable prime ideals of $\overline{R}$ lying over P, in contradiction to what we have proved earlier. Therefore, there are at most 2 prime ideals of A lying over P.   □

The next corollary shows that the property of being finitely stable localizes at prime ideals.

Corollary 5.9. *If R is finitely stable in S and P is a prime ideal of R, then R_P is finitely stable in S_P.*

Proof. Suppose that R is finitely stable in S, and let P be a prime ideal of R. The proof is via Theorem 5.4. We claim first that R_P is tight in S_P. Let $s/b \in S_P$, where $s \in S$ and $b \in R \setminus P$. Then since R is tight in S, there exists an S-invertible ideal I of R such that $sI \subseteq R$. Therefore, $(s/b)I_P \subseteq R_P$, so that the S_P-invertible ideal I_P is in $(R_P :_{R_P} s/b)$. Thus R_P is tight in S_P. Next, by Lemma 3.1, $R_P \subseteq S_P$ is a quadratic extension, and by Lemma 4.5, $\overline{R}_P \subseteq S_P$ is a Prüfer extension. Finally, by Corollary 5.8, there are at most two prime ideals of $\overline{R}$ lying over P, so PR_P has at most 2 maximal ideals in $\overline{R}_P$ lying over it. Therefore, by Theorem 5.4, R_P is finitely stable in S_P. $\quad\square$

The next corollary shows that the converse is also true: A tight subring is finitely stable if and only if it is locally finitely stable.

Corollary 5.10. *The ring R is finitely stable in S if and only if R is tight in S and R_M is finitely stable in S_M for each S-regular maximal ideal M of R.*

Proof. If R is finitely stable in S, then Corollary 5.9 shows that R_M is finitely stable in S_M for each maximal ideal M of R. Conversely, suppose that R is tight in S and R_M is finitely stable in S_M for all S-regular maximal ideals M of R. Let I be a finitely generated S-regular ideal of R, and let M be a maximal ideal of R containing I. Then M is S-regular, so by assumption, I_M is a stable ideal of R_M. Hence by Proposition 2.11 and Lemma 5.3, $1 \in [I_M :_{S_M} I_M^2]I_M = [I :_S I^2]I_M$. On the other hand, if M is a maximal ideal not containing I, then it is clear that $1 \in [I :_S I^2]I_M$. Therefore, $1 \in [I :_S I^2]I$, since this containment holds locally, and hence by Proposition 2.11, I is a stable ideal of R. This proves that R is finitely stable in S. $\quad\square$

Restricting to the case in which $S = \mathrm{Quot}(R)$, we obtain a characterization of finitely stable rings.

Corollary 5.11. *Let R be a ring, and let $\overline{R}$ denote its integral closure in $\mathrm{Quot}(R)$. The ring R is finitely stable if and only if $R \subseteq \overline{R}$ is a quadratic extension, $\overline{R}$ is a Prüfer ring, and each regular maximal ideal of R has at most 2 maximal ideals of $\overline{R}$ lying over it.*

Proof. The ring R is finitely stable if and only if R is finitely stable in $\mathrm{Quot}(R)$, and an ideal of R is regular if and only if it is $\mathrm{Quot}(R)$-regular. Thus the corollary is a consequence of Theorem 5.4. $\quad\square$

Restricting further to domains, we obtain a stronger localization result than Corollary 5.10. If R is not a domain, then since it need not be the case that $\mathrm{Quot}(R)_M = \mathrm{Quot}(R_M)$, we cannot assert something similar for finitely stable rings with zerodivisors; in that case, the best we can assert is that R is finitely stable if and only if R_M is finitely stable in $\mathrm{Quot}(R)_M$ for each maximal ideal M.

Corollary 5.12. *A domain R is a finitely stable ring if and only if R_M is a finitely stable ring for each maximal ideal M of R.*

Proof. Since R is a domain, $\text{Quot}(R)_M = \text{Quot}(R_M)$ for each maximal ideal M of R. Moreover, the domain R is a finitely stable ring if and only if R is finitely stable in $\text{Quot}(R)$, so the claim follows from Corollary 5.10. $\square$

6 Existence and Examples of Finitely Stable Rings

In this section we briefly discuss some constructions and examples of finitely stable rings. Where arguments can be easily adapted, we also prove similar results for stable subrings, where by a *stable subring R of S* we mean a tight subring of S such that every S-regular ideal of R is stable. Stable subrings will be treated in more detail in a future article.

By Theorem 5.4, if R is a finitely stable domain, then $\overline{R}$ is a Prüfer domain and $R \subseteq \overline{R}$ is a quadratic extension. The following proposition shows that every uncountable Prüfer domain S occurs as the integral closure of a finitely stable domain that is properly contained in S.

Proposition 6.1. *If S is an uncountable Prüfer domain with quotient field F, then there exists a finitely stable domain R with quotient field F and integral closure S such that $R \subsetneq S$.*

Proof. Let K be a nonzero free S-module of at most countable rank. By [17, Lemmas 3.3 and 3.4], there exists a subring A of S such that S/A is a torsion-free divisible A-module, and there exists an A-linear derivation $D : F \to FK$ such that $D(S) = FK$. (Here, FK is the divisible hull of K.) Set $R = S \cap D^{-1}(K)$. In the terminology of [17], R is a subring of S twisted along the set of nonzero elements of A, and hence by [17, Theorem 7.1], R is a finitely stable domain with quotient field F and integral closure S. By [17, Lemma 4.4], with Q the quotient field of A, we have $S/R \cong QK/K$ as R-modules. Since K is a nonzero free S-module, then $QK \neq K$, so $R \subsetneq S$. $\square$

Remark 6.2. If in the theorem S has positive characteristic and F is separable of infinite transcendence degree over a countable subfield, then the cited references show that $S/R \cong FK/K$. (Specifically, S is "strongly twisted" by K.) Moreover, the S-module K need not be chosen to be free; we only assumed this in the proof to guarantee that $QK \neq K$. In any case, if $K \neq FK$, then S is not a fractional ideal of R, and the "distance" from S can be prescribed by the choice of K.

When $R \subseteq S$ is a ring extension and I is an ideal of R that is also an ideal of S, then R is Prüfer in S if and only if R/I is Prüfer in S/I [10, Proposition 5.8, p. 52]. In Proposition 6.4 we prove the analogous statements for stable and finitely stable rings.

Lemma 6.3. *Let $R \subseteq S \subseteq Q(R)$ be rings, and suppose that I is an ideal of R that is also an ideal of S. Then an ideal J of R is an S-regular stable ideal if and only if $I \subseteq J$ and J/I is an S/I-regular stable ideal of R/I.*

Proof. Let J be an S-regular stable ideal of R. If $I \nsubseteq J$, then there exists $x \in I \setminus J$, and since $JS = S$, there exist $y_1, \ldots, y_n \in J$ and $s_1, \ldots, s_n \in S$ such that $1 = y_1 s_1 + \cdots + y_n s_n$. But then $x = y_1(s_1 x) + \cdots + y_n(s_n x) \in J$, contrary to the choice of x. Therefore, $I \subseteq J$. Since also $J^2 S = S$, this same argument shows that $I \subseteq J^2$. Now since J is an S-regular stable ideal of R, then by Proposition 2.11, $1 \in [J :_S J^2]J$, and hence $1 + I \in [J/I :_{S/I} J^2/I](J/I)$. Therefore, since J/I is an S/I-regular ideal of R/I, Proposition 2.11 implies that J/I is a stable ideal of R/I.

Conversely, suppose that $I \subseteq J$ and J/I is an S/I-regular stable ideal of R/I. Then $JS = S$, and by Proposition 2.11, $1 + I \in [J/I :_{S/I} J^2/I](J/I)$, and hence there exist $x_1, \ldots, x_n \in J$, $y_1, \ldots, y_n \in [J :_S J^2]$ and $x \in I$ such that $1 = y_1 x_1 + \cdots + y_n x_n + x$. Now $x \in I \subseteq J \subseteq [J :_S J^2]J$, so $1 \in [J :_S J^2]J$, and hence, since $JS = S$ implies that J is dense, Proposition 2.11 implies that J is a stable ideal of R. $\square$

Proposition 6.4. *Let $R \subseteq S$ be an extension of rings, and suppose I is an ideal of R that is also an ideal of S. Then R is stable (resp., finitely stable) in S if and only if R/I is stable (resp., finitely stable) in S/I.*

Proof. First we claim that R is tight in S if and only if R/I is tight in S/I. Suppose that R is tight in S, and let $x \in S$. Then there exists an S-invertible ideal J such that $xJ \subseteq R$. Since J is S-invertible, there exists an R-submodule K of S with $JK = R$. Now $((J + I)/I)((K + I)/I) = R/I$, so J/I is an S/I-invertible ideal of R/I. Moreover, $J/I \subseteq (R/I : x + I)$, which proves that R/I is tight in S/I. Conversely, suppose that R/I is tight in S/I, and let $x \in S$. Then there exists an S/I-invertible ideal J/I of R/I such that $xJ \in R$. Since J/I is S/I-invertible, there exists an R-submodule K of S with $JK + I = R$. But since $I \subseteq J$, this implies that $J(K + R) = JK + J = R$, and hence J is an S-invertible ideal of R.

Now suppose that R is finitely stable in S, and let J be an ideal of R such that J/I is a finitely generated S/I-regular ideal of R. Since J/I is finitely generated, there exist $x_1, \ldots, x_n \in J$ such that $J = (x_1, \ldots, x_n)R + I$, and since $JS = S$, there exists $x \in I$ such that $(x_1, \ldots, x_n, x)S = S$. As in the proof of Lemma 6.3, this implies that $I \subseteq (x_1, \ldots, x_n, x)R$, and hence $J = (x_1, \ldots, x_n, x)R$. As a finitely generated S-regular ideal of R, J is stable, so Lemma 6.3 implies that J/I is a stable ideal of R/I. Therefore, R/I is finitely stable in S/I. The converse follows also from Lemma 6.3 and the fact that $JS = S$ implies $I \subseteq J$, as does the claim that R is stable in S if and only if R/I is stable in S/I. $\square$

Corollary 6.5. *Suppose that the ring R occurs in a commutative diagram of the form below, where A and S are rings. Then R is a stable (resp., finitely stable) subring of S if and only if A is a stable (resp., finitely stable) ring.*

Proof. The diagram allows us to identify R with a subring of S and A with R/I for some ideal I. Necessarily, I is an ideal of S also and we may identify $\text{Quot}(A)$ with S/I. Therefore, the corollary follows from Proposition 6.4. $\square$

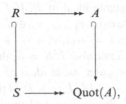

$$S \longrightarrow \mathrm{Quot}(A),$$

Example 6.6. (1) Let A be a domain with quotient field F, let X be an indeterminate for F, and let $R = A + XF[X]$. Then by Corollary 6.5, R is stable (resp., finitely stable) in $F[X]$ if and only if A is a stable (resp., finitely stable) domain.

(2) Let A be a ring, let L be a $\mathrm{Quot}(A)$-module, and let $A \star L$ denote the Nagata idealization of L; that is, as an A-module, $A \star L$ is $A \oplus L$, and multiplication is given by $(a_1, \ell_1) \cdot (a_2, \ell_2) = (a_1 a_2, a_1 \ell_2 + a_2 \ell_1)$ for all $a_1, a_2 \in A$, $\ell_1, \ell_2 \in L$. Then Corollary 6.5 implies that $A \star L$ is a stable (resp., finitely stable) ring if and only if A is a stable (resp., finitely stable) ring.

Remark 6.7. Nagata idealization can also be used to produce finitely stable rings with a more flexible choice of the A-module L but a narrower choice of A. It follows from [18, Lemma 3.3] that when A is a Bézout domain and L is an A-module, then $A \star L$ is a finitely stable ring.

References

1. H. Bass, On the ubiquity of Gorenstein rings. Math. Z. **82**, 8–28 (1963)
2. S. Bazzoni, Finite character of finitely stable domains. J. Pure Appl. Algebra **215**(6), 1127–1132 (2011)
3. N. Bourbaki, *Commutative Algebra*, Chapters 1–7. Translated from the French. Reprint of the 1989 English translation. Elements of Mathematics (Springer, Berlin, 1998)
4. S. El Baghdadi, S. Gabelli, Ring-theoretic properties of PvMDs. Comm. Algebra **35**(5), 1607–1625 (2007)
5. E.G. Evans, A generalization of Zariski's main theorem. Proc. Am. Math. Soc. **26**, 45–48 (1970)
6. S. Gabelli, G. Picozza, Star stable domains. J. Pure Appl. Algebra **208**(3), 853–866 (2007)
7. S. Gabelli, G. Picozza, Stability and Clifford regularity with respect to star operations. Comm. Algebra **40**(9) 3558–3582 (2012)
8. D. Handelman, Propinquity of one-dimensional Gorenstein rings. J. Pure Appl. Algebra **24**(2), 145–150 (1982)
9. S. Kabbaj, A. Mimouni, t-class semigroups of integral domains. J. Reine Angew. Math. **612**, 213–229 (2007)
10. M. Knebusch, D. Zhang, *Manis Valuations and Prüfer Extensions. I. A New Chapter in Commutative Algebra*. Lecture Notes in Mathematics, vol. 1791 (Springer, Berlin, 2002)
11. J. Lambek, *Lectures on Rings and Modules*, 2nd edn. (Chelsea Publishing Co., New York, 1976)
12. M. Larsen, P. McCarthy, *Multiplicative Theory of Ideals* (Academic, New York, 1971)
13. J. Lipman, Stable ideals and Arf rings. Am. J. Math. **93**, 649–685 (1971)

14. A. Mimouni, Ratliff-Rush closure of ideals in integral domains. Glasg. Math. J. **51**(3), 681–689 (2009)
15. B. Olberding, Stability of ideals and its applications. *Ideal Theoretic Methods in Commutative Algebra (Columbia, MO, 1999)*. Lecture Notes in Pure and Applied Mathematics, vol. 220 (Dekker, New York, 2001), pp. 319–341
16. B. Olberding, On the structure of stable domains. Comm. Algebra **30**(2), 877–895 (2002)
17. B. Olberding, A counterpart to Nagata idealization. J. Algebra **365**, 199–221 (2012)
18. B. Olberding, Generic formal fibers and analytically ramified stable rings. Nagoya Math. J. **211**, 109–135 (2013)
19. B. Olberding, One-dimensional bad Noetherian domains. Trans. Amer. Math. Soc. **366**, 4067–4095 (2014)
20. B. Olberding, Prescribed subintegral extensions of local Noetherian doamins. J. Pure Appl. Alg. **218**, 506–521 (2014)
21. G. Picozza, F. Tartarone, Flat ideals and stability in integral domains. J. Algebra **324**(8), 1790–1802 (2010)
22. D. Rush, Rings with two-generated ideals. J. Pure Appl. Algebra **73**(3), 257–275 (1991)
23. D. Rush, Two-generated ideals and representations of abelian groups over valuation rings. J. Algebra **177**(1), 77–101 (1995)
24. J. Sally, W. Vasconcelos, Stable rings. J. Pure Appl. Algebra **4**, 319–336 (1974)
25. J. Sally, W. Vasconcelos, Stable rings and a problem of Bass. Bull. Am. Math. Soc. **79**, 575–576 (1973)
26. L. Sega, Ideal class semigroups of overrings. J. Algebra **311**(2), 702–713 (2007)
27. P. Zanardo, The class semigroup of local one-dimensional domains. J. Pure Appl. Algebra **212**(10), 2259–2270 (2008)
28. P. Zanardo, Algebraic entropy of endomorphisms over local one-dimensional domains. J. Algebra Appl. **8**(6), 759–777 (2009)

Integral Closure of Rings of Integer-Valued Polynomials on Algebras

Giulio Peruginelli and Nicholas J. Werner

Abstract Let D be an integrally closed domain with quotient field K. Let A be a torsion-free D-algebra that is finitely generated as a D-module. For every a in A we consider its minimal polynomial $\mu_a(X) \in D[X]$, i.e. the monic polynomial of least degree such that $\mu_a(a) = 0$. The ring $\mathrm{Int}_K(A)$ consists of polynomials in $K[X]$ that send elements of A back to A under evaluation. If D has finite residue rings, we show that the integral closure of $\mathrm{Int}_K(A)$ is the ring of polynomials in $K[X]$ which map the roots in an algebraic closure of K of all the $\mu_a(X)$, $a \in A$, into elements that are integral over D. The result is obtained by identifying A with a D-subalgebra of the matrix algebra $M_n(K)$ for some n and then considering polynomials which map a matrix to a matrix integral over D. We also obtain information about polynomially dense subsets of these rings of polynomials.

Keywords Integer-valued polynomial • Matrix • Triangular matrix • Integral closure • Pullback • Polynomially dense

MSC(2010) classification: []13B25, 13B22 11C20.

G. Peruginelli (✉)
Institut für Analysis und Comput. Number Theory, Technische University,
Steyrergasse 30, A-8010 Graz, Austria
e-mail: peruginelli@math.tugraz.at

N.J. Werner
The Ohio State University-Newark, 1179 University Drive, Newark, OH 43055, USA
e-mail: nwerner@newark.osu.edu

M. Fontana et al. (eds.), *Commutative Algebra: Recent Advances in Commutative Rings,*
Integer-Valued Polynomials, and Polynomial Functions, DOI 10.1007/978-1-4939-0925-4_17,
© Springer Science+Business Media New York 2014

1 Introduction

Let D be a (commutative) integral domain with quotient field K. The ring $\text{Int}(D)$ of integer-valued polynomials on D consists of polynomials in $K[X]$ that map elements of D back to D. More generally, if $E \subseteq K$, then one may define the ring $\text{Int}(E, D)$ of polynomials that map elements of E into D.

One focus of recent research ([4–6, 9–11]) has been to generalize the notion of integer-valued polynomial to D-algebras. When $A \supseteq D$ is a torsion-free module finite D-algebra we define $\text{Int}_K(A) := \{f \in K[X] \mid f(A) \subseteq A\}$. The set $\text{Int}_K(A)$ forms a commutative ring. If we assume that $K \cap A = D$, then $\text{Int}_K(A)$ is contained in $\text{Int}(D)$ (these two facts are indeed equivalent), and often $\text{Int}_K(A)$ shares properties similar to those of $\text{Int}(D)$ (see the references above, especially [6]).

When $A = M_n(D)$, the ring of $n \times n$ matrices with entries in D, $\text{Int}_K(A)$ has proven to be particularly amenable to investigation. For instance, [9, Thm. 4.6] shows that the integral closure of $\text{Int}_{\mathbb{Q}}(M_n(\mathbb{Z}))$ is $\text{Int}_{\mathbb{Q}}(\mathscr{A}_n)$, where $\mathscr{A}_n$ is the set of algebraic integers of degree at most n, and $\text{Int}_{\mathbb{Q}}(\mathscr{A}_n) = \{f \in \mathbb{Q}[X] \mid f(\mathscr{A}_n) \subseteq \mathscr{A}_n\}$. In this paper, we will generalize this theorem and describe the integral closure of $\text{Int}_K(A)$ when D is an integrally closed domain with finite residue rings. Our description (Theorem 13) may be considered an extension of both [9, Thm. 4.6] and [2, Thm. IV.4.7] (the latter originally proved in [7, Prop. 2.2]), which states that if D is Noetherian and D' is its integral closure in K, then the integral closure of $\text{Int}(D)$ equals $\text{Int}(D, D') = \{f \in K[X] \mid f(D) \subset D'\}$.

Key to our work will be rings of polynomials that we dub *integral-valued polynomials* and which act on certain subsets of $M_n(K)$. Let $\overline{K}$ be an algebraic closure of K. We will establish a close connection between integral-valued polynomials and polynomials that act on elements of $\overline{K}$ that are integral over D. We will also investigate polynomially dense subsets of rings of integral-valued polynomials.

In Sect. 2, we define what we mean by integral-valued polynomials, discuss when sets of such polynomials form a ring, and connect them to the integral elements of $\overline{K}$. In Sect. 3, we apply the results of Sect. 2 to $\text{Int}_K(A)$ and prove the aforementioned theorem about its integral closure. Section 4 covers polynomially dense subsets of rings of integral-valued polynomials. We close by posing several problems for further research.

2 Integral-Valued Polynomials

Throughout, we assume that D is an integrally closed domain with quotient field K. We denote by $\overline{K}$ a fixed algebraic closure of K. When working in $M_n(K)$, we associate K with the scalar matrices, so that we may consider K (and D) to be subsets of $M_n(K)$.

For each matrix $M \in M_n(K)$, we let $\mu_M(X) \in K[X]$ denote the minimal polynomial of M, which is the monic generator of $N_{K[X]}(M) = \{f \in K[X] \mid$

$f(M) = 0\}$, called the null ideal of M. We define Ω_M to be the set of eigenvalues of M considered as a matrix in $M_n(\overline{K})$, which are the roots of μ_M in $\overline{K}$. For a subset $S \subseteq M_n(K)$, we define $\Omega_S := \bigcup_{M \in S} \Omega_M$. Note that a matrix in $M_n(K)$ may have minimal polynomial in $D[X]$ even though the matrix itself is not in $M_n(D)$. A simple example is given by $\left(\begin{smallmatrix} 0 & q \\ 0 & 0 \end{smallmatrix}\right) \in M_2(K)$, where $q \in K \setminus D$.

Definition 1. We say that $M \in M_n(K)$ is *integral over D* (or just *integral*, or is an *integral matrix*) if M solves a monic polynomial in $D[X]$. A subset S of $M_n(K)$ is said to be *integral* if each $M \in S$ is integral over D.

Our first lemma gives equivalent definitions for a matrix to be integral.

Lemma 2. *Let $M \in M_n(K)$. The following are equivalent:*

  (i) *M is integral over D*
  (ii) *$\mu_M \in D[X]$*
 (iii) *Each $\alpha \in \Omega_M$ is integral over D*

Proof. $(i) \Rightarrow (iii)$ Suppose M solves a monic polynomial $f(X)$ with coefficients in D. As $\mu_M(X)$ divides $f(X)$, its roots are then also roots of $f(X)$. Hence, the elements of Ω_M are integral over D.

$(iii) \Rightarrow (ii)$ The coefficients of $\mu_M \in K[X]$ are the elementary symmetric functions of its roots. Assuming (iii) holds, these roots are integral over D; hence, the coefficients of μ_M are integral over D. Since D is integrally closed, we must have $\mu_M \in D[X]$.

$(ii) \Rightarrow (i)$ Obvious. $\qquad\qquad\qquad\qquad\qquad\qquad\qquad\qquad\qquad\qquad$ $\square$

For the rest of this section, we will study polynomials in $K[X]$ that take values on sets of integral matrices. These are the integral-valued polynomials mentioned in the introduction.

Definition 3. Let $S \subseteq M_n(K)$. Let $K[S]$ denote the K-subalgebra of $M_n(K)$ generated by K and the elements of S. Define $S' := \{M \in K[S] \mid M \text{ is integral}\}$ and $\mathrm{Int}_K(S, S') := \{f \in K[X] \mid f(S) \subseteq S'\}$. We call $\mathrm{Int}_K(S, S')$ a set of *integral-valued polynomials*.

Remark 4. In the next lemma, we will prove that forming the set S' is a closure operation in the sense that $(S')' = S'$. We point out that this construction differs from the usual notion of integral closure in several ways. First, if S itself is not integral, then $S \not\subseteq S'$. Second, S' need not have a ring structure. Indeed, if $D = \mathbb{Z}$ and $S = M_2(\mathbb{Z})$, then both $\left(\begin{smallmatrix} 1 & 0 \\ 0 & 0 \end{smallmatrix}\right)$ and $\left(\begin{smallmatrix} 1/2 & 1/2 \\ 1/2 & 1/2 \end{smallmatrix}\right)$ are in S', but neither their sum nor their product is integral. Lastly, if S is a commutative ring, then S' need not be the same as the integral closure of S in $K[S]$, because we insist that the elements of S satisfy a monic polynomial in $D[X]$ rather than $S[X]$.

However, if S is a commutative D-algebra and it is an integral subset of $M_n(K)$, then S' is equal to the integral closure of S in $K[S]$ (see Corollary 1 to Proposition 2 and Proposition 6 of [1, Chapt. V]).

Lemma 5. *Let $S \subseteq M_n(K)$. Then, $(S')' = S'$.*

Proof. We just need to show that $K[S'] = K[S]$. By definition, $S' \subseteq K[S]$, so $K[S'] \subseteq K[S]$. For the other containment, let $M \in K[S]$. Let $d \in D$ be a common multiple for all the denominators of the entries in M. Then, $dM \in S' \subseteq K[S']$. Since $1/d \in K$, we get $M \in K[S']$.   □

An integral subset of $M_n(K)$ need not be closed under addition or multiplication, so at first glance it may not be clear that $\mathrm{Int}_K(S, S')$ is closed under these operations. As we now show, $\mathrm{Int}_K(S, S')$ is in fact a ring.

Proposition 6. *Let $S \subseteq M_n(K)$. Then, $\mathrm{Int}_K(S, S')$ is a ring, and if $D \subseteq S$, then $\mathrm{Int}_K(S, S') \subseteq \mathrm{Int}(D)$.*

Proof. Let $M \in S$ and $f, g \in \mathrm{Int}_K(S, S')$. Then, $f(M), g(M)$ are integral over D. By Corollary 2 after Proposition 4 of [1, Chap. V], the D-algebra generated by $f(M)$ and $g(M)$ is integral over D. So, $f(M) + g(M)$ and $f(M)g(M)$ are both integral over D and are both in $K[S]$. Thus, $f(M) + g(M), f(M)g(M) \in S'$ and $f + g, fg \in \mathrm{Int}_K(S, S')$. Assuming $D \subseteq S$, let $f \in \mathrm{Int}_K(S, S')$ and $d \in D$. Then, $f(d)$ is an integral element of K. Since D is integrally closed, $f(d) \in D$. Thus, $\mathrm{Int}_K(S, S') \subseteq \mathrm{Int}(D)$.   □

We now begin to connect our rings of integral-valued polynomials to rings of polynomials that act on elements of $\overline{K}$ that are integral over D. For each $n > 0$, let

$$\Lambda_n := \{\alpha \in \overline{K} \mid \alpha \text{ solves a monic polynomial in } D[X] \text{ of degree } n\}.$$

In the special case $D = \mathbb{Z}$, we let $\mathscr{A}_n := \Lambda_n = \{ \text{algebraic integers of degree} \}$ $\{\text{at most } n \} \subset \overline{\mathbb{Q}}$.

For any subset $\mathscr{E}$ of Λ_n, define

$$\mathrm{Int}_K(\mathscr{E}, \Lambda_n) := \{f \in K[X] \mid f(\mathscr{E}) \subseteq \Lambda_n\}.$$

to be the set of polynomials in $K[X]$ mapping elements of $\mathscr{E}$ into Λ_n. If $\mathscr{E} = \Lambda_n$, then we write simply $\mathrm{Int}_K(\Lambda_n)$. As with $\mathrm{Int}_K(S, S')$, $\mathrm{Int}_K(\mathscr{E}, \Lambda_n)$ is a ring despite the fact that Λ_n is not.

Proposition 7. *For any $\mathscr{E} \subseteq \Lambda_n$, $\mathrm{Int}_K(\mathscr{E}, \Lambda_n)$ is a ring and is integrally closed.*

Proof. Let Λ_∞ be the integral closure of D in $\overline{K}$. We set $\mathrm{Int}_{\overline{K}}(\mathscr{E}, \Lambda_\infty) = \{f \in \overline{K}[X] \mid f(\mathscr{E}) \subseteq \Lambda_\infty\}$. Then, $\mathrm{Int}_{\overline{K}}(\mathscr{E}, \Lambda_\infty)$ is a ring, and by [2, Prop. IV.4.1] it is integrally closed.

Let $\mathrm{Int}_K(\mathscr{E}, \Lambda_\infty) = \{f \in K[X] \mid f(\mathscr{E}) \subseteq \Lambda_\infty\}$. Clearly, $\mathrm{Int}_K(\mathscr{E}, \Lambda_n) \subseteq \mathrm{Int}_K(\mathscr{E}, \Lambda_\infty)$. However, if $\alpha \in \mathscr{E}$ and $f \in K[X]$, then $[K(f(\alpha)) : K] \leq [K(\alpha) : K] \leq n$, so in fact $\mathrm{Int}_K(\mathscr{E}, \Lambda_n) = \mathrm{Int}_K(\mathscr{E}, \Lambda_\infty)$. Finally, since $\mathrm{Int}_K(\mathscr{E}, \Lambda_\infty) = \mathrm{Int}_{\overline{K}}(\mathscr{E}, \Lambda_\infty) \cap K[X]$ is the contraction of $\mathrm{Int}_{\overline{K}}(\mathscr{E}, \Lambda_\infty)$ to $K[X]$, it is an integrally closed ring, proving the proposition.   □

Theorem 4.6 in [9] shows that the integral closure of $\mathrm{Int}_{\mathbb{Q}}(M_n(\mathbb{Z}))$ equals the ring $\mathrm{Int}_{\mathbb{Q}}(\mathscr{A}_n)$. As we shall see (Theorem 9), this is evidence of a broader connection between the rings of integral-valued polynomials $\mathrm{Int}_K(S, S')$ and rings of polynomials that act on elements of Λ_n. The key to this connection is the observation contained in Lemma 2 that the eigenvalues of an integral matrix in $M_n(K)$ lie in Λ_n and also the well-known fact that if $M \in M_n(K)$ and $f \in K[X]$, then the eigenvalues of $f(M)$ are exactly $f(\alpha)$, where α is an eigenvalue of M. More precisely, if $\chi_M(X) = \prod_{i=1,\ldots,n}(X - \alpha_i)$ is the characteristic polynomial of M (the roots α_i are in $\overline{K}$ and there may be repetitions), then the characteristic polynomial of $f(M)$ is $\chi_{f(M)}(X) = \prod_{i=1,\ldots,n}(X - f(\alpha_i))$. Phrased in terms of our Ω-notation, we have

$$\text{if } M \in M_n(K) \text{ and } f \in K[X], \text{ then } \Omega_{f(M)} = f(\Omega_M) = \{f(\alpha) \mid \alpha \in \Omega_M\}. \quad (8)$$

Using this fact and our previous work, we can equate $\mathrm{Int}_K(S, S')$ with a ring of the form $\mathrm{Int}_K(\mathscr{E}, \Lambda_n)$.

Theorem 9. *Let* $S \subseteq M_n(K)$. *Then,* $\mathrm{Int}_K(S, S') = \mathrm{Int}_K(\Omega_S, \Lambda_n)$, *and in particular* $\mathrm{Int}_K(S, S')$ *is integrally closed.*

Proof. We first prove this for $S = \{M\}$. Using Lemma 2 and (8), for each $f \in K[X]$ we have

$$f(M) \in S' \iff f(M) \text{ is integral } \iff \Omega_{f(M)} \subseteq \Lambda_n \iff f(\Omega_M) \subseteq \Lambda_n.$$

This proves that $\mathrm{Int}_K(\{M\}, \{M\}') = \mathrm{Int}_K(\Omega_M, \Lambda_n)$. For a general subset S of $M_n(K)$, we have

$$\mathrm{Int}_K(S, S') = \bigcap_{M \in S} \mathrm{Int}_K(\{M\}, S') = \bigcap_{M \in S} \mathrm{Int}_K(\Omega_M, \Lambda_n) = \mathrm{Int}_K(\Omega_S, \Lambda_n).$$

$\square$

The above proof shows that if a polynomial is integral-valued on a matrix, then it is also integral-valued on any other matrix with the same set of eigenvalues. Note that for a single integral matrix M we have these inclusions:

$$D \subset D[M] \subseteq \{M\}' \subset K[M].$$

Moreover, $\{M\}'$ is equal to the integral closure of $D[M]$ in $K[M]$ (because $D[M]$ is a commutative algebra).

3 The Case of a D-Algebra

We now use the results from Sect. 2 to gain information about $\mathrm{Int}_K(A)$, where A is a D-algebra. In Theorem 13 below, we shall obtain a description of the integral closure of $\mathrm{Int}_K(A)$.

As mentioned in the introduction, we assume that A is a torsion-free D-algebra that is finitely generated as a D-module. Let $B = A \otimes_D K$ be the extension of A to a K-algebra. Since A is a faithful D-module, B contains copies of D, A, and K. Furthermore, K is contained in the center of B, so we can evaluate polynomials in $K[X]$ at elements of B and define

$$\operatorname{Int}_K(A) := \{f \in K[X] \mid f(A) \subseteq A\}.$$

Letting n be the vector space dimension of B over K, we also have an embedding $B \hookrightarrow M_n(K), b \mapsto M_b$. More precisely, we may embed B into the ring of K-linear endomorphisms of B (which is isomorphic to $M_n(K)$) via the map $B \hookrightarrow \operatorname{End}_K(B)$ sending $b \in B$ to the endomorphism $x \mapsto b \cdot x$. Consequently, starting with just D and A, we obtain a representation of A as a D-subalgebra of $M_n(K)$. Note that n may be less than the minimum number of generators of A as a D-module.

In light of the aforementioned matrix representation of B, several of the definitions and notations we defined in Sect. 2 will carry over to B. Since the concepts of minimal polynomial and eigenvalue are independent of the representation $B \hookrightarrow M_n(K)$, the following are well defined:

- For all $b \in B$, $\mu_b(X) \in K[X]$ is the minimal polynomial of b. So, $\mu_b(X)$ is the monic polynomial of minimal degree in $K[X]$ that kills b. Equivalently, μ_b is the monic generator of the null ideal $N_{K[X]}(b)$ of b. This is the same as the minimal polynomial of $M_b \in M_n(K)$, since for all $f \in K[X]$ we have $f(M_b) = M_{f(b)}$. To ease the notation, from now on we will identify b with M_b.
- By the Cayley-Hamilton Theorem, $\deg(\mu_b) \leq n$, for all $b \in B$.
- For all $b \in B$, $\Omega_b = \{\text{roots of } \mu_b \text{ in } \overline{K}\}$. The elements of Ω_b are nothing else than the eigenvalues of b under any matrix representation $B \hookrightarrow M_n(K)$. If $S \subseteq B$, then $\Omega_S = \bigcup_{b \in S} \Omega_b$.
- $b \in B$ is *integral over* D (or just *integral*) if b solves a monic polynomial in $D[X]$.
- $B = K[A]$, since B is formed by extension of scalars from D to K.
- $A' = \{b \in B \mid b \text{ is integral}\}$. By [1, Theorem 1, Chapt. V] $A \subseteq A'$. In particular, this implies $A \cap K = D$ (because D is integrally closed), so that $\operatorname{Int}_K(A) \subseteq \operatorname{Int}(D)$ (see the remarks in the introduction).
- $\operatorname{Int}_K(A, A') = \{f \in K[X] \mid f(A) \subseteq A'\}$.

Working exactly as in Proposition 6, we find that $\operatorname{Int}_K(A, A')$ is another ring of integral-valued polynomials. Additionally, Lemma 2 and (8) hold for elements of B. Consequently, we have

Theorem 10. $\operatorname{Int}_K(A, A')$ *is an integrally closed ring and is equal to* $\operatorname{Int}_K(\Omega_A, \Lambda_n)$.

By generalizing results from [9], we will show that if D has finite residue rings, then $\operatorname{Int}_K(A, A')$ is the integral closure of $\operatorname{Int}_K(A)$. This establishes the analogue of [2, Thm. IV.4.7] (originally proved in [7, Prop. 2.2]) mentioned in the introduction.

We will actually prove a slightly stronger statement and give a description of $\text{Int}_K(A, A')$ as the integral closure of an intersection of pullbacks. Notice that

$$\bigcap_{a \in A}(D[X] + \mu_a(X) \cdot K[X]) \subseteq \text{Int}_K(A)$$

because if $f \in D[X] + \mu_a(X) \cdot K[X]$, then $f(a) \in D[a] \subseteq A$. We thus have a chain of inclusions:

$$\bigcap_{a \in A}(D[X] + \mu_a(X) \cdot K[X]) \subseteq \text{Int}_K(A) \subseteq \text{Int}_K(A, A') \tag{11}$$

and our work below will show that this is actually a chain of integral ring extensions.

Lemma 12. Let $f \in \text{Int}_K(A, A')$, and write $f(X) = g(X)/d$ for some $g \in D[X]$ and some nonzero $d \in D$. Then, for each $h \in D[X]$, $d^{n-1}h(f(X)) \in \bigcap_{a \in A}(D[X] + \mu_a(X) \cdot K[X])$.

Proof. Let $a \in A$. Since $f \in \text{Int}_K(A, A')$, $m := \mu_{f(a)} \in D[X]$, and $\deg(m) \leq n$.

Now, m is monic, so we can divide h by m to get $h(X) = q(X)m(X) + r(X)$, where $q, r \in D[X]$, and either $r = 0$ or $\deg(r) < n$. Then,

$$d^{n-1}h(f(X)) = d^{n-1}q(f(X))m(f(X)) + d^{n-1}r(f(X))$$

The polynomial $d^{n-1}q(f(X))m(f(X)) \in K[X]$ is divisible by $\mu_a(X)$ because $m(f(a)) = 0$. Since $\deg(r) < n$, $d^{n-1}r(f(X)) \in D[X]$. Thus, $d^{n-1}h(f(X)) \in D[X] + \mu_a(X) \cdot K[X]$, and since a was arbitrary the lemma is true. □

For the next result, we need an additional assumption. Recall that a ring D has finite residue rings if for all proper nonzero ideal $I \subset D$, the residue ring D/I is finite. Clearly, this condition is equivalent to asking that for all nonzero $d \in D$, the residue ring D/dD is finite.

Theorem 13. Assume that D has finite residue rings. Then, $\text{Int}_K(A, A') = \text{Int}_K(\Omega_A, \Lambda_n)$ is the integral closure of both $\bigcap_{a \in A}(D[X] + \mu_a(X) \cdot K[X])$ and $\text{Int}_K(A)$.

Proof. Let $R = \bigcap_{a \in A}(D[X] + \mu_a(X) \cdot K[X])$. By (11), it suffices to prove that $\text{Int}_K(A, A')$ is the integral closure of R. Let $f(X) = g(X)/d \in \text{Int}_K(A, A')$. By Theorem 10, $\text{Int}_K(A, A')$ is integrally closed, so it is enough to find a monic polynomial $\phi \in D[X]$ such that $\phi(f(X)) \in R$.

Let $\mathscr{P} \subseteq D[X]$ be a set of monic residue representatives for $\{\mu_{f(a)}(X)\}_{a \in A}$ modulo $(d^{n-1})^2$. Since D has finite residue rings, $\mathscr{P}$ is finite. Let $\phi(X)$ be the product of all the polynomials in $\mathscr{P}$. Then, ϕ is monic and is in $D[X]$.

Fix $a \in A$ and let $m = \mu_{f(a)}$. There exists $p(X) \in \mathscr{P}$ such that $p(X)$ is equivalent to $m \mod (d^{n-1})^2$, so $p(X) = m(X) + (d^{n-1})^2 r(X)$ for some $r \in D[X]$. Furthermore, $p(X)$ divides $\phi(X)$, so there exists $q(X) \in D[X]$ such that $\phi(X) = p(X)q(X)$. Thus,

$$\phi(f(X)) = p(f(X))q(f(X))$$
$$= m(f(X))q(f(X)) + (d^{n-1})^2 r(f(X))q(f(X))$$
$$= m(f(X))q(f(X)) + d^{n-1} r(f(X)) \cdot d^{n-1} q(f(X))$$

As in Lemma 12, $m(f(X))q(f(X)) \in \mu_a(X) \cdot K[X]$ because $m(f(a)) = 0$. By Lemma 12, $d^{n-1}r(f(X))$ and $d^{n-1}q(f(X))$ are in $D[X] + \mu_a(X) \cdot K[X]$. Hence, $\phi(f(X)) \in D[X] + \mu_a(X) \cdot K[X]$, and since a was arbitrary, $\phi(f(X)) \in R$. $\square$

Theorem 13 says that the integral closure of $\text{Int}_K(A)$ is equal to the ring of polynomials in $K[X]$ which map the eigenvalues of all the elements $a \in A$ to integral elements over D.

Remark 14. By following essentially the same steps as in Lemma 12 and Theorem 13, one may prove that $\text{Int}_K(A, A')$ is the integral closure of $\text{Int}_K(A)$ without the use of the pullbacks $D[X] + \mu_a(X) \cdot K[X]$. However, employing the pullbacks gives a slightly stronger theorem without any additional difficulty.

In the case $A = M_n(D)$, $\text{Int}_K(M_n(D))$ is equal to the intersection of the pullbacks $D[X] + \mu_M(X) \cdot K[X]$, for $M \in M_n(D)$. Indeed, let $f \in \text{Int}_K(M_n(D))$ and $M \in M_n(D)$. By [11, Remark 2.1 & (3)], $\text{Int}_K(M_n(D))$ is equal to the intersection of the pullbacks $D[X] + \chi_M(X)K[X]$, for $M \in M_n(D)$, where $\chi_M(X)$ is the characteristic polynomial of M. By the Cayley-Hamilton Theorem, $\mu_M(X)$ divides $\chi_M(X)$ so that $f \in D[X] + \chi_M(X)K[X] \subseteq D[X] + \mu_M(X)K[X]$ and we are done.

Remark 15. The assumption that D has finite residue rings implies that D is Noetherian (and in fact that D is a Dedekind domain, because D is integrally closed). Given that [2, Thm. IV.4.7] (or [7, Prop. 2.2]) requires only the assumption that D is Noetherian, it is fair to ask if Theorem 13 holds under the weaker condition that D is Noetherian.

Note that $\Omega_{M_n(D)} = \Lambda_n$ (and in particular, $\Omega_{M_n(\mathbb{Z})} = \mathscr{A}_n$). Hence, we obtain the following (which generalizes [9, Thm. 4.6]):

Corollary 16. *If D has finite residue rings, then the integral closure of $\text{Int}_K(M_n(D))$ is $\text{Int}_K(\Lambda_n)$.*

The algebra of upper triangular matrices yields another interesting example.

Corollary 17. *Assume that D has finite residue rings. For each $n > 0$, let $T_n(D)$ be the ring of $n \times n$ upper triangular matrices with entries in D. Then, the integral closure of $\text{Int}_K(T_n(D))$ equals $\text{Int}(D)$.*

Proof. For each $a \in T_n(D)$, μ_a splits completely and has roots in D, so $\Omega_{T_n(D)} = D$. Hence, the integral closure of $\text{Int}_K(T_n(D))$ is $\text{Int}_K(\Omega_{T_n(D)}, \Lambda_n) = \text{Int}_K(D, \Lambda_n)$. But, polynomials in $K[X]$ that move D into Λ_n actually move D into $\Lambda_n \cap K = D$. Thus, $\text{Int}_K(D, \Lambda_n) = \text{Int}(D)$. $\square$

Since $\text{Int}_K(T_n(D)) \subseteq \text{Int}_K(T_{n-1}(D))$ for all $n > 0$, the previous proposition proves that

$$\cdots \subseteq \text{Int}_K(T_n(D)) \subseteq \text{Int}_K(T_{n-1}(D)) \subseteq \cdots \subseteq \text{Int}(D)$$

is a chain of integral ring extensions.

4 Matrix Rings and Polynomially Dense Subsets

For any D-algebra A, we have $(A')' = A'$, so $\text{Int}_K(A')$ is integrally closed by Theorem 10. Furthermore, $\text{Int}_K(A')$ is always contained in $\text{Int}_K(A, A')$. One may then ask: when does $\text{Int}_K(A')$ equal $\text{Int}_K(A, A')$? In this section, we investigate this question and attempt to identify *polynomially dense* subsets of rings of integral-valued polynomials. The theory presented here is far from complete, so we raise several related questions worthy of future research.

Definition 18. Let $S \subseteq T \subseteq M_n(K)$. Define $\text{Int}_K(S, T) := \{f \in K[X] \mid f(S) \subseteq T\}$ and $\text{Int}_K(T) := \text{Int}_K(T, T)$. To say that S is *polynomially dense* in T means that $\text{Int}_K(S, T) = \text{Int}_K(T)$.

Thus, the question posed at the start of this section can be phrased as: is A polynomially dense in A'?

In general, it is not clear how to produce polynomially dense subsets of A', but we can describe some polynomially dense subsets of $M_n(D)'$.

Proposition 19. *For each $\Omega \subset \Lambda_n$ of cardinality at most n, choose $M \in M_n(D)'$ such that $\Omega_M = \Omega$. Let S be the set formed by such matrices. Then, S is polynomially dense in $M_n(D)'$. In particular, the set of companion matrices in $M_n(D)$ is polynomially dense in $M_n(D)'$.*

Proof. We know that $\text{Int}_K(M_n(D)') \subseteq \text{Int}_K(S, M_n(D)')$, so we must show that the other containment holds. Let $f \in \text{Int}_K(S, M_n(D)')$ and $N \in M_n(D)'$. Let $M \in S$ such that $\Omega_M = \Omega_N$. Then, $f(M)$ is integral, so by Lemma 2 and (8), $f(N)$ is also integral.

The proposition holds for the set of companion matrices because for any $\Omega \subset \Lambda_n$, we can find a companion matrix in $M_n(D)$ whose eigenvalues are the elements of Ω. $\qquad\square$

By the proposition, any subset of $M_n(D)$ containing the set of companion matrices is polynomially dense in $M_n(D)'$. In particular, $M_n(D)$ is polynomially dense in $M_n(D)'$.

When $D = \mathbb{Z}$, we can say more. In [10] it is shown that $\text{Int}_{\mathbb{Q}}(\mathscr{A}_n) = \text{Int}_{\mathbb{Q}}(A_n, \mathscr{A}_n)$, where A_n is the set of algebraic integers of degree equal to n. Letting $\mathscr{I}$ be the set of companion matrices in $M_n(\mathbb{Z})$ of irreducible polynomials, we have $\Omega_{\mathscr{I}} = A_n$. Hence, by Corollary 16 and Theorem 9, $\mathscr{I}$ is polynomially dense in $M_n(\mathbb{Z})'$.

Returning to the case of a general D-algebra A, the following diagram summarizes the relationships among the various polynomial rings we have considered:

$$\mathrm{Int}_K(A) \subseteq \mathrm{Int}_K(A, A') = \mathrm{Int}_K(\Omega_A, \Lambda_n)$$
$$\cup\mathsf{I} \qquad\qquad \cup\mathsf{I}$$
$$\mathrm{Int}_K(A') \qquad = \mathrm{Int}_K(\Omega_{A'}, \Lambda_n) \qquad\qquad (20)$$
$$\cup\mathsf{I} \qquad\qquad \cup\mathsf{I}$$
$$\mathrm{Int}_K(M_n(D)') = \mathrm{Int}_K(\Lambda_n)$$

From this diagram, we deduce that A is polynomially dense in A' if and only if Ω_A is polynomially dense in $\Omega_{A'}$.

It is fair to ask what other relationships hold among these rings. We present several examples and a proposition concerning possible equalities in the diagram. Again, we point out that such equalities can be phrased in terms of polynomially dense subsets. First, we show that $\mathrm{Int}_K(A)$ need not equal $\mathrm{Int}_K(A, A')$ (i.e., A need not be polynomially dense in A').

Example 21. Take $D = \mathbb{Z}$ and $A = \mathbb{Z}[\sqrt{-3}]$. Then, $A' = \mathbb{Z}[\theta]$, where $\theta = \frac{1+\sqrt{-3}}{2}$. The ring $\mathrm{Int}_K(A, A')$ contains both $\mathrm{Int}_K(A)$ and $\mathrm{Int}_K(A')$.

If $\mathrm{Int}_K(A, A')$ equaled $\mathrm{Int}_K(A')$, then we would have $\mathrm{Int}_K(A) \subseteq \mathrm{Int}_K(A')$. However, this is not the case. Indeed, working mod 2, we see that for all $\alpha = a + b\sqrt{-3} \in A$, $\alpha^2 \equiv a^2 - 3b^2 \equiv a^2 + b^2$. So, $\alpha^2(\alpha^2 + 1)$ is always divisible by 2, and hence $\frac{x^2(x^2+1)}{2} \in \mathrm{Int}_K(A)$. On the other hand, $\frac{\theta^2(\theta^2+1)}{2} = -\frac{1}{2}$, so $\frac{x^2(x^2+1)}{2} \notin \mathrm{Int}_K(A')$. Thus, we conclude that $\mathrm{Int}_K(A') \subsetneq \mathrm{Int}_K(A, A')$.

The work in the previous example suggests the following proposition.

Proposition 22. *Assume that D has finite residue rings. Then, A is polynomially dense in A' if and only if $\mathrm{Int}_K(A) \subseteq \mathrm{Int}_K(A')$.*

Proof. This is similar to [2, Thm. IV.4.9]. If A is polynomially dense in A', then $\mathrm{Int}_K(A') = \mathrm{Int}_K(A, A')$, and we are done because $\mathrm{Int}_K(A, A')$ always contains $\mathrm{Int}_K(A)$. Conversely, assume that $\mathrm{Int}_K(A) \subseteq \mathrm{Int}_K(A')$. Then, $\mathrm{Int}_K(A) \subseteq \mathrm{Int}_K(A') \subseteq \mathrm{Int}_K(A, A')$. Since $\mathrm{Int}_K(A')$ is integrally closed by Theorem 10 and $\mathrm{Int}_K(A, A')$ is integral over $\mathrm{Int}_K(A)$ by Theorem 13, we must have $\mathrm{Int}_K(A') = \mathrm{Int}_K(A, A')$. $\qquad\square$

By Proposition 19, $\mathrm{Int}_K(M_n(D)') = \mathrm{Int}_K(M_n(D), M_n(D)')$. There exist algebras other than matrix rings for which $\mathrm{Int}_K(A') = \mathrm{Int}_K(A, A')$. We now present two such examples.

Example 23. Let $A = T_n(D)$, the ring of $n \times n$ upper triangular matrices with entries in D. Define $T_n(K)$ similarly. Then, A' consists of the integral matrices in $T_n(K)$, and since D is integrally closed, such matrices must have diagonal entries in D. Thus, $\Omega_{A'} = D = \Omega_A$. It follows that $\mathrm{Int}_K(T_n(D), T_n(D)') = \mathrm{Int}_K(T_n(D)')$.

Example 24. Let $\mathbf{i}$, $\mathbf{j}$, and $\mathbf{k}$ be the standard quaternion units satisfying $\mathbf{i}^2 = \mathbf{j}^2 = \mathbf{k}^2 = -1$ and $\mathbf{ij} = \mathbf{k} = -\mathbf{ji}$ (see, e.g., [8, Ex. 1.1, 1.13] or [3] for basic material on quaternions).

Let A be the $\mathbb{Z}$-algebra consisting of Hurwitz quaternions:

$$A = \{a_0 + a_1\mathbf{i} + a_2\mathbf{j} + a_3\mathbf{k} \mid a_\ell \in \mathbb{Z} \text{ for all } \ell \text{ or } a_\ell \in \mathbb{Z} + \tfrac{1}{2} \text{ for all } \ell\}$$

Then, for B we have

$$B = \{q_0 + q_1\mathbf{i} + q_2\mathbf{j} + q_3\mathbf{k} \mid q_\ell \in \mathbb{Q}\}$$

It is well known that the minimal polynomial of the element $q = q_0 + q_1\mathbf{i} + q_2\mathbf{j} + q_3\mathbf{k} \in B \setminus \mathbb{Q}$ is $\mu_q(X) = X^2 - 2q_0 X + (q_0^2 + q_1^2 + q_2^2 + q_3^2)$, so A' is the set

$$A' = \{q_0 + q_1\mathbf{i} + q_2\mathbf{j} + q_3\mathbf{k} \in B \mid 2q_0, \ q_0^2 + q_1^2 + q_2^2 + q_3^2 \in \mathbb{Z}\}$$

As with the previous example, by (20), it is enough to prove that $\Omega_{A'} = \Omega_A$.

Let $q = q_0 + q_1\mathbf{i} + q_2\mathbf{j} + q_3\mathbf{k} \in A'$ and $N = q_0^2 + q_1^2 + q_2^2 + q_3^2 \in \mathbb{Z}$. Then, $2q_0 \in \mathbb{Z}$, so q_0 is either an integer or a half-integer. If $q_0 \in \mathbb{Z}$, then $q_1^2 + q_2^2 + q_3^2 = N - q_0^2 \in \mathbb{Z}$. It is known (see for instance [12, Lem. B p. 46]) that an integer which is the sum of three rational squares is a sum of three integer squares. Thus, there exist $a_1, a_2, a_3 \in \mathbb{Z}$ such that $a_1^2 + a_2^2 + a_3^2 = N - q_0^2$. Then, $q' = q_0 + a_1\mathbf{i} + a_2\mathbf{j} + a_3\mathbf{k}$ is an element of A such that $\Omega_{q'} = \Omega_q$.

If q_0 is a half-integer, then $q_0 = \frac{t}{2}$ for some odd $t \in \mathbb{Z}$. In this case, $q_1^2 + q_2^2 + q_3^2 = \frac{4N - t^2}{4} = \frac{u}{4}$, where $u \equiv 3 \bmod 4$. Clearing denominators, we get $(2q_1)^2 + (2q_2)^2 + (2q_3)^2 = u$. As before, there exist integers $a_1, a_2,$ and a_3 such that $a_1^2 + a_2^2 + a_3^2 = u$. But since $u \equiv 3 \bmod 4$, each of the a_ℓ must be odd. So, $q' = (t + a_1\mathbf{i} + a_2\mathbf{j} + a_3\mathbf{k})/2 \in A$ is such that $\Omega_{q'} = \Omega_q$.

It follows that $\Omega_{A'} = \Omega_A$ and thus that $\mathrm{Int}_K(A, A') = \mathrm{Int}_K(A')$.

Example 25. In contrast to the last example, the Lipschitz quaternions $A_1 = \mathbb{Z} \oplus \mathbb{Z}\mathbf{i} \oplus \mathbb{Z}\mathbf{j} \oplus \mathbb{Z}\mathbf{k}$ (where we only allow $a_\ell \in \mathbb{Z}$) are not polynomially dense in A_1'. With A as in Example 24, we have $A_1 \subset A$, and both rings have the same B, so $A_1' = A'$. Our proof is identical to Example 21. Working mod 2, the only possible minimal polynomials for elements of $A_1 \setminus \mathbb{Z}$ are X^2 and $X^2 + 1$. It follows that $f(X) = \frac{x^2(x^2+1)}{2} \in \mathrm{Int}_K(A_1)$. Let $\alpha = \frac{1 + \mathbf{i} + \mathbf{j} + \mathbf{k}}{2} \in A'$. Then, the minimal polynomial of α is $X^2 - X + 1$ (note that this minimal polynomial is shared by $\theta = \frac{1 + \sqrt{-3}}{2}$ in Example 21). Just as in Example 21, $f(\alpha) = -\frac{1}{2}$, which is not in A'. Thus, $\mathrm{Int}_K(A_1) \not\subseteq \mathrm{Int}(A')$, so A_1 is not polynomially dense in $A_1' = A'$ by Proposition 22.

5 Further Questions

Here, we list more questions for further investigation.

Question 26. Under what conditions do we have equalities in (20)? In particular, what are necessary and sufficient conditions on A for A to be polynomially dense in A'? In Examples 23 and 24, we exploited the fact that if $\Omega_A = \Omega_{A'}$, then $\mathrm{Int}_K(A, A') = \mathrm{Int}_K(A')$. It is natural to ask whether the converse holds. If $\mathrm{Int}_K(A, A') = \mathrm{Int}_K(A')$, does it follow that $\Omega_A = \Omega_{A'}$? In other words, if A is polynomially dense in A', then is Ω_A equal to $\Omega_{A'}$?

Question 27. By [2, Proposition IV.4.1] it follows that $\mathrm{Int}(D)$ is integrally closed if and only if D is integrally closed. By Theorem 10 we know that if $A = A'$, then $\mathrm{Int}_K(A)$ is integrally closed. Do we have a converse? Namely, if $\mathrm{Int}_K(A)$ is integrally closed, can we deduce that $A = A'$?

Question 28. In our proof (Theorem 13) that $\mathrm{Int}_K(A, A')$ is the integral closure of $\mathrm{Int}_K(A)$, we needed the assumption that D has finite residue rings. Is the theorem true without this assumption? In particular, is it true whenever D is Noetherian?

Question 29. When is $\mathrm{Int}_K(A, A') = \mathrm{Int}_K(\Omega_A, \Lambda_n)$ a Prüfer domain? When $D = \mathbb{Z}$, $\mathrm{Int}_{\mathbb{Q}}(A, A')$ is always Prüfer by [9, Cor. 4.7]. On the other hand, even when $A = D$ is a Prüfer domain, $\mathrm{Int}(D)$ need not be Prüfer (see [2, Sec. IV.4]).

Question 30. In Remark 14, we proved that $\mathrm{Int}_K(M_n(D))$ equals an intersection of pullbacks:

$$\bigcap_{M \in M_n(D)} (D[X] + \mu_M(X) \cdot K[X]) = \mathrm{Int}_K(M_n(D))$$

Does such an equality hold for other algebras?

Acknowledgements The authors wish to thank the referee for his/her suggestions. The first author wishes to thank Daniel Smertnig for useful discussions during the preparation of this paper about integrality in noncommutative settings. The same author was supported by the Austrian Science Foundation (FWF), Project Number P23245-N18.

References

1. N. Bourbaki, *Commutative Algebra* (Addison-Wesley Publishing Co., Reading, Mass, Hermann, Paris 1972)
2. J.-P. Cahen, J.-L. Chabert, *Integer-Valued Polynomials*, vol. 48 (American Mathematical Society Surveys and Monographs, Providence, RI 1997)
3. L.E. Dickson, *Algebras and their Arithmetics* (Dover Publications, New York 1960)
4. S. Evrard, Y. Fares, K. Johnson, Integer valued polynomials on lower triangular integer matrices. Monats. für Math. **170**(2), 147–160 (2013)

5. S. Frisch, Corrigendum to integer-valued polynomials on algebras. J. Algebra **373**, 414–425 (2013), J. Algebra **412** 282 (2014). DOI: 10.1016/j.jalgebra.2013.06.003. http://dx.doi.org/10.1016/j.jalgebra.2013.06.003
6. S. Frisch, Integer-valued polynomials on algebras. J. Algebra **373**, 414–425 (2013)
7. R. Gilmer, W. Heinzer, D. Lantz, The Noetherian property in rings of integer-valued polynomials. Trans. Amer. Math. Soc. **338**(1), 187–199 (1993)
8. T.Y. Lam, *A First Course in Noncommutative Rings* (Springer, New York, 1991)
9. K.A. Loper, N.J. Werner, Generalized rings of integer-valued polynomials. J. Number Theory **132**(11), 2481–2490 (2012)
10. G. Peruginelli, Integral-valued polynomials over sets of algebraic integers of bounded degree. J. Number Theory **137**, 241–255 (2014)
11. G. Peruginelli, Integer-valued polynomials over matrices and divided differences. Monatsh. Math. **173**(4), 559–571 (2014)
12. J.P. Serre, *A Course in Arithmetic* (Springer, New York 1996)

On Monoids and Domains Whose Monadic Submonoids Are Krull

Andreas Reinhart

Abstract A submonoid S of a given monoid H is called monadic if it is a divisor-closed submonoid of H generated by one element (i.e., there is some (non-zero) $b \in H$ such that S is the smallest divisor-closed submonoid of H such that $b \in S$). In this paper we study monoids and domains whose monadic submonoids are Krull monoids. These monoids resp. domains are called monadically Krull. Every Krull monoid is a monadically Krull monoid, but the converse is not true. We provide several types of counterexamples and present a few characterizations for monadically Krull monoids. Furthermore, we show that rings of integer-valued polynomials over factorial domains are monadically Krull. Finally, we investigate the connections between monadically Krull monoids and generalizations of SP-domains.

Keywords Monadically • Integer-valued • Krull monoid • Mori set • SP-domain

2000 Mathematics Subject Classification. 13A15, 13F05, 20M11, 20M12

1 Introduction

The main goal of this paper is to study the so-called monadically Krull monoids (i.e., monoids where every divisor-closed submonoid generated by one element is a Krull monoid). Studying monoids "monadically" (i.e., investigating properties

A. Reinhart (✉)
Institut für Mathematik und wissenschaftliches Rechnen, Karl-Franzens-Universität,
Heinrichstrasse 36, 8010 Graz, Austria
e-mail: andreas.reinhart@uni-graz.at

M. Fontana et al. (eds.), *Commutative Algebra: Recent Advances in Commutative Rings,*   307
Integer-Valued Polynomials, and Polynomial Functions, DOI 10.1007/978-1-4939-0925-4_18,
© Springer Science+Business Media New York 2014

that are satisfied by all divisor-closed submonoids generated by one element) is reasonable, since some types of monoids are better situated in the "local" than in the "global" situation. On the other hand it turns out that there are a lot of monoid theoretical properties that are satisfied by the monoid if and only if they are satisfied "monadically" (e.g., being atomic, completely integrally closed, factorial). However, the Krull property does not behave like this (as pointed out in this work), and thus monadically Krull monoids are of special interest. Being monadically Krull is related to "weak factorization properties" that have been studied in a series of papers (see [9, 10, 24–26]). Moreover, some recent work in studying monoids "monadically" has been done in [16, 23]. Investigating monadically Krull monoids was also motivated by a problem that we want to discuss in more detail. Let R be a (possibly noncommutative) ring and let $\mathcal{C}$ be a class of finitely generated (right) R-modules which is closed under finite direct sums, direct summands, and isomorphisms. Then the set $\mathcal{V}(\mathcal{C})$ of isomorphism classes of modules is a commutative semigroup with operation induced by the direct sum. This semigroup encodes all possible information about direct sum decompositions of modules in $\mathcal{C}$ (see [6, 12]). If the endomorphism ring of each module in $\mathcal{C}$ is semilocal, then $\mathcal{V}(\mathcal{C})$ is a Krull monoid [11, Theorem 3.4]. Moreover, every reduced Krull monoid can be realized by such a monoid of modules [13]. Thus the (global) property that $\mathcal{V}(\mathcal{C})$ is Krull follows from a family of local data, namely that all $\text{End}_R(M)$ are semilocal. Furthermore, the assumption that $\text{End}_R(M)$ is semilocal implies that the smallest divisor-closed submonoid of $\mathcal{V}(\mathcal{C})$ generated by the class of M [denoted by add(M)] is a Krull monoid [4–6].

In the second section we will discuss the most important terminology. We give a brief introduction to finitary ideal systems to simplify and unify the terminology about various types of ideals (e.g., ring ideals and t-ideals).

In the third section we will prove that several interesting properties (like being completely integrally closed, being atomic or being an FF-monoid) can be characterized by using the divisor-closed submonoids generated by one element. Moreover, we provide another characterization of Krull monoids. The main result in this section is a characterization of monadically Krull monoids. It turns out that the monadically Krull monoids are precisely the atomic, completely integrally closed monoids where special sets of atoms are finite up to associates.

In the fourth section we deal with the question whether every monadically Krull monoid is already a Krull monoid. We provide several counterexamples. First we present a ring theoretical counterexample and later we will introduce a counterexample in the monoid setting that is substantially stronger. The second example will show that radical factorial FF-monoids (they are always monadically Krull) also need not be Krull. By the way we answer some questions that have been raised in the literature in the negative. In [10] it has been shown that every atomic IDPF-domain that contains a field of characteristics zero is already completely integrally closed. We will point out that such a domain is not necessarily a Krull domain. Furthermore, we deal with the problem whether the t-dimension of a t-SP-monoid (which is some sort of generalized Krull monoid) is bounded by one and show that t-SP-monoids whose height-one prime t-ideals are divisorial are not necessarily Krull. Moreover, it is well known that an integral domain is a Prüfer domain that does not have non-zero idempotent prime ideals if and only if each of its primary ideals is a power of its radical and its set of prime ideals satisfies the ACC (e.g., see [29, Corollary 5.5]). We show that the "t-analogue" of this statement is not true in the monoid setting.

In the fifth section we investigate rings of integer-valued polynomials. We prove that rings of integer-valued polynomials over factorial domains are monadically Krull. Using this result we are able to provide a large class of monadically Krull domains that are t-Prüfer domains and that fail to be Krull.

In the last section we deal with the question whether every radical factorial FF-domain of Krull dimension one is already a Krull domain. Although we could not solve this problem so far, we will present partial solutions. For example, we will construct a BF-domain that is an SP-domain but not a Krull domain (note that every radical factorial domain of Krull dimension one is an SP-domain; see [29, Proposition 3.11]). The counterexamples in this section are based on a construction used in [20]. Furthermore, we investigate how far SP-domains (and their generalizations) are from being monadically Krull by studying a special necessary property that pops up in the characterization of monadically Krull (in [24] this special property is called pseudo-IDPF).

2 Preliminaries

In the following, a monoid is a commutative semigroup (multiplicatively written if not stated otherwise) that possesses an identity and (if not stated otherwise) a zero element different from the identity such that every non-zero element is cancellative. A quotient monoid of a monoid H is a monoid containing H as a submonoid where every non-zero element is invertible and that is minimal with respect to this property.

Let H be a monoid, K a quotient monoid of H, and $X \subseteq H$. Set $H^\bullet = H \setminus \{0\}$.

- For $A, B \subseteq K$ let $(A :_K B) = \{z \in K \mid zB \subseteq A\}$, $A^{-1} = (H :_K A)$ and $A_v = (A^{-1})^{-1}$.
- X is called $(H\text{-})$divisor-closed if for all $x \in H$ and $y \in H^\bullet$ such that $xy \in X$ it follows that $x \in X$.
- By $[X]_H$ (resp. $[\![X]\!]_H$) we denote the smallest (divisor-closed) submonoid of H that contains X.
- If $a \in H$, set $[a]_H = [\{a\}]_H$.
- X is called an $(H\text{-})$Mori set if for every $F \subseteq X$ there exists some finite $E \subseteq F$ such that $E^{-1} = F^{-1}$.

Note that $[\![a]\!]_H = \{b \in H \mid b \mid_H a^n \text{ for some } n \in \mathbb{N}\} \cup \{0\}$ for all $a \in H^\bullet$.

Since we will use a slightly different version of ideal systems than those dealt within [22], we will recall the definition. The ideal systems in this work will always be ideal systems in the sense of [22] (but not conversely). Let $\mathbb{P}(H)$ be the power set of H and $r : \mathbb{P}(H) \to \mathbb{P}(H)$ be a map. The map r is called a (finitary) ideal system on H if the following properties are satisfied for all $X, Y \subseteq H$ and $c \in H$.

- $XH \cup \{0\} \subseteq r(X)$.
- $r(cX) = cr(X)$.
- If $X \subseteq r(Y)$, then $r(X) \subseteq r(Y)$.
- $(r(X) = \bigcup_{E \subseteq X, |E| < \infty} r(E).)$

Note that if $H^{\bullet} \neq H^{\times}$, then $v : \mathbb{P}(H) \to \mathbb{P}(H)$ defined by $v(X) = X_v$ for all $X \subseteq H$ is an ideal system on H and $t : \mathbb{P}(H) \to \mathbb{P}(H)$ defined by $t(X) = \bigcup_{E \subseteq X, |E| < \infty} E_v$ for all $X \subseteq H$ is a finitary ideal system on H. If R is an integral domain, then $d : \mathbb{P}(R) \to \mathbb{P}(R)$ defined by $d(X) = (X)_R$ for all $X \subseteq R$ is a finitary ideal system on R. Furthermore, $s : \mathbb{P}(H) \to \mathbb{P}(H)$ defined by $s(X) = XH$ if $\emptyset \neq X \subseteq H$ and $s(\emptyset) = \{0\}$ is a finitary ideal system on H.

In the following we will use most of the definitions and notations in [15, 22] without further reference. Especially, we will freely use the following terms: "BF-monoid", "FF-monoid", "ACCP", "atomic", "factorial", "Krull", "completely integrally closed", "valuation monoid", "v-closed", "root-closed", and "GCD-monoid". A monoid is called a Mori monoid if it is v-noetherian in the terminology of [15].

Observe that Mori sets defined in this work differ from those introduced in [30]. Note that if $S \subseteq H$ is a divisor-closed submonoid and H is a Krull monoid (a Mori monoid, a completely integrally closed monoid), then S has the same property by [15, Proposition 2.4.4.2].

3 Monadic Properties and Mori Sets

First we present a simple characterization of being a Mori set. Using this result it is straightforward to prove that H is a Mori monoid if and only if H is a Mori set.

Lemma 3.1. *Let H be a monoid and $X \subseteq H$. Then X is not a Mori set if and only if there is some $(a_i)_{i \in \mathbb{N}} \in X^{\mathbb{N}}$ such that $\{a_i \mid i \in [1, n+1]\}^{-1} \subsetneq \{a_i \mid i \in [1, n]\}^{-1}$ for all $n \in \mathbb{N}$.*

Proof. "$\Rightarrow$": Let X be not a Mori set. Then there is some $F \subseteq X$ such that for every finite $E \subseteq F$ it follows that $F^{-1} \subsetneq E^{-1}$. There exists some $a_1 \in F$. Now let $n \in \mathbb{N}$ and $(a_i)_{i=1}^{n} \in F^{[1,n]}$. Then $F^{-1} \subsetneq \{a_i \mid i \in [1, n]\}^{-1}$, and thus $F \nsubseteq \{a_i \mid i \in [1, n]\}_v$. Consequently, there exists some $a_{n+1} \in F \backslash \{a_i \mid i \in [1, n]\}_v$, and thus $\{a_i \mid i \in [1, n+1]\}^{-1} \subsetneq \{a_i \mid i \in [1, n]\}^{-1}$. Hence there is some $(a_i)_{i \in \mathbb{N}} \in F^{\mathbb{N}}$ such that $\{a_i \mid i \in [1, n+1]\}^{-1} \subsetneq \{a_i \mid i \in [1, n]\}^{-1}$ for all $n \in \mathbb{N}$. "$\Leftarrow$": Let $F = \{a_i \mid i \in \mathbb{N}\}$. Assume that there is some finite $E \subseteq F$ such that $E^{-1} = F^{-1}$. Then there exists some $n \in \mathbb{N}$ such that $E \subseteq \{a_i \mid i \in [1, n]\}$. This implies that $F^{-1} \subseteq \{a_i \mid i \in [1, n+1]\}^{-1} \subsetneq \{a_i \mid i \in [1, n]\}^{-1} \subseteq E^{-1} = F^{-1}$, a contradiction. $\square$

Next we specify Krull monoids by using Mori sets. Note that the equivalence of 1 and 4 is well known.

Proposition 3.2. *Let H be a monoid. The following conditions are equivalent:*

1. H *is a Krull monoid.*
2. H *is atomic, completely integrally closed, and $\mathcal{A}(H)$ is a Mori set.*
3. H *is completely integrally closed and there is some Mori set $F \subseteq H$ such that $H = [F \cup H^\times]_H$.*
4. H *is completely integrally closed and every t-maximal t-ideal of H is divisorial.*

Proof. $1 \Rightarrow 2$: Clear. $2 \Rightarrow 3$: Set $F = \mathcal{A}(H)$. Since H is atomic we have $H = [F \cup H^\times]$. $3 \Rightarrow 4$: Let $F \subseteq H$ be a Mori set such that $H = [F \cup H^\times]_H$, P a t-maximal t-ideal of H and $x \in P^\bullet$. Then there are some $\varepsilon \in H^\times$ and $(\alpha_e)_{e \in F} \in \mathbb{N}_0^{(F)}$ such that $x = \varepsilon \prod_{e \in F} e^{\alpha_e}$. Therefore, there exists some $e \in F$ such that $e \in P$ and $x \in eH$. It follows that $x \in \{e\}_t \subseteq (P \cap F)_t$. Consequently, $P \subseteq (P \cap F)_t$. There is some finite $E \subseteq P \cap F$ such that $E^{-1} = (P \cap F)^{-1}$. This implies that $P \subseteq (P \cap F)_t \subseteq (P \cap F)_v = E_v = E_t \subseteq (P \cap F)_t \subseteq P$; hence $P = E_v$, and thus P is divisorial. $4 \Rightarrow 1$: Let I be a non-zero t-ideal of H. It is sufficient to show that I is t-invertible. Since H is completely integrally closed, it follows that $(II^{-1})_v = H$. Assume that $(II^{-1})_t \subsetneq H$. Then there exists some t-maximal t-ideal P of H such that $(II^{-1})_t \subseteq P$. We have $H = (II^{-1})_v \subseteq ((II^{-1})_t)_v \subseteq P_v = P$, a contradiction. Consequently, $(II^{-1})_t = H$. $\square$

Now we provide a few minor results about Mori sets and divisor-closed submonoids to prepare for the main result in this section.

Lemma 3.3. *Let H be a monoid, $S \subseteq H$ a divisor-closed submonoid, and $X \subseteq S$ a subset. If X is an H-Mori set, then X is an S-Mori set.*

Proof. Let K be a quotient monoid of H and $L \subseteq K$ the quotient monoid of S. First we show that for every $Y \subseteq S$ it follows that $(H :_L Y) = (S :_L Y)$. Let $Y \subseteq S$. "$\subseteq$": Let $x \in (H :_L Y)$, then $xY \subseteq H$. Since $S \subseteq H$ is divisor closed, it follows that $xY \subseteq H \cap L = S$, hence $x \in (S :_L Y)$. "$\supseteq$": trivial. Now let X be an H-Mori set and $F \subseteq X$. Then there exists some finite $E \subseteq F$ such that $(H :_K F) = (H :_K E)$. This implies that $(S :_L F) = (H :_L F) = (H :_K F) \cap L = (H :_K E) \cap L = (H :_L E) = (S :_L E)$, hence X is an S-Mori set. $\square$

Let H be a monoid, $x \in H$, and $n \in \mathbb{N}$ and let $\mathbb{A}$ be some property that can be stated in the language of monoids (e.g., atomic, Krull, Mori).

- Set $\mathcal{D}_n(x) = \{u \in \mathcal{A}(H) \mid u \mid_H x^n\}$.
- A submonoid $S \subseteq H$ is called monadic if $S = [\![a]\!]$ for some $a \in H^\bullet$.
- We say that H is monadically $\mathbb{A}$ (or H is a monadically $\mathbb{A}$ monoid) if every monadic submonoid of H satisfies $\mathbb{A}$.
- The property $\mathbb{A}$ is said to be monadic (for H) if H has property $\mathbb{A}$ if and only if every monadic submonoid of H has property $\mathbb{A}$.
- If H is an integral domain, we say that H is a $\mathbb{A}$ domain if H satisfies $\mathbb{A}$ as a monoid.

Note that H is monadically $\mathbb{A}$ if and only if $[\![E]\!]$ satisfies $\mathbb{A}$ for all non-empty finite $E \subseteq H^\bullet$.

Proposition 3.4. *Let H be a monoid and K a quotient monoid of H:*

1. $H^\times = [a]^\times$ *and* $\mathcal{A}([a]) = \mathcal{A}(H) \cap [a] = \bigcup_{n\in\mathbb{N}} \mathcal{D}_n(a)$ *for all $a \in H^\bullet$.*
2. *H is atomic if and only if H is monadically atomic.*
3. *H is completely integrally closed if and only if H is monadically completely integrally closed.*
4. *H is an FF-monoid if and only if H is a monadically FF-monoid.*

Proof. 1. Let $a \in H^\bullet$. Clearly, $[a]^\times \subseteq H^\times$. If $\varepsilon \in H^\times$, then $\varepsilon\varepsilon^{-1} = 1 \in [a]$; hence $\varepsilon, \varepsilon^{-1} \in [a]$, and thus $\varepsilon \in [a]^\times$. If $x \in \mathcal{A}([a])$ and $b, c \in H$ are such that $x = bc$, then $b, c \in [a]$; hence $b \in [a]^\times = H^\times$ or $c \in [a]^\times = H^\times$. Finally, if $x \in \mathcal{A}(H) \cap [a]$ and $b, c \in [a]$ are such that $x = bc$, then $b \in H^\times = [a]^\times$ or $c \in H^\times = [a]^\times$. Obviously, $\mathcal{A}(H) \cap [a] = \bigcup_{n\in\mathbb{N}} \{u \in \mathcal{A}(H) \mid u\mid_H a^n\} = \bigcup_{n\in\mathbb{N}} \mathcal{D}_n(a)$.

2. This is an immediate consequence of 1.

3. "$\Rightarrow$": trivial "$\Leftarrow$": Let $x \in K^\bullet$ be almost integral over H. There exists some $c \in H^\bullet$ such that $cx^n \in H$ for all $n \in \mathbb{N}$ and there are some $a, b \in H^\bullet$ such that $x = \frac{a}{b}$. Let $L \subseteq K$ be the quotient monoid of $[abc]$. Then $c \in [abc]^\bullet$, $x \in L$ and $cx^n \in H \cap L = [abc]$ for all $n \in \mathbb{N}$. Since $[abc]$ is completely integrally closed, we have $x \in [abc] \subseteq H$.

4. "$\Rightarrow$": This follows from 1 and [15, Theorem 1.5.6.2]. "$\Leftarrow$": Let $x \in H^\bullet$. It is an easy consequence of 1 that $f : \{y[x] \mid y \in [x], y \mid_{[x]} x\} \to \{yH \mid y \in [x], y \mid_{[x]} x\}$ defined by $f(I) = IH$ is a bijective map. Since $\{y[x] \mid y \in [x], y \mid_{[x]} x\}$ is finite, we have $\{yH \mid y \in H, y \mid_H x\} = \{yH \mid y \in [x], y \mid_{[x]} x\}$ is finite. $\square$

Let H be a monoid and K a quotient monoid of H.

- H is called seminormal if for all $x \in K$ such that $x^2, x^3 \in H$ we have $x \in H$.
- H is called a weakly factorial if every $x \in H^\bullet \setminus H^\times$ is a finite product of primary elements of H (i.e., of elements $x \in H^\bullet$ such that xH is primary).
- H is called radical factorial if every $x \in H^\bullet$ is a finite product of radical elements of H (i.e., of elements $x \in H^\bullet$ such that $\sqrt{xH} = xH$).

We leave to the reader to prove that "satisfying the ACCP", "seminormal", "root-closed", "atomic and weakly factorial", "atomic and radical factorial", "being a BF-monoid", "being a valuation monoid", "being a GCD-monoid", and "factorial" are also monadic properties for H. We do not know whether "weakly factorial", "radical factorial", and "v-closed" are monadic properties. If H is weakly factorial (resp. radical factorial), then H is monadically weakly factorial (resp. monadically radical factorial) and the following holds.

Remark 3.5. Let H be a monoid that is monadically v-closed. Then H is v-closed.

Proof. Let K be a quotient monoid of H, $\emptyset \neq E \subseteq H^\bullet$ finite, and $x \in K^\bullet$ such that $xE \subseteq E_v$. There are some $y, z \in H^\bullet$ such that $x = \frac{y}{z}$. Set $S = [E \cup xE \cup \{y, z\}]$. Observe that S is a monadic submonoid of H, and thus S is v_S-closed. Let $L \subseteq K$

be the quotient monoid of S. Clearly, $x \in L$ and it follows by [15, Proposition 2.4.2.3] that $xE \subseteq E_v \cap S \subseteq (E_{vs})_v \cap S = E_{vs}$; hence $x \in S$. Consequently, $x \in H$. □

Now we present the main result in this section. It connects monadically Krull monoids with concepts that are well known in the literature.

Theorem 3.6. *Let H be a monoid. The following conditions are equivalent:*

1. *H is a monadically Krull monoid.*
2. *H is atomic and completely integrally closed and $\{uH \mid u \in \mathcal{A}(\llbracket a \rrbracket)\}$ is finite for all $a \in H^\bullet$.*
3. *H is a completely integrally closed FF-monoid and for all $a \in H^\bullet$, $\mathcal{A}(\llbracket a \rrbracket) \subseteq \mathcal{D}_k(a)$ for some $k \in \mathbb{N}$.*
4. *H is atomic and completely integrally closed and $\mathcal{A}(\llbracket a \rrbracket)$ is an H-Mori set for all $a \in H^\bullet$.*

Proof. $1 \Rightarrow 2$: By Propositions 3.4(2) and 3.4(3) we have H is atomic and completely integrally closed. Let $a \in H^\bullet$. If $P \in \mathfrak{X}(\llbracket a \rrbracket)$, then there is some $b \in P^\bullet$; hence there are some $c \in H$ and $n \in \mathbb{N}$ such that $bc = a^n$. It follows that $c \in \llbracket a \rrbracket$; hence $a^n \in P$, and thus $a \in P$. Therefore, $a \in P$ for all $P \in \mathfrak{X}(\llbracket a \rrbracket)$. It follows by [15, Proposition 2.2.4.2] and [15, Theorem 2.2.5.2] that $\mathfrak{X}(\llbracket a \rrbracket)$ is finite. It can be easily deduced from [15, Theorem 2.7.14] that $\{u\llbracket a \rrbracket \mid u \in \mathcal{A}(\llbracket a \rrbracket)\}$ is finite. It follows by Proposition 3.4(1) that $f : \{u\llbracket a \rrbracket \mid u \in \mathcal{A}(\llbracket a \rrbracket)\} \to \{uH \mid u \in \mathcal{A}(\llbracket a \rrbracket)\}$ defined by $f(I) = IH$ is bijective; hence $\{uH \mid u \in \mathcal{A}(\llbracket a \rrbracket)\}$ is finite.

$2 \Rightarrow 3$: Let $a \in H^\bullet$. It follows that $\{uH \mid u \in \mathcal{D}_1(a)\} \subseteq \{uH \mid u \in \mathcal{A}(\llbracket a \rrbracket)\}$ is finite which implies (together with the fact that H is atomic) that H is an FF-monoid. On the other hand there is some finite $E \subseteq \mathcal{A}(\llbracket a \rrbracket)$ such that $\{uH \mid u \in \mathcal{A}(\llbracket a \rrbracket)\} = \{uH \mid u \in E\}$. There is some $k \in \mathbb{N}$ such that $E \subseteq \mathcal{D}_k(a)$. Let $u \in \mathcal{A}(\llbracket a \rrbracket)$. Then some $\varepsilon \in H^\times$ and some $v \in E$ exist such that $u = \varepsilon v$. Since $v \in \mathcal{D}_k(a)$, we immediately obtain that $u \in \mathcal{D}_k(a)$.

$3 \Rightarrow 4$: Let $a \in H^\bullet$ and $F \subseteq \mathcal{A}(\llbracket a \rrbracket)$. There is some $k \in \mathbb{N}$ such that $F \subseteq \mathcal{D}_k(a)$ and since H is an FF-monoid we have $\{uH \mid u \in F\} \subseteq \{uH \mid u \in \mathcal{D}_k(a)\}$ is finite. Consequently, there is some finite $E \subseteq F$ such that $\{uH \mid u \in F\} = \{uH \mid u \in E\}$. If $x \in K$, then $x \in E^{-1}$ if and only if $xuH \subseteq H$ for all $u \in E$ if and only if $xuH \subseteq H$ for all $u \in F$ if and only if $x \in F^{-1}$. Therefore, $E^{-1} = F^{-1}$.

$4 \Rightarrow 1$: Let $a \in H^\bullet$. By Propositions 3.4(2) and 3.4(3) it follows that $\llbracket a \rrbracket$ is atomic and completely integrally closed. Lemma 3.3 implies that $\mathcal{A}(\llbracket a \rrbracket)$ is an $\llbracket a \rrbracket$-Mori set. By Proposition 3.2 we obtain that $\llbracket a \rrbracket$ is a Krull monoid. □

Using the terminology in [24] we obtain by Theorem 3.6 that H is a monadically Krull monoid if and only if it is an atomic, completely integrally closed IDPF-monoid if and only if it is a completely integrally closed FF-monoid that is a pseudo-IDPF monoid. We will see later that monadically Krull monoids are not necessarily Krull monoids. Especially, we have that monadically Mori monoids are not necessarily Mori monoids. The next result shows that if the Mori property is satisfied by a bigger class of divisor-closed submonoids, then the monoid itself satisfies the Mori property.

Proposition 3.7. *Let H be a monoid. Then H is a Mori monoid if and only if $[\![X]\!]$ is a Mori monoid for every denumerable subset $X \subseteq H^\bullet$.*

Proof. Let K be a quotient monoid of H. "$\Rightarrow$": Trivial. "$\Leftarrow$": Assume that H is not a Mori monoid. Then H is not a Mori set. By Lemma 3.1 there exists some $(a_i)_{i \in \mathbb{N}} \in H^\mathbb{N}$ such that $(H :_K \{a_i \mid i \in [1, n+1]\}) \subsetneqq (H :_K \{a_i \mid i \in [1, n]\})$ for all $n \in \mathbb{N}$. Therefore, there exist some $(x_i)_{i \in \mathbb{N}} \in H^\mathbb{N}$ and $(y_i)_{i \in \mathbb{N}} \in (H^\bullet)^\mathbb{N}$ such that for all $n \in \mathbb{N}$ we have $a_{n+1} \frac{x_n}{y_n} \in K \backslash H$ and $a_i \frac{x_n}{y_n} \in H$ for all $i \in [1, n]$. Let $S = [\![\{a_n x_n y_n \mid n \in \mathbb{N}\}]\!]$ and let $L \subseteq K$ be the quotient monoid of S. Then $(a_i)_{i \in \mathbb{N}} \in S^\mathbb{N}$ and $(\frac{x_i}{y_i})_{i \in \mathbb{N}} \in L^\mathbb{N}$. Moreover, we have for all $n \in \mathbb{N}$ that $a_{n+1} \frac{x_n}{y_n} \in L \backslash S$ and $a_i \frac{x_n}{y_n} \in H \cap L = S$ for all $i \in [1, n]$. This implies that $(S :_L \{a_i \mid i \in [1, n+1]\}) \subsetneqq (S :_L \{a_i \mid i \in [1, n]\})$ for all $n \in \mathbb{N}$. It follows by Lemma 3.1 that S is not an S-Mori set, and thus S is not a Mori monoid, a contradiction. $\square$

4 Counterexamples

It is of interest to know whether every monadically Krull monoid is already a Krull monoid. In this section we prove that this is not necessarily true and show that even strong improvements of monadically Krull can fail to be Krull. For technical reasons we will consider (multiplicatively written) monoids that do not possess a zero element in this section. Moreover, we will use monoids that are additively written (the zero element is their identity and they do not possess an "additive" analogue of a "multiplicative" zero element). All terminology that has been introduced so far can be adapted in an obvious way for these types of monoids. Observe that the label "quotient monoid" will be replaced by "quotient group" for monoids without a zero element. Note that a monoid is root closed if and only if it is integrally closed in terms of [18]. We want to thank F. Kainrath who led our attention to the integral domain constructed in the next example.

Example 4.1. There exists a monadically Krull domain that is not a Krull domain.

Proof. Let R be an integrally closed noetherian domain, $(X_i)_{i \in \mathbb{N}_0}$ a sequence of independent indeterminates over R and K a field of quotients of $R[\{X_i \mid i \in \mathbb{N}_0\}]$. For $n \in \mathbb{N}_0$ set $S_n = R[\{\prod_{i=0}^n X_i^{a_i} \mid (a_i)_{i=0}^n \in \mathbb{N}_0^{[0,n]}, a_0 \geq \sum_{i=1}^n \frac{a_i}{2^i}\}]$. Let $S = \bigcup_{n \in \mathbb{N}} S_n$. We show that S is a monadically Krull domain that is not a Krull domain. Note that S is a subring of $R[\{X_i \mid i \in \mathbb{N}_0\}]$ and K is a field of quotients of S.

First we show that S is not a Krull domain. For $n \in \mathbb{N}_0$ set $a_n = X_0^{n+1}(\prod_{i=1}^n X_i^{2^i}) X_{n+1}^{2^{n+1}-1}$. By Lemma 3.1 it is sufficient to show that $X_{k+1} \in \{a_i \mid i \in [0, k]\}^{-1} \backslash \{a_i \mid i \in [0, k+1]\}^{-1}$ for all $k \in \mathbb{N}_0$. Let $k \in \mathbb{N}_0$ and

$i \in [0, k]$. Since $\sum_{j=1}^{i} \frac{2^j}{2^j} + \frac{2^{i+1}-1}{2^{i+1}} + \frac{1}{2^{k+1}} = i + 1 - \frac{1}{2^{i+1}} + \frac{1}{2^{k+1}} \leq i + 1$
it follows that $X_{k+1} a_i = X_0^{i+1} (\prod_{j=1}^{i} X_j^{2^j}) X_{i+1}^{2^{i+1}-1} X_{k+1} \in S$, and thus
$X_{k+1} \in \{a_j \mid j \in [0, k]\}^{-1}$. Since $\sum_{j=1}^{k} \frac{2^j}{2^j} + \frac{2^{k+1}+1}{2^{k+1}} + \frac{2^{k+2}-1}{2^{k+2}} = k + 2 + \frac{1}{2^{k+2}} >$
$k + 2$, we have $X_{k+1} a_{k+1} = X_0^{k+2} (\prod_{j=1}^{k} X_j^{2^j}) X_{k+1}^{2^{k+1}+1} X_{k+2}^{2^{k+2}-1} \notin S$; hence
$X_{k+1} \notin \{a_j \mid j \in [0, k+1]\}^{-1}$.

Next we prove that S_l is a divisor-closed subring of S that is noetherian and integrally closed for all $l \in \mathbb{N}$. Let $l \in \mathbb{N}$. Clearly, S_l is a subring of S. Let $f, g \in S^{\bullet}$ be such that $fg \in S_l$. There is some $m \in \mathbb{N}$ such that $m \geq l$ and $f, g \in S_m$. Since $fg \in R[\{X_i \mid i \in [0, l]\}]$, it follows that $f \in R[\{X_i \mid i \in [0, l]\}]$; hence $f \in S_m \cap R[\{X_i \mid i \in [0, l]\}] = S_l$. Therefore, S_l is a divisor-closed subring of S.

Set $T = \{\prod_{i=0}^{l} X_i^{a_i} \mid (a_i)_{i=0}^{l} \in \mathbb{N}_0^{[0,l]}, a_0 \geq \sum_{i=1}^{l} \frac{a_i}{2^i}\}$, $U = \{\prod_{i=0}^{l} X_i^{a_i} \mid (a_i)_{i=0}^{l} \in \mathbb{N}_0^{[0,l]}, l \geq a_0 \geq \sum_{i=1}^{l} \frac{a_i}{2^i}\}$ and $V = [U]$. Then T is a submonoid of S_l. We show that T is root closed and $T = V$. Let $L \subseteq K$ be a quotient group of T. Let $x \in L$ and $r \in \mathbb{N}$ be such that $x^r \in T$. Since $\{\prod_{i=0}^{l} X_i^{a_i} \mid (a_i)_{i=0}^{l} \in \mathbb{N}_0^{[0,l]}\}$ is a root closed submonoid of L that contains T it follows that there is some $(b_i)_{i=0}^{l} \in \mathbb{N}_0^{[0,l]}$ such that $x = \prod_{i=0}^{l} X_i^{b_i}$. This implies that $\prod_{i=0}^{l} X_i^{r b_i} \in T$, and thus $r b_0 \geq \sum_{i=1}^{l} \frac{r b_i}{2^i}$. Consequently, $b_0 \geq \sum_{i=1}^{l} \frac{b_i}{2^i}$; hence $x \in T$. Therefore, T is root closed. It remains to prove by induction on k that for all $k \in \mathbb{N}$ and all $(a_i)_{i=1}^{l} \in \mathbb{N}_0^{[1,l]}$ such that $k \geq \sum_{i=1}^{l} \frac{a_i}{2^i}$ it follows that $X_0^k \prod_{i=1}^{l} X_i^{a_i} \in V$. If $k = 1$, then the assertion is clear. Now let $k \in \mathbb{N}$ and $(a_i)_{i=1}^{l} \in \mathbb{N}_0^{[1,l]}$ be such that $k + 1 \geq \sum_{i=1}^{l} \frac{a_i}{2^i}$.
Case 1: There is some $j \in [1, l]$ such that $a_j > 2^j$; we have $k \geq \sum_{i=1,i \neq j}^{l} \frac{a_i}{2^i} + \frac{a_j - 2^j}{2^j}$. It follows by the induction hypothesis that $X_0^k (\prod_{i=1,i \neq j}^{l} X_i^{a_i}) X_j^{a_j - 2^j} \in V$.
Therefore, $X_0^{k+1} \prod_{i=1}^{l} X_i^{a_i} = X_0 X_j^{2^j} X_0^k (\prod_{i=1,i \neq j}^{l} X_i^{a_i}) X_j^{a_j - 2^j} \in V$. Case 2: $a_j \leq 2^j$ for all $j \in [1, l]$: We have $\sum_{i=1}^{l} \frac{a_i}{2^i} \leq l$. If $l \geq k + 1$, then $X_0^{k+1} \prod_{i=1}^{l} X_i^{a_i} \in V$, by definition. Now let $l < k + 1$. Since $X_0 \in V$, it follows that $X_0^{k+1} \prod_{i=1}^{l} X_i^{a_i} = X_0^{k+1-l} X_0^l \prod_{i=1}^{l} X_i^{a_i} \in V$. By [17, Corollary 15.12] it follows that S_l is noetherian and integrally closed.

Let $a \in S^{\bullet}$. There is some $s \in \mathbb{N}$ such that $a \in S_s$. Since S_s is a divisor-closed subring of S it follows that $[\![a]\!]_S$ is a divisor-closed submonoid of S_s. Since S_s is a Krull domain this implies that $[\![a]\!]_S$ is a Krull monoid. Consequently, S is a monadically Krull domain. $\qquad \square$

It has been pointed out in [10] that every atomic IDPF-domain (this notion has been introduced in [24]) that contains a field of characteristics zero is already completely integrally closed. Now let the domain R in the last example be a field of characteristics zero. Then the domain S in the last example is an atomic IDPF-domain that contains a field of characteristics zero and yet S is not a Krull domain. Let H be a monoid and r a finitary ideal system on H. Let $\mathcal{J} \subseteq \mathbb{P}(H)$. We say that $\mathcal{J}$ possesses a length function if there exists some map $\lambda : \mathcal{J} \to \mathbb{N}_0$ such that $\lambda(J) < \lambda(I)$ for all $I, J \in \mathcal{J}$ such that $I \subsetneq J$. Note that if $\mathcal{J}$ possesses a length function, then $\mathcal{J}$ satisfies the ACC. Moreover, if R is an integral domain, then the set of non-zero ideals of R possesses a length function if and only if R is a noetherian domain and $\dim(R) \leq 1$. Observe that a monoid H is a BF-monoid if and only if $\{xH \mid x \in H^\bullet\}$ possesses a length function. Moreover, possessing a length function is in some sense the same as satisfying a strong version of the ACC. We use it in the next example to highlight that it is not only a BF-monoid but also that the set of radicals of principal ideals possesses a length function. Next we introduce several other types of generalizations of the Krull property to study the following example in detail:

- H is called an r-SP-monoid if every r-ideal of H is a finite r-product of radical r-ideals of H.
- H is called an r-Prüfer monoid if every non-zero r-finitely generated r-ideal of H is r-invertible.
- An r-ideal I of H is called r-cancellative if I is cancellative in the r-ideal semigroup of H.

Clearly, H is a Krull monoid if and only if it is a Mori monoid that is a t-Prüfer monoid. Note that if H is a radical factorial monoid, then for all $a \in H^\bullet$ we have $\mathcal{A}(\llbracket a \rrbracket) \subseteq \mathcal{D}_1(a)$. Using Theorem 3.6 and [29, Proposition 2.4] it is straightforward to prove that every radical factorial FF-monoid is a monadically Krull monoid (but even a Krull monoid needs not be radical factorial; see [29, Example 4.3]). In this light we will sharpen our first counterexample in the monoid setting and prove that even radical factorial FF-monoids can fail to be Krull. A sequence $(x_i)_{i \in \mathbb{N}_0}$ of integers is called formally infinite if $\{i \in \mathbb{N}_0 \mid x_i \neq 0\}$ is finite. If H is additively written, $k \in \mathbb{N}$ and $I \subseteq H$, then set $kI = \{\sum_{i=1}^{k} a_i \mid (a_i)_{i=1}^{k} \in I^{[1,k]}\}$.

Example 4.2. Let G be a free abelian group with basis $(e_i)_{i \in \mathbb{N}_0}$. For $x \in G$, let $(x_i)_{i \in \mathbb{N}_0} \in \mathbb{Z}^{(\mathbb{N}_0)}$ denote the unique formally infinite sequence such that $x = \sum_{i \in \mathbb{N}_0} x_i e_i$. Set $H = \{x \in G \mid x_0 \geq x_i \geq 0 \text{ for all } i \in \mathbb{N}_0\}$. Then H is a submonoid of G, G is a quotient group of H, and the following is true:

1. $v\text{-spec}(H)^\bullet = \mathfrak{X}(H)$, $t\text{-spec}(H)^\bullet = \mathfrak{X}(H) \cup \{H \setminus H^\times\}$, every non-empty t-ideal of H is t-cancellative and $(kP)_t$ is P-primary for all $k \in \mathbb{N}$ and $P \in t\text{-spec}(H)^\bullet$.
2. H is a t-SP-monoid, $t\text{-}\dim(H) = 2$, H is an FF-monoid, $\{\sqrt{y + H} \mid y \in H\}$ possesses a length function, and every radical element of H is either an atom or a unit.

In particular, H is a radical factorial monoid that is neither a Mori monoid nor a t-Prüfer monoid.

Proof. Clearly, H is a submonoid of G and $H^\times = \{0\}$. Let $K \subseteq G$ be the quotient group of H and $i \in \mathbb{N}_0$. Obviously, $e_0, e_0 + e_i \in H$; hence $e_i = e_0 + e_i - e_0 \in K$. Therefore, $G = K$ is a quotient group of H. For $r \in \mathbb{N}_0^{\mathbb{N}_0}$ set $I_r = \{x \in G \mid x_0 \geq x_{j+1} + r_{2j+1}, x_j \geq r_{2j} \text{ for all } j \in \mathbb{N}_0\}$. Set $\mathfrak{J} = \{r \in \mathbb{N}_0^{\mathbb{N}_0} \mid |\{j \in \mathbb{N}_0 \mid r_{2j} \neq 0\}| < \infty, r_0 \geq r_{2j+1} + r_{2j+2} \text{ for all } j \in \mathbb{N}_0\}$ and $\mathfrak{L} = \{r \in \mathbb{N}_0^{\mathbb{N}_0} \mid |\{j \in \mathbb{N}_0 \mid r_{2j} \neq 0\}| < \infty, r_0 = \max\{r_{2j+1} + r_{2j+2} \mid j \in \mathbb{N}_0\}\}$. For $i \in \mathbb{N}_0$, let $s^{(i)} \in \mathbb{N}_0^{\mathbb{N}_0}$ be defined by

$s_j^{(i)} = 1$ if $j \in \{0, i\}$ and $s_j^{(i)} = 0$ if $j \in \mathbb{N}_0 \backslash \{0, i\}$. If $r, s \in \mathbb{N}_0^{\mathbb{N}_0}$ and $k \in \mathbb{N}_0$, then we set $r + s = (r_i + s_i)_{i \in \mathbb{N}_0}$, $kr = (kr_i)_{i \in \mathbb{N}_0}$ and $r \leq s$ if $r_j \leq s_j$ for all $j \in \mathbb{N}_0$.

Claim 1: For all $x, y \in H$ it follows that $x \in \sqrt{y + H}$ if and only if $\{i \in \mathbb{N}_0 \mid y_i > 0\} \subseteq \{i \in \mathbb{N}_0 \mid x_i > 0\}$ and $\{i \in \mathbb{N}_0 \mid y_0 > y_i\} \subseteq \{i \in \mathbb{N}_0 \mid x_0 > x_i\}$. Let $x, y \in H$. Observe that $x \in \sqrt{y + H}$ if and only if there is some $k \in \mathbb{N}$ such that $kx_i \geq y_i$ and $k(x_0 - x_i) \geq y_0 - y_i$ for all $i \in \mathbb{N}_0$. "$\Rightarrow$": Let $k \in \mathbb{N}$ be such that $kx_j \geq y_j$ and $k(x_0 - x_j) \geq y_0 - y_j$ for all $j \in \mathbb{N}_0$. Let $i \in \mathbb{N}_0$. If $y_i > 0$, then $kx_i \geq y_i > 0$, and thus $x_i > 0$. If $y_0 > y_i$, then $k(x_0 - x_i) \geq y_0 - y_i > 0$, hence $x_0 > x_i$. "$\Leftarrow$": Let $\{i \in \mathbb{N}_0 \mid y_i > 0\} \subseteq \{i \in \mathbb{N}_0 \mid x_i > 0\}$ and $\{i \in \mathbb{N}_0 \mid y_0 > y_i\} \subseteq \{i \in \mathbb{N}_0 \mid x_0 > x_i\}$. Set $k = 1 + \max\{y_i \mid i \in \mathbb{N}_0\}$. Then $k \in \mathbb{N}$ and it is clear that $kx_i \geq y_i$ and $k(x_0 - x_i) \geq y_0 - y_i$ for all $i \in \mathbb{N}_0$.

Claim 2: For all $\emptyset \neq A \subseteq H$ we have $A_v = \{x \in G \mid x_0 \geq x_i + \min\{a_0 - a_i \mid a \in A\}$ and $x_i \geq \min\{a_i \mid a \in A\}$ for all $i \in \mathbb{N}\}$. Let $\emptyset \neq A \subseteq H$. Set $m^{(i)} = \min\{a_0 - a_i \mid a \in A\}$ and $n^{(i)} = \min\{a_i \mid a \in A\}$ for all $i \in \mathbb{N}$. Observe that $A^{-1} = \{x \in G \mid x + a \in H \text{ for all } a \in A\} = \{x \in G \mid \text{for all } a \in A \text{ and } i \in \mathbb{N}, x_0 + a_0 \geq x_i + a_i \geq 0\} = \{x \in G \mid x_0 + m^{(i)} \geq x_i \text{ and } x_i + n^{(i)} \geq 0 \text{ for all } i \in \mathbb{N}\}$. "$\subseteq$": Let $x \in A_v$ and $i \in \mathbb{N}$. Set $y = m^{(i)}e_i$ and $z = -n^{(i)}e_i$. We have $y, z \in A^{-1}$; hence $x + y \in H$ and $x + z \in H$. Therefore, $x_0 + y_0 \geq x_i + y_i$ and $x_i + z_i \geq 0$. This implies that $x_0 \geq x_i + m^{(i)}$ and $x_i \geq n^{(i)}$. "$\supseteq$": Let $x \in G$ be such that $x_0 \geq x_j + m^{(j)}$ and $x_j \geq n^{(j)}$ for all $j \in \mathbb{N}$. Let $y \in A^{-1}$ and $i \in \mathbb{N}$. Then $x_0 \geq x_i + m^{(i)}$, $y_0 + m^{(i)} \geq y_i$, $x_i \geq n^{(i)}$ and $y_i + n^{(i)} \geq 0$; hence $x_0 + y_0 + m^{(i)} \geq x_i + y_i + m^{(i)}$ and $x_i + y_i + n^{(i)} \geq n^{(i)}$. Therefore, $x_0 + y_0 \geq x_i + y_i \geq 0$. This implies that $x + y \in H$. Consequently, $x \in A_v$.

As usual we denote by $\mathcal{I}_t(H)^\bullet$ (resp. $\mathcal{I}_v(H)^\bullet$) the set of non-empty t-ideals of H (resp. the set of non-empty divisorial ideals of H).

Claim 3: $\mathcal{I}_t(H)^\bullet = \{I_r \mid r \in \mathfrak{J}\}$ and $\mathcal{I}_v(H)^\bullet = \{I_r \mid r \in \mathfrak{L}\}$. First we prove that $\mathcal{I}_t(H)^\bullet = \{I_r \mid r \in \mathfrak{J}\}$. "$\subseteq$": Let $I \in \mathcal{I}_t(H)^\bullet$. For $i \in \mathbb{N}_0$ set $r_{2i} = \min\{y_i \mid y \in I\}$ and $r_{2i+1} = \min\{y_0 - y_{i+1} \mid y \in I\}$. There is some sequence $(z^{(j)})_{j \in \mathbb{N}_0} \in I^{\mathbb{N}_0}$ such that $z_i^{(2i)} = r_{2i}$ and $z_0^{(2i+1)} - z_{i+1}^{(2i+1)} = r_{2i+1}$ for all $i \in \mathbb{N}_0$. If $j \in \mathbb{N}_0$, then since $z_0^{(0)} - z_{j+1}^{(0)} \geq r_{2j+1}$ and $z_{j+1}^{(0)} \geq r_{2j+2}$ we obtain that $r_0 = z_0^{(0)} \geq r_{2j+1} + r_{2j+2}$. Moreover, $|\{j \in \mathbb{N}_0 \mid r_{2j} \neq 0\}| \leq |\{j \in \mathbb{N}_0 \mid z_j^{(0)} \neq 0\}| < \infty$. Therefore, $r \in \mathfrak{J}$. It remains to show that $I = I_r$. "$\subseteq$": trivial. "$\supseteq$": Let $x \in I_r$. Set $E = \{i \in \mathbb{N} \mid x_i \neq 0 \text{ or } z_i^{(0)} \neq 0\}$. Then E is finite. It is sufficient to prove that $x_0 - x_j \geq \min(\{z_0^{(2i-1)} - z_j^{(2i-1)}, z_0^{(2i)} - z_j^{(2i)} \mid i \in E\} \cup \{z_0^{(0)} - z_j^{(0)}\})$ and $x_j \geq \min(\{z_j^{(2i-1)}, z_j^{(2i)} \mid i \in E\} \cup \{z_j^{(0)}\})$ for all $j \in \mathbb{N}$, because then $x \in (\{z^{(2i-1)}, z^{(2i)} \mid i \in E\} \cup \{z^{(0)}\})_v$ by Claim 2; hence $x \in I$. Let $j \in \mathbb{N}$. Case 1a: $x_j \neq 0$. It follows that $x_0 - x_j \geq r_{2j-1} = z_0^{(2j-1)} - z_j^{(2j-1)}$. Case 1b: $x_j = 0$. We have $x_0 - x_j = x_0 \geq r_0 = z_0^{(0)} \geq z_0^{(0)} - z_j^{(0)}$. Case 2a: $j \in E$. It follows that $x_j \geq r_{2j} = z_j^{(2j)}$. Case 2b: $j \notin E$. We have $x_j = 0 = z_j^{(0)}$. "$\supseteq$": Let $r \in \mathfrak{J}$ and $x \in (I_r)_t$. Then there is some finite $\emptyset \neq A \subseteq I_r$ such that $x \in A_v$. It is an immediate consequence of Claim 2 that $x_0 \geq x_j + r_{2j-1}$ and $x_j \geq r_{2j}$ for all

$j \in \mathbb{N}$. Since A is finite, there is some $l \in \mathbb{N}$ such that $x_l = 0$ and $a_l = 0$ for all $a \in A$. It follows by Claim 2 that $x_0 \geq x_l + \min\{a_0 - a_l \mid a \in A\} = \min\{a_0 \mid a \in A\} \geq r_0$. Consequently, $x_0 \geq x_{j+1} + r_{2j+1}$ and $x_j \geq r_{2j}$ for all $j \in \mathbb{N}_0$, and thus $x \in I_r$. Observe that $\sum_{i \in \mathbb{N}_0} r_{2i} e_i \in I_r$; hence $I_r \in \mathcal{I}_t(H)^\bullet$.

Next we show that $\mathcal{I}_v(H)^\bullet = \{I_r \mid r \in \mathfrak{L}\}$. "$\subseteq$": Let $I \in \mathcal{I}_v(H)^\bullet$. As in the preceding part of the proof there is some $r \in \mathfrak{J}$ such that $I = I_r$, $r_{2i} = \min\{y_i \mid y \in I\}$ and $r_{2i+1} = \min\{y_0 - y_{i+1} \mid y \in I\}$ for all $i \in \mathbb{N}_0$. Set $s = \max\{r_{2i+1} + r_{2i+2} \mid i \in \mathbb{N}_0\}$. It remains to show that $s \geq r_0$, because then $r \in \mathfrak{L}$. Set $x = s e_0 + \sum_{i \in \mathbb{N}} r_{2i} e_i$. If $i \in \mathbb{N}$, then $x_0 = s \geq r_{2i-1} + r_{2i} = x_i + r_{2i-1}$ and $x_i = r_{2i}$. Therefore, Claim 2 implies that $x \in (I_r)_v = I_r$, and thus $s = x_0 \geq r_0$. "$\supseteq$": Let $r \in \mathfrak{L}$ and $x \in (I_r)_v$. It follows by Claim 2 that $x_0 \geq x_i + r_{2i-1}$ and $x_i \geq r_{2i}$ for all $i \in \mathbb{N}$. There is some $j \in \mathbb{N}$ such that $r_0 = r_{2j-1} + r_{2j}$; hence $x_0 \geq x_j + r_{2j-1} \geq r_{2j} + r_{2j-1} = r_0$. This implies that $x \in I_r$.

Claim 4: For all $a, b \in \mathfrak{J}$, $I_{a+b} = (I_a + I_b)_t$ and $I_a \subseteq I_b$ if and only if $b \leq a$. Let $a, b \in \mathfrak{J}$. Set $y^{(0)} = \sum_{i \in \mathbb{N}_0} a_{2i} e_i$, $z^{(0)} = \sum_{i \in \mathbb{N}_0} b_{2i} e_i$ and for $j \in \mathbb{N}$ set $y^{(j)} = \sum_{i \in \mathbb{N}_0, i \neq j} a_{2i} e_i + (a_0 - a_{2j-1}) e_j$ and $z^{(j)} = \sum_{i \in \mathbb{N}_0, i \neq j} b_{2i} e_i + (b_0 - b_{2j-1}) e_j$. Observe that $y^{(j)} \in I_a$ and $z^{(j)} \in I_b$ for all $j \in \mathbb{N}_0$. "$\subseteq$": Let $x \in I_{a+b}$. We prove that $x_0 \geq x_j + \min\{y_0^{(l)} + z_0^{(l)} - y_j^{(l)} - z_j^{(l)} \mid l = 0 \text{ or } l \in \mathbb{N}, x_l \neq 0\}$ and $x_j \geq \min\{y_j^{(l)} + z_j^{(l)} \mid l = 0 \text{ or } l \in \mathbb{N}, x_l \neq 0\}$ for all $j \in \mathbb{N}$, because then $x \in \{y^{(l)} + z^{(l)} \mid l = 0 \text{ or } l \in \mathbb{N}, x_l \neq 0\}_v$ by Claim 2; hence $x \in (I_a + I_b)_t$. Let $j \in \mathbb{N}$. Clearly, $x_j \geq a_{2j} + b_{2j} = y_j^{(0)} + z_j^{(0)}$. Case 1: $x_j \neq 0$. We have $x_j + y_0^{(j)} + z_0^{(j)} - y_j^{(j)} - z_j^{(j)} = x_j + a_0 + b_0 - (a_0 - a_{2j-1}) - (b_0 - b_{2j-1}) = x_j + a_{2j-1} + b_{2j-1} \leq x_0$. Case 2: $x_j = 0$. It follows that $x_j + y_0^{(0)} + z_0^{(0)} - y_j^{(0)} - z_j^{(0)} = a_0 + b_0 - a_{2j} - b_{2j} \leq a_0 + b_0 \leq x_0$. "$\supseteq$": Obviously, $I_a + I_b \subseteq I_{a+b}$ and $a + b \in \mathfrak{J}$. Therefore, Claim 3 implies that $(I_a + I_b)_t \subseteq I_{a+b}$.

Clearly, if $b \leq a$, then $I_a \subseteq I_b$. Now let $I_a \subseteq I_b$. Note that $y^{(i)} \in I_b$ for all $i \in \mathbb{N}_0$; hence $y_0^{(i)} \geq y_{j+1}^{(i)} + b_{2j+1}$ and $y_j^{(i)} \geq b_{2j}$ for all $i, j \in \mathbb{N}_0$. If $j \in \mathbb{N}_0$, then $y_0^{(j+1)} \geq y_{j+1}^{(j+1)} + b_{2j+1}$ and $y_j^{(0)} \geq b_{2j}$; hence $a_0 \geq a_0 - a_{2j+1} + b_{2j+1}$ and $a_{2j} \geq b_{2j}$. Consequently, $a_i \geq b_i$ for all $i \in \mathbb{N}_0$, and thus $b \leq a$.

Claim 5: $t\text{-spec}(H)^\bullet = \{I_{s^{(i)}} \mid i \in \mathbb{N}_0\}$ and $\mathfrak{X}(H) = \{I_{s^{(i)}} \mid i \in \mathbb{N}\}$. First we show that $t\text{-spec}(H)^\bullet = \{I_{s^{(i)}} \mid i \in \mathbb{N}_0\}$. "$\subseteq$": Let $P \in t\text{-spec}(H)^\bullet$. By Claim 3 there is some $r \in \mathfrak{J}$ such that $P = I_r$. Case 1: $r_j = 0$ for all $j \in \mathbb{N}$. Since $P \neq H$, we have $r_0 \neq 0$. This implies that $r = k s^{(0)}$ for some $k \in \mathbb{N}$, and thus $P = (k I_{s^{(0)}})_t$ by Claim 4. Therefore, $P = I_{s^{(0)}}$. Case 2: $r_j \neq 0$ for some $j \in \mathbb{N}$. Let $a \in \mathbb{N}_0^{\mathbb{N}_0}$ be defined by $a_i = r_i$ if $i \in \mathbb{N}_0, i \neq j$ and $a_i = r_i - 1$ otherwise. Then $a \in \mathfrak{J}$, $r \leq s^{(j)} + a$ and $r \not\leq a$. Therefore, Claim 4 implies that $(I_{s^{(j)}} + I_a)_t \subseteq P$ and $I_a \not\subseteq P$; hence $I_{s^{(j)}} \subseteq P$. Note that $s^{(j)} \leq r$, and thus $P = I_{s^{(j)}}$ by Claim 4. "$\supseteq$": Observe that $I_{s^{(2i)}} = \{x \in H \mid x_i \geq 1\}$ and $I_{s^{(2i+1)}} = \{x \in H \mid x_0 \geq x_{i+1} + 1\}$ for all $i \in \mathbb{N}_0$. Using this and Claim 3 it is straightforward to prove that $I_{s^{(i)}} \in t\text{-spec}(H)^\bullet$ for all $i \in \mathbb{N}_0$.

Next we show that $\mathfrak{X}(H) = \{I_{s^{(i)}} \mid i \in \mathbb{N}\}$. "$\subseteq$": Let $P \in \mathfrak{X}(H)$. Then $P \in t\text{-spec}(H)^\bullet$; hence $P = I_{s^{(i)}}$ for some $i \in \mathbb{N}_0$. Since $s^{(0)} < s^{(1)}$, it follows by Claim 4 that $I_{s^{(1)}} \subsetneqq I_{s^{(0)}}$, and thus $i \in \mathbb{N}$. "$\supseteq$": Let $i \in \mathbb{N}$ and $P \in t\text{-spec}(H)^\bullet$ be such that $P \subseteq I_{s^{(i)}}$. There is some $j \in \mathbb{N}_0$ such that $P = I_{s^{(j)}}$. It follows by Claim 4 that $s^{(i)} \le s^{(j)}$; hence $s^{(i)} = s^{(j)}$. Therefore, $P = I_{s^{(i)}}$, and thus $I_{s^{(i)}} \in \mathfrak{X}(H)$.

Claim 6: I_r is a radical t-ideal of H for all $r \in \mathfrak{I}$ such that $r_0 = 1$. Let $r \in \mathfrak{I}$ be such that $r_0 = 1$. By Claims 4 and 5 we have $\sqrt{I_r} = \bigcap_{P \in t\text{-spec}(H)^\bullet, P \supseteq I_r} P = \bigcap_{i \in \mathbb{N}_0, s^{(i)} \le r} I_{s^{(i)}} = \{x \in G \mid x_0 \ge x_{j+1} + \max\{s^{(i)}_{2j+1} \mid i \in \mathbb{N}_0, s^{(i)} \le r\}, x_j \ge \max\{s^{(i)}_{2j} \mid i \in \mathbb{N}_0, s^{(i)} \le r\}$ for all $j \in \mathbb{N}_0\} = \{x \in G \mid x_0 \ge x_{j+1} + r_{2j+1}, x_j \ge r_{2j}$ for all $j \in \mathbb{N}_0\} = I_r$.

1. By Claims 3 and 5 we have $v\text{-spec}(H)^\bullet = t\text{-spec}(H)^\bullet \cap \mathcal{I}_v(H)^\bullet = \{I_{s^{(i)}} \mid i \in \mathbb{N}_0\} \cap \{I_r \mid r \in \mathcal{L}\} = \{I_{s^{(i)}} \mid i \in \mathbb{N}\} = \mathfrak{X}(H)$ and $t\text{-spec}(H)^\bullet = \mathfrak{X}(H) \cup \{I_{s^{(0)}}\} = \mathfrak{X}(H) \cup \{H \setminus H^\times\}$. Let $A, B, C \in \mathcal{I}_t(H)^\bullet$ be such that $(A + B)_t = (A + C)_t$. By Claim 3 there exist some $a, b, c \in \mathfrak{I}$ such that $A = I_a$, $B = I_b$, and $C = I_c$. It follows by Claim 4 that $I_{a+b} = (A + B)_t = (A + C)_t = I_{a+c}$, and thus $a + b = a + c$ by Claim 4. Consequently, $b = c$; hence $B = C$. Now let $k \in \mathbb{N}$ and $P \in t\text{-spec}(H)^\bullet$. By Claim 4 we have $(kI_{s^{(2i)}})_t = I_{ks^{(2i)}} = \{x \in H \mid x_i \ge k\}$ and $(kI_{s^{(2i+1)}})_t = I_{ks^{(2i+1)}} = \{x \in H \mid x_0 \ge x_{i+1} + k\}$ for all $i \in \mathbb{N}_0$. Using this it is straightforward to prove that $(kP)_t$ is P-primary.

2. It follows by 1 that $t\text{-dim}(H) = 2$, since $H \setminus H^\times$ is not divisorial. Let $I \in \mathcal{I}_t(H)^\bullet$. By Claim 3 there is some $r \in \mathfrak{I}$ such that $I = I_r$. For $i \in [1, r_0]$ and $j \in \mathbb{N}_0$ set $r^{(i)}_j = 1$, if $[(j$ is even and $r_j \ge i)$ or $(j$ is odd and $r_{j+1} < i \le r_j + r_{j+1})]$ and set $r^{(i)}_j = 0$ otherwise. Observe that $(r^{(i)})^{r_0}_{i=1} \in \{a \in \mathfrak{I} \mid a_0 = 1\}^{[1, r_0]}$ and $r = \sum_{i=1}^{r_0} r^{(i)}$. By Claim 4 we have $I = (\sum_{i=1}^{r_0} I_{r^{(i)}})_t$. Moreover, $I_{r^{(i)}}$ is a radical t-ideal for all $i \in [1, r_0]$ by Claim 6. Therefore, H is a t-SP-monoid. Set $F = \{x \in G \mid x_i \ge 0$ for all $i \in \mathbb{N}_0\}$. Obviously, F is a free abelian monoid and $H \subseteq F$ is a submonoid. Consequently, H is an FF-monoid.

Set $\mathcal{M} = \{\sqrt{y + H} \mid y \in H\}$ and let $I \in \mathcal{M}$. Then $I = \sqrt{x + H}$ for some $x \in H$. Set $m = |\{i \in \mathbb{N}_0 \mid x_i > 0\}|$ and $l = (m + 1)^2$. Let $\mathcal{K} \subseteq \mathcal{M}$ be a chain such that $\min(\mathcal{K}) = I$ (where $\min(\mathcal{K})$ denotes the smallest element of $\mathcal{K}$ with respect to inclusion). There is some sequence $(x^{(L)})_{L \in \mathcal{K}} \in H^{\mathcal{K}}$ such that $J = \sqrt{x^{(J)} + H}$ for all $J \in \mathcal{K}$. Let $f : \mathcal{K} \to [0, m] \times [0, m]$ be defined by $f(J) = (|\{i \in \mathbb{N}_0 \mid x^{(J)}_i > 0\}|, |\{i \in \mathbb{N}_0 \mid x^{(J)}_0 > x^{(J)}_i > 0\}|)$. Using Claim 1 and the fact that $\min(\mathcal{K}) = I$ it is easy to prove that f is well defined. We show that f is injective. Let $J, L \in \mathcal{K}$ be such that $f(J) = f(L)$. Without restriction let $J \subseteq L$. By Claim 1 we have $\{i \in \mathbb{N}_0 \mid x^{(L)}_i > 0\} \subseteq \{i \in \mathbb{N}_0 \mid x^{(J)}_i > 0\}$ and $\{i \in \mathbb{N}_0 \mid x^{(L)}_0 > x^{(L)}_i > 0\} \subseteq \{i \in \mathbb{N}_0 \mid x^{(J)}_0 > x^{(J)}_i > 0\}$. Since $f(J) = f(L)$, this implies that $\{i \in \mathbb{N}_0 \mid x^{(J)}_i > 0\} = \{i \in \mathbb{N}_0 \mid x^{(L)}_i > 0\}$ and $\{i \in \mathbb{N}_0 \mid x^{(J)}_0 > x^{(J)}_i > 0\} = \{i \in \mathbb{N}_0 \mid x^{(L)}_0 > x^{(L)}_i > 0\}$. Consequently, $\{i \in \mathbb{N}_0 \mid x^{(J)}_0 > x^{(J)}_i\} = \{i \in \mathbb{N}_0 \mid x^{(L)}_0 > x^{(L)}_i\}$, and thus

$J = \sqrt{x^{(J)} + H} = \sqrt{x^{(L)} + H} = L$ by Claim 1. Since f is injective we have $|\mathcal{K}| \leq l$. Let $\lambda : \mathcal{M} \to \mathbb{N}_0$ be defined by $\lambda(J) = \max\{|\mathcal{K}| \mid \mathcal{K} \subseteq \mathcal{M}$ is a chain and $\min(\mathcal{K}) = J\}$. Using the previous it is easy to prove that λ is a well-defined map and $\lambda(J) < \lambda(L)$ for all $J, L \in \mathcal{M}$ such that $L \subsetneq J$. Consequently, $\mathcal{M}$ possesses a length function.

Now let y be a radical element of H. There is some $k \in \mathbb{N}$ such that $y_k = 0$. Set $x = 2e_0 + e_k + \sum_{i \in \mathbb{N}, y_i > 0} e_i$. It follows by Claim 1 that $x \in \sqrt{y + H} = y + H$; hence $x_0 - y_0 \geq x_i - y_i \geq 0$ for all $i \in \mathbb{N}_0$. Therefore, $2 - y_0 = x_0 - y_0 \geq x_k - y_k = 1$, and thus $y_0 \leq 1$. Consequently, $y \in \mathcal{A}(H) \cup H^{\times}$.

Since $H \setminus H^{\times}$ is a t-ideal it follows that $\mathcal{C}_t(H)$ is trivial, and thus we have H is radical factorial by [29, Proposition 3.10.2]. Moreover, since $t\text{-dim}(H) = 2$, we have H is not a Krull monoid. Therefore, H is not a Mori monoid by [29, Proposition 2.6]. It follows by [29, Proposition 3.9] and [29, Corollary 3.14] that H is not a t-Prüfer monoid.   □

Note that if H is a discrete valuation monoid (i.e., an atomic valuation monoid H with $H^{\bullet} \neq H^{\times}$), then every radical element of H is either an atom or a unit. The last example also shares this property with discrete valuation monoids. An integral domain is called an almost Krull domain if all its localizations at prime ideals are Krull domains. The following question has been raised by Pirtle (see [28]): Is every almost Krull domain whose height-one prime ideals are divisorial already a Krull domain? Arnold and Matsuda answered Pirtle's question in the negative (see [3]). Note that our last example is of similar type, since it shows that a (radical factorial) t-SP-monoid whose height-one prime t-ideals are divisorial is not necessarily a Krull monoid. This also answers the questions raised after Proposition 2.6 in [29] in the negative. Finally, Example 4.2 shows that being a t-Prüfer monoid is not a monadic property and being "primary r-ideal inclusive" in [29, Corollary 5.3 and Theorem 5.4] cannot be omitted.

5 Connections with Rings of Integer-Valued Polynomials

In this section we investigate the connections between rings of integer-valued polynomials and monadically Krull monoids. In particular, we continue our search for examples of monadically Krull domains that are not Krull. As in Sect. 4, we will consider additively written monoids that do not possess an "additive" analogue of a "multiplicative" zero element.

Let R be an integral domain, K a field of quotients of R, and X an indeterminate over K. If $a, b \in R$, then we write $a \simeq_R b$ if $b = ac$ for some $c \in R^{\times}$. Set $\text{Int}(R) = \{f \in K[X] \mid f(c) \in R$ for all $c \in R\}$, called the ring of integer-valued polynomials over R. Observe that $R \subseteq R[X]$ and $R \subseteq \text{Int}(R)$ are divisor closed,

$\text{Int}(R)^\times = R[X]^\times = R^\times$ and $\mathcal{A}(\text{Int}(R)) \cap R = \mathcal{A}(R[X]) \cap R = \mathcal{A}(R)$. Especially, if $R[X]$ is monadically Krull or $\text{Int}(R)$ is monadically Krull, then R is monadically Krull.

Now let R be factorial and Q a system of representatives of prime elements of R. Recall that $R[X]$ is factorial. If $T \subseteq R$, then let $\text{GCD}_R(T)$ be the set of all greatest common divisors of T (in R). If $g \in R[X] \backslash R$, then g is called primitive if $\text{GCD}_{R[X]}(g, c) = R[X]^\times$ for all $c \in R^\bullet$ (equivalently: $\text{GCD}_R(\{a_i \mid i \in [0, k]\}) = R^\times$ for all $k \in \mathbb{N}_0$ and $(a_i)_{i=0}^k \in R^{[0,k]}$ such that $g = \sum_{i=0}^k a_i X^i$). If $q \in Q$, then let $v_q : R \to \mathbb{N}_0 \cup \{\infty\}$ denote the q-adic valuation of R. Let $d_Q : \text{Int}(R)^\bullet \to R^\bullet$ be defined by $d_Q(g) = \prod_{p \in Q} p^{\min\{v_p(g(c))| c \in R\}}$ for all $g \in \text{Int}(R)^\bullet$. Set $d = d_Q$. Note that $d(g) \in \text{GCD}_R(\{g(c) \mid c \in R\})$ and $\frac{g}{d(g)} \in \text{Int}(R)$ for all $g \in \text{Int}(R)^\bullet$.

If M is a set and $l \in \mathbb{N}$, then a finite sequence $(a_i)_{i=1}^l \in M^l$ will be denoted by $\underline{a}$ (i.e., $\underline{a} = (a_i)_{i=1}^l$). Let $n \in \mathbb{N}$, $\underline{f} \in (\text{Int}(R)^\bullet)^n$ and $\underline{x} \in \mathbb{N}_0^n \backslash \{\underline{0}\}$. Then $\underline{x}$ is called $\underline{f}$-irreducible if for all $\underline{y}, \underline{z} \in \mathbb{N}_0^n$ such that $\underline{x} = \underline{y} + \underline{z}$ and $d(\prod_{i=1}^n f_i^{x_i}) = d(\prod_{i=1}^n f_i^{y_i}) d(\prod_{i=1}^n f_i^{z_i})$ it follows that $\underline{y} = \underline{0}$ or $\underline{z} = \underline{0}$ (this definition does not depend on the choice of Q). In the next lemma we will use [15, Definition 1.5.2] and Dickson's Theorem (see [15, Theorem 1.5.3]) without further citation.

Lemma 5.1. *Let R be a factorial domain, $n \in \mathbb{N}$, and $\underline{f} \in (\text{Int}(R)^\bullet)^n$. Then $\{\underline{x} \in \mathbb{N}_0^n \mid \underline{x} \text{ is } \underline{f}\text{-irreducible}\}$ is finite.*

Proof. Let Q be a system of representatives of prime elements of R and $T = \{w \in R \mid (\prod_{i=1}^n f_i)(w) \neq 0\}$. We prove that $\min\{v_q(g(w)) \mid w \in R\} = \min\{v_q(g(w)) \mid w \in T\}$ for all $q \in Q$ and $g \in \text{Int}(R)^\bullet$. Let $q \in Q$ and $g \in \text{Int}(R)^\bullet$. Then $\min\{v_q(g(w)) \mid w \in R\} = v_q(g(v))$ for some $v \in R$. Observe that there is some $k \in \mathbb{N}$ such that $v_q(g(v)) = v_q(g(v + q^l))$ for all $l \in \mathbb{N}_{\geq k}$. Since $R \backslash T$ is finite, there is some $m \in \mathbb{N}_{\geq k}$ such that $v + q^m \in T$. We have $\min\{v_q(g(w)) \mid w \in R\} = v_q(g(v + q^m))$, and thus $\min\{v_q(g(w)) \mid w \in R\} = \min\{v_q(g(w)) \mid w \in T\}$.

Set $P = \{p \in Q \mid \min\{v_p((\prod_{i=1}^n f_i)(w)) \mid w \in R\} > 0\}$. Clearly, P is finite. If $p \in P$, then there is some finite $S_p \subseteq T$ such that $\text{Min}(\{(v_p(f_i(w)))_{i=1}^n \mid w \in T\}) = \{(v_p(f_i(w)))_{i=1}^n \mid w \in S_p\}$. Set $S = \bigcup_{p \in P} S_p$. Then S is finite. For $\gamma \in S^P$ set $\Omega_\gamma = \{\underline{u} \in \mathbb{N}_0^n \mid \sum_{i=1}^n (v_p(f_i(w)) - v_p(f_i(\gamma(p)))) u_i \geq 0$ for all $p \in P$ and $w \in S\}$. If $\gamma \in S^P$, then Ω_γ is an additive monoid and by [15, Theorem 2.7.14] and [15, Proposition 1.1.7.2] we have $\mathcal{A}(\Omega_\gamma)$ is finite. It suffices to show that $\{\underline{x} \in \mathbb{N}_0^n \mid \underline{x} \text{ is } \underline{f}\text{-irreducible}\} \subseteq \bigcup_{\gamma \in S^P} \mathcal{A}(\Omega_\gamma)$. Let $\underline{x} \in \mathbb{N}_0^n$ be $\underline{f}$-irreducible. There is some $\delta \in S^P$ such that $\min\{\sum_{i=1}^n v_p(f_i(w)) x_i \mid w \in S\} = \sum_{i=1}^n v_p(f_i(\delta(p))) x_i$ for all $p \in P$; hence $\underline{x} \in \Omega_\delta \backslash \{\underline{0}\}$. Let $\underline{u} \in \Omega_\delta$. If $p \in P$, then $\min\{v_p((\prod_{i=1}^n f_i^{u_i})(w)) \mid w \in R\} = \min\{\sum_{i=1}^n v_p(f_i(w)) u_i \mid w \in T\} = \min\{\sum_{i=1}^n v_p(f_i(w)) u_i \mid w \in S\} = \sum_{i=1}^n v_p(f_i(\delta(p))) u_i$, and if $p \in Q \backslash P$, then $\min\{v_p((\prod_{i=1}^n f_i^{u_i})(w)) \mid w \in R\} = 0$. Let $\underline{y}, \underline{z} \in \Omega_\delta$ be such that $\underline{x} = \underline{y} + \underline{z}$. If $p \in P$, then $\min\{v_p((\prod_{i=1}^n f_i^{x_i})(w)) \mid w \in R\} = \sum_{i=1}^n v_p(f_i(\delta(p))) x_i = \sum_{i=1}^n v_p(f_i(\delta(p))) y_i + \sum_{i=1}^n v_p(f_i(\delta(p))) z_i = \min\{v_p((\prod_{i=1}^n f_i^{y_i})(w)) \mid w \in R\} + \min\{v_p((\prod_{i=1}^n f_i^{z_i})(w)) \mid w \in R\}$. This implies that $d_Q(\prod_{i=1}^n f_i^{x_i}) = d_Q(\prod_{i=1}^n f_i^{y_i}) d_Q(\prod_{i=1}^n f_i^{z_i})$; hence $\underline{y} = \underline{0}$ or $\underline{z} = \underline{0}$. Therefore, $\underline{x} \in \mathcal{A}(\Omega_\delta)$. $\square$

Now we present the main result of this section.

Theorem 5.2. *Let R be a factorial domain. Then* $\mathrm{Int}(R)$ *is monadically Krull.*

Proof. Let K be a field of quotients of R, X an indeterminate over K, Q a system of representatives of prime elements of R, and $d = d_Q$. Set $S = R[X]$ and $T = \mathrm{Int}(R)$. It is well known that T is atomic and completely integrally closed (see [8, Propositions VI.2.1 and VI.2.9]). By Theorem 3.6 we need to prove that $\{yT \mid y \in \mathcal{A}(\llbracket g \rrbracket_T)\}$ is finite for all $g \in T^\bullet$. Let $g \in T^\bullet$. Some $a, b \in R^\bullet$, $n \in \mathbb{N}$, $\underline{v} \in \mathbb{N}_0^n$ and $\underline{f} \in (\mathcal{A}(S) \backslash R)^n$ exist such that $g = \frac{a}{b} \prod_{i=1}^n f_i^{v_i}$ and $f_j \not\simeq_S f_k$ for all different $j, k \in [1, n]$. By Lemma 5.1 it is sufficient to show that $\{yT \mid y \in \mathcal{A}(\llbracket g \rrbracket_T)\} \subseteq \{yT \mid y \in \mathcal{A}(R), y \mid_R d(g)\} \cup \{\frac{\prod_{i=1}^n f_i^{\alpha_i}}{d(\prod_{i=1}^n f_i^{\alpha_i})} T \mid \underline{\alpha} \in \mathbb{N}_0^n, \underline{\alpha} \text{ is } \underline{f}\text{-irreducible}\}$. Let $y \in \mathcal{A}(\llbracket g \rrbracket_T)$. Then $y \in \mathcal{A}(T)$ and $y \mid_T g^l$ for some $l \in \mathbb{N}$.

Case 1: $y \in R$. We have $y \in \mathcal{A}(R)$ and $y \mid_R d(g^l) = d(g)^l$. Therefore, $y \mid_R d(g)$.

Case 2: $y \notin R$. Some primitive $t \in S$ and some $c, e \in R^\bullet$ exist such that $\mathrm{GCD}_S(c, et) = S^\times$ and $y = \frac{et}{c}$. This implies that $c \mid_R d(t)$. Observe that $y = \frac{ed(t)}{c} \cdot \frac{t}{d(t)}$, $\frac{ed(t)}{c} \in T$ and $\frac{t}{d(t)} \in T \backslash T^\times$. Consequently, $y \simeq_T \frac{t}{d(t)}$. There are some $w \in S$ and $u \in R^\bullet$ such that $y \frac{w}{u} = g^l$. Therefore, $etwb^l = cua^l \prod_{i=1}^n f_i^{lv_i}$ and since t is primitive it follows that $t \mid_S \prod_{i=1}^n f_i^{lv_i}$. Hence, there is some $\underline{\alpha} \in \mathbb{N}_0^n \backslash \{\underline{0}\}$ such that $t \simeq_S \prod_{i=1}^n f_i^{\alpha_i}$. This implies that $y \simeq_T \frac{\prod_{i=1}^n f_i^{\alpha_i}}{d(\prod_{i=1}^n f_i^{\alpha_i})}$, and thus $yT = \frac{\prod_{i=1}^n f_i^{\alpha_i}}{d(\prod_{i=1}^n f_i^{\alpha_i})} T$. Let $\underline{\beta}, \underline{\gamma} \in \mathbb{N}_0^n$ be such that $\underline{\alpha} = \underline{\beta} + \underline{\gamma}$ and $d(\prod_{i=1}^n f_i^{\alpha_i}) = d(\prod_{i=1}^n f_i^{\beta_i}) d(\prod_{i=1}^n f_i^{\gamma_i})$. Note that $\frac{\prod_{i=1}^n f_i^{\beta_i}}{d(\prod_{i=1}^n f_i^{\beta_i})}, \frac{\prod_{i=1}^n f_i^{\gamma_i}}{d(\prod_{i=1}^n f_i^{\gamma_i})} \in T$ and $\frac{\prod_{i=1}^n f_i^{\beta_i}}{d(\prod_{i=1}^n f_i^{\beta_i})} \cdot \frac{\prod_{i=1}^n f_i^{\gamma_i}}{d(\prod_{i=1}^n f_i^{\gamma_i})} = \frac{\prod_{i=1}^n f_i^{\alpha_i}}{d(\prod_{i=1}^n f_i^{\alpha_i})} \in \mathcal{A}(T)$. Therefore, $\frac{\prod_{i=1}^n f_i^{\beta_i}}{d(\prod_{i=1}^n f_i^{\beta_i})} \in T^\times$ or $\frac{\prod_{i=1}^n f_i^{\gamma_i}}{d(\prod_{i=1}^n f_i^{\gamma_i})} \in T^\times$; hence $\underline{\beta} = \underline{0}$ or $\underline{\gamma} = \underline{0}$. Consequently, $\underline{\alpha}$ is $\underline{f}$-irreducible. $\square$

Theorem 5.2 is also interesting from a purely factorization theoretical point of view, since it provides a class of Krull monoids whose arithmetic is not fully understood by now. The arithmetic of the Krull monoids involved may also differ from the arithmetic of monadic submonoids of principal orders in algebraic number fields.

Corollary 5.3. *Let R be a factorial domain. Then* $\mathrm{Int}(R)$ *is an FF-domain.*

Proof. This follows from Theorems 3.6 and 5.2. $\square$

In [14] it has been shown that $\mathrm{Int}(\mathbb{Z})$ is an FF-domain. Corollary 5.3 is a generalization of this result. By Theorem 5.2, [8, Theorem VI.1.7] and [8, Remark VI.2.10] we obtain that $\mathrm{Int}(\mathbb{Z})$ and $\mathrm{Int}(\mathbb{Z}_{(p)})$ for $p \in \mathbb{P}$ are monadically Krull domains and Prüfer domains (and thus t-Prüfer domains) that are no Krull domains.

6 Further Remarks

In Sect. 4 we showed that a radical factorial FF-monoid is not necessarily a Krull monoid. So far we do not know whether every radical factorial, 1-dimensional FF-domain is a Krull domain. In this last section we investigate special types of examples that have been introduced in [20] to construct atomic Prüfer domains that are no Dedekind domains. We study these examples in greater detail and generality to obtain an interesting class of examples that are not "too far away" from being examples of radical factorial 1-dimensional FF-domains that are not Krull.

Let H be a monoid. We say that H is a weakly Krull monoid if $\bigcap_{P \in \mathfrak{X}(H)} H_P = H$ and $\{P \in \mathfrak{X}(H) \mid a \in P\}$ is finite for all $a \in H^\bullet$. Note that H is a Krull monoid if and only if H is a weakly Krull monoid and H_P is a Krull monoid for all $P \in \mathfrak{X}(H)$. It follows from Example 4.2 that being a weakly Krull monoid is not a monadic property (since the monoid in this example is monadically Krull, hence monadically weakly Krull, but by [29, Proposition 2.6] it fails to be a weakly Krull). By [1, Theorem 1] and [2, Theorem 5.1] we have H is an FF-monoid if and only if H is atomic and $\{uH \mid u \in \mathcal{A}(H), u|_H x\}$ is finite for all $x \in H^\bullet$ (such monoids are called IDF-monoids; e.g., see [24]) if and only if H is a BF-monoid and $\{uH \mid u \in \mathcal{A}(H), u|_H x\}$ is finite for all $x \in H^\bullet$. Clearly, if H is a BF-monoid, then H satisfies the ACCP. First we start with a simple lemma that might be of independent interest. It gives a hint how to construct monoids H where $\{uH \mid u \in \mathcal{A}(H), u|_H x\}$ is finite for all $x \in H^\bullet$, but that fail to be FF-monoids.

Lemma 6.1. *[(cf. [21])] Let K be a monoid, S a submonoid of K that is an FF-monoid, $T \subseteq K$ a submonoid of K that is a valuation monoid and $H = S \cap T$. Then $\{uH \mid u \in \mathcal{A}(H), u|_H x\}$ is finite for all $x \in H^\bullet$.*

Proof. Let $x \in H^\bullet$ and $\mathcal{D}(x) = \{u \in \mathcal{A}(H) \mid u|_H x\}$. Then $\{uS \mid u \in \mathcal{D}(x)\} \subseteq \{uS \mid u \in S \text{ and } u|_S x\}$. Since S is an FF-monoid, it follows that $\{uS \mid u \in \mathcal{D}(x)\}$ is finite. Let $v, w \in \mathcal{A}(H)$ be such that $vS = wS$. We have $vT \subseteq wT$ or $wT \subseteq vT$. Therefore, $vH = vS \cap vT \subseteq wS \cap wT = wH$ or $wH = wS \cap wT \subseteq vS \cap vT = vH$; hence $vH = wH$. Consequently, $f : \{uH \mid u \in \mathcal{D}(x)\} \to \{uS \mid u \in \mathcal{D}(x)\}$ defined by $f(I) = IS$ for all $I \in \{uH \mid u \in \mathcal{D}(x)\}$ is an injective map. This implies that $\{uH \mid u \in \mathcal{D}(x)\}$ is finite. $\qquad\square$

Proposition 6.2. *Let H be a monoid, K a quotient monoid of H, $\mathcal{U}$ a set of overmonoids of H that are FF-monoids, and $\mathcal{V}$ a set of overmonoids of H that are valuation monoids such that $H = \bigcap_{S \in \mathcal{U} \cup \mathcal{V}} S$. Let $(\mathcal{N}_T)_{T \in \mathcal{V}} \in \mathbb{P}(\mathcal{U})^{\mathcal{V}}$ be such that $T \cap \bigcap_{S \in \mathcal{N}_T} S$ is atomic for all $T \in \mathcal{V}$ and $\{T \in \mathcal{V} \mid S \in \mathcal{N}_T\}$ is finite for all $S \in \mathcal{U}$. If $\{S \in \mathcal{U} \cup \mathcal{V} \mid a \notin S^\times\}$ is finite for all $a \in H^\bullet$, then H is an FF-monoid.*

Proof. Let $\{S \in \mathcal{U} \cup \mathcal{V} \mid a \notin S^\times\}$ be finite for all $a \in H^\bullet$ and $\mathcal{M} = \mathcal{U} \cup \{T \cap \bigcap_{S \in \mathcal{N}_T} S \mid T \in \mathcal{V}\}$.

Claim 1: For all $U \in \mathcal{M}$ it follows that U is an FF-monoid. Let $U \in \mathcal{M}$ and $T \in \mathcal{V}$ be such that $U = T \cap \bigcap_{S \in \mathcal{N}_T} S$. We show that $\bigcap_{S \in \mathcal{N}_T} S$ is an FF-monoid.

If $\mathcal{N}_T = \emptyset$, then $\bigcap_{S \in \mathcal{N}_T} S = K$; hence $\bigcap_{S \in \mathcal{N}_T} S$ is an FF-monoid. Since $\{S \in \mathcal{N}_T \mid a \notin S^\times\}$ is finite for all $a \in H^\bullet$, we have $\{S \in \mathcal{N}_T \mid a \notin S^\times\}$ is finite for all $a \in K^\bullet$; hence $\{S \in \mathcal{N}_T \mid a \notin S^\times\}$ is finite for all $a \in (\bigcap_{S \in \mathcal{N}_T} S)^\bullet$. Therefore, [1, Theorem 2] implies that $\bigcap_{S \in \mathcal{N}_T} S$ is an FF-monoid. It follows by Lemma 6.1 that U is an FF-monoid.

Claim 2: For every $a \in H^\bullet$, $\{S \in \mathcal{M} \mid a \notin S^\times\}$ is finite. Let $a \in H^\bullet$. We have $\{T \in \mathcal{V} \mid a \notin (T \cap \bigcap_{S \in \mathcal{N}_T} S)^\times\} \subseteq \{T \in \mathcal{V} \mid a \notin T^\times\} \cup \bigcup_{S \in \mathcal{U}, a \notin S^\times} \{T \in \mathcal{V} \mid S \in \mathcal{N}_T\}$, and thus $\{T \in \mathcal{V} \mid a \notin (T \cap \bigcap_{S \in \mathcal{N}_T} S)^\times\}$ is finite. Therefore, $\{S \in \mathcal{M} \mid a \notin S^\times\} = \{S \in \mathcal{U} \mid a \notin S^\times\} \cup \{T \cap \bigcap_{S \in \mathcal{N}_T} S \mid T \in \mathcal{V}, a \notin (T \cap \bigcap_{S \in \mathcal{N}_T} S)^\times\}$ is finite.

Since $H = \bigcap_{S \in \mathcal{M}} S$, it follows by Claim 1, Claim 2 and [1, Theorem 2] that H is an FF-monoid. $\qquad\square$

Proposition 6.3. *Let K be a monoid, H a submonoid of K and Λ a set of intermediate monoids of H and K such that $\{S \in \Lambda \mid a \notin S^\times\}$ is finite for all $a \in H^\bullet$ and $H \cap \bigcap_{S \in \Lambda} S^\times = H^\times$:*

1. *If S satisfies the ACCP for all $S \in \Lambda$, then H satisfies the ACCP.*
2. *If S is a BF-monoid for all $S \in \Lambda$, then H is a BF-monoid.*

Proof. 1. Let S satisfy the ACCP for all $S \in \Lambda$. Let $(a_i)_{i \in \mathbb{N}} \in H^{\mathbb{N}}$ be such that $a_i H \subseteq a_{i+1} H$ for all $i \in \mathbb{N}$. Without restriction let $a_1 \neq 0$. Let $\mathcal{A} = \{S \in \Lambda \mid a_1 \notin S^\times\}$. Since $\mathcal{A}$ is finite there is some $r \in \mathbb{N}$ such that $a_k S = a_r S$ for all $S \in \mathcal{A}$ and $k \in \mathbb{N}_{\geq r}$. It is sufficient to show that $a_k H = a_r H$ for all $k \in \mathbb{N}_{\geq r}$. Let $k \in \mathbb{N}_{\geq r}$ and $T \in \Lambda$. If $a_1 \in T^\times$, then $a_r, a_k \in T^\times$, and thus $\frac{a_r}{a_k} \in T^\times$. If $a_1 \notin T^\times$, then $a_r T = a_k T$; hence $\frac{a_r}{a_k} \in T^\times$. Consequently, $\frac{a_r}{a_k} \in H \cap \bigcap_{S \in \Lambda} S^\times = H^\times$, and thus $a_r H = a_k H$.

2. Let S be a BF-monoid for all $S \in \Lambda$ and set $\mathcal{M} = \{(S \setminus S^\times) \cap H \mid S \in \Lambda\}$. It follows by [15, Proposition 1.3.2] that $\bigcap_{n \in \mathbb{N}} (S \setminus S^\times)^n = \{0\}$ for all $S \in \Lambda$. Therefore, $\bigcap_{n \in \mathbb{N}} M^n = \{0\}$ for all $M \in \mathcal{M}$. Let $a \in H^\bullet \setminus H^\times$. Then $\{M \in \mathcal{M} \mid a \in M\} \subseteq \{(S \setminus S^\times) \cap H \mid S \in \Lambda, a \notin S^\times\}$; hence $\{M \in \mathcal{M} \mid a \in M\}$ is finite. Since $a \notin H^\times$, there is some $T \in \Lambda$ such that $a \notin T^\times$, hence $(T \setminus T^\times) \cap H \in \{M \in \mathcal{M} \mid a \in M\}$. Consequently, [15, Theorem 1.3.4] implies that H is a BF-monoid. $\qquad\square$

Corollary 6.4. *Let H be a monoid and $\mathcal{M} \subseteq s\text{-spec}(H)$ such that $\bigcup_{M \in \mathcal{M}} M = H \setminus H^\times$ and $\{M \in \mathcal{M} \mid a \in M\}$ is finite for all $a \in H^\bullet$.*

1. *If H_M satisfies the ACCP for all $M \in \mathcal{M}$, then H satisfies the ACCP.*
2. *If H_M is a BF-monoid for all $M \in \mathcal{M}$, then H is a BF-monoid.*

Proof. Let $\Lambda = \{H_M \mid M \in \mathcal{M}\}$. We have $H \cap \bigcap_{S \in \Lambda} S^\times = H \cap \bigcap_{M \in \mathcal{M}} (H_M \setminus M_M) = \bigcap_{M \in \mathcal{M}} (H \setminus M) = H \setminus \bigcup_{M \in \mathcal{M}} M = H^\times$. Let $a \in H^\bullet \setminus H^\times$. Then $\{S \in \Lambda \mid a \notin S^\times\} = \{H_M \mid M \in \mathcal{M}, a \in M\}$ is finite. Consequently, the assertions follow from Proposition 6.3. $\qquad\square$

If S is an integral domain and $R \subseteq S$ is a subring, then let $\mathrm{cl}_S(R)$ denote the integral closure of R in S. We say that $M \in \max(S)$ is critical if for each finite $E \subseteq M$, there exists $Q \in \max(S)$ such that $E \subseteq Q^2$.

Proposition 6.5. *[(cf. [7])] Let R be a Dedekind domain that is not a field, K a field of quotients of R, L/K a field extension, $S = \mathrm{cl}_L(R)$, $(L_i)_{i \in \mathbb{N}}$ a sequence of intermediate fields of K and L such that $L_i \subseteq L_{i+1}$ and $[L_i : K] < \infty$ for all $i \in \mathbb{N}$ and $L = \bigcup_{j \in \mathbb{N}} L_j$. Let $(\mathcal{A}_i)_{i \in \mathbb{N}} \in \mathbb{P}(\max(R))^{\mathbb{N}}$ be such that $\mathcal{A}_i \subseteq \mathcal{A}_{i+1}$, for all $i \in \mathbb{N}$. Set $\mathcal{A} = \bigcup_{i \in \mathbb{N}} \mathcal{A}_i$ and $\mathcal{N} = \{M \in \max(S) \mid M \cap R \in \mathcal{A}\}$. For $i \in \mathbb{N}$ set $S_i = \mathrm{cl}_{L_i}(R)$ and $\mathcal{N}_i = \{M \in \max(S) \mid M \cap R \in \mathcal{A}_i\}$:*

1. *S_i is a Dedekind domain for all $i \in \mathbb{N}$, S is a 1-dimensional Prüfer domain, and $S = \bigcup_{i \in \mathbb{N}} S_i$.*
2. *Let for all $i \in \mathbb{N}$ and $P \in \max(S_i)$ such that $P \cap R \in \mathcal{A}_i$ be $P \not\subseteq Q^2$ for all $Q \in \max(S_{i+1})$. Then for all $M \in \mathcal{N}$ we have M is not critical and if $\mathcal{A} = \max(R)$, then S is an SP-domain.*
3. *Let for all $i \in \mathbb{N}$ and $P \in \max(S_i)$ such that $P \cap R \in \mathcal{A}_i$ be $^{s_i+}\!\sqrt{PS_{i+1}} \in \max(S_{i+1})$. Then for all $a \in S^\bullet$ it follows that $\{M \in \mathcal{N} \mid a \in M\}$ is finite and if $\mathcal{A} = \max(R)$, then S is a weakly Krull.*
4. *Let $\bigcup_{P \in \mathcal{A}} P = R \backslash R^\times$ and let for all $i \in \mathbb{N}$ and $P \in \max(S_i)$ such that $P \cap R \in \mathcal{A}_i$ be $PS_{i+1} \in \max(S_{i+1})$. Then S is a BF-domain and if $|\max(S)\backslash\mathcal{N}| \leq 1$, then S is an FF-domain.*
5. *If there is some sequence $(M_i)_{i \in \mathbb{N}}$ such that $M_i \in \max(S_i)$ and $M_{i+1} \cap S_i = M_i$ for all $i \in \mathbb{N}$, and $\{j \in \mathbb{N} \mid M_j S_{j+1} \not\in \max(S_{j+1})\}$ is infinite, then S is not a Dedekind domain.*

Proof.
1. Clearly, S is 1-dimensional and $S = \bigcup_{i \in \mathbb{N}} S_i$. By the Theorem of Krull-Akizuki we have S_i is a Dedekind domain for all $i \in \mathbb{N}$. Since L/K is an algebraic field extension we have S is a Prüfer domain.
2. Claim 1: For all $j \in \mathbb{N}$ and $M \in \mathcal{N}_j$ it follows that $M \cap S_j \not\subseteq (M \cap S_k)^2$ for all $k \in \mathbb{N}_{\geq j}$. Let $j \in \mathbb{N}$ and $M \in \mathcal{N}_j$. We show by induction on k that $M \cap S_j \not\subseteq (M \cap S_k)^2$ for all $k \in \mathbb{N}_{\geq j}$. Obviously, $M \cap S_j \not\subseteq (M \cap S_j)^2$. Now let $k \in \mathbb{N}_{\geq j}$ be such that $M \cap S_j \not\subseteq (M \cap S_k)^2$. Since $(M \cap S_j)S_k \subseteq M \cap S_k$, there is some ideal I of S_k such that $(M \cap S_j)S_k = (M \cap S_k)I$. Since $M \cap S_j \not\subseteq (M \cap S_k)^2$, it follows that $I \not\subseteq M \cap S_k$; hence $IS_{k+1} \not\subseteq M \cap S_{k+1}$. We have $M \cap S_k \in \max(S_k)$, $M \cap S_{k+1} \in \max(S_{k+1})$ and $(M \cap S_k) \cap R = M \cap R \in \mathcal{A}_j \subseteq \mathcal{A}_k$, and thus $(M \cap S_k)S_{k+1} \not\subseteq (M \cap S_{k+1})^2$. Since $(M \cap S_{k+1})^2$ is $M \cap S_{k+1}$-primary it follows that $(M \cap S_j)S_{k+1} = (M \cap S_k)S_{k+1}IS_{k+1} \not\subseteq (M \cap S_{k+1})^2$; hence $M \cap S_j \not\subseteq (M \cap S_{k+1})^2$.
Claim 2: For every $M \in \max(S)$, we have $M^2 = \bigcup_{i \in \mathbb{N}}(M \cap S_i)^2$. Let $M \in \max(S)$. "$\subseteq$": Let $x \in M^2$. There exist some $r \in \mathbb{N}$ and $(x_i)_{i=1}^r, (y_i)_{i=1}^r \in M^{[1,r]}$ such that $x = \sum_{i=1}^r x_i y_i$. There is some $l \in \mathbb{N}$ such that $x_i, y_i \in S_l$ for all $i \in [1, r]$. Consequently, $x \in (M \cap S_l)^2$. "$\supseteq$": Trivial.
Now let $Q \in \mathcal{N}$. There is some $j \in \mathbb{N}$ such that $Q \in \mathcal{N}_j$. Assume that there is some $M \in \max(S)$ such that $Q \cap S_j \subseteq M^2$. Then $Q \cap S_j = M \cap S_j$ and $M \cap R = Q \cap R \in \mathcal{A}_j$; hence $M \in \mathcal{N}_j$. It follows by Claim 2 that there exists some $k \in \mathbb{N}_{\geq j}$ such that $M \cap S_j \subseteq (M \cap S_k)^2$ which is a contradiction

to Claim 1. Consequently, $(Q \cap S_j)S \not\subseteq M^2$ for all $M \in \max(S)$. Since $(Q \cap S_j)S$ is a finitely generated ideal of S we have Q is not critical.

Now let $\mathcal{A} = \max(R)$. Then $\mathcal{N} = \max(S)$; hence every $M \in \max(S)$ is not critical. It follows by 1 that S is a 1-dimensional Prüfer domain. Consequently, S is an SP-domain by [27, Corollary 2.2].

3. Claim: For all $i \in \mathbb{N}$ we have $f_i : \mathcal{N}_i \to \{M \cap S_i \mid M \in \mathcal{N}_i\}$ defined by $f_i(M) = M \cap S_i$ is a bijective map. Let $i \in \mathbb{N}$. Obviously, f_i is a surjective map. Let $M, Q \in \mathcal{N}_i$ be such that $M \cap S_i = Q \cap S_i$. We show by induction on j that for all $j \in \mathbb{N}_{\geq i}$, $M \cap S_j = Q \cap S_j$. Let $j \in \mathbb{N}_{\geq i}$. The assertion holds for $j = i$. Now let $M \cap S_j = Q \cap S_j$. We have $(M \cap S_j)S_{j+1} \subseteq M \cap S_{j+1}$; hence $\sqrt[S_{j+1}]{(M \cap S_j)S_{j+1}} = M \cap S_{j+1}$. Analogously $\sqrt[S_{j+1}]{(Q \cap S_j)S_{j+1}} = Q \cap S_{j+1}$; hence $M \cap S_{j+1} = Q \cap S_{j+1}$. Finally, it follows that $M = \bigcup_{j \in \mathbb{N}_{\geq i}}(M \cap S_j) = \bigcup_{j \in \mathbb{N}_{\geq i}}(Q \cap S_j) = Q$.

Let $a \in S^\bullet$. There is some $l \in \mathbb{N}$ such that $a \in S_l$. Obviously, there is some surjective map from $\{M \cap S_l \mid M \in \mathcal{N}, a \in M\}$ to $\{M \cap R \mid M \in \mathcal{N}, a \in M\}$. Since $\{M \cap S_l \mid M \in \mathcal{N}, a \in M\} \subseteq \{Q \in \max(S_l) \mid a \in Q\}$ and $\{Q \in \max(S_l) \mid a \in Q\}$ is finite we have $\{M \cap R \mid M \in \mathcal{N}, a \in M\}$ is finite. Therefore, there is some $k \in \mathbb{N}_{\geq l}$ such that $\{M \cap R \mid M \in \mathcal{N}, a \in M\} \subseteq \mathcal{A}_k$. Since $f_k(\{M \in \mathcal{N}_k \mid a \in M\}) = \{M \cap S_k \mid M \in \mathcal{N}_k, a \in M\} \subseteq \{Q \in \max(S_k) \mid a \in Q\}$, it follows by the claim that $\{M \in \mathcal{N} \mid a \in M\} = \{M \in \mathcal{N}_k \mid a \in M\}$ is finite.

Now let $\mathcal{A} = \max(R)$. Then $\mathcal{N} = \max(S) = \mathfrak{X}(S)$ by 1, and thus S is a weakly Krull domain.

4. It follows by 3 that $\{M \in \mathcal{N} \mid a \in S\}$ is finite for all $a \in S^\bullet$. By [19, Proposition 4] we have $\bigcup_{M \in \mathcal{N}} M = S \setminus S^\times$. Let $M \in \mathcal{N}$. By 2 it follows that M is not critical; hence $M \neq M^2$. Since M^2 is M-primary we have $M_M^2 \neq M_M$. Therefore, 1 implies that S_M is a valuation domain, and thus M_M is a principal ideal of S_M. This implies that S_M is a Dedekind domain; hence S_M is an FF-domain and a BF-domain. Consequently, Corollary 6.4(2) implies that S is a BF-domain. Now let $|\max(S) \setminus \mathcal{N}| \leq 1$. Set $\mathcal{U} = \{S_M \mid M \in \mathcal{N}\}$ and $\mathcal{V} = \{S_M \mid M \in \max(S) \setminus \mathcal{N}\}$. Every $T \in \mathcal{U}$ is an FF-domain and by 1 we have that every $T \in \mathcal{V}$ is a valuation domain. Obviously, $\mathcal{U} \cup \mathcal{V} = \{S_M \mid M \in \max(S)\}$; hence $\bigcap_{T \in \mathcal{U} \cup \mathcal{V}} T = S$. For $T \in \mathcal{V}$ set $\mathcal{N}_T = \mathcal{U}$. Since $|\mathcal{V}| \leq 1$, we have $T \cap \bigcap_{U \in \mathcal{N}_T} U = S$ is atomic for all $T \in \mathcal{V}$. It follows that $\{T \in \mathcal{U} \cup \mathcal{V} \mid a \notin T^\times\}$ is finite for all $a \in S^\bullet$, and thus Proposition 6.2 implies that S is an FF-domain.

5. Let $(M_i)_{i \in \mathbb{N}}$ be such that $M_i \in \max(S_i)$ and $M_{i+1} \cap S_i = M_i$ for all $i \in \mathbb{N}$ and $\{j \in \mathbb{N} \mid M_j S_{j+1} \not\subseteq \max(S_{j+1})\}$ is infinite. Let $M = \bigcup_{i \in \mathbb{N}} M_i$. Observe that $M \in \max(S)$. Assume that S is a Dedekind domain, then there is some finite $E \subseteq M$ such that $M = (E)_S$. There is some $i \in \mathbb{N}$ such that $E \subseteq M_i$. There is some $j \in \mathbb{N}_{\geq i}$ such that $M_j S_{j+1} \not\subseteq \max(S_{j+1})$, and thus there are some $Q, Q' \in \max(S_{j+1})$ such that $M_j S_{j+1} \subseteq QQ'$. This implies that $M = QS = Q'S$ and $M^2 = QSQ'S = M$; hence $M = \{0\}$, a contradiction. $\qquad\square$

Proposition 6.6. *Let R be a Dedekind domain such that $\max(R)$ is infinite, K a field of quotients of R, $(\mathcal{A}_i)_{i \in \mathbb{N}}, (\mathcal{B}_i)_{i \in \mathbb{N}}, (\mathcal{C}_i)_{i \in \mathbb{N}} \in \mathbb{P}(\max(R))^{\mathbb{N}}$ such that $\mathcal{A}_i$, $\mathcal{B}_i$, $\mathcal{C}_i$ are finite and $\mathcal{A}_i \subseteq \mathcal{A}_{i+1}, \mathcal{B}_i \subseteq \mathcal{B}_{i+1}, \mathcal{C}_i \subseteq \mathcal{C}_{i+1}$ for all $i \in \mathbb{N}$. Set $\mathcal{A} = \bigcup_{i \in \mathbb{N}} \mathcal{A}_i$, $\mathcal{B} = \bigcup_{i \in \mathbb{N}} \mathcal{B}_i$ and $\mathcal{C} = \bigcup_{i \in \mathbb{N}} \mathcal{C}_i$. Assume that $\mathcal{A} \cap \mathcal{B} = \mathcal{A} \cap \mathcal{C} = \mathcal{B} \cap \mathcal{C} = \emptyset$ and let R/M be finite for all $M \in \max(R)$. Then there exists some sequence $(L_i)_{i \in \mathbb{N}}$ of extension fields of K such that $L_1 = K$, $L_i \subseteq L_{i+1}$, $[L_i : K] < \infty$ and $S_i = \mathrm{cl}_{L_i}(R)$ for all $i \in \mathbb{N}$ and such that the following conditions are satisfied:*

1. *For all $i \in \mathbb{N}$ and $M \in \max(S_i)$ such that $M \cap R \in \mathcal{A}_i$ we have $M S_{i+1} \in \max(S_{i+1})$.*
2. *For all $i \in \mathbb{N}$ and $M \in \max(S_i)$ such that $M \cap R \in \mathcal{B}_i$ we have $M S_{i+1} \notin \max(S_{i+1})$ and $M \not\subseteq Q^2$ for all $Q \in \max(S_{i+1})$.*
3. *For all $i \in \mathbb{N}$ and $M \in \max(S_i)$ such that $M \cap R \in \mathcal{C}_i$ we have $M S_{i+1} \notin \max(S_{i+1})$ and $\sqrt[s_{i+1}]{M S_{i+1}} \in \max(S_{i+1})$.*

Proof. This follows by induction from [18, Theorem 42.5]. $\qquad\square$

By [29, Example 4.3] there is some Dedekind domain R such that $\max(R)$ is countable, R/M is finite for all $M \in \max(R)$, and $\mathrm{Pic}(R)$ is torsion-free. Let $M : \mathbb{N}_0 \to \max(R)$ be a bijection. Note that since $\mathrm{Pic}(R)$ is torsion-free we obtain that $\bigcup_{M \in \max(R) \setminus \{M_0\}} M = R \setminus R^\times$ (since every non-unit of R is contained in at least two different maximal ideals of R). For $j \in \mathbb{N}$ set $\mathcal{A}_j = \{M_i \mid i \in [1, j]\}$.

First set $\mathcal{B}_j = \{M_0\}$ and $\mathcal{C}_j = \emptyset$ for all $j \in \mathbb{N}$. Let $(L_i)_{i \in \mathbb{N}}$ be the sequence in Proposition 6.6, $L = \bigcup_{i \in \mathbb{N}} L_i$ and $S = \mathrm{cl}_L(R)$. Then S is an SP-domain that is a BF-domain but not Krull by Proposition 6.5.

Next set $\mathcal{B}_j = \emptyset$ and $\mathcal{C}_j = \{M_0\}$ for all $j \in \mathbb{N}$. Let $(L_i)_{i \in \mathbb{N}}$ be the sequence in Proposition 6.6, $L = \bigcup_{i \in \mathbb{N}} L_i$ and $S = \mathrm{cl}_L(R)$. Then S is a completely integrally closed FF-domain that is a weakly Krull domain but not a Krull domain by Proposition 6.5.

Proposition 6.7. *Let R be a Prüfer domain, K a field of quotients of R, L/K an algebraic field extension, and $S = \mathrm{cl}_L(R)$:*

1. *If for all intermediate fields $K \subseteq M \subseteq L$ such that $[M : K] < \infty$ it follows that $\mathrm{Pic}(\mathrm{cl}_M(R))$ is a torsion group, then $\mathrm{Pic}(S)$ is a torsion group.*
2. *If for all $a \in L$ and $n \in \mathbb{N}$ there is some $b \in L$ such that $b^n = a$, then $\mathrm{Pic}(S)$ is torsion-free.*

Proof. 1. Let I be an invertible ideal of S. Then there are some $m \in \mathbb{N}$ and some sequence $(a_i)_{i=1}^m \in I^{[1,m]}$ such that $I = \sum_{i=1}^m a_i S$. Set $M = K(\{a_i \mid i \in [1, m]\})$, $T = \mathrm{cl}_M(R)$ and $J = \sum_{i=1}^m a_i T$. Note that $\{a_i \mid i \in [1, m]\} \subseteq M \cap S = T$, and thus J is a non-zero finitely generated ideal of T. Since T is a Prüfer domain we have J is an invertible ideal of T. Since $\mathrm{Pic}(\mathrm{cl}_M(R))$ is a torsion group, there are some $n \in \mathbb{N}$ and $a \in T$ such that $J^n = aT$. Therefore, $I^n = J^n S = aS$.

2. Let I be an invertible ideal of S, $n \in \mathbb{N}$ and $a \in S$ such that $I^n = aS$. There is some $b \in L$ such that $b^n = a$. Observe that $b \in S$ and $I^n = b^n S$. Let $M \in \max(S)$. Since S is a Prüfer domain it follows that S_M is a valuation domain; hence there is some $c \in S$ such that $I_M = cS_M$. This implies that $c^n S_M = I_M^n = b^n S_M$, and thus there exists some $\varepsilon \in S_M^\times$ such that $c^n = \varepsilon b^n$. Since S_M is a valuation domain it follows that $bS_M \subseteq cS_M$ or $cS_M \subseteq bS_M$. Case 1: $bS_M \subseteq cS_M$. There exists some $v \in S_M$ such that $b = cv$. This implies that $b^n = c^n v^n = \varepsilon b^n v^n$; hence $1 = \varepsilon v^n$. Consequently, $v \in S_M^\times$, and thus $I_M = cS_M = cvS_M = bS_M$. Case 2: $cS_M \subseteq bS_M$. There is some $v \in S_M$ such that $c = bv$. We have $c^n = b^n v^n = \varepsilon^{-1} c^n v^n$; hence $v^n = \varepsilon$. This implies that $v \in S_M^\times$, and thus $I_M = cS_M = bvS_M = bS_M$. Therefore, $I_Q = bS_Q$ for all $Q \in \max(S)$; hence $I = bS$. $\qquad\square$

Let H be a monoid. So far we said little about the additional property that popped up in Theorem 3.6(3) (i.e., that for every $a \in H^\bullet$, $\mathcal{A}(\llbracket a \rrbracket) \subseteq \mathcal{D}_k(a)$ for some $k \in \mathbb{N}$). Note that this additional property is equivalent to the notion of being pseudo-IDPF introduced in [24]. Since we are interested in studying monadically Krull monoids and their specializations, we investigate how to control the r-ideal class group of an r-SP-monoid to obtain this additional property. This is reasonable since there are non-trivial situations using the construction in Proposition 6.5 where SP-domains that are BF-domains can show up (as pointed out before). Moreover, Proposition 6.7 indicates that the class group of domains in this construction can behave nicely. If G is an abelian group, then let $\exp(G)$ be the exponent of G (i.e., if 1 is the identity of G, then $\exp(G) = \inf(\{n \in \mathbb{N} \mid x^n = 1 \text{ for all } x \in G\}))$. The group G is called bounded if $\exp(G) < \infty$.

Proposition 6.8. *Let H be a monoid and r a finitary ideal system on H such that H is an r-SP-monoid:*

1. *If $\mathcal{C}_r(H)$ is finite, then for all $a \in H^\bullet$, $\mathcal{A}(\llbracket a \rrbracket) \subseteq \mathcal{D}_k(a)$ for some $k \in \mathbb{N}$.*
2. *If H is an FF-monoid and $\mathcal{C}_r(H)$ is bounded, then for all $a \in H^\bullet$, $\mathcal{A}(\llbracket a \rrbracket) \subseteq \mathcal{D}_k(a)$ for some $k \in \mathbb{N}$.*
3. *If H is r-Prüfer and $\mathcal{C}_r(H)$ is bounded, then for all $a \in H^\bullet$, $\mathcal{A}(\llbracket a \rrbracket) \subseteq \mathcal{D}_k(a)$ for some $k \in \mathbb{N}$.*

Proof. 1. Let $a \in H^\bullet$. Set $k = |\mathcal{C}_r(H)|$. We prove that $\mathcal{A}(\llbracket a \rrbracket) \subseteq \mathcal{D}_k(a)$. Let $u \in \mathcal{A}(\llbracket a \rrbracket)$. There are some $l, s \in \mathbb{N}$ and some sequence $(I_i)_{i=1}^s$ of proper radical r-ideals of H such that $a^l \in uH = (\prod_{i=1}^s I_i)_r$. Observe that $a^l \in I_i$ for all $i \in [1, s]$; hence $a \in I_i$ for all $i \in [1, s]$. This implies that $a^s \in uH$. If $s \le k$, then $a^k \in uH$, and thus $u \in \mathcal{D}_k(a)$. Now let $s > k$. There is some $\emptyset \ne E \subseteq [1, s]$ such that $|E| \le k$ and $(\prod_{i \in E} I_i)_r$ is principal. Since $uH = (\prod_{i=1}^s I_i)_r \subseteq (\prod_{i \in E} I_i)_r \subsetneq H$, we have $uH = (\prod_{i \in E} I_i)_r$. Therefore, $a^k \in a^{|E|} H \subseteq (\prod_{i \in E} I_i)_r = uH$. Consequently, $u \in \mathcal{D}_k(a)$.

2. Let H be an FF-monoid, $\mathcal{C}_r(H)$ bounded, and $a \in H^\bullet$. Set $l = \exp(\mathcal{C}_r(H))$, $\mathcal{M} = \{I \mid I \text{ is an } r\text{-invertible radical } r\text{-ideal of } H, a \in I\}$ and $\mathcal{N} = \{bH \mid b \in H, a^l \in bH\}$. Let $f : \mathcal{M} \to \mathcal{N}$ be defined by $f(I) = (I^l)_r$ for all $I \in \mathcal{M}$.

If $I \in \mathcal{M}$, then there is some $b \in H$ such that $(I^l)_r = bH$. Set $J = aI^{-1}$. Then $J \in \mathcal{I}_r(H)$ and $aH = (IJ)_r$. This implies that $a^l \in a^l H = (I^l J^l)_r = b(J^l)_r \subseteq bH$, and thus f is well defined. Now let $I, J \in \mathcal{M}$ be such that $f(I) = f(J)$. It follows that $I = \sqrt{(I^l)_r} = \sqrt{f(I)} = \sqrt{f(J)} = \sqrt{(J^l)_r} = J$. Therefore, f is injective. Since H is an FF-monoid we have $|\mathcal{M}| \leq |\mathcal{N}| < \infty$. Set $k = l|\mathcal{M}|$. We show that $\mathcal{A}([\![a]\!]) \subseteq \mathcal{D}_k(a)$. Let $u \in \mathcal{A}([\![a]\!])$. There are some $m, n \in \mathbb{N}$, some sequence $(\alpha_i)_{i=1}^n \in \mathbb{N}^{[1,n]}$ and some sequence $(I_i)_{i=1}^n$ of distinct proper radical r-ideals of H such that $a^m \in uH = (\prod_{i=1}^n I_i^{\alpha_i})_r$. Note that $I_i \in \mathcal{M}$ for all $i \in [1,n]$, hence $n \leq |\mathcal{M}|$. If $\alpha_j > l$ for some $j \in [1,n]$, then $uH \subseteq (I_j^{\alpha_j})_r \subseteq (I_j^l)_r \subseteq I \subsetneq H$, and thus $uH = (I_j^l)_r = (I_j^{\alpha_j})_r$ which implies that $I_j = H$, a contradiction. Therefore, $a^l \in (I_j^{\alpha_j})_r$ for all $j \in [1,n]$. It follows that $a^k \in (\prod_{i=1}^n I_i^{\alpha_i})_r = uH$, hence $u \in \mathcal{D}_k(a)$.

3. Let H be an r-Prüfer monoid, $\mathcal{C}_r(H)$ bounded and $a \in H^{\bullet}$. Set $m = \exp(\mathcal{C}_r(H))$ and $k = 2m$. It is sufficient to show that $\mathcal{A}([\![a]\!]) \subseteq \mathcal{D}_k(a)$. Let $u \in \mathcal{A}([\![a]\!])$. By [29, Proposition 3.9], [29, Theorem 3.13] and [29, Theorem 3.3.2] there are some $l \in \mathbb{N}$ and some ascending sequence $(I_i)_{i=1}^l$ of proper radical r-ideals of H such that $uH = (\prod_{i=1}^l I_i)_r$. Set $F = I_l$. Then F is r-invertible. Clearly, $uH \subseteq (F^l)_r$ and there is some $b \in H \setminus H^{\times}$ such that $(F^m)_r = bH$. Assume that $l > k$. We have $uH \subseteq (F^l)_r \subseteq (F^k)_r = b^2 H \subseteq bH$. Since $u \in \mathcal{A}(H)$, this implies that $uH = b^2 H = bH$, hence $b \in H^{\times}$, a contradiction. Therefore, $l \leq k$, and thus $a^k \in a^l H \subseteq (\prod_{i=1}^l I_i)_r = uH$. Consequently, $u \in \mathcal{D}_k(a)$.   □

Acknowledgements We want to thank A. Geroldinger, F. Halter-Koch, F. Kainrath and the referee for their comments and suggestions. This work was supported by the Austrian Science Fund FWF, Project Number P21576-N18.

References

1. D.D. Anderson, B. Mullins, Finite factorization domains. Proc. Am. Math. Soc. **124**, 389–396 (1996)
2. D.D. Anderson, D.F. Anderson, M. Zafrullah, Factorization in integral domains. J. Pure Appl. Algebra **69**, 1–19 (1990)
3. J.T. Arnold, R. Matsuda, An almost Krull domain with divisorial height one primes. Canad. Math. Bull. **29**, 50–53 (1986)
4. N.R. Baeth, A. Geroldinger, Monoids of Modules and Arithmetic of Direct-Sum Decompositions. Pacific J. Math. (to appear)
5. N.R. Baeth, R. Wiegand, Factorization theory and decompositions of modules. Am. Math. Monthly **120**, 3–34 (2013)
6. N.R. Baeth, A. Geroldinger, D.J. Grynkiewicz, D. Smertnig, A Semigroup-Theoretical View of Direct-Sum Decompositions and Associated Combinatorial Problems. J. Algebra Appl. (to appear)
7. H.S. Butts, R.W. Yeagy, Finite bases for integral closures. J. Reine Angew. Math. **282**, 114–125 (1976)

8. P.J. Cahen, J.L. Chabert, *Integer-Valued Polynomials*. Mathematical Surveys and Monographs, vol. 48 (American Mathematical Society, Providence, 1997)

9. J. Coykendall, P. Malcolmson, F. Okoh, On fragility of generalizations of factoriality. Comm. Algebra **41**, 3355–3375 (2013)

10. P. Etingof, P. Malcolmson, F. Okoh, Root extensions and factorization in affine domains. Canad. Math. Bull. **53**, 247–255 (2010)

11. A. Facchini, Direct sum decompositions of modules, semilocal endomorphism rings, and Krull monoids. J. Algebra **256**, 280–307 (2002)

12. A. Facchini, Direct-sum decompositions of modules with semilocal endomorphism rings. Bull. Math. Sci. **2**, 225–279 (2012)

13. A. Facchini, R. Wiegand, Direct-sum decompositions of modules with semilocal endomorphism rings. J. Algebra **274**, 689–707 (2004)

14. S. Frisch, A construction of integer-valued polynomials with prescribed sets of lengths of factorizations. Monatsh. Math. **171**, 341–350 (2013)

15. A. Geroldinger, F. Halter-Koch, *Non-Unique Factorizations: Algebraic, Combinatorial and Analytic Theory*. Pure and Applied Mathematics (Chapman and Hall/CRC, Boca Raton, 2006)

16. A. Geroldinger, F. Halter-Koch, W. Hassler, F. Kainrath, Finitary monoids. Semigroup Forum **67**, 1–21 (2003)

17. R. Gilmer, *Commutative Semigroup Rings* (University of Chicago Press, Chicago, 1984)

18. R. Gilmer, *Multiplicative Ideal Theory*. Queen's Papers in Pure and Applied Mathematics, vol. 90 (Queen's University, Kingston, 1992)

19. R. Gilmer, W.J. Heinzer, Overrings of Prüfer domains. II. J. Algebra **7**, 281–302 (1967)

20. A. Grams, Atomic rings and the ascending chain condition for principal ideals. Proc. Cambridge Philos. Soc. **75**, 321–329 (1974)

21. A. Grams, H. Warner, Irreducible divisors in domains of finite character. Duke Math. J. **42**, 271–284 (1975)

22. F. Halter-Koch, *Ideal Systems. An Introduction to Multiplicative Ideal Theory* (Marcel, New York, 1998)

23. W. Hassler, Arithmetic of weakly Krull domains. Comm. Algebra **32**, 955–968 (2004)

24. P. Malcolmson, F. Okoh, A class of integral domains between factorial domains and IDF-domains. Houston J. Math. **32**, 399–421 (2006)

25. P. Malcolmson, F. Okoh, Factorization in subalgebras of the polynomial algebra. Houston J. Math. **35**, 991–1012 (2009)

26. P. Malcolmson, F. Okoh, Polynomial extensions of IDF-domains and of IDPF-domains. Proc. Am. Math. Soc. **137**, 431–437 (2009)

27. B. Olberding, Factorization into radical ideals, in *Arithmetical Properties of Commutative Rings and Monoids*. Lecture Notes in Pure and Applied Mathematics, vol. 241 (Chapman and Hall/CRC, Boca Raton, 2005), pp. 363–377

28. E.M. Pirtle, Families of valuations and semigroups of fractionary ideal classes. Trans. Am. Math. Soc. **144**, 427–439 (1969)

29. A. Reinhart, Radical factorial monoids and domains. Ann. Sci. Math. Québec **36**, 193–229 (2012)

30. A. Reinhart, On integral domains that are C-monoids. Houston J. Math. **39**, 1095–1116 (2013)

Integral Closure

Irena Swanson

Abstract Since 2006, when the book on integral closures with Huneke and Swanson (Integral Closure of Ideals, Rings, and Modules. Cambridge University Press, Cambridge, 2006) was published, there has been more development in the area. This chapter is an attempt at catching up with that development as well as to fill in a few omissions. Some topics are worked out in detail whereas others are only outlined or mentioned.

Keywords Integral closure • Rees valuations • Computing integral closure • Lipman–Sathaye theorem • Multiplicity • j-multiplicity • Epsilon multiplicity • Monomial ideals • Goto numbers

Subject Classifications: 13B22; Secondary: 13P05, 13P25, 13H15

1 Rees Valuations

This section is an update of Chap. 10 of [19].

The constructions in Chap. 10 show that the set of the Rees valuations of I is contained in the union of the sets of the Rees valuations of I modulo each minimal prime ideal; the following shows that the other inclusion holds as well.

Proposition 1.1 ([28]). *Let R be a Noetherian ring and I an ideal in R not contained in any minimal prime ideal. For each $P \in \mathrm{Min}\,(R)$, let T_P be the set*

I. Swanson (✉)
Reed College, 3203 SE Woodstock Blvd, Portland, OR 97202, USA
e-mail: iswanson@reed.edu

M. Fontana et al. (eds.), *Commutative Algebra: Recent Advances in Commutative Rings,*   331
Integer-Valued Polynomials, and Polynomial Functions, DOI 10.1007/978-1-4939-0925-4__19,
© Springer Science+Business Media New York 2014

of the Rees valuations of $I(R/P)$. By abuse of notation, these valuations are also valuations on R, with $\{r \in R : v(r) = \infty\} = P$. Then $\cup_P T_P$ is the set of the Rees valuations of I.

Proof. The standard proofs of the existence of the Rees valuations show that the set of the Rees valuations of I is contained in $\cup_P T_P$. We need to prove that no valuation in $\cup_P T_P$ is redundant.

Let $Q \in \text{Min}(R)$ and $v \in T_Q$. By the minimality of the Rees valuations of $I(R/Q)$, there exist $n \in \mathbb{N}$ and $r \in R$ such that for all $w \in T_Q \setminus \{v\}$, $w(r) \geq nw(I)$, yet $r \notin \overline{I^n}(R/Q)$ (i.e., $v(r) < nv(I)$). Let r' be an element of R that lies in precisely those minimal prime ideals that do not contain r. Then $r + r'$ is not contained in any minimal prime ideal, for all $w \in T_Q \setminus \{v\}$, $w(r + r') \geq nw(I)$, and $v(r+r') < nv(I)$). Let J' be the intersection of all the minimal primes other than Q, let J'' be the intersection of all the centers of $w \in T_Q$, and let $s \in J' \cap J'' \setminus Q$. By assumption on r, there exists a positive integer k such that for all $w \in T_Q \setminus \{v\}$,

$$\frac{v(s)}{v(I)} - \frac{w(s)}{w(I)} + 1 < k \left(\frac{w(r + r')}{w(I)} - \frac{v(r + r')}{v(I)} \right).$$

Note that for all $w \in \cup_{P \neq Q} S_P$, $w(s) = \infty$. Thus for all $w \in \cup_P T_P \setminus \{v\}$, $\frac{v(s)}{v(I)} - \frac{w(s)}{w(I)} + 1 < k \left(\frac{w(r+r')}{w(I)} - \frac{v(r+r')}{v(I)} \right)$. Then with $m = \lfloor \frac{v(s(r+r')^k)}{v(I)} \rfloor$, $s(r + r')^k \notin \overline{I^{m+1}}$, yet for all $w \in \cup_P T_P \setminus \{v\}$, $w(s(r + r')^k) \geq (m + 1)w(I)$. This proves that v is not redundant. $\square$

Another basic new result is due to Katz and Validashti [21].

Theorem 1.2 (Katz and Validashti [21, Proposition 3.1]). *Let (R, m) be a Noetherian local ring and $I \subseteq R$ an ideal that is not contained in any minimal prime ideal. Then $\ell(I) = \dim R$ if and only if some Rees valuation v of I has center on m and $\text{trdeg}_{R/m}(\kappa(m_v)) = \dim R - 1$.*

Proof. Suppose that $\ell(I) = \dim R$. By Proposition 5.1.7 in [19], there exists a minimal prime ideal P in R such that $\ell(I(R/P)) = \ell(I) = \dim R$. Then by Burch's theorem [19, Proposition 5.4.7], m/P is associated to $\overline{I^n}(R/P)$ for all large n, so that by Discussion 10.1.3 in [19], m/P is the center of some Rees valuation of $I(R/P)$. Hence by Proposition 1.1, m is the center of some Rees valuation of I.

Conversely, suppose that m is the center of some Rees valuation of I and that $\text{trdeg}_{R/m}(\kappa(m_v)) = \dim R - 1$. Let $P = \{r \in R : v(r) = \infty\}$. By the definition of the Rees valuations, P is a minimal prime ideal in R. By [19, Theorem 6.6.7], $\dim R = 1 + \text{trdeg}_{R/m}(\kappa(m_v)) = 1 + \text{trdeg}_{(R/P)/(m/P)}(\kappa(m_v)) \leq \dim(R/P)$, so that $\dim(R/P) = \dim R$. If we can show that the theorem holds for domains, then $\ell(I(R/P)) = \dim(R/P) = \dim R$, and since $\dim R \geq \ell(I) \geq \ell(I(R/P))$, it follows that $\ell(I) = \dim R$. Thus it suffices to prove this direction in case R is a domain. By [19, Proposition 10.4.3], v is the contraction of a valuation w on $\hat{R}$, actually on $\hat{R}/Q$, where Q is a minimal prime ideal in $\hat{R}$ with $\dim(\hat{R}/Q) = \mathrm{trdeg}_{R/m}(\kappa(m_v)) + 1 = \dim R$. Furthermore, this w is a Rees valuation of $I\hat{R}$ and of $I\hat{R}/Q$. Since $\hat{R}/Q$ is formally equidimensional, by [19, Theorem 10.4.2], we have that $\ell(I(\hat{R}/Q)) = \dim(\hat{R}/Q)$. But then $\dim R \geq \ell(I) \geq \ell(I(\hat{R}/Q)) = \dim(\hat{R}/Q) = \dim R$, so we are done.   □

The following in particular proves that if (R, m) is a Noetherian local ring of positive dimension, the number of the Rees valuations of an ideal I in R is the same as the number of the Rees valuations of $I\hat{R}$.

Proposition 1.3 (Katz and Validashti [21, Lemma 5.1(b)]). *Let (R, m) be a Noetherian local ring with completion $\hat{R}$. Let $Q \in \mathrm{Min}\ \hat{R}$ be such that the integral closure of $\hat{R}/Q$ is a discrete valuation ring. Let W be this valuation ring. Then the intersection of W with $\kappa(Q \cap R)$ is the localization of the integral closure of $R/Q \cap R$ at the contraction of the maximal ideal of W.*

Proof. The flow of the proof below, as well as of the proof in [21], uses some techniques from [24]. Neither the conclusion nor the hypotheses change if we replace R with $R/Q \cap R$, so that we may assume that R is a Noetherian local domain and $Q \cap R = (0)$. By [19, Proposition 6.4.7], $V = W \cap \kappa(0_R)$ is a discrete valuation ring, and it contains the integral closure $\overline{R}$ of R. Let m_W be the maximal ideal in W, and $M = m_W \cap \overline{R}$.

If Q is the only associated prime ideal of $\hat{R}$, then $\dim \overline{R} = \dim R = \dim \hat{R} = 1$. Then M is a maximal ideal in $\overline{R}$, so of height 1, and by the Mori–Nagata Theorem [19, Theorem 4.10.5], $\overline{R}_M \subseteq V$ are both discrete valuation rings; hence $\overline{R}_M = V$. This proves the proposition in case Q is the only associated prime of $\hat{R}$.

For the rest of this proof we assume that there are other associated primes of $\hat{R}$. Let q be the Q-primary component of $0\hat{R}$. By prime avoidance there exists d' in the intersection of all other primary components of $0\hat{R}$ that is not in Q. Since the height of $m\hat{R}/Q$ is 1, necessarily $q + d'\hat{R}$ is $m\hat{R}$-primary. Thus $(q + d'\hat{R}) \cap R$ is m-primary. Let b be any non-zero element of this intersection. Then b is a non-zerodivisor in R and hence also in $\hat{R}$. We can write $b = a + sd'$ for some $a \in q$ and $s \in \hat{R}$. Let $d = sd'$. Then $b = a + d$. Since $Q \cap R = (0)$, it follows that $d \neq 0$.

If $d = rb$ for some $r \in \hat{R}$, then $b(1 - r) = b - d = a \in q$, and since b is a non-zerodivisor in $\hat{R}$, necessarily $1 - r \in q$, so that r is a unit in $\hat{R}$. Thus $d = rb$ is also a non-zerodivisor in $\hat{R}$, which is a contradiction. So $d \notin b\hat{R}$.

Since there are no prime ideals strictly between Q and $m\hat{R}$, any other prime ideal P in $\hat{R}$ does not contain Q. Since d is annihilated by a power of each element in Q, it follows that d is contained in every P-primary component. In particular, if m is not associated to R/bR, then $b\hat{R} = (bR :_R m^\infty)\hat{R} = (b\hat{R} :_{\hat{R}} m^\infty)$ contains d, which contradicts the previous paragraph. Thus m is associated to bR.

We write $bR = I \cap J$, where J is an m-primary ideal and I is the intersection of all other primary components. Since d is in every P-primary component, where P is a prime ideal in $\hat{R}$ different from Q and $m\hat{R}$, then $d \in I\hat{R}$. As J is m-primary, we have $(I\hat{R} + J\hat{R})/J\hat{R} = (I + J)\hat{R}/J\hat{R} = (I + J)/J$, so that we can write

$d = i - j$ for some $i \in I$ and $j \in J$. Then $i - d = j \in J\hat{R}$, and both i and d are in $I\hat{R}$, so that $i - d \in I\hat{R} \cap J\hat{R} = b\hat{R}$. Thus $(b, d)\hat{R} = (b, i)\hat{R}$. Since $b - d = a \in q$, then $(b - d)d \in qd = (0)$. It follows that

$$i^2 = (i - d)^2 + 2(i - d)d + d^2 \in (b^2, bd, d^2)\hat{R} = b(b, d)\hat{R} = b(b, i)\hat{R},$$

so that $i^2 \in b(b, i)\hat{R} \cap R = b(b, i)$, and so i/b is integral over R.

We now let $D = R[\frac{i}{b}]$. Since D is module-finite over R, the completion of D (in the m- or mD-adic topology) is $\hat{D} = \hat{R}[\frac{i}{b}]$. Let $\hat{K}$ be the total ring of fractions of $\hat{R}$. Since $i - d \in \hat{R}$, then $\hat{R} \subseteq \hat{D} = \hat{R}[\frac{i}{b}] = \hat{R}[\frac{d}{b}] \subseteq \overline{\hat{R}}$. Since $\hat{K}$ is the localization of both $\hat{R}$ and $\hat{D}$ at the set of non-zerodivisors in $\hat{R}$, there is a natural one-to-one correspondence between associated primes of $\hat{R}$ and $\hat{D}$: an associated prime ideal P in $\hat{R}$ corresponds to $P\hat{K} \cap \hat{D}$. Also, by integral dependence, $\dim(\hat{R}/P) = \dim(\hat{D}/P\hat{K} \cap \hat{D})$. Since d/b is integral over $\hat{R}/Q$, W contains $\hat{D}/Q\hat{K} \cap \hat{D}$. Let $\hat{P}$ be the prime ideal in $\hat{D}$ which is the center of W modulo $Q\hat{K} \cap \hat{D}$. In particular, $\hat{P}$ properly contains $Q\hat{K} \cap \hat{D}$. Note that $1 - d/b = (b - d)/b \in Q\hat{K} \cap \hat{D}$ and that d/b is contained in all the other associated primes of $\hat{D}$. Thus $Q\hat{K} \cap \hat{D}$ is the only minimal prime ideal in $\hat{D}$ that is contained in $\hat{P}$. Hence by the established dimension equalities, $\hat{P}$ is a maximal ideal in $\hat{D}$ and the height of $\hat{P}$ is 1. Set $P = \hat{P} \cap D$. Since $\hat{D}$ is the completion of D and $\hat{P}$ is a maximal ideal in $\hat{D}$, $\hat{P} = P\hat{D}$. Hence P is a height one maximal ideal in D.

Recall that $M = m_W \cap \overline{R}$. Necessarily $M \cap D = P$, so that as $R \subseteq D = R[\frac{i}{b}] \subseteq \overline{R}$ are integral extensions, M is a maximal ideal. Furthermore, $D_P \subseteq \overline{R}_{D\backslash P} \subseteq \overline{R}_M \subseteq V$, the first inclusion is an integral extension, D_P is one-dimensional, so that $\overline{R}_{D\backslash P}$ is one-dimensional and integrally closed, hence a discrete valuation ring. Thus $\overline{R}_{D\backslash P} = \overline{R}_M = V$.   □

Theorem 1.4 (Katz and Validashti [21, Theorem 5.3]). *Let I be an ideal in a Noetherian local ring (R, m) that is not contained in any minimal prime ideal. Let w be a Rees valuation of $I\hat{R}$ with center $m\hat{R}$, and let Q be the corresponding minimal prime ideal in $\hat{R}$ such that w is a valuation on $\kappa(Q)$. Then w restricted to $\kappa(Q \cap R)$ is a Rees valuation of I with center m. The function*

$$w \mapsto w|_{\kappa(\{r \in R: w(r) = \infty\})}$$

from the Rees valuations of $I\hat{R}$ with center on $m\hat{R}$ to the Rees valuations of I with center on m is a one-to-one and onto function.

Proof. The function is onto by [19, Proposition 10.4.3]. By faithful flatness, $q = Q \cap R$ is a minimal prime ideal in R. Set $R' = R/q$. Since Q is a minimal prime ideal in $\widehat{R'} = \hat{R}/q\hat{R}$, then by Proposition 1.1, w is a Rees valuation of $I\widehat{R'}$. If we can prove that the restriction of w to R' is a Rees valuation of IR', then by Proposition 1.1 the declared function is well defined. So by replacing R with R' we may assume that R is a domain, and it remains to prove that the function has the designated codomain and that the function is one-to-one.

Let $e = \dim(\hat{R}/Q)$. Since $\hat{R}$ is formally equidimensional, by [19, Theorem 10.4.2], $\ell(I\hat{R}/Q) = e$. Since the set of the Rees valuations of an ideal is the same as the set of the Rees valuations of any power of that ideal, by replacing I by its power we may assume by [19, Proposition 8.3.8] that $I\hat{R}/Q$ has a minimal reduction generated by e elements. By vector space avoidance, since reductions correspond to subspaces of I/mI, we may assume that $I = (a_1, \ldots, a_s)$ such that $(a_1, \ldots, a_e)\hat{R}/Q$ is a minimal reduction of $I\hat{R}/Q$ and even that the images of $a_1, \ldots, a_e$ in the fiber ring of I in R are algebraically independent. Since $(a_1, \ldots, a_e)$ is a reduction of I, $w(I) = w(a_1, \ldots, a_e)$. Let A be any of $R, \hat{R}, \hat{R}/Q$. By [19, Corollary 8.3.5], the fiber ring of $(a_1, \ldots, a_e)A$ is a polynomial ring in the images of $a_1, \ldots, a_e$ over A/mA. Thus the image of m in the Rees algebra $A[(a_1, \ldots, a_e)t]$ is a prime ideal. By possibly reindexing we may assume that $w(a_1, \ldots, a_e) = w(a_1)$. By analytic independence, $a_1 t \notin mA[(a_1, \ldots, a_e)t]$, so that in the localization $A[(a_1, \ldots, a_e)t]_{a_1 t}$, the image of m remains a prime ideal. The homogeneous component of this localization of t-degree 0 is the ring $A[\frac{a_2}{a_1}, \ldots, \frac{a_e}{a_1}]$, and necessarily the image of m in that is still a prime ideal. Thus we can define $U_A = A[\frac{a_2}{a_1}, \ldots, \frac{a_e}{a_1}]_{mA[\frac{a_2}{a_1}, \ldots, \frac{a_e}{a_1}]}$. When $A = \hat{R}/Q$, $a_1, \ldots, a_e$ generate an $m\hat{R}/Q$-primary ideal, so that the prime ideal $m(\hat{R}/Q)[\frac{a_2}{a_1}, \ldots, \frac{a_e}{a_1}]$ is minimal over the principal ideal generated by a_1, so that $U_{\hat{R}/Q}$ has dimension 1. By the set-up, W is the localization of the integral closure of $U_{\hat{R}/Q}$. In particular, the transcendence degree of $\kappa(m_W)$ over $\kappa(mU_{\hat{R}/Q})$ is zero. Other consequences are that W is a Rees valuation of $(a_1, \ldots, a_e)U_{\hat{R}/Q}$ and thus also of $IU_{\hat{R}/Q}$. Since $U_{\hat{R}/Q}$ is the quotient of $U_{\hat{R}}$ by the minimal prime ideal $Q\hat{K} \cap U_{\hat{R}}$, where $\hat{K}$ is the total ring of fractions of $\hat{R}$, it follows that W is a Rees valuation of $(a_1, \ldots, a_e)U_{\hat{R}}$ and of $IU_{\hat{R}}$, with the transcendence degree of $\kappa(m_W)$ over $\kappa(mU_{\hat{R}})$ being zero. Observe that the completions of U_R and $U_{\hat{R}}$ are identical. By [19, Proposition 10.4.3], there exist a minimal prime ideal $\hat{Q}$ in $\widehat{U_R}$ and a valuation ring $\hat{W}$ in the field of fractions of $\widehat{U_R}/\hat{Q}$ such that $\hat{W}$ is a Rees valuation of $(a_1, \ldots, a_e)\widehat{U_R}$ such that W is the contraction of $\hat{W}$ and such that $\dim(\widehat{U_R}/\hat{Q}) - 1$ equals the transcendence degree of $\kappa(m_W)$ over $\kappa(mU_{\hat{R}})$, namely zero. Thus $\hat{Q}$ is a minimal prime ideal in $\widehat{U_R}$ such that $\dim(\widehat{U_R}/\hat{Q}) = 1$. Since $\widehat{U_R}/\hat{Q}$ is complete, its integral closure is local, so that $\hat{W}$ is the integral closure of $\widehat{U_R}/\hat{Q}$. Hence by Proposition 1.3, the intersection of $\hat{W}$ and hence of W with K is the localization of the integral closure of U_R. In particular, V is the localization of the integral closure of $R[\frac{a_2}{a_1}, \ldots, \frac{a_e}{a_1}]$ at a prime ideal necessarily containing a_1, so that V is a Rees valuation of $(a_1, \ldots, a_e)$. Furthermore, since $w(I) = w(a_1)$,

$$U_R = R\left[\frac{a_2}{a_1}, \ldots, \frac{a_e}{a_1}\right] \subseteq U'_R = R\left[\frac{a_2}{a_1}, \ldots, \frac{a_s}{a_1}\right] \subseteq V,$$

so that

$$V = (\overline{U_R})_{m_W \cap \overline{U_R}} \subseteq (\overline{U'_R})_{m_W \cap \overline{U'_R}} \subseteq V,$$

and so by construction of the Rees valuations, V is a Rees valuation of I as well. $\square$

2 The Lipman–Sathaye Theorem for Reduced Rings

In Sect. 12.3 of [19] we proved a version of the Lipman–Sathaye theorem for domains. This section here proves a generalized version for reduced equidimensional rings. In places the exposition here is almost a verbatim repetition of Sect. 12.3, with parts taken from Hochster's generalization in [17]. One motivation for working out the details of the generalization is that in the literature there are some faulty applications of the domain case to reduced rings, with claims that $J_{S/R}\overline{S} \subseteq S$ whereas only $J_{S/R}\overline{S} \subseteq \prod_{P \in \mathrm{Min}\, S} S/P$ can be concluded from the domain case.

Throughout this section, let R be a universally catenary Cohen–Macaulay Noetherian domain of positive dimension, and let K be its field of fractions. Let $X_1, \ldots, X_n$ be variables over R, $T = R[X_1, \ldots, X_n]$, and $S = T/I$ for some radical ideal I, all of whose minimal primes have the same height n. We assume that $R \subseteq S$, that all non-zero elements of R are non-zerodivisors on S, and that each direct summand of the total ring of fractions L of S is finite and separable over K. (Hochster [17] calls the last property *generically étale*.) By [19, Proposition 4.4.4], the Jacobian ideal $J_{S/R}$ of S over R is well defined. If $g_1, \ldots, g_n \in I$, we let g_X denote the determinant of the matrix whose (i, j)th entry is $\frac{\partial g_i}{\partial X_j}$. Thus $J_{S/R} = \{g_X \mid g_1, \ldots, g_n \in I\}S$.

Remark 2.1. The goal of this section is to prove that

$$(\overline{S} :_L J_{\overline{S}/R}) \subseteq S :_L J_{S/R}.$$

The following reduction is from [17]. Let $t_1, \ldots, t_k$ be variables over T, where k is very large. Let U be the subset of $R[t_1, \ldots, t_k]$ generated by all polynomials in $t_1, \ldots, t_k$ whose coefficients generate the unit ideal in R. Then U is a multiplicatively closed set. Then $R' = U^{-1}R[t_1, \ldots, t_k]$ is a faithfully flat extension of R that is a universally catenary Cohen–Macaulay Noetherian domain of positive dimension and with field of fractions $K' = K(t_1, \ldots, t_k)$. The polynomial ring $T' = R'[X_1, \ldots, X_n]$ over R' is faithfully flat over T; for any ideal H in T, the primary decomposition of H extends to a primary decomposition of HT', so that in particular IT' is reduced and all prime ideals in T' minimal over IT' have height n. Furthermore, $S' = T'/IT' = (T/I) \otimes_R R' = S \otimes_R R'$ is a reduced ring. The inclusion $R' \subseteq S'$ still holds, and every non-zero element of R' is a non-zerodivisor on S'. Every direct summand of the total ring of fractions L' of S' is finite and separable over K'. If we can prove the displayed formula for S', L', R' in place of S, L, R, then by the structure of these extensions,

$$(\overline{S} :_L J_{\overline{S}/R}) \subseteq (\overline{S'} :_{L'} J_{\overline{S'}/R'}) \cap L \subseteq (S' :_{L'} J_{S'/R'}) \cap L \subseteq S :_L J_{S/R}.$$

Thus it suffices to prove the goal for R', T', S'. Let $I = (a_1, \ldots, a_l)$. For $i = 1, \ldots, l$, set $f_i = \sum_{j=1}^{l} u_{ij} a_j$, where u_{ij} are distinct elements of $\{t_1, \ldots, t_k\}$ (so $k \geq l^2$). Then by genericity, $(f_1, \ldots, f_l) T' = I T'$, and any n of $f_1, \ldots, f_l$ form a regular sequence in T'. Furthermore, let Q be a prime ideal in T' minimal over $I T'$. Since all elements of R' are non-zerodivisors on S', it follows that $Q \cap R' = 0$, so that $T_{Q'}$ is a localization of $K'[X_1, \ldots, X_n]$ at a prime ideal of height n, thus at a maximal ideal. Thus $Q T'_Q$ is generated by n elements (even if K is not infinite, e.g., Exercise 2.28 in [19]). But $Q T'_Q = I T'_Q$, so that again by genericity, any n of $f_1, \ldots, f_l$ generate $I T'_Q$.

Proposition 2.2. *1. $S \otimes_R K$ is a direct sum of fields and for every prime ideal Q in T minimal over I, $I_Q = Q_Q$ has height n and T_Q is a regular local ring of dimension n.*

2. Let $Q \in \operatorname{Spec} T$ with $I \subseteq Q$. If $g_1, \ldots, g_n \in I$ and $(g_1, \ldots, g_n) : I \nsubseteq Q$, then $(J_{S/R})_Q = g_X S_Q$.

3. If $g_1, \ldots, g_n \in I$, then g_X is not in a prime ideal Q minimal over I if and only if $(g_1, \ldots, g_n) : I \nsubseteq Q$.

Proof. Note that $S \otimes_R K$ is a localization of S at the set of all non-zero elements of R, so that $K \subseteq S \otimes_R K \subseteq L$. Since L is module-finite over K, so is $S \otimes_R K$, so that $S \otimes_R K$ is a reduced zero-dimensional ring, hence a direct sum of fields. Necessarily $S \otimes_R K = L$. A prime ideal Q in T minimal over I corresponds to a minimal prime ideal in S, and since S_Q is a field, necessarily $I_Q = Q_Q$. By assumption this has height n. Since non-zero elements of R are non-zerodivisors on S, necessarily $Q \cap R = (0)$, so that T_Q is a localization of $K[X_1, \ldots, X_n]$ and is hence regular.

To prove (2), note that $(J_{S/R})_Q = J_{S_Q/R}$ by Corollary 4.4.5 in [19]. By assumption $(g_1, \ldots, g_n) : I \nsubseteq Q$ it follows that $I_Q = (g_1, \ldots, g_n)_Q$, so that by independence of Jacobian ideals of the presentation (Proposition 4.4.4 in [19]), $(J_{S/R})_Q = g_X S_Q$.

Suppose that $(g_1, \ldots, g_n) : I$ is not contained in Q. Then by (2), $(J_{S/R})_Q = g_X S_Q$. By the Jacobian criterion [19, Theorem 4.4.9], since S_Q is a domain, $(J_{S/R})_Q$ is non-zero, so that g_X is not contained in Q. This proves one direction in (3). Conversely, suppose that g_X is not contained in a prime ideal Q minimal over I. Since $\operatorname{ht}(g_1, \ldots, g_n) \leq n$, the Jacobian ideal of $T/(g_1, \ldots, g_n)$ contains $\det\left(\frac{\partial g_i}{\partial X_j}\right)$. Thus by the Jacobian criterion [19, Theorem 4.4.9], $(T/(g_1, \ldots, g_n))_Q$ is regular. However, $(T/(g_1, \ldots, g_n))_Q$ is the localization of $K[X_1, \ldots, X_n]/(g_1, \ldots, g_n)$ at the image of Q, so that as $K[X_1, \ldots, X_n]$ is regular, necessarily $(g_1, \ldots, g_n)_Q$ is generated by part of a minimal generating set of Q_Q. But then if the height of $(g_1, \ldots, g_n)_Q$ is strictly smaller than n, we have that $g_X = \det\left(\frac{\partial g_i}{\partial X_j}\right)$ has zero image in S_Q, so that $g_X \in Q$, which is a contradiction. This proves (3). $\square$

Proposition 2.3. *With notation as in Remark 2.1, the Jacobian ideal $J_{S'/R'}$ is generated by elements g_X such that $g_1, \ldots, g_n$ is a regular sequence in $I T'$ and such that g_X is not in any prime ideal minimal over $I T'$.*

Proof. As in Remark 2.1, $IT' = (f_1, \ldots, f_l)T'$, and any n of $f_1, \ldots, f_l$ form a regular sequence in IT' and generate IT' generically (after localizing at each prime ideal minimal over IT'). Certainly $J_{S'/R'}$ is generated by g_X as $g_1, \ldots, g_n$ vary over n elements of $\{f_1, \ldots, f_l\}$. Since locally at each prime ideal Q minimal over IT', $(g_1, \ldots, g_n)_Q = IT'_Q$, it follows that $(g_1, \ldots, g_n) : I \not\subseteq Q$, so that by Proposition 2.2(3), g_X is not in Q. $\square$

Definition 2.4. We say that $\underline{g} = g_1, \ldots, g_n \in I$ is an *acceptable sequence* if $g_1, \ldots, g_n$ form a regular sequence and if g_X is a non-zerodivisor modulo I. For an acceptable $g_1, \ldots, g_n \in I$ we define an S-homomorphism $\varphi_{\underline{g}} : ((g_1, \ldots, g_n) : I)/(g_1, \ldots, g_n) \to L$ as follows: if $u \in (g_1, \ldots, g_n) : I$ represents a class $\bar{u} \in ((g_1, \ldots, g_n) : I)/(g_1, \ldots, g_n)$, define (by abuse of notation) $\varphi_{\underline{g}}(\bar{u}) = \frac{u}{g_X}$, where this fraction is in L. We set $M_{\underline{g}}$ to be the image of $\varphi_{\underline{g}}$. Similarly we define an acceptable sequence in IT'.

A key point in the theorem of Lipman and Sathaye is the following:

Proposition 2.5. *If $\underline{g}$ and $\underline{h}$ are acceptable sequences in IT' (where k is large enough), then $M_{\underline{g}} = M_{\underline{h}}$.*

Proof. For two acceptable sequences $\underline{g}$ and $\underline{h}$ in T' we define a distance $\rho(\underline{g}, \underline{h})$ between them as the minimum integer s such that for some invertible matrices E_1 and E_2 with coefficients in T' and for some $w_{s+1}, \ldots, w_n$ in T', $\underline{g} E_1 = (g'_1, \ldots, g'_s, w_{s+1}, \ldots, w_n)$ and $\underline{h} E_2 = (h'_1, \ldots, h'_s, w_{s+1}, \ldots, w_n)$.

We prove the proposition by induction on $\rho(\underline{g}, \underline{h})$. If $\rho(\underline{g}, \underline{h}) = 0$, then it is easy to check that $M_{\underline{g}} = M_{\underline{h}}$ as there is an invertible matrix E such that $\underline{g} = \underline{h} E$.

We need to do the case $\rho(\underline{g}, \underline{h}) = 1$ separately. By possibly multiplying by invertible matrices, we may assume without loss of generality that $\underline{g} = (y_1, \ldots, y_{n-1}, g)$ and $\underline{h} = (y_1, \ldots, y_{n-1}, h)$. Let $u \in (\underline{g}) : I$, and write

$$uh = \sum_{i=1}^{n-1} r_i y_i + vg.$$

It is straightforward to check that $\frac{u}{g_X} = \frac{v}{h_X}$. We claim that $v \in (\underline{h}) : I$. Multiplying the displayed equation by an arbitrary element $z \in I$ yields $zuh = \sum_{i=1}^{n-1} r_i y_i z + vgz$. As $z \in I$ and $u \in (\underline{g}) : I$, there is an equation $zu = \sum_{i=1}^{n-1} s_i y_i + sg$, and upon substitution in the preceding equation one obtains that

$$g(sh - vz) \in (y_1, \ldots, y_{n-1}).$$

Since $\underline{g} = y_1, \ldots, y_{n-1}, g$ is acceptable; this is a regular sequence in T', so that $sh - vz \in (y_1, \ldots, y_{n-1})$. It follows that $vz \in (\underline{h})$ and hence that $v \in (\underline{h}) : I$. Note that $M_{\underline{g}}$ is generated by elements $\frac{u}{g_X}$ as u ranges over elements of $(\underline{g}) : I$. Since $\frac{u}{g_X} = \frac{v}{h_X}$ and $v \in (\underline{h}) : I$, this proves that $M_{\underline{g}} \subseteq M_{\underline{h}}$. By symmetry we obtain that $M_{\underline{h}} = M_{\underline{g}}$. This finishes the case $\rho(\underline{g}, \underline{h}) = 1$.

Suppose that $\rho(\underline{g}, \underline{h}) = m > 1$. We may assume that $g_i = h_i$ for $i > m$. Let $I = (a_1, \ldots, a_l)$. Let $b = \sum_{i=1}^n u_i a_i$, where $u_1, \ldots, u_n \in \{t_1, \ldots, t_k\}$ such that these variables do not appear in any g_i, h_i. Then $\underline{g}' = b, g_2, \ldots, g_n$ and $\underline{h}' = b, h_2, \ldots, h_n$ are regular sequences, and since g_X, h_X are not in any minimal prime ideals over I, the same must be true for g'_X an h'_X. Thus $\underline{g}'$ and $\underline{h}'$ are acceptable sequences. Since $\rho(\underline{g}', \underline{g}) \leq 1$, we have proved that $M_{\underline{g}} = M_{\underline{g}'}$. Similarly $M_{\underline{h}} = M_{\underline{h}'}$. By induction on ρ, $M_{\underline{g}'} = M_{\underline{h}'}$, so that $M_{\underline{g}} = M_{\underline{h}}$. □

Definition 2.6. Whenever R, S, T are such that the S-module $M_{\underline{g}}$ is independent of the acceptable sequence, we emphasize that with writing $M_{\underline{g}}$ as $\overline{K}_{S/R}$.

The notation $K_{S/R}$ is meant to suggest a relative canonical module, which is the role this module plays in the proofs.

Proposition 2.7. *Let $\underline{g}$ be an acceptable sequence. Then*

1. *$((\underline{g}) :_T I) \cap I = (\underline{g})$.*
2. *The map $\varphi_{\underline{g}} : (\underline{g} :_T I)/(\underline{g}) \to L$ is injective. In particular, $(\underline{g} :_T I)/(\underline{g}) \cong K_{S/R}$.*
3. *$K_{S'/R'} \subseteq (S' :_{L'} J_{S'/R'})$.*
4. *If S is normal and $\underline{g}$ is acceptable, then $M_{\underline{g}}$ is a reflexive S-module.*
5. *Assume that S is normal and that for every height one prime Q of S, $R_{Q \cap R}$ is regular. Then $K_{S'/R'} = S' :_{L'} J_{S'/R'}$.*

Proof. Certainly $(\underline{g}) \subseteq ((\underline{g}) :_T I) \cap I$. Let Q be associated to $(\underline{g})$. Since $\underline{g}$ is a regular sequence and T is Cohen–Macaulay, ht $Q \leq n$. If $I \subseteq Q$, then Q is minimal over I, so that by acceptability, $(((\underline{g}) :_T I) \cap I)_Q = I_Q = (\underline{g})_Q$. If $I \not\subseteq Q$, then $(((\underline{g}) :_T I) \cap I)_Q = (\underline{g})_Q$. Thus (1) holds.

Let $u \in (\underline{g} :_T I)$ represent the class of an element $\overline{u}$ in $((\underline{g}) :_T I)/(\underline{g})$. If $\varphi_{\underline{g}}(\overline{u}) = 0$, then $u' = 0$ and hence $u \in I$. Hence $u \in ((\underline{g}) :_T I) \cap I$. By (1) this means that $u \in (\underline{g})$, so that $\overline{u} = 0$. Thus $\varphi_{\underline{g}}$ is injective. The last part is immediate.

To prove $K_{S'/R'} \subseteq (S' :_{L'} J_{S'/R'})$, by Remark 2.1 it suffices to prove that for any g_X where $\underline{g}$ is acceptable, $g_X K_{S'/R'} \subseteq S'$. But by Proposition 2.5, $K_{S'/R'} = M_{\underline{g}}$, and then (3) follows trivially.

Assume that S is normal. The S-module $M_{\underline{g}}$ is reflexive if and only if it is reflexive after localization at all prime ideals P (in S) with depth $S_P \leq 1$ and if depth$(M_{\underline{g}})_P \geq 2$ for all prime ideals P with depth $S_P \geq 2$. Since S is normal, depth $S_P \leq 1$ means that ht $P \leq 1$; whence S_P is a regular local ring of dimension 0 or 1, so that every torsion-free S_P-module is reflexive. But $(M_{\underline{g}})_P$ is a subset of the quotient field L_P of S_P; whence it is torsion-free, hence free, and hence reflexive. Now assume that depth $S_P \geq 2$. Let Q be a prime ideal in T such that $Q/I = P$. Then ht$(Q/I) \geq 2$, and for any acceptable sequence $\underline{g}$, depth$(T/((\underline{g}) :_T I))_Q \geq 1$ since $\underline{g}$ is generated by a regular sequence and T is Cohen–Macaulay. Likewise, depth$(T/\underline{g})_Q \geq 2$. The exact sequence

$$0 \to ((\underline{g} :_T I)/(\underline{g}))_Q \to (T/\underline{g})_Q \to (T/(\underline{g} :_T I))_Q \to 0$$

then gives depth $M_{\underline{g}} = \text{depth}(\underline{g} :_T I)/(\underline{g}))_Q \geq 2$. Hence $M_{\underline{g}}$ is reflexive.

By (3), $K_{S'/R'} \subseteq (S' :_{L'} J_{S'/R'})$, and by (3), $K_{S'/R'}$ is a reflexive S'-module. Suppose that the two modules are not equal. Then there exists a prime ideal Q in S' minimal over the quotient module $M = (S' :_{L'} J_{S'/R'})/K_{S'/R'}$, so that M_Q has finite length. If depth $S'_Q \geq 2$, we can choose a regular sequence $a, b \in S'_Q$ in S'_Q and $u \in (S' :_{L'} J_{S'/R'})_Q \setminus (K_{S'/R'})_Q$, such that $au, bu \in (K_{S'/R'})_Q$. By reflexivity, $K_{S'/R'} = \text{Hom}_{S'}(\text{Hom}_{S'}(K_{S'/R'}, S'), S')$, and since a, b is regular on S'_Q, it is also regular on $(K_{S'/R'})_Q$. Thus $b(au) = a(bu) \in (K_{S'/R'})_Q$ implies that $au \in a(K_{S'/R'})_Q$; whence $u \in (K_{S'/R'})_Q$, which is a contradiction. So we may assume that depth $S'_Q \leq 1$. Since S' is normal, this means that ht $Q \leq 1$. Set $q = Q \cap R'$. By assumption, R'_q is regular, and as S' is normal, S'_Q is also regular. Lift Q to a prime ideal Q' in T'_q. Since $T'_{Q'}$ and $S'_Q = T'_{Q'}/I_{Q'}$ are regular, $I_{Q'}$ is generated by a regular sequence, say $g_1, \ldots, g_n$, which we may assume are in I. Then $(g_1, \ldots, g_n) : I \nsubseteq Q'$, and Proposition 2.2(2) shows that $(J_{S'/R'})_Q = g_X S'_Q$. Hence $(S' :_{L'} J_{S'/R'})_Q = (S'_Q :_{L'} (J_{S'/R'})_Q) = \left(\frac{1}{g_X}\right) S'_Q$, and since $(g_1, \ldots, g_n) :$ I contains an element not in Q', it follows that $M_{\underline{g}} = \left(\frac{1}{g_X}\right) S'_Q$, proving the desired equality. $\square$

We next compare $K_{S/R}$ and $K_{B/R}$ where B is a finite extension of S with the same field of fractions.

Remark 2.8. Let $y \in L \setminus S$ be integral over S. Set $B = S[y]$. Extend the map of T onto S to an epimorphism $\varphi : T[Y] \to B$. Let H be the kernel of φ. Clearly $H \cap T = I$. Since the fields of fractions are the same, H contains an element of the form $aY - b$, where $a, b \in T$, $a \notin I$, and by integrality H contains a monic polynomial in Y with coefficients in T. Let h be such a monic polynomial of least possible degree m. We suppose below that S has sufficiently many units (say S is actually S' from Remark 2.1).

Suppose that $\underline{g}$ is an acceptable sequence for I. The sequence $\underline{g}^* = g_1, \ldots, g_n, h$ is clearly a sequence in Q having height $n + 1$ and the Jacobian $g^*_{X,Y} = g_X \frac{\partial h}{\partial Y}$. If $\frac{\partial h}{\partial Y}$ is not contained in any prime ideal minimal over H, then $\underline{g}^*$ is an acceptable sequence for H. Now suppose that $\frac{\partial h}{\partial Y}$ is contained in some prime ideals minimal over H. Let $c \in T[Y]$ be contained precisely in those prime ideals minimal over H that do not contain $\frac{\partial h}{\partial Y}$. Since a is not contained in any prime ideal minimal over I or over H, then for some sufficiently general unit u in S (this is the assumption), $\frac{\partial(h+cu(aY-b))}{\partial Y} = \frac{\partial h}{\partial Y} + ca$ is not contained in any prime ideal minimal over H. Since $y \notin S$, it follows that $m \geq 2$, so by replacing h with $h + cu(aY - b))$ we assume that $\frac{\partial h}{\partial Y}$ is not contained in any prime ideal in $T[Y]$ minimal over H.

We now have the map from $((\underline{g}^*) : H)/(\underline{g}^*) \to L$ given by $v \in ((\underline{g}^*) : H)$ goes to $\frac{v}{(g^*_{X,Y})}$. We denote the image of the map by $M_{\underline{g}^*}$.

Lemma 2.9. *Let the notation be as above. Then $M_{\underline{g}^*} \subseteq M_{\underline{g}}$. Precisely, for every* $v \in (g_1, \ldots, g_n, h)T[Y] :_{T[Y]} H$, *there is an element $u \in (g_1, \ldots, g_n)T :_T I$ such* *that*

$$v \equiv u \frac{\partial h}{\partial Y} \bmod Q.$$

Proof. By polynomial division we may write $v = hw + a_1 Y^{m-1} + a_2 Y^{m-2} + \cdots + a_m$ for some $w \in T[Y]$ and $a_i \in T$. Since $vI \subseteq vH \subseteq (g_1, \ldots, g_n, h)T[Y]$, this forces $(a_1 Y^{m-1} + a_2 Y^{m-2} + \cdots + a_m)I \subseteq (g_1, \ldots, g_n, h)T[Y]$. By degree count, $(a_1 Y^{m-1} + a_2 Y^{m-2} + \cdots + a_m)I \subseteq (g_1, \ldots, g_n)T[Y]$, and since $g_1, \ldots, g_n \in T$, it follows that $a_1, \ldots, a_m \in (g_1, \ldots, g_n)$.

We must also have that $v(aY - b) \in (g_1, \ldots, g_n, h)T[Y]$. Thus $(a_1 Y^{m-1} + a_2 Y^{m-2} + \cdots + a_m)(aY - b) = (v - hw)(aY - b) \in (g_1, \ldots, g_n, h)T[Y]$. Since $(a_1 Y^{m-1} + a_2 Y^{m-2} + \cdots + a_m)(aY - b)$ is a polynomial in Y of degree m with leading coefficient aa_1, we have that $(v - hw)(aY - b) - aa_1 h \in (g_1, \ldots, g_n)T[Y]$ is a polynomial of Y-degree at most $m - 1$. Differentiating with respect to Y gives that $(v - hw)a - a_1 a \frac{\partial h}{\partial Y} \equiv 0$ modulo H or even that $va - a_1 a \frac{\partial h}{\partial Y} \equiv 0$ modulo H. As a is a non-zerodivisor in $S[y]$, it follows that $v \equiv a_1 \frac{\partial h}{\partial Y}$ modulo Q. □

Theorem 2.10 (Lipman–Sathaye Theorem). *Let R be a universally catenary Cohen–Macaulay Noetherian domain of positive dimension, and let K be its field of fractions. Let $X_1, \ldots, X_n$ be variables over R, $T = R[X_1, \ldots, X_n]$, and I a radical ideal in T, all of whose minimal primes have the same height n. Set $S = T/I$. We assume that $R \subseteq S$, that all non-zero elements of R are non-zerodivisors on S, that each direct summand of the total ring of fractions L of S is finite and separable over K, and that for all prime ideals Q in S of height one, $R_{Q \cap R}$ is a regular local ring. Furthermore we assume that the integral closure $\overline{S}$ of S is a finitely generated S-module. Then*

$$(\overline{S} :_L J_{\overline{S}/R}) \subseteq S :_L J_{S/R}.$$

In particular, $J_{S/R} \overline{S} \subseteq S$.

Proof. By Remark 2.1 we may switch to R', T', S' in place of R, T, S. Then $K = S/R = M_{\underline{g}}$ is independent of acceptable sequence $\underline{g}$ for S. Fix one such $\underline{g}$. We write $\overline{S} = S[y_1, \ldots, y_l]$ and use Lemma 2.9 repeatedly. This lemma shows that there is an acceptable sequence $\underline{g}^*$ for $\overline{S}$ such that $M_{\underline{g}*} \subseteq M_{\underline{g}}$. By Proposition 2.7 we have an equality $\overline{S} :_L J_{\overline{S}/R} = K_{\overline{S}/R} = M_{\underline{g}*}$, and by definition $K_{S/R} = M_{\underline{g}}$. Hence $M_{\underline{g}*} \subseteq M_{\underline{g}}$ gives that

$$\overline{S} :_L J_{\overline{S}/R} \subseteq K_{S/R} \subseteq S :_L J_{S/R}.$$

The last containment follows since $\overline{S} \subseteq (\overline{S} :_L J_{\overline{S}/R})$. □

3   Improvements for Computing the Integral Closure of Rings

Chapter 15 of [19] presents de Jong's [8] algorithm for computing the integral closure, as well as some modifications due to Vasconcelos and to Lipman. Those algorithms work by successively finding more and more elements in the integral closure until the ring is integrally closed. Since publication, there have been two new developments: an improvement of de Jong's algorithm due to Greuel et al. [13] and a very different algorithm due to Leonard and Pellikaan [22] and Singh and Swanson [27]. We present the two in chronological order.

Leonard and Pellikaan [22] devised an algorithm for computing the integral closure of weighted rings that are finitely generated over finite fields, and Singh and Swanson [27] generalized the method to affine equidimensional reduced rings over perfect fields in positive prime characteristic. This new algorithm starts with a special module containing the integral closure and then successively constructs submodules that eventually stabilize in the integral closure. In general the descending chain condition does not hold between the integral closure and the initial module, but the particular descending chain of submodules does stabilize. The algorithm is now implemented both in Macaulay2 and in Singular, and it sometimes terminates much faster than the other implementations.

The algorithm is based on the following theorem:

Theorem 3.1 ([27, Theorem 1.1]). *Let R be a reduced ring that is finitely generated over a computable field of characteristic $p > 0$. Set $\overline{R}$ to be the integral closure of R in its total ring of fractions. Suppose that D is a non-zerodivisor in the conductor ideal of R, i.e., that D is a non-zerodivisor with $D\overline{R} \subseteq R$.*

1. Set $V_0 = \frac{1}{D}R$, and for $e \geq 0$ inductively define

$$V_{e+1} = \{f \in V_e \mid f^p \in V_e\}.$$

Then the V_e are algorithmically constructible modules.
2. The descending chain

$$V_0 \supseteq V_1 \supseteq V_2 \supseteq V_3 \supseteq \cdots$$

stabilizes.

If $V_e = V_{e+1}$, then V_e equals $\overline{R}$.

Proof. For every $e \geq 0$, the module DV_e is a submodule of DV_0 and hence of R. Thus $U_e = DV_e$ is an ideal in R. Certainly $U_0 = R$, and for any $e \geq 0$,

$$U_{e+1} = \{r \in U_e \mid r^p \in D^{p-1}U_e\}.$$

Let $F : R \to R$ be the Frobenius homomorphism taking r to r^p, and let $\pi : R \to R/D^{p-1}U_e$ be the canonical surjection. Then the kernel of $\pi \circ F$ is computable, and as $U_{e+1} = U_e \cap \ker(\pi \circ F)$ is computable. Thus V_{e+1} is computable.

Certainly the chain is a descending chain, and if $V_e = V_{e+1}$, then $V_e = V_s$ for all $s \ge e$.

By assumption, $\overline{R} \subseteq V_0$. Suppose that $\overline{R} \subseteq V_e$. Let $f \in \overline{R}$. By assumption $f \in V_e$, and since $\overline{R}$ is a ring, also $f^p \in \overline{R} \subseteq V_e$. Thus $f \in V_{e+1}$, which proves that $\overline{R} \subseteq V_{e+1}$.

Let $v_1, \ldots, v_s$ be the Rees valuations of the ideal DR, i.e., v_i are valuations such that for each $n \in \mathbf{N}$, $\overline{D^n R} = \{r \in R \mid v_i(r) \ge n v_i(D) \text{ for each } i\}$. Let e be a positive integer such that $p^e > v_i(D)$ for each i. We claim that $V_e \subseteq \overline{R}$, i.e., that $V_e = V_{e+1}\overline{R}$. Let $f \in V_e$. We can write $f = r/D$. By definition, $(r/D)^{p^e} \in V_0 = \frac{1}{D}R$, so that $r^{p^e} \in D^{p^e-1}R$. Thus for all $i = 1, \ldots, s$,

$$v_i(r) = \frac{1}{p^e} v_i(r^{p^e}) \ge \frac{1}{p^e} v_i(D^{p^e-1}) = \frac{p^e - 1}{p^e} v_i(D) > v_i(D) - 1.$$

Since $v_i(r)$ and $v_i(D)$ are integers, it follows that $v_i(r) \ge v_i(D)$ for each i. Thus $r \in \overline{DR} \subseteq \overline{D\overline{R}}$. In integrally closed rings, principal ideals generated by non-zerodivisors are integrally closed, so that $r \in D\overline{R}$; whence $f \in \overline{R}$. $\square$

How does one make an algorithm out of this theorem if the underlying field k is perfect? Once D is found as in the hypotheses of the theorem, the proof shows how the rest is straightforwardly algorithmic. Thus the question is how to find such a D algorithmically.

Write $R = k[x_1, \ldots, x_n]/(f_1, \ldots, f_m)$. Let $h = \mathrm{ht}(f_1, \ldots, f_m)$. Let $J_{R/k}$ be the Jacobian ideal of R over k, i.e., $J_{R/k}$ is an ideal in R generated by the determinants of all the $h \times h$ submatrices of the Jacobian matrix $(\partial f_i/\partial x_j)$. By the Jacobian criterion (Theorem 4.4.9 in [19]), since R is reduced, $J_{R/k}$ contains a non-zerodivisor. Let D be the determinant of some $h \times h$ submatrix of the Jacobian matrix. By sampling random $h \times h$ submatrices or approaching them in order, we eventually get to a non-zero D. Note that $0 :_R D$ is non-zero if and only if D is a zero divisor. In case that D is a zero divisor and not zero, both $0 :_R D$ and $0 :_R (0 :_R D)$ are non-zero radical ideals of height zero, and $\overline{R} = \overline{R/(0 :_R D)} \times \overline{R/(0 :_R (0 :_R D))}$, and it suffices to compute the integral closures of $R/(0 :_R D)$ and $R/(0 :_R (0 :_R D))$ that are equidimensional and have strictly smaller fewer minimal prime ideals. Thus we have reduced to the case where D is not a zero divisor. By Theorem 2.10, D multiplies the integral closure of R into R. Thus by applying Theorem 3.1 we construct $V_0, V_1, \ldots$ with this D, to eventually get the stable value $V_e = \overline{R}$.

We now turn to another new development in the computation of integral closure. The improvement due to Greuel et al. [13] is as follows. Start with an affine domain R over a perfect field k. Compute the Jacobian ideal $J_{R/k}$ of R over k. With the help of Serre's conditions (Theorem 4.5.7 in [19]) determine if R is integrally closed. If it is not, then compute $J = \sqrt{J_{R/k}}$, and $R_1 = \mathrm{Hom}_R(J, J)$. This R_1 is a ring strictly larger than R and contained in $\overline{R}$. So far this is the same as de Jong's original algorithm. But the big saving comes next:

1. The Jacobian ideal of R_1 over k is up to radical the same as JR_1, so that by Serre's conditions (Theorem 4.5.7 in [19]), R_1 is integrally closed if and only if JR_1 has grade at least two.
2. If R_1 is not integrally closed, rather than repeat the procedure that we did before on R now on R_1, we can instead use $J_1 = \sqrt{JR_1}$, and compute $R_2 = \text{Hom}_{R_1}(J_1, J_1)$.

Note that this improvement allows us to skip the very time-consuming step of computing the Jacobian ideal for R_1.

Similarly, once R_n is computed, it is integrally closed if and only if JR_n has grade at least 2, and if it is not integrally closed, then we compute $J_{n+1} = \sqrt{JR_n}$ and $R_{n+1} = \text{Hom}_{R_n}(J_{n+1}, J_{n+1})$.

Greuel et al. [13] make further improvements. Namely, to recognize $\text{Hom}_R(J, J)$ as a ring rather than as an R-module, the standard algorithms use the following identification: $\text{Hom}_R(J, J) = \frac{1}{d}(dJ :_R J)$ for any non-zerodivisor $d \in J$ (Lemma 2.4.3 in [19]). Thus in particular for all n, $R_{n+1} = \frac{1}{d}(dJ_n :_{R_n} J_n)$. This is of course computed in the ring R_n. But by expressing $R_n = \frac{1}{e}U$ and $J_n = \frac{1}{e}H$ for some non-zerodivisor e and some ideals U and H in R, [13] gives the following:

Theorem 3.2 (Theorem 3.5 in [13]). *With notation as above,*

$$R_{n+1} = \frac{1}{d}(dJ_n :_{R_n} J_n) = \frac{1}{ed}(edH :_R H).$$

Thus R_{n+1} is computed in R rather than in R_n.

Proof. Let $r \in (dJ_n :_{R_n} J_n)$. Write $r = \frac{a}{e}$ for some $a \in U$. Then $\frac{a}{e}\frac{H}{e} = \frac{a}{e}J_n \subseteq dJ_n = d\frac{H}{e}$, so that $aH \subseteq edH$; whence $r = \frac{a}{e} \in \frac{1}{e}(edH :_R H)$. This proves that $\frac{1}{d}(dJ_n :_{R_n} J_n) \subseteq \frac{1}{ed}(edH :_R H)$.

Now let $r \in (edH :_R H)$. Then $rH \subseteq edH$, so that $\frac{r}{e}\frac{H}{e} \subseteq d\frac{H}{e}$, i.e., that $\frac{r}{e}J_n \subseteq dJ_n$. Since $d \in J_n$, it follows that $\frac{r}{e}d \in dJ_n$, and as d is a non-zerodivisor, $\frac{r}{e} \in J_n \subseteq R_n$. Hence $\frac{r}{e} \in (dJ_n :_{R_n} J_n)$. This proves that $\frac{1}{ed}(edH :_R H) \subseteq \frac{1}{d}(dJ_n :_{R_n} J_n)$. $\square$

Thus the computation of R_{n+1} can be done in R rather than in R_n; however, the computation of $J_n = \sqrt{JR_n}$ must still be done in R_n.

In positive prime characteristic p, to compute the radical of JR_n, Greuel et al. [13] use the following. Write $R_n = \frac{1}{e}U$ as before. Then

$$J_n = \{f \in R_n : f^m \in JR_n \text{ for some positive integer } m\}$$

$$= \{f \in R_n : f^m \in JR_n \text{ for some power } m \text{ of } p\}$$

$$= \left\{\frac{r}{e} : r \in U, r^m \in e^{m-1}JU \text{ for some power } m \text{ of } p\right\}.$$

For any power m of p, $\{\frac{r}{e} : r \in U, r^m \in e^{m-1} J U\}$ can be computed as the kernel of a composition of a Frobenius map with a surjection to $R/(e^{m-1} J U)$. Thus J_n is a union of these computable ideals, as m varies. Since R is Noetherian, only finitely many m are needed; by Hermann [16] or Seidenberg [26], there is a huge a priori upper bound on m depending on the number of variables in R and the degrees and the number of generators of J and of the presenting ideal of R. However, in practice it may still be best to compute J_n in R_n rather than in R.

4 Integral Closure with a View Towards Constructive Mathematics

This section is motivated by Lombardi's paper [23] and Grinberg's post [14] and only imports a few of the results written by them.

It is of interest that both papers generalize the definition of integrality. The following version is taken from [14]. Let $R \subseteq S$ be an inclusion of rings. For each $m \in \mathbf{N}_0$ let I_m be an ideal in R, and assume that $I_0 = R$ and that for all m, m', $I_m I'_m \subseteq I_{m+m'}$. Then $x \in S$ is **n-*integral over* $(\mathbf{R}, \{\mathbf{I_m}\}_\mathbf{m})$** if there exists an equation (of integral dependence) $x^n + a_1 x^{n-1} + \cdots + a_n$ with $a_j \in I_j$ for each j. This notion of integrality is, after some translation of language, the same as the old notion of integrality of the ring $S[t]$ over the Rees ring over R associated to the filtration $\{I_m\}_m$; thus we do not describe these notions further here. It should be mentioned that some of the proofs about basic facts on integral closures are more streamlined with the new definition. Interested reader can go directly to the sources.

The rest of the section proves some interesting results encountered in the two papers.

Theorem 4.1 ([14, Theorem 2]). *Let $R \subseteq S$ be rings; let $a_0, \ldots, a_n \in R$, $x \in S$, such that $\sum_{i=0}^n a_i x^i = 0$. Then for any $k \subset \{0, \ldots, n\}$, $\sum_{i=0}^{n-k} a_{i+k} x^i$ satisfies an equation of integral dependence of degree n over R.*

Proof. Let $u = \sum_{i=0}^{n-k} a_{i+k} x^i$. Then

$$x^k u = \sum_{i=0}^{n-k} a_{i+k} x^{i+k} = \sum_{i=k}^{n} a_i x^i = -\sum_{i=0}^{k-1} a_i x^i.$$

Thus for $t \in \{k, \ldots, n\}$, $x^t u = -\sum_{i=0}^{k-1} a_i x^{i+(t-k)}$, and for $t \in \{0, \ldots, k-1\}$, $x^t u = \sum_{i=0}^{n-k} a_{i+k} x^{i+t}$, which shows that for $t = 0, \ldots, n$, $x^t u \in \sum_{i=0}^n x^i R$. Note that $U = \sum_{i=0}^n x^i R$ is a faithful finitely generated R-module; we just showed that $uU \subseteq U$, so that by [19, Lemma 2.1.8], u is integral over R and it satisfies an equation of integral dependence of degree n. $\square$

The following is due to the classical Dedekind–Mertens Lemma (see Sect. 1.7 in [19], where it should have been stated explicitly). It is also related to [19, Lemma 2.1.19].

Theorem 4.2 (Kronecker's Theorem). *Let R be a ring and X a variable over R. If $f \in R[X]$ factors as*

$$f = \left(\sum_{i=0}^{m} a_i X^i \right) \left(\sum_{i=j}^{n} b_j X^j \right),$$

then for all $i \in \{0, \ldots, m\}$ and all $j \in \{0, \ldots, n\}$, $a_i b_j$ is integral over A.

Theorem 4.3 (Kronecker's Theorem). *Let k be a field, let $X_0, \ldots, X_m, Y_0, \ldots, Y_n$ be indeterminates over k, and for $k = 0, \ldots, m + n$, let $Z_k = \sum_{i+j=k} X_i Y_j$. Then for all $i \in \{0, \ldots, m\}$ and all $j \in \{0, \ldots, n\}$, $X_i Y_j$ is integral over $k[Z_0, \ldots, Z_{m+n}]$.*

Theorem 4.4 (Theorem 22 in [14]). *Let $R \subseteq S$ be rings, and let $x, y, z \in S$. Suppose that z is integral over $R[x]$ and over $R[y]$. Then z is integral over $R[xy]$.*

Proof. Reasoning very similar to the one for [19, Proposition 2.1.16] shows that we may assume that R and S are domains. Let v be a valuation on the field of fractions of $R[xy]$ that is non-negative on $R[xy]$. Then v is non-negative either on $R[x]$ or on $R[y]$. Hence by assumption $v(x) \geq 0$. Since v is an arbitrary valuation, by [19, Proposition 6.8.14], z is integral over $R[xy]$.   □

Remark 4.5. Another related paper in the constructive spirit is [1] due to Barhoumi and Lombardi. The Traverso–Swan theorem says that a reduced ring R is seminormal if and only if the canonical map $\operatorname{Pic} R \rightarrow \operatorname{Pic} R[X]$ of Picard groups is an isomorphism. The paper [1] gives an explicit algorithm for obtaining from a given set of generators a sequence of elements $c_1, \ldots, c_m$ that generate the overring and such that for all i, $c_i^2, c_i^3 \in R[c_1, \ldots, c_{i-1}]$.

5 Multiplicity and Monomial Ideals

In this section, let k be a field, $X_1, \ldots, X_d$ variables over k, and let R be either $k[[X_1, \ldots, X_d]]$ or $k[X_1, \ldots, X_d]$.

In [19, Proposition 1.4.2] it was proved that the integral closure of a monomial ideal is monomial. Not all monomial ideals have monomial reductions, however, say for example (X^4, XY^2, Y^9). In this section we address the question of when the integral closure of a not-necessarily monomial ideal in R monomial. We restrict our attention to zero-dimensional ideals. A characterization for such ideals was given by Saia in [25] for the ring of convergent power series over $\mathbf{C}$. An algebraic proof for $\mathbf{C}[[X_1, \ldots, X_d]]$ was given by Biviá-Ausina in [2]. The treatment below follows that of [2], using multiplicities.

We start with a geometric result for multiplicity of monomial ideals.

Theorem 5.1. *Let I be a zero-dimensional monomial ideal in R. Then $e(I) = d!$ vol I, where $\mathrm{vol}\ I$ is the complement in $\mathbf{R}_{\geq 0}^d$ of the Newton polyhedron $\mathrm{NP}(I)$ of I.*

Proof. For any monomial ideal J, define $E'(J)$ to be the set in $\mathbf{Z}_{\geq 0}^d$ of all exponent vectors of monomial elements of J, and let $E'(J) = E(J) + \mathbf{R}_{\geq 0}^d \subseteq \mathbf{R}_{\geq 0}^d$. By the structure of the integral closure of monomial ideals, say by discussion on page 11 of [19], $\lim_n \frac{1}{n} E(I^n) = \mathrm{NP}(I)$, $\frac{1}{n} E(I^n) \subseteq \mathrm{NP}(I)$ for all n, and for sufficiently divisible n, $\frac{1}{n} E(I^n) \cap \mathbf{Z}_{\geq 0}^d = \mathrm{NP}(I) \cap \mathbf{Z}_{\geq 0}^d$. By the definition of multiplicity,

$$e(I) = \lim_n \frac{d!}{n^d} \lambda\left(\frac{R}{I^n}\right) = \lim_n \frac{d!}{n^d} \lambda\left(\frac{R}{\overline{I^n}}\right) \tag{1}$$

$$= \lim_n \frac{d!}{n^d}\ (\text{number of integer lattice points in } \mathbf{R}_{\geq 0}^d \setminus E(\overline{I^n})) \tag{2}$$

$$= \lim_n \frac{d!}{n^d}\ \left(\text{number of } \tfrac{1}{n}\text{-integer lattice points in } \frac{\mathbf{R}_{\geq 0}^d \setminus E(\overline{I^n})}{n}\right) \tag{3}$$

$$= d!\ \lim_n \frac{\text{number of } \tfrac{1}{n}\text{-integer lattice points in } \mathbf{R}_{\geq 0}^d \setminus \frac{E(\overline{I^n})}{n}}{n^d}. \tag{4}$$

Any $\frac{1}{n}$-integer lattice point $\frac{1}{n}(\alpha_1, \ldots, \alpha_d)$ in $\mathbf{R}_{\geq 0}^d \setminus \frac{E(\overline{I^n})}{n}$ determines a unique d-dimensional box whose 2^d vertices are $\frac{1}{n}(\alpha_1 + \varepsilon_1, \ldots, \alpha_d + \varepsilon_d)$, as $\varepsilon_1, \ldots, \varepsilon_d$ vary over the set $\{0, 1\}$. This box has volume $\frac{1}{n^d}$. The union of all such boxes contains $\mathbf{R}_{\geq 0}^d \setminus \frac{E(\overline{I^n})}{n}$, and for sufficiently divisible n, the union of all such boxes contains $\mathrm{NP}(I) \cap \mathbf{Z}_{\geq 0}^d$. Thus vol I, the volume of $\mathbf{R}_{\geq 0}^d \setminus \mathrm{NP}(I)$, is at most (and approximately)

$$\frac{\text{number of } \tfrac{1}{n}\text{-integer lattice points in } \mathbf{R}_{\geq 0}^d \setminus \frac{E(\overline{I^n})}{n}}{n^d}.$$

But $\lim_n \frac{1}{n} E(\overline{I^n}) = \mathrm{NP}(I)$, so that by Riemann integrals, the conclusion follows. $\square$

Theorem 5.2 (Saia [25], Biviá-Ausina [2]). *Let I be a $(X_1, \ldots, X_d)$-primary ideal in R. Let $\mathrm{NP}(I)$ be the convex hull in $\mathbf{R}^d$ of the set of all d-tuples $(e_1, \ldots, e_d)$ such that $X_1^{e_1} \cdots X_d^{e_d}$ appears with a non-zero coefficient in some element of I. (This set is called the Newton polyhedron of I and generalizes the same notion for monomial ideals.) Then the complement of $\mathrm{NP}(I)$ in $\mathbf{R}_{\geq 0}^d$ has finite volume $\mathrm{vol}(I)$. If $d!\ \mathrm{vol}(I) = e(I)$, then $\overline{I}$ is a monomial ideal.*

Proof. Let J be the monomial ideal in R generated by all the monomials that appear with a non-zero coefficient in some element of I. Then $I \subseteq J$, and $\mathrm{NP}(I) =$

NP(J). By Theorem 5.1, $d!$ vol(J) $= e(J)$, so that by assumption $e(J) = e(I)$. If R is local (if it is a power series ring), then by the Rees theorem (11.3.1 in [19]), $\overline{I} = \overline{J}$, which is monomial by [19, Proposition 1.4.2]. If R is not local, say if it is the polynomial ring, then $\overline{I} = \overline{J}$ holds locally after localizing at the unique the maximal ideal that contains I, and hence it holds globally. $\square$

In Saia, the condition for the integral closure of a $(X_1, \ldots, X_d)$-primary ideal being monomial is phrased in terms of non-degeneracy of its Newton polyhedron. We do not discuss degeneracy of Newton polyhedra.

6 Epsilon and j-Multiplicities

At the end of Sect. 11.3 in [19], we defined j-multiplicity and stated one theorem without proof. Since then, there have been more activity on j-multiplicities and a generalization to ε-multiplicities. We briefly indicate some of that development.

An equivalent formulation of j-multiplicity of an ideal I in a Noetherian local ring (R, m) of dimension d is as

$$j(I) = \lim_{n \to \infty} \frac{(d-1)!}{n^{d-1}} \lambda_R \left(H_m^0 \left(\frac{I^n}{I^{n+1}} \right) \right),$$

where $H_m^0(M)$ is the 0th local cohomology of a module M with support in m; namely it is the submodule of the module M generated by all elements that are annihilated by a power of m.

Katz and Validashti [21, Theorem 3.9] proved that if I has analytic spread equal to the dimension of the ring, then $j(I)$ can be expressed as an integer-linear combination of the v-values of I, as v varies over all normalized Rees valuations of I. They also prove in [21, Corollary 3.11] that for all positive integers n, $j(\overline{I^n}) = n^{\dim R} \cdot j(I)$.

Jeffries and Montaño [20] recently proved that with the proper definition of a truncated volume of a not-necessarily bounded complement of the Newton polyhedron of an ideal in $\mathbf{R}_{\geq 0}^d$, for any monomial (not-necessarily zero-dimensional) ideal I in a polynomial ring, $j(I)$ equals $d!$ times this volume. This generalizes Theorem 5.1.

Ulrich and Validashti generalized the j-multiplicity to modules in [29]: if E is a submodule of a free finitely generated R-module $F = R^e$, then the j-multiplicity of E is

$$j(E) = \lim_{n} \frac{(d+e-1)!}{n^{d+e-1}} \cdot \sum_{i=0}^{n-1} \lambda_R \left(H_m^0 \left(\frac{E^i F^{n-i}}{E^{i+1} F^{n-i-1}} \right) \right),$$

where the products are as in Sect. 16.5 of [19]. (That section is about the Buchsbaum–Rim multiplicity.) In fact, j-multiplicity of modules contains both the j-multiplicity of ideals and the Buchsbaum–Rim multiplicity to modules, and correspondingly [29] contains characterizations of integral dependence via j-multiplicity.

In [21] Katz and Validashti pursue a modification of j-multiplicity for ideals, and in [29], Ulrich and Validashti pursue a modification of j-multiplicity for modules as above: the ε multiplicity of E is

$$\varepsilon(E) = \limsup_n \frac{(d+e-1)!}{n^{d+e-1}} \cdot \lambda_R \left(H_m^0 \left(\frac{F^n}{E^n} \right) \right).$$

Both [21, 30] prove another characterization of integral dependence (under certain assumptions) using this new multiplicity, and non-vanishing theorems. Jeffries and Montaño [20] present the ε multiplicity of a monomial ideal in terms of a certain volume.

Cutkosky [7] proved that under some assumptions, such as depth $R \geq 2$ and R is essentially of finite type over a field of characteristic zero, in the definition of ε multiplicity, "limsup" can be replaced with "lim".

7 Grading

Huneke and Swanson [19, Theorem 2.3.2] proves that the integral closure of a $\mathbf{N}^d \times \mathbf{Z}^e$-graded ring R inside a compatibly $\mathbf{N}^d \times \mathbf{Z}^e$-graded ring S is graded as well. Shiro Goto pointed out the following shorter proof.

For this proof we assume (or one could give a proof simpler than that of Theorem 2.3.2) that if A is a subring of B, and if $A' = A[X_1, \ldots, X_{d+e}, X_{d+1}^{-1}, \ldots, X_{d+e}^{-1}]$ and $B' = B[X_1, \ldots, X_{d+e}, X_{d+1}^{-1}, \ldots, X_{d+e}^{-1}]$ are $\mathbf{N}^d \times \mathbf{Z}^e$-graded by monomials in the variables $X_1, \ldots, X_{d+e}$, then the integral closure of a homogeneous subring of A' in B' is $\mathbf{N}^d \times \mathbf{Z}^e$-graded.

Now let $G = \mathbf{N}^d \times \mathbf{Z}^e$. Then the group $R[G]$ is $R[X_1, \ldots, X_{d+e}, X_{d+1}^{-1}, \ldots, X_{d+e}^{-1}]$. If R is also G-graded, define $\varphi_R : R \to R[G]$ to be the homomorphism that takes a homogeneous $r \in R$ of degree g to $r \cdot g \in R[G]$. Similarly define φ_S.

By a previous paragraph, when we think of $R[G]$ as G-graded by monomials in $X_1, \ldots, X_{d+e}$, then the integral closure of $R[G]$ in $S[G]$ is G-graded. Now let $s \in S$ be integral over R. Applying φ_S to the equation of integral dependence of s over R shows that $\varphi_S(s)$ is integral over $R[G]$. Thus all the homogeneous components of $\varphi_S(s)$ are integral over $R[G]$. But the homogeneous components of $\varphi_S(s)$ are precisely of the form $h \cdot m$, where h is a homogeneous component of s in S and m is a monomial in $X_1, \ldots, X_{d+e}, X_{d+1}^{-1}, \ldots, X_{d+e}^{-1}$. But then by writing out the equation of integral dependence of $h \cdot m$ over $R[G]$ we get that h, an arbitrary homogeneous component of s, is integral over R.

8 Goto Numbers

In this section (R, m) is a Noetherian local ring. The *Goto number* of a parameter ideal Q is the largest integer q such that $Q : m^q$ is integral over Q. These numbers were explored in [3, 12, 15, 31]. The motivation for Goto numbers came from the following result that predates [19]:

Theorem 8.1 (Corso et al. [6], Corso and Polini [4], Corso et al. [5], Goto [9]). *Let (R, m) be a Cohen–Macaulay local ring of positive dimension. Let Q be a parameter ideal in R and let $I = Q : m$. Then the following are equivalent:*

1. $I^2 \neq QI$.
2. *The integral closure of Q is Q.*
3. *R is a regular local ring and $\mu(m/Q) \leq 1$.*

Thus $I^2 = QI$ if (R, m) is a Cohen–Macaulay local ring that is not regular. Furthermore, when this is the case, the associated graded ring $\mathrm{gr}_I(R) = R[It]/IR[It]$ and the fiber ring $R[It]/mR[It]$ are both Cohen–Macaulay. If in addition $\dim R > 1$, it follows that the Rees algebra $R[It]$ is a Cohen–Macaulay ring. Several papers explored more generally $I : m^q$ for various positive integers q, keeping in mind related questions on good properties of the Rees algebras and fiber rings. See for example Goto et al. [10], Goto et al. [11], Horiuchi [18], and so on.

References

1. S. Barhoumi, H. Lombardi, An algorithm for the Traverso-Swan theorem on seminormal rings. J. Algebra **320**, 1531–1542 (2008)
2. C. Biviá-Ausina, Nondegenerate ideals in formal power series rings. Rocky Mt. J. Math. **34**, 495–511 (2004)
3. L. Bryant, Goto numbers of a numerical semigroup ring and the Gorensteiness of associated graded rings. Comm. Algebra **38**, 2092–2128 (2010)
4. A. Corso, C. Polini, Links of prime ideals and their Rees algebras. J. Algebra **178**, 224–238 (1995)
5. A. Corso, C. Polini, W. Vasconcelos, Links of prime ideals. Math. Proc. Camb. Phil. Soc. **115**, 431–436 (1995)
6. A. Corso, C. Huneke, W. Vasconcelos, On the integral closure of ideals . Manuscripta Math. **95**, 331–347 (1998)
7. S.D. Cutkosky, Asymptotic growth of saturated powers and epsilon multiplicity. Math. Res. Lett. **18**, 93–106 (2011)
8. T. de Jong, An algorithm for computing the integral closure. J. Symbolic Comput. **26**, 273–277 (1998)
9. S. Goto, Integral closedness of complete intersection ideals. J. Algebra **108**, 151–160 (1987)
10. S. Goto, N. Matsuoka, R. Takahashi, Quasi-socle ideals in a Gorenstein local ring. J. Pure Appl. Algebra **212**, 969–980 (2008)
11. S. Goto, S. Kimura, N. Matsuoka, Quasi-socle ideals in Gorenstein numerical semigroup rings. J. Algebra **320**, 276–293 (2008)
12. S. Goto, S. Kimura, T.T. Phuong, H.L Truong, Quasi-socle ideals in Goto numbers of parameters. J. Pure Appl. Algebra **214**, 501–511 (2010)

13. G.-M. Greuel, S. Laplagne, F. Seelisch, Normalization of rings. J. Symbolic Comput. **45**(9), 887–901 (2010)
14. D. Grinberg, A few facts on integrality DETAILED VERSION. http://www.cip.ifi.lmu.de/~grinberg/Integrality.pdf
15. W. Heinzer, I. Swanson, Goto numbers of parameter ideals. J. Algebra **321**, 152–166 (2009)
16. G. Hermann, Die Frage der endlich vielen Schritte in der Theorie der Polynomideale. Math. Ann. **95**, 736–788 (1926)
17. M. Hochster, Presentation depth and the Lipman-Sathaye Jacobian theorem, in *The Roos Festschrift*, vol. 2, Homology Homotopy Appl. **4**, 295–314 (2002)
18. J. Horiuchi, Stability of quasi-socle ideals. J. Commut. Algebra **4**, 269–279 (2012)
19. C. Huneke, I. Swanson, *Integral Closure of Ideals, Rings, and Modules*. London Mathematical Society Lecture Note Series, vol. 336 (Cambridge University Press, Cambridge, 2006)
20. J. Jeffries, J. Montaño, The j-multiplicity of monomial ideals. arXiv:math.AC/1212.1419 (preprint)
21. D. Katz, J. Validashti, Multiplicities and Rees valuations. Collect. Math. **61**, 1–24 (2010) 2009.
22. D.A. Leonard, R. Pellikaan, Integral closures and weight functions over finite fields. Finite Fields Appl. **9**, 479–504 (2003)
23. H. Lombardi, Hidden constructions in abstract algebra. I. Integral dependence. J. Pure Appl. Algebra **167**, 259–267 (2002)
24. L.J. Ratliff, Jr., On quasi-unmixed local domains, the altitude formula, and the chain condition for prime ideals, (I). Am. J. Math. **91**, 508–528 (1969)
25. M. J. Saia, The integral closure of ideals and the Newton filtration.J. Algebraic Geom. **5**, 1–11 (1996)
26. A. Seidenberg, Constructions in algebra. Trans. Am. Math. Soc. **197**, 273–313 (1974)
27. A.K. Singh, I. Swanson, Associated primes of local cohomology modules and of Frobenius powers. Int. Math. Res. Notices **30**, 1703–1733 (2004)
28. I. Swanson, Rees valuations, in *Commutative Algebra: Noetherian and Non-Noetherian Perspectives*, ed. by M. Fontana, S. Kabbaj, B. Olberding, I. Swanson (Springer, New York, 2010), pp. 421–440
29. B. Ulrich, J. Validashti, A criterion for integral dependence of modules. Math. Res. Lett. **15**, 149–162 (2008)
30. B. Ulrich, J. Validashti, Numerical criteria for integral dependence. Math. Proc. Camb. Phil. Soc. **151**, 95–102 (2011)
31. K.-i. Watanabe, K.-i. Yoshida, A variant of Wang's theorem. J. Algebra **369**, 129–145 (2012)

Open Problems in Commutative Ring Theory

Paul-Jean Cahen, Marco Fontana, Sophie Frisch, and Sarah Glaz

Abstract This chapter consists of a collection of open problems in commutative algebra. The collection covers a wide range of topics from both Noetherian and non-Noetherian ring theory and exhibits a variety of research approaches, including the use of homological algebra, ring theoretic methods, and star and semistar operation techniques. The problems were contributed by the authors and editors of this volume, as well as other researchers in the area.

Keywords Prüfer ring • Homological dimensions • Integral closure • Group ring • Grade • Complete ring • McCoy ring • Straight domain • Divided domain • Integer-valued polynomials • Factorial • Density • Matrix ring • Overring • Absorbing ideal • Kronecker function ring • Stable ring • Divisorial domain • Mori domain • Finite character • PvMD • Semistar operation • Star operation • Jaffard domain • Locally tame domain • Factorization • Spectrum of a ring • Integral closure of an ideal • Rees algebra • Rees valuation

P.-J. Cahen (✉)
12 Traverse du Lavoir de Grand-Mère, 13100 Aix en Provence, France
e-mail: pauljean.cahen@gmail.com

M. Fontana
Dipartimento di Matematica e Fisica, Università degli Studi Roma Tre, Largo San Leonardo Murialdo 1, 00146 Roma, Italy
e-mail: fontana@mat.uniroma3.it

S. Frisch
Mathematics Department, Graz University of Technology, Steyrergasse 30, 8010 Graz, Austria
e-mail: frisch@tugraz.at

S. Glaz
Department of Mathematics, University of Connecticut, Storrs, CT 06269, USA
e-mail: Sarah.Glaz@uconn.edu

M. Fontana et al. (eds.), *Commutative Algebra: Recent Advances in Commutative Rings,*   353
Integer-Valued Polynomials, and Polynomial Functions, DOI 10.1007/978-1-4939-0925-4_20,
© Springer Science+Business Media New York 2014

Mathematics Subject Classification (2010): 13-02; 13A05; 13A15; 13A18; 13B22; 13C15; 13D05; 13D99; 13E05; 13F05; 13F20; 13F30; 13G05

1 Introduction

This chapter consists of a collection of open problems in commutative algebra. The collection covers a wide range of topics from both Noetherian and non-Noetherian ring theory and exhibits a variety of research approaches, including the use of homological algebra, ring theoretic methods, and star and semistar operation techniques. The problems were sent to us, in response to our request, by the authors of the articles in this volume and several other researchers in the area. We also included our own contributions. Some of these problems appear in other chapters of this volume, while others are unrelated to any of them, but were considered important by their proposers. The problems were gathered by the contributors from a variety of sources and include long-standing conjectures as well as questions that were generated by the most recent research in the field. Definitions and clarifying comments were added to make the problems more self-contained, but, as a rule, if unidentified notions are used, the reader can find the relevant definitions in the cited references. The purpose of this chapter is to generate research, and we hope that the problems proposed here will keep researchers happily busy for many years.

The underlying assumption is that all rings are commutative with 1 (and $1 \neq 0$), all modules are unital, all groups are abelian, and the term "local ring" refers to a not-necessarily Noetherian ring with a unique maximal ideal. Several notions and ring constructions appear in a number of the proposed problems. For the reader's convenience, we mention a few definitions and sources of information here:

Let R be a commutative ring and let G be an abelian group written multiplicatively. The group ring RG is the free R module on the elements of G with multiplication induced by G. An element x in RG has a unique expression $x = \sum_{g \in G} x_g g$, where $x_g \in R$ and all but finitely many x_g are zero. Addition and multiplication in RG are defined analogously to the standard polynomial operations. Basic information about commutative group rings may be found in [60, 84] and [63, Chap. 8 (Sect. 2)].

Let D be a domain with quotient field K; Int(D) denotes the ring of integer-valued polynomials, that is, Int$(D) = \{f \in K[X] \mid f(D) \subseteq D\}$. More generally, for a subset E of K, Int$(E, D) = \{f \in K[X] \mid f(E) \subseteq D\}$, and thus, Int$(D) =$ Int(D, D). For several indeterminates, Int$(D^n) = \{f \in K[X_1, \ldots, X_n] \mid f(D^n) \subseteq D\}$. For a D-algebra A, containing D, Int$_K(A) = \{f \in K[X] \mid f(A) \subseteq A\}$. Note that Int$_K(A)$ is contained in Int(D) if and only if $A \cap K = D$. Basic information about integer-valued polynomials may be found in [19].

Basic information on the star and semistar operations that appear in some of these problems may be found in [59, Sects. 32 and 34] and [95].

Basic information on the integral closure of ideals in Noetherian rings that is used in some of these problems may be found in [81].

Finally, for interested readers, we provide the list of contributors in the Acknowledgment section at the end of this chapter.

2 Open Problems

Problem 1. Glaz [67] and Bazzoni and Glaz [14] consider, among other properties, the finitistic and weak global dimensions of rings satisfying various Prüfer conditions. The Prüfer conditions under considerations are:

(1) R is *semihereditary* (i.e., finitely generated ideals of R are projective).
(2) w.dim $R \leq 1$.
(3) R is *arithmetical* (i.e., ideals of R_m are totally ordered by inclusion for all maximal ideals m of R).
(4) R is *Gaussian* (i.e., $c(fg) = c(f)c(g)$ for all $f, g \in R[x]$).
(5) R is *locally Prüfer* (i.e., R_p is Prüfer for every prime ideal p of R).
(6) R is *Prüfer* (i.e., every finitely generated regular ideal of R is invertible).

Let mod R be the set of all R-modules admitting a projective resolution consisting of finitely generated projective modules. The *finitistic projective dimension of* R is defined as fp.dim $R = \sup \{\text{proj.dim}_R M : M \in \text{mod } R \text{ and proj.dim}_R M < \infty\}$. In general, fp.dim $R \leq$ w.dim R, and if R is a local coherent regular ring, then fp.dim $R =$ w.dim R [67, Lemma 3.1]:

Problem 1a. Let R be a Prüfer ring. Is fp.dim $R \leq 1$?
  The answer is affirmative for Gaussian rings [67, Theorems 3.2 and 14, Proposition 5.3]. It is also clearly true for Prüfer domains.

Problem 1b. Let R be a total ring of quotients. Is fp.dim $R = 0$?
  Note that a total ring of quotients is always a Prüfer ring, so this question asks if for this particular kind of Prüfer ring, fp.dim R is not only at most equal to 1, but is actually equal to 0. This is true for a local Gaussian total ring of quotients [67, Theorem 3.2 (Case 1)]. More information and a detailed bibliography on the subject may be found in [66, 68].

Problem 2. Using the results obtained for the finitistic projective dimension of rings (see previous problem) with various Prüfer conditions, it is possible to determine the values of the weak global dimensions of rings under certain Prüfer conditions [14, 67], but many questions are not yet answered:

Problem 2a. If R is a Gaussian ring, then is w.dim $R = 0, 1$, or ∞?
  This is the case for coherent Gaussian rings [67, Theorem 3.3] (and actually, more generally, for coherent Prüfer rings [14, Proposition 6.1]), arithmetical rings [97 and 14, remark in the last paragraph], and a particular case of Gaussian rings [14, Theorem 6.4].

Problem 2b. If R is a total ring of quotients, is w.dim $R = 0, 1$, or ∞?

This is true if R is coherent [68, Corollary 6.7], in which case w.dim $R = 0$ or ∞, and also holds for one example of a non-coherent ring [68, Example 6.8].

In general:

Problem 2c. What are the values of w.dim R when R is a Prüfer ring?

Additional information and references on the subject may be found in [68].

Problem 3. Is the integral closure of a one-dimensional coherent domain in its field of quotients a Prüfer domain?

This question, which also appears in [23, Problem 65], was posed by Vasconcelos. It had been answered positively in many, but not all, cases. A useful reference that will lead to many other useful references is [63, Chap. 5 (Sect. 3) and Chap. 7 (Sect. 4)].

Problem 4. A ring R is a *finite conductor ring* if $aR \cap bR$ and $(0 : c)$ are finitely generated ideals of R for all elements a, b, and c in R. A ring R is a *quasi-coherent ring* if $a_1 R \cap \ldots \cap a_n R$ and $(0 : c)$ are finitely generated ideals of R for all elements $a_1, \ldots, a_n$ and c in R. Examples of both classes of rings include all coherent rings, UFDs, GCD domains, G-GCD domains (i.e., domains in which the intersection of two invertible ideals is an invertible ideal), and the still more general G-GCD rings (i.e., rings in which principal ideals are projective and the intersection of two finitely generated flat ideals is a finitely generated flat ideal). For more information on these classes of rings see references [64, 65]. Let G be a multiplicative abelian group and let RG be the group ring of G over R. In the group ring setting where R is a domain, characterizations of group rings as UFDs and GCD domains were obtained in [61, Theorems 6.1, 6.4, and 7.17]. In the case where R is a ring with zero divisors, however, the behavior of the finite conductor and quasi-coherent properties has been only partially described. Specifically, in the general ring setting, both properties descend from RG to R [65, Proposition 3.2], and the question of ascent from R to RG reduces to the situation where G is finitely generated [65, Proposition 3.1]. This, however, does not solve the problem of ascent for either property. Even in the case where R is a G-GCD ring and G is an infinite cyclic group, ascent is unknown.

Problem 4a. Assume that R is a G-GCD ring and G is a finitely generated abelian group. Do the finite conductor and the quasi-coherent properties ascend from R to RG?

Further explorations of these conditions in the group ring setting may shed light on a more general problem:

Problem 4b. Are the finite conductor and quasi-coherent properties for rings distinct?

Other useful references include [51, 69].

Problem 5. Let R be a commutative ring and let G be an abelian group with the property that the order of every element of G is invertible in R. Then w.dim $RG =$ w.dim $R + \mathrm{rank}\, G$ [33, Theorem] and [62, Theorem 2]. With the aid of this formula, it is possible to characterize von Neumann regular and semihereditary group rings RG [62, Corollaries 1 and 2].

Problem 5a. Is there a similar formula that relates the global dimension of RG to the global dimension of R in combination with some invariant of the group G?

In this direction, there is one classical result, Maschke's theorem [84, page 47], which pertains to a special case of semisimple rings (i.e., rings of global dimension zero): *Let G be a finite group and let k be a field. Then kG is a semisimple ring if and only if the characteristic of k does not divide the order of the group G.* Beyond this, characterizations of group rings of finite global dimension are not known, even for the simple cases of global dimension zero (semisimple rings) or one (hereditary rings).

Problem 5b. Find characterizations of semisimple and hereditary group rings.

Other useful references include [69, 103].

Problem 6. The generalization of the notion of a Cohen–Macaulay and a Gorenstein ring from the Noetherian to the non-Noetherian context is a recent development. As such, the questions of when a group ring RG, where R is a commutative ring and G is an abelian group, is Cohen–Macaulay or Gorenstein have yet to be investigated. The articles [72, 80] introduce the notions of non-Noetherian Cohen–Macaulay and non-Noetherian Gorenstein, respectively. An excellent survey of these theories which discusses the underlying homological framework is found in [79].

Problem 7. This problem arises within the context of work done by Hamilton and Marley [72] to characterize non-Noetherian Cohen–Macaulay rings, as well as work of Hummel and Marley [80] to characterize non-Noetherian Gorenstein rings. In light of the role that homological dimensions, infinite finite free resolutions of modules, and local (co)homology played in the development of these theories for local coherent rings, Hummel [79] posed the question:

Is there a non-Noetherian characterization of local complete intersection rings such that (coherent) local Gorenstein rings are complete intersections and such that (coherent) local complete intersection rings are Cohen–Macaulay?

André [10] provided a characterization of non-Noetherian complete intersection rings parallel to its following Noetherian counterpart using André–Quillen homology: *Let (R, m) be a local Noetherian ring with residue field k, and let $\beta_i = \dim_k \operatorname{Tor}_i^R(k, k)$ be the ith Betti number. R is a complete intersection if and only if the Betti numbers appear in the following equality of power series:* $\sum \beta_i x^i = \frac{(1+x)^r}{(1-x^2)^s}$ *with $0 < r - s = \dim R$.*

While able to prove a similar characterization for non-Noetherian rings, André showed there was no relation between the integers r and s above in the non-Noetherian case.

In [79] one can find more background on this problem, as well as potential definitions or characterizations of non-Noetherian complete intersection rings.

Problem 8. Let (R, M) be a Noetherian local ring. R is said to be *quasi-complete* if for any decreasing sequence $\{A_n\}_{n=1}^{\infty}$ of ideals of R and each natural number k, there exists a natural number s_k with $A_{s_k} \subseteq (\bigcap_{n=1}^{\infty} A_n) + M^k$. If this condition holds for any decreasing sequence $\{A_n\}_{n=1}^{\infty}$ of ideals of R with $\bigcap_{n=1}^{\infty} A_n = 0$, then R is called *weakly quasi-complete* (in which case, we actually have $A_{s_k} \subseteq M^k$). Now, if R is complete, then R is quasi-complete, which implies that R is weakly quasi-complete. Also R is quasi-complete if and only if each homomorphic image of R is weakly quasi-complete. The implication "R complete implies that R is quasi-complete" was first proved by Chevalley [24, Lemma 7]; for a proof in this volume see [4, Theorem 1.3]. Note that a DVR is quasi-complete, but need not be complete. More generally, a one-dimensional Noetherian local domain is (weakly) quasi-complete if and only if it is analytically irreducible [4, Corollary 2.2].

Problem 8a. Is a weakly quasi-complete ring quasi-complete?

Problem 8b. Let k be a field and $R = k[X_1, \ldots, X_n]_{(X_1, \ldots, X_n)}, n \geq 2$. Is R (weakly) quasi-complete?

Regarding Problem 8b, in [4, Conjecture 1] it is conjectured that R is not weakly quasi-complete and in [4, Example 2.1] this is shown to be the case if k is countable. Note that R not being quasi-complete is equivalent to the existence of a height-one prime ideal P of $k[[X_1, \ldots, X_n]]$ with $P \cap k[X_1, \ldots, X_n] = 0$. For additional information see [4].

Problem 9. A commutative ring R is said to be a *McCoy ring* if each finitely generated ideal $I \subseteq Z(R)$ (where $Z(R)$ denotes the set of zero divisors of R) has a nonzero annihilator. In 1980, Akiba proved that if R is an integrally closed reduced McCoy ring, then the polynomial ring $R[X]$ is also integrally closed [2, Theorem 3.2]. He also proved that if R_M is an integrally closed domain for each maximal ideal M, then $R[X]$ is integrally closed [2, Corollary 1.3]. In addition he provided an example of a reduced ring R that is not a McCoy ring but locally is an integrally closed domain [2, Example]. Combining his results, one has that if R is a reduced ring such that R_M is an integrally closed McCoy ring for each maximal ideal M, then $R[X]$ is integrally closed [78, page 103].

Does there exist an integrally closed reduced ring R such that R_M is an integrally closed McCoy ring for each maximal ideal M, but R is not a McCoy ring and it is not locally a domain?

Problem 10. With the notation and definitions of the previous problem, a ring R has (A_n) if each ideal $I \subseteq Z(R)$ that can be generated by n (or fewer) elements has a nonzero annihilator. Example 2.5 in [90] shows that for each $n \geq 2$, there are reduced rings which have (A_n) but not (A_{n+1}). An alternate restriction on zero divisors is (a.c.): R has (a.c.) if for each pair of elements $r, s \in R$, there is an element $t \in R$ such that $\text{Ann}(r, s) = \text{Ann}(t)$. The rings in Examples 2.2 and 2.4 of [90] show that there are reduced McCoy rings that do not have (a.c.), and reduced rings with (a.c.) that are not McCoy.

Do there exist reduced rings that have both (a.c.) and (A_n) for some $n \geq 2$ that are not McCoy rings?

Problem 11. Let P be a nonzero prime ideal of a Prüfer domain R. Then, it is known that PS is a divisorial prime in each overring $R \subseteq S \subseteq R_P$ if and only if PR_P is principal and each P-primary ideal (of R) is a divisorial ideal of R.

Problem 11a. Characterize when P (and R) is such that $P \cap T$ is a divisorial prime of T for each Prüfer domain $T \subseteq R$ with the same quotient field as R.

Note that no assumption has been made about PS being, or not being, divisorial for $R \subsetneq S \subseteq R_P$. This is true for each P if $\mathbb{Z} \subseteq R \subseteq \mathbb{Q}$.

Problem 11b. Characterize when P (and R) is such that there is no Prüfer domain $T \subseteq R$ with the same quotient field as R such that $P \cap T$ is a divisorial prime of T.

More information about the problem may be found in [45].

Problem 12. Let D be an almost Dedekind domain with a non-invertible maximal ideal M. Then $D(X)$ is also an almost Dedekind domain and $MD(X)$ is a non-invertible maximal ideal with corresponding residue field $F(X)$ where $F = D/M$. Let R be the pullback of $F[X]_{(X)}$ over $MD(X)$. Then R is a Prüfer domain, $R \subsetneq D(X) \subsetneq D(X)_{MD(X)}$, $MD(X)$ is a divisorial prime ideal of R, it is not a divisorial (prime) ideal of $D(X)$, but $MD(X)_{MD(X)}$ is a divisorial prime ideal of $D(X)_{MD(X)}$. In contrast, if P is a nonzero non-maximal prime of the ring of entire functions E and N is a maximal ideal that contains P, then $E \subsetneq E_N \subsetneq E_P$, $P(= P^2)$ is not a divisorial ideal of E, PE_N is a divisorial prime ideal of E_N, and PE_P is not a divisorial (prime) ideal of E_P.

Problem 12a. Does there exist a Prüfer domain R with a nonzero prime P such that there is a countably infinite chain of overrings $R = R_0 \subsetneq R_1 \subsetneq R_2 \subsetneq \cdots \subsetneq R_P$ where for all $n \geq 0$, PR_{2n} is a divisorial ideal of R_{2n} and PR_{2n+1} is an ideal of R_{2n+1} that is not divisorial? If such a chain exists, is PS divisorial or not divisorial as an ideal of $S = \bigcup R_m (\subseteq R_P)$?

Problem 12b. Does there exist a Prüfer domain R with a nonzero prime P such that there is a countably infinite chain of underrings $R = R_0 \supsetneq R_1 \supsetneq R_2 \supsetneq \cdots \supsetneq T = \bigcap R_m$ with T having the same quotient field as R where for all $n \geq 0$, $P \cap R_{2n}$ is a divisorial ideal of R_{2n} and $P \cap R_{2n+1}$ is not a divisorial ideal of R_{2n+1}? If such a chain exists, is $P \cap T$ divisorial or not divisorial as an ideal of T?

More information about the problem may be found in [45].

Problem 13. A domain D with field of quotients K is called a *straight domain* if for every overring S of D, S/PS is torsion-free over D/P, for every prime ideal P of D.

Problem 13a. If R_P and R/P are straight domains, does this imply that $R + PR_P$ is a straight domain?

A domain D is called *divided* if $PD_P = P$ for every prime ideal P of D; D is called *locally divided* if D_M is divided for every maximal ideal M of D. An answer to Question 13a may shed light on an open question posed in [30]:

Problem 13b. Does there exist a straight domain which is not locally divided?

Extending these definitions, analogously, to general rings, we note that there exists a straight ring which is not a domain and not locally divided. For more details, see [32, Sect. 2].

Problem 14. Comparing the Krull dimensions of the ring $\text{Int}(D)$ of integer-valued polynomials and of the classical ring $D[X]$ of polynomials with coefficients in D, it is known that $\dim(\text{Int}(D)) \geq \dim(D[X]) - 1$, with possibility of equality [19, Example V.1.12]. The question of an upper bound remains open: It is conjectured to be equal to $\dim(D[X])$.

Problem 15. Considering a one-dimensional local Noetherian domain D with finite residue field, when does $\text{Int}(D)$ satisfy the almost strong Skolem property?

See [19, Chap. VII] for a survey of Skolem properties. It is known that it is enough that D be analytically irreducible and was recently shown that D must be unibranched [20] leaving open the question of a necessary and sufficient condition.

Problem 16. A sequence $\{a_n\}_{n\geq 0}$ of integers is said to be *self* (*simultaneously*) *ordered* if for all positive integers m, n : $\prod_{k=0}^{n-1}(a_n - a_k)$ divides $\prod_{k=0}^{n-1}(a_m - a_k)$. If f is a nonconstant polynomial of $\mathbb{Z}[X]$, distinct from $\pm X$, then, for every $x \in \mathbb{Z}$, the sequence $\{f^{*n}(x)\}_{n\geq 0}$ (where f^{*n} denotes the nth iterate of f) is self-ordered [1, Proposition 18]. For instance, the sequence $\{q^n\}_{n\geq 0}$, with $q \neq 0, \pm 1$, obtained with $f(X) = qX$ and $x = 1$, is self-ordered. Aside from the (infinitely many) sequences obtained with such a dynamical construction are the three following "natural" sequences: $\{(-1)^n \left[\frac{n}{2}\right]\}_{n\geq 1}$, $\{n^2\}_{n\geq 0}$, and $\left\{\frac{n(n+1)}{2}\right\}_{n\geq 0}$. Also, if the sequence $\{a_n\}_{n\geq 0}$ is self-ordered, then so is $\{ba_n + c\}_{n\geq 0}$, for all integers $b \neq 0$ and c. We note the importance of the word "natural" (although not defined), for instance, the "natural candidates" $\{n^2\}_{n\geq 1}$ and $\{n^k\}_{n\geq 0}$ for $k \geq 3$ are not self-ordered. On the other hand, one can construct infinitely many "artificial" self-ordered sequences by the following ad hoc construction: choose two distinct integers a_0 and a_1 and, for $n \geq 1$, define inductively a_{n+1} to be any integer, distinct from the a_k's for $0 \leq k \leq n$, such that $\prod_{k=0}^{n-1}(a_n - a_k)$ divides $a_{n+1} - a_0$. These integers can be chosen to be prime numbers, thanks to Dirichlet's theorem, and hence, there are infinitely many self-ordered sequences contained in the set $\mathbb{P}$ of prime numbers, although $\mathbb{P}$ itself cannot be self ordered. The question is:

Are there any other "natural" self-ordered sequences neither obtained by a dynamical construction nor by an affine map applied to the three previous examples? More details can be found in [22, Sect. 5.2, Q.2].

Problem 17. The nth Bhargava factorial [15] associated to an infinite subset E of $\mathbb{Z}$ is the integer $n!_E$ such that $\frac{1}{n!_E}$ is the generator of the fractional ideal formed by the leading coefficients of the polynomials $f(X) \in \text{Int}(E, \mathbb{Z})$ with degree $\leq n$ [22, Sect. 3]. These factorials have the following properties:

1. $0!_E = 1$.
2. $\forall n \geq 0, n!$ *divides* $n!_E$.
3. $\forall n, m \geq 0, n!_E \times m!_E$ divides $(n + m)!_E$.

Mingarelli [92] called *abstract factorials* a sequence $\{n!_a\}_{n\geq 0}$ of positive integers satisfying these three properties. Clearly, such sequences are non-decreasing, but although there cannot be three consecutive equal terms [92, Lemma 8], one may have $k!_a = (k+1)!_a$ for infinitely many k [92, Proposition 12].

Are there subsets E of $\mathbb{Z}$ such that the sequence of Bhargava's factorials $\{n!_E\}_{n\geq 0}$ is not ultimately strictly increasing? [22, Sect. 4.3, Q.1].

Note that, for $E = \{n^3 \mid n \geq 0\}$, one has $3!_E = 4!_E = 504$, yet the sequence $\{n!_E\}_{n\geq 0}$ is ultimately strictly increasing.

Problem 18. Given an abstract factorial $\{n!_a\}_{n\geq 0}$ as in the previous question, one may define a generalization of the constant e [92, Definition 17] by $e_a = \sum_{n\geq 0} \frac{1}{n!_a}$. For Bhargava's factorial associated to a subset E one denotes this number by e_E. For example, if $E = \mathbb{N}^{(2)} = \{n^2 \mid n \geq 0\}$, then $e_{\mathbb{N}^{(2)}} = e + \frac{1}{e}$. The constant e_a is always irrational [92, Theorem 28].

For which infinite subset E is e_E a transcendental number [22, Sect. 6.5]?

Problem 19. Int(D) is known to be a free D module if D is a Dedekind domain. *TV* domains, defined by Houston and Zafrullah in [77], are domains in which t-ideals coincide with v-ideals. *TV PvMD* domains were extensively studied in [82]. For a Krull domain, or more generally for a TV PvMD domain, Int(D) is known to be locally free and hence a flat D module [37].

Problem 19a. Is Int(D) a flat D module for any domain D?

Problem 19b. Is Int(D) a free D-module for any domain D?

Problem 20. Let D be an integral domain. The canonical D-algebra homomorphism Int$(D)^{\otimes D^n} \longrightarrow$ Int(D^n) is known to be an isomorphism if Int(D) is free or if D is locally free and Int$(D_{\mathfrak{m}}) = $ Int$(D)_{\mathfrak{m}}$ for every maximal ideal $\mathfrak{m}$ of D [35].

Is this canonical morphism always injective? surjective?

Problem 21. Let D be an integral domain. Does Int(D) always have a unique structure of a D-D-biring such that the inclusion $D[X] \longrightarrow $ Int(D) is a homomorphism of D-D-birings?

This is the case if the canonical D-algebra homomorphism Int$(D)^{\otimes D^n} \longrightarrow$ Int(D^n) is an isomorphism for all n [36].

Problem 22. Let $D \subseteq A$ be an extension of domains. Let B be the quotient field of A and Int$(A) = \{f \in B[X] \mid f(A) \subseteq A\}$ be the ring of integer-valued polynomials of A and similarly Int(A^n) be the ring of integer-valued polynomials in several indeterminates. If Int$(D^n) \subseteq $ Int(A^n) for all positive integer n, then the extension $D \subseteq A$ is said to be *almost polynomially complete* [35].

If Int$(D) \subseteq $ Int(A), does it follow that the extension $D \subseteq A$ is almost polynomially complete?

Problem 23. Recall that, for a set S of nonnegative integers, the *natural density* $\delta(S)$ of S is defined to be $\delta(S) = \lim_{n\to\infty} \frac{|\{a\in S : a < n\}|}{n}$, provided the limit exists.

Let K be a number field and $\mathcal{O}_K$ be the corresponding ring of algebraic integers. Consider the natural density $\delta(\text{Int}(\mathcal{O}_K))$ of the set of nonnegative integers n such that $\text{Int}_n(\mathcal{O}_K)$ is free ($\text{Int}_n(\mathcal{O}_K)$ denotes the $\mathcal{O}_K$-module formed by the integer-valued polynomials of degree at most n).

Prove or disprove the following conjecture [38]: $\delta(\text{Int}(\mathcal{O}_K))$ exists, is rational, and is at least $1/\text{Card}(\text{PO}(\mathcal{O}_K))$ (where $\text{PO}(\mathcal{O}_K)$) denotes the Pólya group of $\mathcal{O}_K$). Then compute $\delta(\text{Int}(\mathcal{O}_K))$.

Problem 24. Consider the ring $\text{Int}(E, \mathbb{Z})$ where E is the set formed by the elements of a sequence $\{u_n\}_{n \geq 0}$ of integers determined by a recursion $u_{n+1} = au_n + bu_{n-1}$ and initial values u_0, u_1.

Problem 24a. Compute a regular basis (i.e., a basis with nth term of degree n) of the $\mathbb{Z}$-module $\text{Int}(E, \mathbb{Z})$.

Problem 24b. Compute the characteristic sequence of $\text{Int}(E, \mathbb{Z})$ with respect to each prime p, that is, the sequence $\{\alpha_p(n)\}_{n \geq 0}$, where $\alpha_p(n)$ is the p-adic valuation of the fractional ideal consisting of 0 and the leading coefficients of elements of $\text{Int}(E, \mathbb{Z})$ of degree no more than n.

Problem 24c. Determine the asymptotic behavior of this sequence, that is, compute the limit $\lim_{n \to \infty} \alpha(n)/n$ (the limit exists, by Fekete's lemma, since the sequence $\{\alpha(n)\}$ is superadditive).

Based on the results of Coelho and Parry [25] answers are known if $b = \pm 1$, which includes the cases of the Fibonacci and Lucas numbers [83, 102]; however the method used there does not seem to extend to general a, b. This question can, of course, be extended to higher-order recursive sequences.

Problem 25. There has been much recent progress in understanding $\text{Int}_{\mathbb{Q}}(M_n(\mathbb{Z}))$ where $M_n(\mathbb{Z})$ is the ring of $n \times n$ matrices with integer coefficients, beginning with Frisch's contribution [48] and more recently those of Peruginelli [98] and of Peruginelli and Werner [99]. These descriptions relate $\text{Int}_{\mathbb{Q}}(M_n(\mathbb{Z}))$ to polynomials integer valued on algebraic integers or divisible by irreducibles modulo $d\mathbb{Z}[x]$. There remains however, as in the previous problem, to:

Problem 25a. Compute a regular basis of $\text{Int}_{\mathbb{Q}}(M_n(\mathbb{Z}))$.

Problem 25b. Compute its characteristic sequence with respect to each prime p.

Problem 25c. Determine the asymptotic behavior of this sequence.

Problem 26. Let D be an integral domain and let A be a torsion-free D-algebra that is finitely generated as a D-module. We consider the ring $\text{Int}_K(A)$ of polynomials, with coefficients in the quotient field K of D, that are integer-valued over A, that is, $\text{Int}_K(A) = \{f \in K[X] \mid f(A) \subseteq A\}$. For every $a \in A$, we denote by $\mu_a(X) \in D[X]$ the minimal polynomial of a over D. Each polynomial f in the pullback $D[X] + \mu_a(X)K[X]$ is obviously such that $f(a) \in A$. Hence, $\bigcap_{a \in A}(D[X] + \mu_a(X)K[X]) \subseteq \text{Int}_K(A)$.

For which algebras A as above does this inclusion become an equality?

Note that equality holds for the D-algebra $A = M_n(D)$ of $n \times n$ matrices over D [99, Remark 3.4]. The ring $\text{Int}_K (M_n(D))$ has been studied in several places; see for instance [46, 98].

Problem 27. Let D be an integral domain with quotient field K. Let A be a torsion-free D-algebra, containing D, finitely generated as a D-module, and such that $K \cap A = D$. Letting $B = K \otimes_D A$ (i.e., the ring of fractions $\frac{a}{d}$ with $a \in A$ and $d \in D, d \neq 0$), one can consider the set $\text{Int}(A) = \{f \in B[X] \mid f(A) \subseteq A\}$. Working with polynomials in $B[X]$, one assumes that the indeterminate X commutes with all elements of B and that polynomials are evaluated with X on the right; see [85] for more details on polynomials with non-commuting coefficients. Note that $\text{Int}_K(A) = \text{Int}(A) \cap K[X]$. The set $\text{Int}(A)$ is always a left A-module, but it is not clear whether it has a ring structure when A is not commutative. A sufficient condition for $\text{Int}(A)$ to be a ring is that each element of A can be written as a finite sum $\sum_i c_i u_i$, where each u_i is a unit of A and each c_i is central in B [106]. Examples of algebras that meet this condition include the matrix rings $M_n(D)$ and group rings DG where G is a finite group. Yet this condition is not necessary [107].

Problem 27a. Give an example of a D-algebra A such that $\text{Int}(A)$ is not a ring (possibly relaxing some of the conditions on A, for instance, that A is finitely generated as a D-module).

Problem 27b. In [107], it is conjectured that $\text{Int}(A)$ is always a ring when D has finite residue rings. Prove this conjecture or give a counterexample in this case.

Problem 28. In [89], it is shown that if A is the ring of integers of an algebraic number field, then $\text{Int}_\mathbb{Q}(A)$ is Prüfer. More generally, let D be an integral domain, let K be its quotient field, and let A be a D-algebra. If $\text{Int}_K(A)$ is a Prüfer domain contained in $\text{Int}(D)$, then $\text{Int}(D)$ must also be Prüfer, and hence D must be integrally closed. This condition is not sufficient: $\text{Int}_K(A)$ need not even be integrally closed and [99] gives some general theorems regarding the integral closure of $\text{Int}_K(A)$.

Determine when this integral closure is Prüfer.

Note that Loper [87] determined such a criterion for the classical ring of integer-valued polynomials.

Problem 29. In the early 1990s, J.D. Sally gave expository talks on the question of which rings lie between a Noetherian domain D and its quotient field F [74]. The manuscript [74] provides abundant evidence that when the dimension of D is greater than one, the class of rings between D and F is rich in interesting Noetherian and non-Noetherian rings. A narrower problem, which remains open, is the following:

Describe the integrally closed rings between a two-dimensional Noetherian domain and its quotient field.

Work on this problem is surveyed in [96]. The evidence suggests that this class of rings is quite complicated. A framework for describing the integrally closed rings between $\mathbb{Z}[X]$ and $\mathbb{Q}[X]$ is given in [88].

Problem 30. Let R be a ring and let n be a positive integer. A proper ideal I of R is called an *n-absorbing ideal* if whenever the product $x_1 x_2 \cdots x_{n+1} \in I$ for $x_1, x_2, \ldots, x_{n+1} \in R$, then there are n of the x_i's whose product is in I. Clearly, a 1-absorbing ideal is just a prime ideal. A proper ideal I of R is called a *strongly n-absorbing ideal* if whenever $I_1 I_2 \cdots I_{n+1} \subseteq I$ for ideals $I_1, I_1, \ldots, I_{n+1}$ of R, then there are n of the ideals I_i's whose product is a subset of I. Obviously, a strongly n-absorbing ideal is an n-absorbing ideal.

Problem 30a. Let I be an n-absorbing ideal of R. Is I a strongly n-absorbing ideal of R?

Note that if $n = 1$ the answer is obviously positive and if $n = 2$ a positive answer is contained in [11, Theorem 2.13].

Problem 30b. Let I be an n-absorbing ideal of R. Is $\mathrm{rad}(I)^n \subseteq I$?

Note that if I is an n-absorbing ideal, then so is $\mathrm{rad}(I)$. For $n = 2$, the positive answer to this question is proved in [11, Theorem 2.4].

Problem 30c. Let I be an n-absorbing ideal of R. Is $I[X]$ an n-absorbing ideal of the polynomial ring $R[X]$?

Note that if $n = 2$, then the answer is affirmative [7, Theorem 4.15].

Problem 31. Let F/K be a transcendental field extension, and let X be an indeterminate over F. By a result of Halter-Koch [70, Theorem 2.2], $A = \bigcap_V V(X) = \bigcap_V V[X]_{M_V[X]}$, (where V ranges over the valuation rings of F/K and M_V is the maximal ideal of V), is a Bézout domain. However, A is not the Kronecker function ring of a domain with quotient field F (and hence does not arise directly from an e.a.b. star operation). Fabbri and Heubo-Kwegna [39] show this issue can be circumvented when $F = K(X_0, X_1, \ldots, X_n)$ is a purely transcendental extension of K by introducing the notion of projective star operations, which are glued together from traditional star operations on affine subsets of projective n-space. The ring A is then a "projective" Kronecker function ring of the projective analogue of the b-operation. More generally, e.a.b. projective star operations give rise to projective Kronecker function rings. The theory of projective star operations is worked out in [39] for projective n-space, but the following problem remains open:

Develop projective star operations for projective varieties (i.e., for the case where F/K is a finitely generated field extension that need not be purely transcendental)

Problem 32. Let D be a domain and let I be a nonzero ideal of D. Recall that I is called *stable* if I is invertible in its endomorphism ring $E(I) = (I : I)$, and the integral domain D is called *(finitely) stable* if each nonzero (finitely generated) ideal is stable. As usual, the ideal I is called *divisorial* if $I = I^v = (D : (D : I))$, and D is called a *divisorial domain* if each nonzero ideal is divisorial. A *Mori domain* is a domain satisfying the ascending chain condition on divisorial ideals. Clearly Noetherian and Krull domains are Mori. It has been proved in [54] that a stable domain is one-dimensional if and only if it is Mori. Since Mori domains satisfy the ascending chain condition on principal ideals, a Mori domain D is *Archimedean,* that is, $\bigcap_{n\geq 0} x^n D = (0)$, for each non-unit $x \in D$. The class of Archimedean domains includes also completely integrally closed domains and one-dimensional domains.

Problem 32a. Is a stable Archimedean domain one-dimensional?

The answer is positive in the semilocal case [54], so that a semilocal stable Archimedean domain is Mori (see also [50, Theorem 2.17]). Hence, a way of approaching this problem is trying to see if for stable domains the Archimedean property localizes. In general the Archimedean property does not pass to localizations. For example, the ring of entire functions is an infinite-dimensional completely integrally closed (hence, Archimedean) Bézout domain which is not locally Archimedean, because an Archimedean valuation domain is one-dimensional.

If D is a Mori stable domain and I is a nonzero ideal of D, we have $I^v = (x, y)^v$, for some $x, y \in I$ [50, Theorem 2.17]. Therefore, we can say that in a stable Mori domain, each divisorial ideal is 2-v-*generated*. Since a divisorial Mori domain is Noetherian, this result generalizes the fact that in a Noetherian domain that is stable and divisorial each ideal can be generated by two elements [50, Theorem 3.6].

Problem 32b. Let D be a one-dimensional Mori domain such that each divisorial ideal is 2-v-generated. Is it true that each divisorial ideal is stable?

Note that the answer to this question is negative when D has dimension greater than one. For example, let D be a Krull domain. Then, each divisorial ideal of D is 2-v-generated [93, Proposition 1.2] and stability coincides with invertibility [50, Proposition 2.3]. Hence, each divisorial ideal of D is stable (i.e., invertible) if and only if D is locally factorial [17, Lemma 1.1].

Recall that a nonzero ideal I of an integral domain D is v-*invertible* if $(I(D : I))^v = D$, and I is called v-*stable* if I^v is v-invertible as an ideal of $E(I^v)$, that is, $(I^v(E(I^v) : I^v))^v = E(I^v)$. Clearly, v-invertibility implies v-stability. If each nonzero ideal of D is v-stable, we say that D is v-*stable* [52]. Each nonzero ideal of a Krull domain is v-invertible; thus a Krull domain is v-stable. However, a Krull domain is stable if and only if it is a Dedekind domain, that is, it has dimension one. An example of a one-dimensional Mori domain that is v-stable but not stable is given in [53, Example 2.6].

Problem 32c. Let D be a Mori domain such that each divisorial ideal is 2-v-generated. Is it true that D is v-stable?

Problem 33. It is well known that if an integral domain D has finite character, then a locally invertible ideal is invertible. Conversely, if each locally invertible ideal is invertible, D need not have finite character (e.g., a Noetherian domain need not have finite character). However, a Prüfer domain such that each locally invertible ideal is invertible does have finite character. This fact was conjectured by Bazzoni [12, p 630] and proved by Holland et al. in [76]. (A simplified proof appears in [91]). Halter-Koch gave independently another proof in the more general context of ideal systems [71]. Other contributions were made by Zafrullah in [111] and by Finocchiaro et al. in [40]. Following Anderson and Zafrullah [6], an integral domain D is called an *LPI domain* if each locally principal nonzero ideal of D is invertible. Since a Prüfer domain is precisely a finitely stable integrally closed domain [50, Proposition 2.5], one is led to ask the following more general question (which appears in [13, Question 4.6]):

Assume that D is a finitely stable LPI domain. Is it true that D has finite character?

The answer to this question is positive if and only if the LPI property extends to fractional overrings [49, Corollary 15]. In particular, this holds when D is Mori or integrally closed. For an exhaustive discussion of this problem, see [49].

Problem 34. Let D be an integral domain, let S be a multiplicatively closed set in D, and let D_S be the ring of fractions of D with respect to S. Consider $R = D+XD_S[X]$, the ring of polynomials over D_S in indeterminate X and with constant terms in D.

Recall that a *GCD domain* is an integral domain with the property that any two nonzero elements have a greatest common divisor and a *PvMD* is an integral domain with the property that every nonzero finitely generated ideal is t-invertible. It was shown in [26, Theorem 1.1] that R is a GCD domain if and only if D is a GCD domain and for all $d \in D\setminus\{0\}$ GCD(d, X) exists. In [9, Theorem 2.5] it was shown that R is a Prüfer v-multiplication domain (for short, PvMD) if and only if D is a PvMD and the ideal (d, X) is t-invertible for all $d \in D\setminus\{0\}$. Next, recall that an integral domain is a *v-domain* if every nonzero finitely generated ideal is v-invertible. Note that if D is a v-domain, it is possible that D_S may not be a v-domain. For more information about v-domains consult [44].

Problem 34a. Find necessary and sufficient conditions for $R = D + XD_S[X]$ to be a v-domain. Prove or disprove that R is a v-domain if and only if D is a v-domain and (d, X) is v-invertible for all $d \in D\setminus\{0\}$.

More generally:

Problem 34b. Let $A \subseteq B$ be an extension of integral domains, and let X be an indeterminate over B. Find necessary and sufficient conditions for the integral domain $A + XB[X]$ to be a v-domain.

Problem 35. An integral domain D is called an *almost GCD* (for short, AGCD) if for each pair $x, y \in D\setminus\{0\}$, there is an integer $n = n(x, y)$ (depending on x and y) such that $x^n D \cap y^n D$ is principal. The theory of AGCD domains runs along lines similar to that of GCD domains; see [5] for more information and a list of references on the topic. An integral domain D is a *domain of finite t-character* if every nonzero non-unit of D belongs to at most a finite number of maximal t-ideals of D. AGCD domains of finite t-character were characterized in [34]. An ideal A of D is t-locally principal if AD_P is principal for every maximal t-ideal P of D. In [111] it was shown that if D is a PvMD, then D is of finite t-character if and only if every nonzero t-locally principal ideal of D is t-invertible. A GCD domain is a PvMD because the v-closure of every nonzero finitely generated ideal in a GCD domain is principal. We can therefore conclude that a GCD domain is of finite t-character if and only if every nonzero t-locally principal ideal A of D is t-invertible.

Now, we are in a position to state an open problem:

Let D be an almost GCD domain such that every nonzero t-locally principal ideal is t-invertible. Is D of finite t-character?

Problem 36. Let D be an integral domain, let K be the field of quotients of D, and let $\overline{F}(D)$ denote the set of all the nonzero D-submodules of K.

Let $\mathcal{T}$ denote a nonempty collection of overrings of D and, for any $T \in \mathcal{T}$, let $\star_T$ be a semistar operation on T. An interesting question posed in [23, Problem 44] is the following:

Problem 36a. Find conditions on $\mathcal{T}$ and on the semistar operations $\star_T$ under which the semistar operation $\star_{\mathcal{T}}$ on D defined by $E^{\star_{\mathcal{T}}} := \bigcap \{(ET)^{\star_T} \mid T \in \mathcal{T}\}$, for all $E \in \overline{F}(D)$, is of finite type.

Note that, if $D = \bigcap \{T \mid T \in \mathcal{T}\}$ is locally finite and each $\star_T$ is a star operation on T of finite type, then Anderson in [3, Theorem 2] proved that $\star_{\mathcal{T}}$ is a star operation on D of finite type. Through the years, several partial answers to this question were given and they are mainly topological in nature. For example, in [42, Corollary 4.6], a description of when the semistar operation $\star_{\mathcal{T}}$ is of finite type was given when $\mathcal{T}$ is a family of localizations of D and $\star_T$ is the identity semistar operation on T, for each $T \in \mathcal{T}$. More recently, in [41], it was proved that if $\mathcal{T}$ is a quasi-compact subspace of the (Zariski–Riemann) space of all valuation overrings of D (endowed with the Zariski topology) and $\star_T$ is the identity (semi)star operation on T, for each $T \in \mathcal{T}$, then $\star_{\mathcal{T}}$ is of finite type. Another more natural way to see the problem stated above is the following.

Problem 36b. Let $\mathcal{S}$ be any nonempty collection of semistar operations on D and let $\wedge_{\mathcal{S}}$ be the semistar operation defined by $E^{\wedge_{\mathcal{S}}} := \bigcap \{E^\star \mid \star \in \mathcal{S}\}$ for all $E \in \overline{F}(D)$. Find conditions on the set $\mathcal{S}$ for the semistar operation $\wedge_{\mathcal{S}}$ on D to be of finite type.

Note that it is not so difficult to show that the constructions of the semistar operations of the type $\star_{\mathcal{T}}$ and $\wedge_{\mathcal{S}}$ are essentially equivalent, in the sense that every semistar operation $\star_{\mathcal{T}}$ can be interpreted as one of the type $\wedge_{\mathcal{S}}$, and conversely.

Problem 37. A finite-dimensional integral domain D is said to be a *Jaffard domain* if $\dim(D[X_1, X_2, \ldots, X_n]) = n + \dim(D)$ for all $n \geq 1$ and, equivalently, if $\dim(D) = \dim_v(D)$, where $\dim(D)$ denotes the (Krull) dimension of D and $\dim_v(D)$ its valuative dimension (i.e., the supremum of dimensions of the valuation overrings of D). As this notion does not carry over to localizations, D is said to be a *locally Jaffard domain* if D_P is Jaffard for each prime ideal P of D. The class of (locally) Jaffard domains contains most of the well-known classes of (locally) finite-dimensional rings involved in dimension theory such as Noetherian domains, Prüfer domains, universally catenarian domains, and universally strong S(eidenberg) domains. It is an open problem to compute the dimension of polynomial rings over Krull domains in general. In this vein, Bouvier conjectured that "finite-dimensional Krull (or, more particularly, factorial) domains need not be Jaffard" [18, 43]. Bouvier's conjecture makes sense beyond the Noetherian context. Explicit finite-dimensional non-Noetherian Krull domains are scarce in the literature and one needs to test them and their localizations as well for the Jaffard property. In [16], the authors scanned all known families of examples of non-Noetherian finite-dimensional Krull (or factorial) domains existing in the literature. They showed that all these examples—except two—are in fact locally Jaffard domains. The two exceptions are addressed below in the open Problems 37b and 37c. Bouvier's conjecture is still elusively open and one may reformulate it in the following simple terms:

Problem 37a. Is there a Krull (or, more particularly, factorial) domain D such that $1 + \dim(D) \lneqq \dim(D[X])$?

In [16], Bouchiba and Kabbaj examined David's second construction described in [29] of a three-dimensional factorial domain which arises as an ascending union of three-dimensional polynomial rings J_n in three variables over a field k; namely, $J = \bigcup J_n$ with $J_n = k[X, \beta_{n-1}, \beta_n]$ for each positive integer n, where the variables β_n satisfy the following condition: For $n \geq 2$, $\beta_n = \frac{-\beta_{n-1}^{s(n)} + \beta_{n-2}}{X}$, where the $s(n)$ are positive integers. We have $J_n \subseteq J \subseteq J_n[X^{-1}]$ for each positive integer n. Therefore, by [31, Theorem 2.3], J is a Jaffard domain (since the J_n's are affine domains, i.e., finitely generated k-algebras), but it is not known if J is locally Jaffard. So the following question is open:

Problem 37b. Is J a locally Jaffard domain?

Clearly, a negative answer to Problem 37b will solve (affirmatively) Bouvier's conjecture for factorial domains (cf. Problem 37a).

In [16], the authors also investigated the known family of examples which stem from the generalized 14th problem of Hilbert (also called Hilbert–Zariski problem): Let k be a field of characteristic zero, T a normal affine domain (i.e., an integrally closed domain which is a finitely generated algebra) over k, and F a subfield of the quotient field of T. The Hilbert–Zariski problem asks whether $D = F \cap T$ is an affine domain over k. Counterexamples for this problem were constructed by Rees [100], Nagata [94], and Roberts [101], where D wasn't even Noetherian. In this vein, Anderson et al. [8] asked whether D and its localizations inherit from T the Noetherian-like main behavior of having Krull and valuative dimensions coincide (i.e., whether D is locally Jaffard). In [18], the authors proved that D is Jaffard, but were not able to determine whether D is locally Jaffard. In fact, they addressed this problem in the more general context of subalgebras of affine domains over Noetherian domains and the following question remains open:

Problem 37c. Let $A \subseteq D$ be an extension of integral domains, where A is Noetherian domain and D is a subalgebra of an affine domain T over A. Is D a locally Jaffard domain?

Clearly, a negative answer to Problem 37c will solve (affirmatively) Bouvier's conjecture for Krull domains.

Problem 38. A one-dimensional local Mori domain R is called *locally tame* if for every irreducible element $u \in R$ there is a constant $t \in \mathbb{N}_0$ with the following property: For every $a \in uR$ and every factorization $v_1 \cdot \ldots \cdot v_m$ of a (where $m \in \mathbb{N}$

and $v_1, \ldots, v_m$ are irreducibles of R), there is a subproduct, say $v_1 \cdot \ldots \cdot v_l$, which is a multiple of u and which has a factorization containing u, say $v_1 \cdot \ldots \cdot v_l = uu_2 \cdot \ldots \cdot u_k$ (where $k \in \mathbb{N}_0$ and $u_2, \ldots, u_k$ are irreducibles of R), such that $\max\{k, l\} \leq t$.

Let R be a one-dimensional local Mori domain. Is R locally tame?

Denote by $\hat{R}$ the complete integral closure of R. It is known that R is locally tame in each of the following cases: The conductor $(R : \hat{R}) \neq \{0\}$ [56, Proposition 2.10.7]; R is Noetherian [73, Theorem 3.3]; $\hat{R}$ is Krull and $|\mathfrak{X}(\hat{R})| \geq 2$ [58, Theorem 3.5 and Corollary 3.6].

Assume that R is locally tame. Then this implies the finiteness of several arithmetical invariants, such as the catenary degree and the set of distances of R. Also, the sets of lengths in R have a well-defined structure [55], [57], and [56, Theorem 3.1.1].

Problem 39. Analyze and describe non-unique factorization in $\mathrm{Int}(D)$, where D is a DVR with finite residue field.

We remark that the results obtained by Frisch [47] for $\mathrm{Int}(\mathbb{Z})$ rely heavily on the fact that $\mathbb{Z}$ has prime ideals of arbitrarily large index.

Problem 40. Let $\mathbb{Z}$ (respectively, $\mathbb{Q}$) denote the integers (respectively, the rationals), and let F be a field. It is known that $\mathrm{Spec}(F[x, y])$ is order-isomorphic to $\mathrm{Spec}(\mathbb{Z}[y])$ if and only if F is contained in the algebraic closure of a finite field [109, Theorem 2.10]; and in that case $\mathrm{Spec}(A)$ is order-isomorphic to $\mathrm{Spec}(\mathbb{Z}[y])$ for every two-dimensional domain that is finitely generated as an F-algebra. Moreover, the poset $\mathrm{Spec}(\mathbb{Z}[y])$ is characterized, among posets, by five specific axioms [21, Theorem 2.9]. For more background information, see references [21, 108–110].

Problem 40a. Find axioms characterizing the poset $\mathbb{Q}[x, y]$.

See [21, Remark 2.11.3].

Problem 40b. Are $\mathrm{Spec}(\mathbb{Q}[x, y])$ and $\mathrm{Spec}(\mathbb{Q}(\sqrt{2})[x, y])$ order-isomorphic?

Problem 40c. More generally, let F and K be algebraic number fields. If A is a two-dimensional affine domain over F, is $\mathrm{Spec}(A)$ order-isomorphic to $\mathrm{Spec}(K[x, y])$?

This question is close to Question 2.15.1 of [110]. It is also observed in Example 2.14 of [110] and Corollary 7 of [108] that if L is an algebraically closed field of infinite transcendence degree over $\mathbb{Q}$, then the spectra of $L[x, y, z]/(x^4 + y^4 + z^4 - 1)$ and $L[x, y]$ are known to be non-isomorphic.

Problem 40d. At the other extreme, let F and K be fields, neither of them algebraic over a finite field. If $\mathrm{Spec}(F[x, y]$ and $\mathrm{Spec}(K[x, y])$ are order-isomorphic, are F and K necessarily isomorphic fields?

This is Question 2 of [110].

Problem 41. Let I be the integral closure of the ideal (x^a, y^b, z^c) in a polynomial ring in variables x, y, z over a field.

Classify all triples (a, b, c) for which all powers of I are integrally closed.

For basic information on the integral closure of monomial ideals, see [81, Chap. 1, Sect. 1.4]. The case of two variables is known by Zariski's theory of integral closure of ideals in a two-dimensional regular ring [81, Chap. 14], and for three variables some work was done by Coughlin in [27].

Problem 42. For some special ideals, such as for monomial ideals, there are fast algorithms for computing their integral closure. For computing the integral closure of general ideals, however, the current algorithms reduce to computing the integral closure of the Rees algebra and then reading off the graded components of the integral closure This computes simultaneously the integral closures of all powers of the ideal, which is doing more than necessary. This excess of work makes the computation sometimes unwieldy.

Problem 42a. Is there a more direct algorithm for computing the integral closure of a general ideal?

Problem 42b. In particular, can the Leonard–Pelikaan [86] and Singh–Swanson [104] algorithms be modified for computing the integral closure of ideals?

More information and references on the topic may be found in [105, Sect. 2].

Problem 43. The following is a general form of the Lipman–Sathaye Theorem, as found in Theorem 2.1 of [75]: *Let R be a Noetherian domain with field of fractions K. Assume that the S_2-locus is open in all algebras essentially of finite type over R. Let S be an extension algebra essentially of finite type over R such that S is torsion-free and generically étale over R, and such that for every maximal ideal M of S, $R_{M \cap R}$ is normal, and S_M has a relatively S_2-presentation over $R_{M \cap R}$. Let $L = K \otimes_R S$ and let $\overline{S}$ be the integral closure of S in L. Assume that $\overline{S}$ is module-finite over S, and that for every height-one prime ideal Q of $\overline{S}$, $R_{Q \cap R}$ is regular. Then $(\overline{S} :_L J_{\overline{S}/R}) \subseteq (S :_L J_{S/R})$.*

The suggested problem is to find how tight is the statement of this theorem: relax some assumption and either prove the theorem or find a counterexample for that relaxation.

Problem 44. Let R be a ring and let I be an ideal in R. A set of *Rees valuation rings* of I is a set $\{V_1, .., V_r\}$ consisting of valuation rings, subject to the following conditions:

(1) Each V_i is Noetherian and is not a field.
(2) For each $i = 1, \ldots, r$ there exists a minimal prime ideal P_i of R such that V_i is a ring between R/P_i and the field of fractions of R/P_i.
(3) For each natural number n, $\overline{I^n} = \cap_{i=1}^{r}(I^n V_i) \cap R$, where $\overline{I^n}$ denotes the integral closure of I^n.
(4) No set of valuation rings of cardinality smaller than r satisfies conditions (1)–(3).

Basic information on Rees valuations of ideals may be found in [81, Chap. 10]. Ideals that have only one Rees valuation have several good properties.

Cutkosky proved in [28] the existence of a Noetherian, two-dimensional, complete, integrally closed local domain (R, m) in which every m-primary ideal has more than one Rees valuation. Give a construction of such a ring.

Acknowledgements We thank all the commutative algebraists who contributed open problems to this chapter. The list of contributors is as follows: D.D. Anderson (Problem 8), A. Badawi (Problem 30), P.-J. Cahen (Problems 14 and 15), J.-L. Chabert (Problems 16–18), J. Elliott (Problems 19–23), C.A. Finocchiaro and M. Fontana (Problem 36), S. Frisch (Problems 28 and 39), S. Gabelli (Problems 32 and 33), A. Geroldinger (Problem 38), S. Glaz (Problems 1–3), L. Hummel (Problem 7), K. Johnson (Problems 24 and 25), S. Kabbaj (Problem 37), T.G. Lucas (Problems 9–12), B. Olberding (Problems 29 and 31), G. Peruginelli (Problem 26), G. Picavet and M. Picavet-L'Hermitte (Problem 13), R. Schwarz (Problems 4–6), I. Swanson (Problems 41–44), N.J. Werner (Problems 27 and 28), S. Wiegand and R. Wiegand (Problem 40), M. Zafrullah (Problems 34 and 35).

Note that while this chapter and reference [54] were in proof, Problem 32a has been answered negatively. A counter example is given in [54, Example 3.9] with a 2-dimensional Prüfer domain.

References

1. D. Adam, J.-L. Chabert, Y. Fares, Subsets of Z with simultaneous ordering. Integers **10**, 437–451 (2010)
2. T. Akiba, Integrally-closedness of polynomial rings. Jpn. J. Math. **6**, 67–75 (1980)
3. D.D. Anderson, Star operations induced by overrings. Comm. Algebra **16**, 2535–2553 (1988)
4. D.D. Anderson, Quasi-complete semilocal rings and modules, in *Commutative Algebra: Recent Advances in Commutative Rings, Integer-Valued Polynomials, and Polynomial Functions*, ed. by M. Fontana, S. Frisch, S. Glaz (Springer, New York, 2014, this volume)
5. D.D. Anderson, M. Zafrullah, Almost Bézout domains. J. Algebra **142**, 285–309 (1991)
6. D.D. Anderson, M. Zafrullah, Integral domains in which nonzero locally principal ideals are invertible. Comm. Algebra **39**, 933–941 (2011)
7. D.F. Anderson, A. Badawi, On n-absorbing ideals of commutative rings. Comm. Algebra **39**, 1646–1672 (2011)
8. D.F. Anderson, D.E. Dobbs, P.M. Eakin, W.J. Heinzer, On the generalized principal ideal theorem and Krull domains. Pacific J. Math. **146**(2), 201–215 (1990)
9. D.D. Anderson, D.F. Anderson, M. Zafrullah, The ring $D + XD_S[X]$ and t-splitting sets. Comm. Algebra Arab. J. Sci. Eng. Sect. C Theme Issues **26**(1), 3–16 (2001)
10. M. André, Non-Noetherian complete intersections. Bull. Am. Math. Soc. **78**, 724–729 (1972)
11. A. Badawi, On 2-absorbing ideals of commutative rings. Bull. Austral. Math. Soc. **75**, 417–429 (2007)
12. S. Bazzoni, Class semigroups of Prüfer domains. J. Algebra **184**, 613–631 (1996)
13. S. Bazzoni, Finite character of finitely stable domains. J. Pure Appl. Algebra **215**, 1127–1132 (2011)
14. S. Bazzoni, S. Glaz, Gaussian properties of total rings of quotients. J. Algebra **310**, 180–193 (2007)
15. M. Bhargava, The factorial function and generalizations. Am. Math. Monthly **107**, 783–799 (2000)
16. S. Bouchiba, S. Kabbaj, *Bouvier's Conjecture*. Commutative Algebra and its Applications (de Gruyter, Berlin, 2009), pp. 79–88
17. A. Bouvier, The local class group of a Krull domain. Canad. Math. Bull. **26**, 13–19 (1983)
18. A. Bouvier, S. Kabbaj, Examples of Jaffard domains. J. Pure Appl. Algebra **54**(2–3), 155–165 (1988)

19. P.-J. Cahen, J.-L. Chabert, *Integer-Valued Polynomials*. Mathematical Surveys and Monographs, vol. 48 (American Mathematical Society, Providence, 1997)
20. P.-J. Cahen, R. Rissner, Finiteness and Skolem closure of ideals for non unibranched domains. Comm. Algebra, in Press
21. E. Celikbas, C. Eubanks-Turner, S. Wiegand, Prime ideals in polynomial and power series rings over Noetherian domains, in *Commutative Algebra: Recent Advances in Commutative Rings, Integer-Valued Polynomials, and Polynomial Functions*, ed. by M. Fontana, S. Frisch, S. Glaz (Springer, New York, 2014, this volume)
22. J.-L. Chabert, Integer-valued polynomials: looking for regular bases (a survey), in *Commutative Algebra: Recent Advances in Commutative Rings, Integer-Valued Polynomials, and Polynomial Functions*, ed. by M. Fontana, S. Frisch, S. Glaz (Springer, New York, 2014, this volume)
23. S. Chapman, S. Glaz, One hundred problems in commutative ring theory, in *Non-Noetherian Commutative Ring Theory*. Mathematics and Its Applications, vol. 520 (Kluwer Academic Publishers, Dordrecht, 2000), pp. 459–476
24. C. Chevalley, On the theory of local rings. Ann. Math. **44**, 690–708 (1943)
25. Z. Coelho, W. Parry, Ergodicity of p-adic multiplications and the distribution of Fibonacci numbers. Am. Math. Soc. Translations **202**, 51–70 (2001)
26. D. Costa, J.L. Mott, M. Zafrullah, The construction $D + XD_S[X]$. J. Algebra **53**, 423–439 (1978)
27. H. Coughlin, Classes of normal monomial ideals, Ph.D. thesis, University of Oregon, 2004
28. S.D. Cutkosky, On unique and almost unique factorization of complete ideals II. Invent. Math. **98**, 59–74 (1989)
29. J. David, A characteristic zero non-Noetherian factorial ring of dimension three. Trans. Am. Math. Soc. **180**, 315–325 (1973)
30. D. Dobbs, G. Picavet, Straight rings. Comm. Algebra **37**(3), 757–793 (2009)
31. D.E. Dobbs, M. Fontana, S. Kabbaj, Direct limits of Jaffard domains and S-domains. Comment. Math. Univ. St. Pauli **39**(2), 143–155 (1990)
32. D. Dobbs, G. Picavet, M. Picavet-L'Hermitte, On a new class of integral domains with the portable property, in *Commutative Algebra: Recent Advances in Commutative Rings, Integer-Valued Polynomials, and Polynomial Functions*, ed. by M. Fontana, S. Frisch, S. Glaz (Springer, New York, 2014, this volume)
33. A. Douglas, The weak global dimension of the group rings of abelian groups. J. London Math. Soc. **36**, 371–381 (1961)
34. T. Dumitrescu, Y. Lequain, J. Mott, M. Zafrullah, Almost GCD domains of finite t-character. J. Algebra **245**, 161–181 (2001)
35. J. Elliott, Universal properties of integer-valued polynomial rings. J. Algebra **318**, 68–92 (2007)
36. J. Elliott, Birings and plethories of integer-valued polynomials, in *Third International Meeting on Integer-Valued Polynomials, 2010*. Actes des Rencontres du CIRM, vol. 2(2) (CIRM, Marseille, 2010), pp. 53–58
37. J. Elliott, Integer-valued polynomial rings, t-closure, and associated primes. Comm. Algebra **39**(11), 4128–4147 (2011)
38. J. Elliott, The probability that $Int_n(D)$ is free, in *Commutative Algebra: Recent Advances in Commutative Rings, Integer-Valued Polynomials, and Polynomial Functions*, ed. by M. Fontana, S. Frisch, S. Glaz (Springer, New York, 2014, this volume)
39. A. Fabbri, O. Heubo-Kwegna, Projective star operations on polynomial rings over a field. J. Comm. Algebra **4**(3), 387–412 (2012)
40. C.A. Finocchiaro, G. Picozza, F. Tartarone, Star-invertibility and t-finite character in integral domains. J. Algebra Appl. **10**, 755–769 (2011)
41. C.A. Finocchiaro, M. Fontana, K.A. Loper, The constructible topology on spaces of valuation domains. Trans. Am. Math. Soc. **365**(12), 6199–6216 (2013)

42. M. Fontana, J. Huckaba, Localizing systems and semistar operations, in *Non-Noetherian Commutative Ring Theory*. Mathematics and Its Applications, vol. 520 (Kluwer Academic Publishers, Dordrecht, 2000), pp. 169–197
43. M. Fontana, S. Kabbaj, Essential domains and two conjectures in dimension theory. Proc. Am. Math. Soc. **132**, 2529–2535 (2004)
44. M. Fontana, M. Zafrullah, On v-domains: a survey, in *Commutative Algebra: Noetherian and Non-Noetherian Perspectives* (Springer, New York, 2011), pp. 145–179
45. M. Fontana, E. Houston, T.G. Lucas, Toward a classification of prime Ideals in Prüfer domains. Forum Mathematicum, **22**, 741–766 (2010)
46. S. Frisch, Polynomial separation of points in algebras, in *Arithmetical Properties of Commutative Rings and Modules* (Chapel Hill Conference Proceedings) (Dekker, New York, 2005), pp. 249–254
47. S. Frisch, A Construction of integer-valued polynomials with prescribed sets of lengths of factorizations. Monatsh. Math. **171**(3–4), 341–350 (2013)
48. S. Frisch, Corrigendum to Integer-valued polynomials on algebras. [J. Algebra **373**, 414–425] (2013), J. Algebra **412** 282 (2014)
49. S. Gabelli, Locally principal ideals and finite character. Bull. Math. Soc. Sci. Math. Roumanie (N.S.) **56**(104), 99–108 (2013)
50. S. Gabelli, Ten problems on stability of domains, in *Commutative Algebra: Recent Advances in Commutative Rings, Integer-Valued Polynomials, and Polynomial Functions*, ed. by M. Fontana, S. Frisch, S. Glaz (Springer, New York, 2014, this volume)
51. S. Gabelli, E. Houston, Coherent-like conditions in pullbacks. Michigan Math. J. **44**, 99–123 (1997)
52. S. Gabelli, G. Picozza, Star-stable domains. J. Pure Appl. Algebra **208**, 853–866 (2007)
53. S. Gabelli, G. Picozza, Star stability and star regularity for Mori domains. Rend. Semin. Mat. Padova, **126**, 107–125 (2011)
54. S. Gabelli, M. Roitman, On finitely stable domains (preprint)
55. W. Gao, A. Geroldinger, W.A. Schmid, Local and global tameness in Krull monoids. Comm. Algebra (to appear) http://arxiv.org/abs/1302.3078
56. A. Geroldinger, F. Halter-Koch, *Non-Unique Factorizations: Algebraic, Combinatorial and Analytic Theory*. Pure and Applied Mathematics, vol. 278 (Chapman & Hall/CRC Press, Boca Raton, 2006)
57. A. Geroldinger, W. Hassler, Local tameness of v-Noetherian monoids. J. Pure Appl. Algebra **212**, 1509–1524 (2008)
58. A. Geroldinger, W. Hassler, G. Lettl, On the arithmetic of strongly primary monoids. Semigroup Forum **75**, 567–587 (2007)
59. R. Gilmer, *Multiplicative Ideal Theory* (Dekker, New York, 1972)
60. R. Gilmer, *Commutative Semigroup Rings*. Chicago Lecture Notes in Mathematics (University of Chicago Press, Chicago, 1984)
61. R. Gilmer, T. Parker, Divisibility properties in semigroup rings. Mich. Math. J. **21**, 65–86 (1974)
62. S. Glaz, On the weak dimension of coherent group rings. Comm. Algebra. **15**, 1841–1858 (1987)
63. S. Glaz, *Commutative Coherent Rings*. Lecture Notes in Mathematics, vol. 1371 (Springer, Berlin, 1989)
64. S. Glaz, Finite conductor rings. Proc. Am. Math. Soc. **129**, 2833–2843 (2000)
65. S. Glaz, Finite conductor rings with zero divisors, in *Non-Noetherian Commutative Ring Theory*. Mathematics and Its Applications, vol. 520 (Kluwer Academic Publishers, Dordrecht, 2000), pp. 251–270
66. S. Glaz, *Prüfer Conditions in Rings with Zero-Divisors*. Lecture Notes in Pure and Applied Mathematics, vol. 241 (CRC Press, London, 2005), pp. 272–281
67. S. Glaz, The weak dimension of Gaussian rings. Proc. Am. Math. Soc. **133**, 2507–2513 (2005)
68. S. Glaz, R. Schwarz, Prüfer conditions in commutative rings. Arabian J. Sci. Eng. **36**, 967–983 (2011)

69. S. Glaz, R. Schwarz, Finiteness and homological conditions in commutative group rings, in *Progress in Commutative Algebra 2* (De Gruyter, Berlin, 2012), pp. 129–143
70. F. Halter-Koch, Kronecker function rings and generalized integral closures. Comm. Algebra **31**, 45–59 (2003)
71. F. Halter-Koch, Clifford semigroups of ideals in monoids and domains. Forum Math. **21**, 1001–1020 (2009)
72. T. Hamilton, T. Marley, Non-Noetherian Cohen-Macaulay rings. J. Algebra **307**, 343–360 (2007)
73. W. Hassler, Arithmetical properties of one-dimensional, analytically ramified local domains. J. Algebra **250**, 517–532 (2002)
74. W. Heinzer, C. Rotthaus, S. Wiegand, Examples using power series over Noetherian integral domains. http://www.math.purdue.edu/~heinzer/preprints/preprints.html (in preparation)
75. M. Hochster, Presentation depth and the Lipman-Sathaye Jacobian theorem, *The Roos Festschrift*, vol. 2, Homology Homotopy Appl. **4**, 295–314 (2002)
76. W.C. Holland, J. Martinez, W.Wm. McGovern, M. Tesemma, Bazzoni's conjecture. J. Algebra **320**(4), 1764–1768 (2008)
77. E. Houston, M. Zafrullah, Integral domains in which each t-ideal is divisorial. Michigan Math J. **35**(2), 291–300 (1988)
78. J.A. Huckaba, *Commutative Rings with Zero Divisors* (Dekker, New York, 1988)
79. L. Hummel, Recent progress in coherent rings: a homological perspective, in *Progress in Commutative Algebra 1* (De Gruyter, Berlin, 2012), pp. 271–292
80. L. Hummel, T. Marley, The Auslander-Bridger formula and the Gorenstein property for coherent rings. J. Comm. Algebra. **1**, 283–314 (2009)
81. C. Huneke, I. Swanson, *Integral Closure of Ideals, Rings, and Modules*. London Mathematical Society Lecture Note Series, vol. 336 (Cambridge University Press, Cambridge, 2006)
82. C.J. Hwang, G.W. Chang, Prüfer v-multiplication domains in which each t-ideal is divisorial. Bull. Korean Math. Soc. **35**(2), 259–268 (1998)
83. K. Johnson, K. Scheibelhut, Polynomials that are integer valued on the Fibonacci numbers (to appear)
84. G. Karpilovsky, *Commutative Group Algebras*. Lecture Notes in Pure and Applied Mathematics, vol. 78. (Dekker, New York, 1983)
85. T.Y. Lam, *A First Course in Non-Commutative Rings* (Springer, New York, 1991)
86. D.A. Leonard, R. Pellikaan, Integral closures and weight functions over finite fields. Finite Fields Appl. **9**, 479–504 (2003)
87. A.K. Loper, A classification of all D such that Int(D) is a Prüfer domain. Proc. Am. Math Soc. **126**(3), 657–660 (1998)
88. K.A. Loper, F. Tartarone, A classification of the integrally closed rings of polynomials containing $Z[X]$. J. Comm. Algebra **1**(1), 91–157 (2009)
89. K.A. Loper, N.J. Werner, Generalized rings of integer-valued polynomials. J. Num. Theory **132**(11), 2481–2490 (2012)
90. T.G. Lucas, Two annihilator conditions: property (A) and (a.c.). Comm. Algebra **14**, 557–580 (1986)
91. W.Wm. McGovern, Prüfer domains with Clifford class semigroup. J. Comm. Algebra **3**, 551–559 (2011)
92. A. Mingarelli, Abstract factorials. arXiv:0705.4299v3[math.NT]. Accessed 10 July 2012
93. J.L. Mott, M. Zafrullah, On Krull domains. Arch. Math. **56**, 559–568 (1991)
94. M. Nagata, On the fourteenth problem of Hilbert, in *Proceedings of the International Congress of Mathematicians, 1958* (Cambridge University Press, London-New York, 1960), pp. 459–462
95. A. Okabe, R. Matsuda, Semistar operations on integral domains. Math. J. Toyama Univ. **17**, 1–21 (1994)
96. B. Olberding, *Intersections of Valuation Overrings of Two-Dimensional Noetherian Domains*. Commutative Algebra–Noetherian and Non-Noetherian Perspectives (Springer, New York, 2011), pp. 459–462

97. B. Osofsky, Global dimensions of commutative rings with linearly ordered ideals. J. Lond. Math. Soc. **44**, 183–185 (1969)

98. G. Peruginelli, Integer-valued polynomials over matrices and divided differences. Monatshefte für Mathematik. http://dx.doi.org/10.1007/s00605-013-0519-9 (to appear)

99. G. Peruginelli, N.J. Werner, Integral closure of rings of integer-valued polynomials on algebras, in *Commutative Algebra: Recent Advances in Commutative Rings, Integer-Valued Polynomials, and Polynomial Functions*, ed. by M. Fontana, S. Frisch, S. Glaz (Springer, New York, 2014, this volume)

100. D. Rees, On a problem of Zariski. Illinois J. Math. **2**, 145–149 (1958)

101. P. Roberts, An infinitely generated symbolic blow-up in a power series ring and a new counterexample to Hilbert's fourteenth problem. J. Algebra **132**, 461–473 (1990)

102. K. Scheibelhut, Polynomials that are integer valued on the Fibonacci numbers, M.Sc. Thesis, Dalhousie University, 2013

103. R. Schwarz, S. Glaz, Commutative group rings with von Neumann regular total rings of quotients. J. Algebra **388**, 287–293 (2013)

104. A.K. Singh, I. Swanson, Associated primes of local cohomology modules and of Frobenius powers. Int. Math. Res. Notices **30**, 1703–1733 (2004)

105. I. Swanson, Integral closure, in *Commutative Algebra: Recent Advances in Commutative Rings, Integer-Valued Polynomials, and Polynomial Functions*, ed. by M. Fontana, S. Frisch, S. Glaz (Springer, New York, 2014, this volume)

106. N.J. Werner, Integer-valued polynomials over matrix rings. Comm. Algebra **40**(12), 4717–4726 (2012)

107. N.J. Werner, Polynomials that kill each element of a finite ring. J. Algebra Appl. **13**, 1350111 (2014) http://dx.doi.org/10.1142/S0219498813501119

108. R. Wiegand, The prime spectrum of a two-dimensional affine domain. J. Pure Appl. Algebra **40**, 209–214 (1986)

109. R. Wiegand, S. Wiegand, Prime ideals and decompositions of modules, in *Non-Noetherian Commutative Ring Theory*. Mathematics and Its Applications, vol. 520 (Kluwer Academic Publishers, Dordrecht, 2000), pp. 403–428

110. R. Wiegand, S. Wiegand, *Prime Ideals in Noetherian Rings: A Survey*. Ring and Module Theory (Birkhauser, Boston, 2010), pp. 175–193

111. M. Zafrullah, *t*-invertibility and Bazzoni-like statements. J. Pure Appl. Algebra **214**, 654–657 (2010)

Printed in the United States
By Bookmasters